Quantum Theory of High-Energy Ion-Atom Collisions

Quantum Theory of High-Energy Ion-Atom Collisions

Dževad Belkić
Karolinska Institute
Stockholm, Sweden

CRC Press is an imprint of the
Taylor & Francis Group, an **informa** business
A TAYLOR & FRANCIS BOOK

CRC Press
Taylor & Francis Group
6000 Broken Sound Parkway NW, Suite 300
Boca Raton, FL 33487-2742

Printed in the United States of America on acid-free paper
10 9 8 7 6 5 4 3 2 1

International Standard Book Number-13: 978-1-58488-728-7 (Hardcover)

Visit the Taylor & Francis Web site at
http://www.taylorandfrancis.com

and the CRC Press Web site at
http://www.crcpress.com

Contents

About the Author

Dževad Belkić is a theoretical physicist. He is Professor of Mathematical Radiation Physics at the Karolinska Institute in Stockholm, Sweden. His current research activities are in atomic collision physics, radiation physics, radiobiology, magnetic resonance physics and mathematical physics.

In atomic heavy-particle collision physics, his past and current work encompass many problems ranging from major challenges in this research field, such as theory of charge exchange and ionization at high non-relativistic energies. He is one of the leading world's experts on Coulomb asymptotic convergence problem, distorted wave representations and perturbations expansions methods. He is known for furthering the powerful and versatile continuum distorted wave method and its derivatives that were advantageously exported to ion-atom and photon-atom collisions and also found their useful applications in medical physics.

In radiation physics, Professor Belkić works on the passage of fast electrons and multiply charged ions through tissue, as needed in radiation therapy in medicine. Here he furthers both deterministic methods through the Boltzmann equation and stochastic simulations via Monte Carlo computations.

In radiobiology, his research entails mathematical modeling for cell survival, with the main emphasis on mechanistic approaches by including the chief pathways for survival of cells under irradiation during radiotherapy.

In magnetic resonance physics, Professor Belkić works on magnetic resonance spectroscopy with the main applications to medical diagnostics, aided critically by high-resolution parametric signal processors that go beyond the conventional shape estimations of spectra and fitting approaches to quantification. The leading processor here is his fast Padé transform for exact spectral analysis of generic time signals and unequivocal signal-noise separation via Froissart doublets or pole-zero cancellations in response functions.

In mathematical physics, he works on many problems including the derivation of analytical expressions for scattering integrals or bound-free form factors, for rational response functions in signal processing, and for coupling parameters in the nearest neighbor approximation, which is one of the most frequently used methods in physics and chemistry.

Professor Belkić has published more than 180 scientific works, which have received over 2500 citations. His two books: "Principles of Quantum Scattering Theory" and "Quantum Mechanical Signal Processing and Spectral Analysis" were published by the Institute of Physics Publishing in 2004 and 2005, respectively.

He has received numerous international awards, including the triple Guest Professorship in Physics from the Nobel Foundation and the Royal Swedish Academy of Sciences for theory of atomic heavy-particle high-energy collisions.

Preface

The main purpose of this book is to provide state-of-the-art coverage of the non-relativistic theory of high-energy four-body ion-atom collisions with four active particles (two electrons and two nuclei), such that the presented content is suitable simultaneously for graduate students and experienced researchers, as well as for experts on the subject.

Regarding these energetic ion-atom collisions, over the last 15 years, the literature has witnessed a remarkable resurgence of interest and activity in the development and wide applications of four-body quantum mechanical methods. This was and still is vigorously coupled to investigations with a similar intensity in the accompanying field of related measurements.

Yet, this subject of four-body interactive dynamics in all its generality has not been presented in full detail in any of the available books. Therefore, it is both timely and necessary to present this field in its full capacity and this is the principal goal of this book. This has been achieved by a thorough exposition and critical review of the most important theoretical methods available to date by scrutinizing their foundation and practical usefulness relative to virtually all the available experimental data.

The reasons for resurgence of interest in energetic ion-atom collisions are multi-faceted and easy to explain. They become apparent as soon as one asks the question: who is in need of cross sections, collision rates and other related observables from this field?

Versatile information and data bases on high-energy ion-atom collisions are of paramount importance in many high-priority branches of science and technology e.g. accelerator based physics, the search for new sources of energy, controlled thermonuclear fusion, weak as well as strong lasers (laser-assisted fast ion-atom collisions), plasma research, astrophysics of upper atmosphere, the earth's environment, solar-terrestrial relations, space research, particle transport physics, medical storage ring accelerators (increasing in number worldwide) with radio-therapeutic ions, etc.

All these inter-disciplinary fields are in need of knowledge about a large variety of cross sections and collisional rates for different kinds of fast ion-atom collisions, such as single ionization, excitation, charge exchange and various combinations thereof. These include two-electron transitions e.g. double ionization, excitation or capture as well as simultaneous electron transfer and ionization or excitation and the like, as analyzed thoroughly in the present book.

The physics of high-energy ion-atom collisions is not restricted to providing its rich data bases to the neighboring branches. Most importantly, experience and expertize from heavy-particle energetic ion-atom collisions through translational research often play pivotal roles both in design of many important measurements and interpretation of obtained experimental results in these multi-disciplinary fields. There are many examples to support this fact by going from basic via applied research to technology e.g. high-temperature fusion energy research and technology (charge exchange spectroscopy for plasma diagnostics), ionosphere research (recombination and absorption processes, solar continuous spectrum), hadron radiotherapy in medicine (usage of energetic ions to help treat deep-seated tumors by reliance upon R.R. Wilson's visionary concept of optimally deposited dose though the Bragg peak at the targeted lesions), etc.

There are several books dealing with high-energy non-relativistic three-body collisions of ions and atoms. Theory in these books is restricted only to one actively participating electron. For multi-electron targets such treatments adopt various forms of the frozen-core approximation that effectively reduce the problem to three-body collisions without the possibility to access the key dielectronic correlations and their interactive dynamics. Some of these books on formalisms with merely one active electron are very recent, whereas the others were published 1-3 decades ago, so that the active and intensive research on this topic of high-energy four-body ion-atom collisions during the past 15 years still remains without adequate coverage on the level of books. The present book is aimed at bridging this gap.

This book covers all the essential aspects of non-relativistic high-energy ion-atom collisions ranging from fundamental principles, the most successful methods and their algorithmic capabilities as well as limitations, all of which is needed for the present self-contained and comprehensive reference. Such a thorough coverage is necessary to match the growing demand from both theory and practice, given that the subject under study is of high relevance for physics *per se*, as well as for other applied fields such as fusion, medicine, etc. For example, very reliable data bases of cross sections and other related observables are required in e.g. Monte Carlo simulations of transport of ions through tissue encountered in radiotherapy. Likewise, thermonuclear fusion requires very accurate cross sections in fusion reactors. We also systematize the analysis on the available cross section data bases for versatile usage in these two very important applications of high-energy ion-atom collisions.

The overall emphasis of the present book is not on providing a catalogue of the existing methods nor on compilation of exhaustive data bases for cross sections, but rather on multi-faceted mechanisms of four-body collisional phenomena with ions and atoms at non-relativistic high energies. This is indispensable for active researchers, since work in this field by the leading experts in the last 15 years needs to be put into a comparative perspective by resetting the standards and paving the way for the main directions of further developments resulting from coherent, synergistic efforts by various groups and centers.

Additionally, graduate students and beginners in this and related fields are also in need of the methodology expounded in this book. The rationale is that currently there is no self-contained and comprehensive source of information in a single publication covering all the relevant and essential aspects, such as rigorous formal theory, design and implementation of most adequate methods with their key analytical/numerical features and their consistent validation by reliance to the first principles of physics, as well as comparison with virtually all the available experimental data.

This is also important from the viewpoint of educating young researchers, especially in the current stage of the computer era, in which often too much emphasis is placed on visualization of the available data bases and overwhelming usage of ready-made algorithms and software, thus depriving them from the invaluable step-by-step development of work from operators to numbers with the ensuing incomparably higher degree of control and insight as well as the enhanced possibility to create novel methods and data bases.

Dževad Belkić
Professor of Mathematical Radiation Physics
Karolinska Institute
Stockholm, Sweden
September 2008

Acknowledgments

The author is grateful to the Royal Swedish Academy of Sciences for the Nobel research grants over a period of three years on quantum mechanical theory of high-energy ion-atom collisions. He also wishes to express his gratitude to the King Gustav the 5th Research Fund in Stockholm for support. Thanks are also due to Professor Ivan Mančev for thorough checking of the presented material and to Professor Karen Belkić for language editing. The author appreciates a fruitful cooperation with Mr. John Navas, Ms Amber Donley and Ms Suzanne Lassandro from Taylor & Francis.

1

Basic notions and main observables in scattering problems

1.1 Observables and elementary processes

In this chapter, we shall devote our attention to the main observables in scattering problems in light of the critical criteria for comparisons between theoretical and experimental data. Hereafter, any physical quantity which can be experimentally measured will be called an observable. We emphasize immediately that we shall not study collision experiments from the viewpoint of their realistic performance, which could certainly be very sophisticated, especially for modern measurements. Our primary concern will be focused on the exposition of the main concepts of scattering experiments regarding the boundary conditions and the corresponding requirements imposed by the theory. Here, we primarily think of the requirements that are a direct consequence of the limitations of the theory itself. Namely, scattering theory is based on the first principles of physics without reliance on any free or adjustable parameters. Therefore, such a theory is complete. However, the price paid for this completeness is an obvious limitation to relatively clear-cut situations dealing with elementary processes of the type of scattering of one given particle on a single particle from the target. These are the so-called binary collisions. Of course, this is an idealization, since in any realistic collision experiment, we encounter a beam of incident particles (projectiles) that scatter on the target which is itself comprized of many particles representing the elementary centers of scattering[1]. However, comparisons with experiments represent the ultimate test for the final validation of any theory in physics. In other words, no matter how apparently attractive and compelling a given theory might appear, from both the physical and mathematical standpoints, it would still be considered as inadequate if it fails to describe properly the available experimental data. But, how to fulfill such a stringent demand when the proper theories[2] deal mainly with idealized problems relative to complex

[1] In a given target, scattering centers are all the individual, structureless particles that collide with the projectiles.

[2] Here, the adjective 'proper' serves to highlight our interest in mathematically rigorous theories that stem from the first principles of physics, as opposed to phenomenological

realistic experiments? The answer to this fundamental question, in fact, defines the necessary initial conditions that the experiment must fulfill in order to make meaningful comparisons between the same theoretical and measured quantities. For example, if a theoretically calculated physical quantity relates to an arbitrary and typical scattering center, then the only thing which could be done *a posteriori* is a certain convenient way of averaging over mutually independent scattering centers that experience the same kind of phenomena as the typical scattering center. An experiment must match this theoretical requirement as closely as possible. The way in which this is usually achieved will be explained in the present chapter. To this end, it is necessary to introduce several basic notations, observables and initial conditions that are of primary concern to typical scattering experiments.

1.2 Energy as the most important physical property

One of the most important characteristics of any given physical system is its total energy E. Already the sign of E provides the first and vital information, since for e.g. $E < 0$ the system is in a bound state. Negative values of the total energy E of the system of particles have certain extraordinarily universal repercussions. For example, on the atomic level of the structure of matter for a given ground state, the fact $E < 0$ signifies stability of the given atom. Positive values $E > 0$ of the total energy of the system imply that the system as a whole is not bound. For instance, in collisional systems, the total energy of the whole system always lies in the continuum ($E > 0$). Of course, such a situation does not preclude that one or more subsystems of the whole system could be in their bound states. As an example of such a bound state is a nitrogen target colliding with an α particle as studied in e.g. the well-known Rutherford experiment [1]–[3]

$$\alpha + {}^{14}_{7}\mathrm{N} \longrightarrow p + {}^{17}_{8}\mathrm{O}. \tag{1.1}$$

In this nuclear transmutation, α particles are absorbed by ${}^{14}_{7}\mathrm{N}$ which subsequently decays into a free proton (p) and an oxygen ${}^{17}_{8}\mathrm{O}$. Hamiltonians of two-particle collisional systems could occasionally exhibit some apparently unexpected characteristics, since they could temporarily possess certain quasi-discrete eigen-energies associated with nearly bound normalizable states of the total system (projectile + target). Such states are proper i.e. physical

models whose primary goal is, at best, to interpret experiments. In so doing, such models inevitably resort to free or adjustable parameters through various fitting recipes and, as such, they cannot qualify to be a theory, let alone explain the underlying mechanisms of the studied phenomenon.

since they belong to the Hilbert space $\mathcal{H}$ despite the fact that we are dealing with continuous spectra. In some measurements of total cross sections $Q(E)$ as a function of energy E, such a situation which is encountered in various branches of physics (nuclear physics, atomic physics, etc) is characterized by the appearance of peaks[3] at certain energies $E_1, E_2, ...$, etc. When a projectile is sent to a target with an impact energy which coincides with one of these so-called resonant energies $E_1, E_2, ...$, then it could easily happen that the incident particle is temporarily captured by the target in a metastable state of the ensemble (projectile+target). Such a possibility is considered as the main cause for the observed sudden and strong variations of cross sections $Q(E)$. A corresponding simple quantitative explanation for this so-called resonant scattering can be established from the analytical properties of the scattering S-matrix in the second Riemann sheet of the complex energy plane. As opposed to stable stationary states, resonances formed in collisions do not have sharp energies. Rather, a given resonance possesses an energy distribution of a certain width Γ whose inverse is the lifetime of the associated state. Such states are called quasi-stationary states. In other words, a resonant state of a nearly bound complex projectile-target is unstable, or metastable. Such a metastable state, after a certain finite lifetime $\tau \sim 1/\Gamma$, will decay to either the previous two aggregates (projectile and target) or to some of their subsystems. This means that on the plot of the cross section $Q(E)$ as a function of the incident energy, the resonant effect must be observed within the energy interval of length $\Gamma \sim 1/\tau$. Simultaneously, the resonance width Γ represents the probability of the decay of the given resonance. In the vicinity of a resonant energy E_r, the cross section $Q(E)$ is given by the Breit-Wigner dispersion formula representing a single Lorentzian

$$Q(E) \propto \frac{\Gamma^2/4}{(E - E_r)^2 + \Gamma^2/4} \tag{1.2}$$

where E is the incident energy[4]. In this case, the curve which depicts the cross section $Q(E)$ as a function of energy E has a pronounced structure of a sharp local peak which is symmetrically distributed around $E = E_r$. The width of this peak taken at the half of the ordinate $Q(E)$ represents the resonance width Γ. It is also called the full width at the half maximum (FWHM)[5]. Let us emphasize that this introductory discussion relates to those resonances that are experimentally detectable as the realistic maximal values of cross section $Q(E)$, and not to some possible, but unmeasurable, poles of the S-matrix in the complex energy plane.

---

[3]The word 'peak' in physics is customarily used in the context of local maximae of a given function for some values of its independent variable.

[4]The incident energy is equal to the total energy when the target is at rest, as is the case in most scattering experiments.

[5]All acronyms used are listed at the end of the book (see pp. 368–370).

Thus far, we have frequently used the term scattering states to which considerable attention will be devoted in this book. In the case of a one-channel collision, by scattering states we shall understand stationary or non-stationary two-particle states of the whole system (projectile+target) with the positive total energy $E > 0$. A given spatial ensemble of a quantum system at rest in the usual laboratory coordinate frame will be called the target which can be any general particle (atoms, ions, molecules, elementary particles, etc) with or without structure. The target can be in a solid, liquid or gaseous state. Projectile is a collective name for the incident beam of particles (with or without mass) that collide with the target. Here, under the notion of a channel we understand every possible state of colliding particles in the initial and/or final configuration. In other words, a channel is one of the possible ways of fragmentation of the given total system[6]. Under the notion of a fragment we understand a group of particles. A channel could be open if the collision is allowed relative to the standard conservation laws of e.g. energy, charge, etc. In the opposite case, a channel is said to be closed. Stated equivalently, for a given value of the total energy $E > 0$ e.g. excitation of one or both particles could take place. In such a case, certain of these internal levels labelled collectively by E_n could be excited, but the remaining energy E might still be insufficient for exciting other internal levels $E_m (m \neq n)$. If such a process does not exhaust the entire initial energy E, then according to the conservation law of the total energy, we must have a positive energy remainder ΔE called the energy defect. This amount ΔE is distributed between the colliding particles as their kinetic energies. Channels that correspond to the first mentioned group of states of colliding particles associated with the collective label E_n for their internal excited levels are open. Likewise, the other channels that correspond to the second mentioned states related to the energies $E_m (m \neq n)$ are closed i.e. inaccessible to the given energy E. The value of energy E, which is sufficient for opening a given channel is called the energy threshold. For instance, such a threshold for ionization of atomic hydrogen from its ground state is 13.6 eV. This means that an incident electron whose kinetic energy is smaller than this threshold will be unable to eject the bound electron from its ground state in an atomic hydrogen. Likewise, we can give another example from nuclear collisions such as in (1.1). In this scattering, the state of the system $\alpha + \mathrm{N}$ in its initial configuration is called the entrance channel. For such an entrance channel, there could be many different possibilities in the so-called exit channel, each of which is associated with certain groups of particles in their final configuration e.g. $\alpha + \mathrm{N}$, $p + \mathrm{O}$, $\alpha + \alpha + \mathrm{B}$, etc. These latter states of the system also represent the open exit channels. A similar situation is encountered in another example from nuclear reactions

[6]By a composite particle, or a particle with structure, we shall call a particle which is comprized of at least two other particles in a bound state. Similarly, a structureless particle is a physical system which does not contain any bound state of other particles.

where a π° meson is produced in a collision of two protons via

$$p + p \longrightarrow p + p + \pi^\circ. \tag{1.3}$$

This is the well-known collision studied experimentally by Lettes *et al.* [4, 5] who confirmed Yukawa's [6] theory of mesons for strong interactions among nucleons. Here also, in addition to $p+p+\pi^\circ$, one could have other exit channels such as $p + n + \pi^+$ or $d + \pi^+$, etc, where d stands for a deuteron (the nucleus of heavy hydrogen atom: deuterium). Processes of this type where a given entrance channel at a selected energy of the projectile potentially corresponds to several exit channels are called many-channel collisions. However, if for a given initial configuration and a fixed incident energy, there could exist only one possible exit channel, then we speak about the one-channel collision. The simplest example of a one-channel scattering is a potential scattering for which the same particles are seen in both the entrance and exit channel.

1.3 Classification of collisions

Collisions are classified according the values of the incident velocity v_i. Thus, e.g. in atomic physics, collisions are said to be slow and fast if $v_i \ll v_0$ and $v_i \gg v_0$, respectively. Here, v_0 is the velocity of the electron in the ground state of the hydrogen atom. Further, we could have elastic collisions between structureless or composite particles A and B without any change of their internal structure as symbolized by

$$\mathrm{A}(i) + \mathrm{B}(i') \longrightarrow \mathrm{A}(i) + \mathrm{B}(i') \tag{1.4}$$

where the labels i and i' denote the sets of the usual quantum numbers that describe the states of A and B, respectively. As opposed to this, there are also inelastic collisions with a change in the internal structure of A and/or B or both e.g. $\mathrm{A}(i) + \mathrm{B}(i') \longrightarrow \mathrm{A}(f) + \mathrm{B}(i')$, $\mathrm{A}(i) + \mathrm{B}(i') \longrightarrow \mathrm{A}(i) + \mathrm{B}(f')$, $\mathrm{A}(i)+\mathrm{B}(i') \longrightarrow \mathrm{A}(f)+\mathrm{B}(f')$, etc. The kinetic energy of the projectile remains unaltered in elastic collisions. This is in contrast to inelastic collisions where the impact kinetic energy is altered, so that a part of it is spent on changes in the internal structures of A and/or B. In such a circumstance, a part of the incident kinetic energy of A could lead to a recoil of the target B. This recoil might, however, be negligible if the masses M_{A} and M_{B} of A and B are very different from each other e.g. $M_{\mathrm{A}} \ll M_{\mathrm{B}}$. There are also direct collisions where the internal structure of the projectile A and target B could change or remain unaltered, but no new particles are produced in the exit channel which still contains only A and B. In contrast to this, there are rearranging collisions where A and/or B decay by creating two or more particles that are different from A and B e.g.

$$\mathrm{A}(i) + \mathrm{B}(i') \longrightarrow \mathrm{C}(f) + \mathrm{D}(f') \tag{1.5}$$

or $\mathrm{A}(i) + \mathrm{B}(i') \longrightarrow \mathrm{C}(f) + \mathrm{D}(i')$, $\mathrm{A}(i) + \mathrm{B}(i') \longrightarrow \mathrm{A}(i) + \mathrm{D}(f')$, etc. The newly created particles C and D could also be certain aggregates of two or more structureless or composite free particles such as $\mathrm{C} = \mathrm{C}_1 + \mathrm{C}_2 + \cdots + \mathrm{C}_n$ and/or $\mathrm{D} = \mathrm{D}_1 + \mathrm{D}_2 + \cdots + \mathrm{D}_n$. Within quantum mechanics, rearranging collisions are obviously more difficult to study than direct scatterings. Both direct and rearranging collisions are frequently encountered in versatile applications in physics and applied sciences as well as in technologies. Rearranging collisions also include chemical reactions as well as many nuclear, atomic or molecular scatterings where A and B exchange one or more of their constituent particles. We could have the same particles in the entrance and the exit channel and still speak of a rearranging collision e.g.

$$p + \mathrm{H}(i) \longrightarrow \mathrm{H}(f) + p. \tag{1.6}$$

This is called charge exchange where an electron (e) is captured by the incident proton to form another hydrogen atom $\mathrm{H}(f)$ which is different from the target $\mathrm{H}(i)$. Even when $f = i$, we still have a rearranging collision in (1.6) which is called resonant electron capture (resonant charge exchange or resonant electron transfer). Since there is no energy change for the case $f = i$, such a process from (1.1) could also be called the rearranging elastic collision, especially when viewed as two intermediary elastic scatterings of the electron on the projectile and target protons (the so-called Thomas double scattering) [7]. When two or three free particles appear in the exit channel, we are talking about binary or tertiary collisions, respectively. In fact, it would be more precise to introduce a symmetric nomenclature for the entrance and exit channel by talking about binary-binary or binary-tertiary collisions, etc. Of course, it is also possible to have an even more general class of collisions of the type tertiary-binary or tertiary-tertiary collisions, or the like, where in the entrance channel three or more particles could collide with each other [8]–[15]. An example along these lines is the so-called three-body recombination process which could also take place in plasma physics or during the passage of e.g. protons p or deuterons $d = (p, n)$ through cooling electrons in cooler storage ring experiments [16, 17]

$$p + p + e \longrightarrow p + \mathrm{H} \qquad d + d + e \longrightarrow d + \mathrm{D} \tag{1.7}$$

where D denotes deuterium $\mathrm{D} = (d, e)$ i.e. heavy hydrogen atom. In this book we shall limit ourselves to binary constellations in the entrance channels. Certain useful aspects of a formal theory of tertiary collisions that encompass three free particles in the entrance channel are given in [8, 9].

After scattering on a target, a projectile or other newly created particles can be scattered to all possible angles from $0°$ to $180°$. The angle between the direction of the incident and scattered projectile is called the scattering angle and is usually denoted by θ. The two values of this angle $\theta = 0^\circ$ and $\theta = 180^\circ$ correspond to forward and backward scattering, respectively. Obviously, for $\theta = 0^\circ$, the projectile is not scattered at all, whereas $\theta = 180^\circ$ corresponds to the situation where a projectile, after colliding with a target, returns back in the same direction from which it originally came. This is the so-called 'head-on' collision which was detected for the first time in the mentioned experiment of Rutherford [1]. It is precisely this head-on collision which has provided the first evidence of the planetary structure of the atom with the nucleus being localized to a small center around which the electrons revolve. Before these relatively rare head-on collisions were detected by Rutherford [1], the peculiar Thomson model was largely unchallenged in viewing the electrons as being stuck like plums in a 'pudding' of positively charged protons occupying the whole atom. The forward and backward collisions are very important from a theoretical viewpoint. Thus, for example, one can encounter the problem of non-integrability of the transition or T$-$amplitudes at $\theta = 0^\circ$.

In order to find out what happens with the incident beam of particles, one uses a detector which should be placed in an asymptotic spatial region behind the target[7]. Such a detector could be fixed at a given point in space in a given direction θ or it could be a kind of a movable measuring device. We could also place a number of detectors at different values of θ to measure the angular distribution of scattered particles. Each detector would be said to correspond to a channel and the measurement would provide counts per channel from which differential cross sections could be extracted. Scattering of particles from the incident beam that hit the detector will be registered through the emergence of a voltage impulse which could also be converted into a signal. Clearly, such an event is not guaranteed to happen. Rather, its occurrence is a probabilistic phenomenon. This means that, at best, the theory can ascribe only a certain probability to each given collisional event. Of course, before the act of measurement has been made, a number of experimental conditions must be fulfilled for a reliable preparation of both the projectile and the target. First, it is customary to assume that the target is at rest, so that the relative velocity of the colliding particles is equal to the incident velocity v_i of the projectile. This is not the only possibility, since there are experiments with merging beams (or crossed beams) that involve two moving beams of the projectile and target [18]. In an experiment with a stationary target, we imagine that the target is localized in the center of the laboratory coordinate system of reference. The $Z-$axis in this laboratory system is chosen to coincide with the incident beam, whereas the spatial positions of the scattered particles are given by the usual spherical coordinates $\Omega = (\theta, \phi)$ where Ω is the solid angle. Further, we ought to know the quantum state of the target

[7]In this book, we shall not go into details of a description of experimental aspects of the process of detection of particles. Instead, the detecting system will simply be conceived as a counter which registers e.g. the outgoing scattered particles.

and prepare it in e.g. the ground state as in most experiments, but metastable states are also occasionally used.

1.4 The role of wave packets

In scattering experiments, one does not deal with an idealized collision between two isolated particles. Rather, the incident beam contains a huge number of stable particles ($\sim 10^{15}$ or more) that usually have sufficiently precise kinetic energies. On the other hand, the momentum vectors of projectiles i.e. the incident wave vectors $\boldsymbol{k}_i$, do not necessarily possess their sharp values. To reflect this latter fact appropriately, an adequate theoretical description should employ a wave packet instead of a plane wave $\phi_{\boldsymbol{k}_i} \propto \exp{(i\boldsymbol{k}_i \cdot \boldsymbol{r}\,)}$. This is usually achieved by forming a time-dependent travelling wave packet $\phi(t)$ through an integration over all the possible intermediate momenta $\boldsymbol{k}$ as

$$\phi(t) = \int \mathrm{d}\,\boldsymbol{k}\, w(\boldsymbol{k}\,)\mathrm{e}^{-iE_k t}\phi_{\boldsymbol{k}} \tag{1.8}$$

where $E_k = k^2/(2m)$ is the classical kinetic energy[8] of the particle of mass m and the linear momentum $\boldsymbol{k} = m\boldsymbol{v}$. Thus, a wave packet is a linear superposition of plane waves $\phi_{\boldsymbol{k}} = (2\pi)^{-3/2} \exp{(i\boldsymbol{k} \cdot \boldsymbol{r}\,)}$ with a certain weight $w(\boldsymbol{k}\,)$ of a peaked form such as a Gaussian, Lorentzian or Voigtian [19]–[22]. The role of the weight function $w(\boldsymbol{k}\,)$ is to cut off the individual superimposing plane waves $\phi_{\boldsymbol{k}}$ beyond a finite portion of the space. Such weight functions are said to be peaked in a given limited range (e.g. $\boldsymbol{k} \approx \boldsymbol{k}_i$), where they attain their maximum values, whereas outside this domain they are practically equal to zero. In this way, the whole wave packet $\phi(t)$ is conceived as being concentrated around a momentum value which is usually the impact momentum $\boldsymbol{k}_i$. As opposed to the corresponding plane wave $\phi_{\boldsymbol{k}_i}$, the wave packet $\phi(t)$ formed in this manner is normalizable i.e. it has a finite norm ($||\phi(t)|| < \infty$). This means that $\phi(t)$ belongs to the Hilbert state space $\mathcal{H}$ and, as such, can be physically interpreted despite the fact that we are dealing with the elements of the continuous spectrum of the Hamiltonian operator $\hat{\mathrm{H}}$ of the considered system. Otherwise, it is obvious that the continuous spectra of Hamiltonians have their most important application precisely in scattering theory [7, 23], as analyzed in chapter 3.

[8]In non-relativistic collisions the velocity of colliding particles is negligible relative to the speed of light c. This justifies the usage of the classical expression for the kinetic energy and momentum.

1.5 Adiabatic switching of interaction potentials

The introduction of wave packets into scattering theory is done not only to more adequately match the experimental situation characterized by the lack of purely mono-energetic incident beams. There is another reason for such a concept and that is the well-known asymptotic convergence of scattering states [7]. Here, given the total $\Psi^{\pm}(t)$ and unperturnbed $\Psi_0(t)$ states, matching between the two strong limits $\lim_{t\to\mp\infty} \Psi^{\pm}(t)$ and $\lim_{t\to\mp\infty} \Psi_0(t)$ is accomplished by establishing convergence of the so-called scattering integrals when plane waves are replaced by wave packets. Yet, wave packets are utterly inconvenient in practical calculations. This is because in e.g. stationary scattering theory all the relevant quantities must be subjected to an additional integration over $\boldsymbol{k}$ with the weight function $w(\boldsymbol{k})$. Alternatively, plane waves can be retained in the analysis with e.g. short-range potentials by using the concept of the so-called adiabatic theorem from chapter 6. Within such a strategy, convergence of the mentioned scattering integrals can be secured in a much more attractive manner for applications. This is achieved via adiabatically turning on and off the interaction potential V which leads to scattering. In so doing, we simply imagine that the potential V is screened through multiplication by the Dyson damping factor $\exp(-\varepsilon|t|)$ such that $V\exp(-\varepsilon|t|)$ vanishes when $t \to \mp\infty$, where ε is an infinitesimally small positive number. Then all the limiting procedures on convergence of scattering states in the asymptotic region exist and can be carried out. In the end of such an analysis, the attenuation factor ε is let to tend to zero through positive numbers. Nevertheless, one of the inconvenient consequences of using wave functions of mono-energetic incident beams is the necessity for the introduction of generalized functions, such as the Dirac $\delta-$function. These are, in fact, not functions in a strict sense of the word, but rather they represent distributions [24]–[26]. Moreover, expressions containing such distributions acquire their meaning only after they are integrated with the help of certain weight functions and this is just another equivalent way of creating wave packets. This is the way by which vectors of continuum 'states' should be understood when they are normalized to the $\delta-$function, as is the customary procedure in the literature [27].

1.6 Collimation of beams of projectiles

The constituent particles in an experimentally well-prepared incident beam are presumed not to interact with each other on their way to the target. For this to happen, it is necessary that the incident beam is not overly narrow and intense. The reason for which this beam cannot be entirely mono-energetic is that each instrument, including the energy analyzer of projectiles, has a finite resolving power. This means that, in practice, the impact energy can be determined up to some reasonable and acceptable uncertainty. In addition to the term 'mono-energetic', one also uses the term mono-chromatic, which refers to an incident beam which contains waves that are regularly repeated with some precisely determined frequencies and wavelengths. If a given wave is exactly mono-chromatic, then it is automatically coherent, meaning that the manner and the form of its propagation are maintained constant in large spatial and time intervals. Certainly, we want a detector (located at a fixed angle $\theta > 0°$) to register only those particles from the scattered beam that hit the detector after they have indeed collided with the target. In other words, the incident beam ought to be sufficiently narrow i.e. collimated in order to eliminate a contribution due to angles $\theta > 0°$ from the particles passing by the target without collision ($\theta = 0°$). This is achieved by letting the incident beam pass through one or more collimating holes on a fixed collimating diaphragm before collision with the target[9]. A collimating diaphragm could be visualized as a sufficiently flat plaque which, throughout its surface, represents a virtually insurmountable barrier to the incident beam, except at a centrally located circular collimating hole. The center of the source, the collimating hole and the target must be properly aligned along the direction which is perpendicular to the plane of the collimating diaphragm. Thus, only a part of the incident beam will pass through this collimating hole on its way from the source (e.g. an accelerator) to the target. For an optimized beam collimation, several parallel collimating planes with aligned collimating holes are customarily used in scattering experiments. A well collimated incident beam is characterized by a wave packet whose longitudinal length should coincide with the diameter of the collimating hole. We then say that after being passed through one or more collimators, the incident beam is considered as being sufficiently collimated. This means that all the particles in the projectile beam move like a rather uniform and narrow stream in approximately parallel directions along the incident wave vector $\boldsymbol{k}_i$. This is how we view the collimated incident wave packet. Due to a finite experimental energy resolution of an energy analyzer as well as because of the statistical fluctuations in energies of individual particles, the profile of the beam is not like a sharp line, but rather it exhibits a predominantly Gaussian shaped form.

Customary collimation discussed thus far is typical of single-pass experiments (projectiles go through the target only once) performed using ion beams

[9]Of course, before being sent to the target, the incident beam must be passed through an energy analyzer to determine the impact energy with some error bars. However, this and similar technical, but nevertheless important details, will not be addressed in the present book. This is because the final goal of our dealing with scattering experiments *per se* is not in a thorough analysis of all the relevant technical characteristics of realistic measurements, but rather our focus is on basic conceptual principles, as in Ref. [7].

from e.g. linear accelerators. However, there are also multiple-pass experiments (target traversed more than once by projectiles) at cooler storage rings [28, 29]. With these accelerators, a much better overall beam quality, including collimation can be achieved by the so-called adiabatic cooling of the initially hot ion beam with a short and parallel beam of cold electrons that are injected into one small ring section of the length of a typical bending magnet and diverted immediately afterwards. Initially, electrons emitted from a cathode at temperature $T \approx 800\,^{\circ}\mathrm{K}$ have energies around $kT \approx 0.1$ meV. Subsequently, the gas of electrons is adiabatically expanded in order to lower its temperature and the corresponding energies can be in the range $kT \approx 0.01 - 0.001$ meV prior to injection into the ring via an electron gun [16, 28].

In kinetic theory of gases, an adiabatic process is the one in which a system exchanges no heat with its environment, as expressed by $\Delta T \approx 0$ where $\Delta T = T_1 - T_2$. Here, T_1 and T_2 are the temperatures of the given system and its environment, respectively (or the temperatures of two gases brought in contact). Temperature T is connected with the average molecular kinetic energy $\bar{E} = \langle Mv^2/2 \rangle$ via $\bar{E} = 3kT/2$, where M is molecular mass and k is the Boltzmann constant[10]. For gases comprized of heavy particles (ions, atoms), the averaging in the stated definition of $\bar{E}$ is taken over the Boltzmann distribution. In the Boltzmann energy distribution, which is a Gaussian function, $\langle Mv^2/2 \rangle$ is proportional to the FWHM. This link between the average energy and temperature of a gas gives the meaning of the mentioned terminology of hot or cold ion beams and the related aspect of collimation in multi-pass collision experiments. Thus, a hot or cold ion beam has a high or low T i.e. a large or small value of $T = 2\bar{E}/(3k)$ pointing to a broad or narrow Gaussian as a characteristic of a poor or good energy collimation, respectively. Further, when two gases of different temperatures are brought into contact, they exchange heat maximally when the vectors of the average velocities of the two gases are nearly equal to each other (adiabatic cooling). Such an effect is experimentally achieved by tuning the electron beam velocity, so that the vector of relative ion-electron velocity becomes close to zero. When this occurs, a hot ion beam is extremely quickly cooled by merely a few of its passages through the electron beam section. This is seen by monitoring the Schottky signal, which by completion of cooling shows that the ion beam profile is drastically narrowed. Hence collimation by electron adiabatic cooling. The invoked mechanism of the electron-ion velocity matching is the signature of a nearly exact resonance effect. At the beginning of ion cooling, the conditions are non-adiabatic due to the existing heat exchange between the two merged parallel beams. Cooling is completed when the two beams are in heat equilibrium, in which case no further heat exchange takes place between the two merged beams. This latter phenomenon is the signature of the adiabatic conditions as per the definition of an adiabatic process.

---

[10] In this context, temperature T is often called "kinetic temperature".

1.7 General waves and quantum mechanical waves

At this point of the exposition, it is instructive to make a slight degression of the analysis and discuss several notions related to customary waves. This is pedagogical because the features which will be mentioned have a similar meaning and significance for quantum mechanical waves. Waves appear in nearly every branch of physics, so that we have mechanical waves, sound waves, light waves, electro-magnetic waves, etc. For example, mechanical waves stem from moving of a part of a given elastic medium away from its normal constellation leading to some oscillations of the medium around a certain equilibrium position. In this phenomenon, we encounter elastic forces among adjacent layers of the examined medium leading to disturbances that are transferred as waves from one layer to another. A similar phenomenon also takes place with waves of other kinds, but as opposed to mechanical waves, it is not always necessary to have a material medium for transmission of waves. For instance, electro-magnetic waves could pass through an ideal vacuum as well. If we draw a surface through all the points that at a given time undergo similar disturbances due to the passage of waves (i.e. they are in the same phase of motion), we shall obtain the so-called wave front. If the medium is homogeneous and isotropic, the direction of propagation of the wave should always be perpendicular to the wave front and to this line we associate the wave vector $\boldsymbol{k}$. Wave fronts can possess very different shapes and forms. If disturbances move through the medium in only one given direction, then we shall have plane waves. This is because at a given time, the conditions are nearly the same throughout any plane orthogonal to the propagation direction. In such a case, the wave fronts are the parallel planes and the wave vectors are the parallel straight lines orthogonal to these planes. Two arbitrary adjacent planes of such a plane wave are separated by the wavelength λ. Another simple case is given by spherical waves. Here, disturbances propagate in all radial directions from a point source of the wave and the wave fronts are concentric spheres. Any two such adjacent spheres are separated by the wavelength λ. Wave vectors of a spherical wave extend along a radius of a sphere with the direction pointed away from the source of the wave. Far away from their source, spherical waves have a very small curvature, so that they can be approximated by plane waves in a limited part of the wave front. This reminder about the mentioned particular waves also applies to general waves, including corpuscular waves i.e. waves comprized of particles. Namely, in quantum mechanics, there exists the so-called de Broglie particle-wave dualism for description of matter. According to this concept, every particle under certain conditions can behave like a wave and *vice versa.* If $p = mv$ is the classical linear impulse or momentum of a particle of mass m and velocity v, then to this particle we can associate a wave of the wavelength $\lambda = h/p$ where h is the Planck universal constant. This de Broglie relation contains the characteristics of both a particle and a wave through p and λ, respectively. The reason for which matter has its dual manifestation in nature, being waves or particles, lies in the Heisenberg uncertainty principle on the impossibility to simultaneously measure any two conjugate variables associated with two non-commuting operators. Thus, for example, this unique principle of quantum mechanics states that it is impossible to simultaneously measure both the position and velocity of a particle in motion due to the following uncertainty relationship

$$\Delta p_\zeta \, \Delta r_\zeta \geq \hbar \qquad (\zeta = x, y, z) \tag{1.9}$$

where $\hbar = h/(2\pi)$, $\boldsymbol{p} = (p_x, p_y, p_z)$ and $\boldsymbol{r} = (r_x, r_y, r_z) \equiv (x, y, z)$. Here, as indeed no place else in physics, the Planck constant h, although exceedingly small in its absolute value ($h = 6.63 \times 10^{-34}$Js), has an exceptionally far reaching significance. Had the product of Δp_ζ with Δr_ζ by some chance been zero instead of $\hbar$, the classical ideas of particles and orbits of electrons in atoms would be correct. This would make simultaneous measurements of position and velocity of a given moving particle possible with unlimited precision, determined solely by the technical resolving power of our instruments. In this way the notion of a trajectory and force would also be justified in quantum mechanics as it is legitimate in classical physics. In classical mechanics, due to the lack of the existence of the inequality (1.9), a state of the given system is determined precisely by the numerical values of all the coordinates and impulses for all the constituent particles at any fixed time t_0. This, under the condition of a free evolution of a classical system, accurately determines the state of the system at any earlier ($t < t_0$) and/or later ($t > t_0$) time. An entirely analogous conclusion is also valid for the case when a classical system is exposed to an arbitrary external influence whose action upon each constituent particle of the system is known. However, the fact that the product $\Delta p_\zeta \, \Delta r_\zeta$ is greater than or equal to $\hbar$ tells us that the outlined classical ideas are wrong. The absolute value of the constant $\hbar$ indicates under which conditions we must resort to quantum mechanical concepts where the notions of e.g. trajectories of particles or forces have no physical meaning whatsoever and, as such, cannot be used for description of quantum systems. Thus, for example, in quantum wave mechanics, one cannot say that e.g. the electron in its ground state described by the wave function Ψ_0 in the atomic hydrogen circulates around the proton on the first orbit of the diameter of the Bohr radius a_0, as has initially been put forward in the Bohr planetary model of orbits [30] before the establishment of the Heisenberg uncertainty principle (1.9). Rather, after the emergence of the principle (1.9), all one is justified to say is that there is a chance i.e. a finite probability $|\Psi_0|^2$ of finding the electron at a given distance from the proton. In other words, not the trajectory of the electron, but rather the quantity $|\Psi_0|^2$ has a physical meaning at any particular point as a measure of the probability that the electron is located near the given position in space. Thus, quantum mechanics can say where it would be most probable to find the electron, but not at which precise position the electron would reside. Such an interpretation of the quantity $|\Psi_0|^2$ as a probability distribution provides a statistical connection between the wave function Ψ and the corresponding particle. This interpretation is the basis of the so-called probabilistic foundation of quantum mechanical description of general phenomena in nature. How dramatic this difference between the classical and the quantum concepts really is, can be illustrated by the example of the stability of the atom. In both the classical (Rutherford [1]) and the quantum (Bohr [30]) models of the atom, the electrons revolve around the nucleus. However, only the quantum model is capable of predicting and describing the stability of the atom. For the classical model, the concept of electronic orbits is valid with the underlying significance of the possibility of simultaneous and precise measurements of all the coordinates and momenta of all the electrons in the atom at every given time. But conceived in this way, the electrons are neither at rest nor in uniform motion and, therefore, cannot be in a stationary state in the atom. This is because according to the laws of electro-magnetism, the electrons will be accelerated towards the nucleus by the attractive electrostatic Coulomb potential and, hence, they must radiate their energy. Losing the energy while moving, the electronic trajectories will be diminished, leading ultimately to a collapse of the electrons into the nucleus. Hence instability of the atom described by classical physics. By contrast, in quantum physics a given stationary electronic state in the atom is possible, since it is described by a wave function as an amplitude of the probability distribution of all the possible positions and velocities. Under such quantum circumstances, the average current can vanish, implying the absence of spontaneous emission of energy. Yet, simultaneously, electronic velocities themselves could still have non-zero values due to non-vanishing average (expected) values, thus yielding non-zero average kinetic energies of the electrons.

1.8 Probability character of quantum collisions

Like quantum mechanics itself, descriptions of collisional systems and their interpretation by means of quantum scattering theory are fundamentally probabilistic. This is because no answer can be given to the questions of the type: 'for the given physical conditions, what exactly is going to happen to the examined colliding particles'? Instead, quantum scattering theory can answer, in principle, the corresponding more realistic and more relevant questions such as 'for the given physical conditions, what is the probability of each of the possible outcomes of the examined colliding particles'? As mentioned, the Heisenberg uncertainty relation (1.9) shows why it is possible that light and non-zero mass particles exhibit the dual wave-corpuscular behavior. Specifically, this is because these two equivalent aspects of matter are opposite to each other in such a profound way that their simultaneous manifestation is impossible if attempted to be detected in the same experiment. For instance, if we imagine an experiment which would force an electron to unequivocally display its wave character, then its corpuscular nature would remain inherently fuzzy (and *vice versa* in another experiment). In general, matter can be manifested in either of its two fundamental and real appearances as waves or particles, but never in both simultaneously. It is something like an impossible situation in which the same person attempts to see simultaneously both sides of the same coin[11]. Bohr used his complementarity principle to conceive wave-particle duality as the two features that complement rather than contradict each other. Of capital importance is the fact that the de Broglie idea about the wave nature of matter has been successfully confirmed in the landmark experiment of Davisson and Germer [31]. They carried out an experiment of collision of an electron beam with the crystal of nickel (Ni). The incident beam was created simply via acceleration of electrons by means of an alternating potential difference U_e. Atoms of Ni served as a three-dimensional filter of diffraction centers for the incident beam of electrons. If the electrons were to display their wave nature in an unambiguous manner, then they should exhibit very pronounced diffraction peaks in certain characteristic directions, similarly to the well-known diffraction of X-rays. Indeed, a remarkable peak has been observed at the scattering angle $\theta = 50°$ and for the kinetic energy 54 eV. This phenomenon can be qualitatively explained in terms of the so-called Bragg scattering provided that the electrons are conceived as waves that for the given conditions have the theoretically calculated de Broglie wavelength $\lambda = h/p = 1.64 \times 10^{-10}$cm. Such electronic waves undergo the diffraction pattern on a sequence of atomic planes, so that every two adjacent planes are separated by the distance $r_0 = 0.91 \times 10^{-10}$cm, as established via X-rays. Notice that the wavelength λ of the electron and the atomic inter-planar distance r_0 are of the same order of magnitude (a spatial resonance). The corresponding experimental data for the wavelength of the electron is $\lambda = 1.6 \times 10^{-10}$cm. This extraordinary agreement between the theory and experiment represents one of the most convincing proofs that, under special conditions, electrons are indeed waves. Similar conclusions have also been reached in experiments with other particles e.g. neutrons, etc. Interestingly, due to neutron diffraction, slow neutrons from nuclear reactors can be used for examination of atomic structure of solid bodies. All told, particles from a beam of projectiles can alternatively be called incident waves. The previously mentioned plane wave used extensively throughout physics has entirely the same meaning as a generic plane from general wave phenomena. Moreover, in one of the formulations of mechanics of atomic and subatomic particles, known as wave quantum mechanics, the general terminology from

[11]Of course, the two sides of a coin exist, although they are not both simultaneously visible to the same observer.

wave phenomena can also be used for quantum mechanical wave functions that describe states of the examined system. This is clear already from the well-known asymptotic form for the stationary wave function of the scattering state $\Psi_{\boldsymbol{k}}^{+}(\boldsymbol{r})$ [7]

$$\Psi_{\boldsymbol{k}}^{+}(\boldsymbol{r}) \underset{r\to\infty}{\longrightarrow} (2\pi)^{-3/2}\left[\mathrm{e}^{i\boldsymbol{k}\cdot\boldsymbol{r}} + f(\theta,\phi)\frac{\mathrm{e}^{ikr}}{r}\right] \tag{1.10}$$

which is a linear superposition of the plane wave $(2\pi)^{-3/2}\mathrm{e}^{i\boldsymbol{k}\cdot\boldsymbol{r}}$ and the spherically scattered outgoing wave $(2\pi)^{-3/2}f(\theta,\phi)\mathrm{e}^{ikr}/r$ in the configuration space. The function $f(\theta,\phi)$ is the scattering amplitude which is essential for the definition of the differential cross section $dQ/d\Omega_i$ given by

$$\frac{dQ}{d\Omega_i} = \frac{\text{scattered flux / solid angle}}{\text{incident flux / area}} = |f(\theta,\phi)|^2. \tag{1.11}$$

The quantity $\Omega_i = (\theta,\phi)$ is a solid angle (expressed in dimensionless units called steradians) around the incident wave vector $\boldsymbol{k}_i \equiv \boldsymbol{k}$ encompassing the polar (θ) and azimuthal (ϕ) angles.

In a measurement of differential cross sections for various values of angle θ, one regularly encounters a problem regarding the forward scattering ($\theta = 0°$). Specifically, it is impossible to distinguish elastic scattering of particles in the forward direction from the transmitted particles that did not collide at all with the target and, as such, are also detected at $\theta = 0°$. Therefore, the differential cross section $[\mathrm{d}Q/\mathrm{d}\Omega_i]_{\theta=0°}$ remains experimentally undetermined. One of the possible ways to partially overcome this difficulty is to perform several successive measurements in the close vicinity of the forward cone ($\theta \approx 0°$) and subsequently to carry out e.g. a spline extrapolation to $\theta = 0°$. The differential cross section $\mathrm{d}Q/\mathrm{d}\Omega_i$ is introduced in (1.11) as a quotient of the spatial density of flux of the outgoing particles around the solid angle Ω_i and the incident flux of particles per unit surface. Obviously, the quantity $\mathrm{d}Q/\mathrm{d}\Omega_i$ has the dimension of surface. In atomic physics, cross sections are customarily given as multipliers of the basic 'geometric' cross section $\pi a_0^2 \approx 8.8\times10^{-17}\mathrm{cm}^2$. In nuclear physics, the standard unit for cross sections is called barn (b) and its numerical value is $1\,b = 10^{-24}\mathrm{cm}^2$. In the majority of collisions, there is an axial symmetry around the $Z-$axis, so that differential cross sections are independent of ϕ i.e. $\mathrm{d}Q/\mathrm{d}\Omega_i = (2\pi/\sin\theta_i)\mathrm{d}Q/\mathrm{d}\theta_i$. This is the case in e.g. a collision involving only a central potential $V(\boldsymbol{r}) = V(r)$ which has the spherical symmetry. An integration of $\mathrm{d}Q/\mathrm{d}\Omega_i$ over the solid angle Ω_i in the interval $\theta \in [0,\pi]$ and $\phi \in [0,2\pi]$ yields the so-called total cross section Q which, of course, has the unit of surface just like $\mathrm{d}Q/\mathrm{d}\Omega_i$. When there is an axial symmetry, the integration over ϕ becomes trivial, giving the overall multiplicative factor 2π. This is usually anticipated already on the level of $\mathrm{d}Q/\mathrm{d}\Omega_i$ by writing $\mathrm{d}Q/\mathrm{d}\Omega_i = (2\pi/\sin\theta_i)\mathrm{d}Q/\mathrm{d}\theta_i$ as we did.

2

Requirements of the theory for the experiment

An interpretation of measured data in terms of simple physical terms, as well as direct and adequate comparisons with corresponding theoretical predictions assumes that a number of important experimental conditions are fulfilled. For example, if the target is comprized of N identical particles, then it is natural to suppose that the final experimental result will be equal to the corresponding N fold data for collisions on one particle of the target. Roughly speaking, this condition in the mentioned Rutherford experiment [1] on collisions of α particles on Au has been achieved by preparing the target in the form of very thin gold slices ($\approx$ 1 micron $\equiv 1\mu$). A detailed inspection of a typical scattering experiment reveals that the following eight most important conditions are necessary to secure a relatively solid basis which would justify subsequent meaningful comparison with theoretical predictions:

• (i) The source of the projectiles must secure that the incident beam is not of an overly weak intensity. This is because any statistically significant measurement must detect a large number of collisional events. The incident beam must not be too strong either, or otherwise interactions among individual particles in the beam would not be small. The existence of a non-negligible effect of the latter interactions would preclude the unequivocal definition of the initial state of the projectile-target system.

• (ii) The dimension of the hole on the collimating diaphragms must not be too small in order: (a) to eliminate as much as possible the undesirable diffraction effects of particles of the incident beam on the edges of the hole and (b) to describe the middle part of the impact beam via plane waves. Clearly, due to diffraction, there would be a number of particles from the incident beam that would scatter without having any collision whatsoever on the target. This would lead to unwanted enhancement of scattering angles because of diffraction. Obviously, one of the basic purposes of collimators is to prevent scattering of the incident beam prior to its arrival to the target. Likewise, collimating holes must not be too wide in order to avoid formation of a wide incident wave front. A consequence of such a wide wave front is that some particles from the projectile beam that are too far from the target would hit the detector without undergoing any collision.

• (iii) The target B must contain a large number $N_{\rm B}$ of scattering centers in order to intensify the overall collisional effects in the sense of accumulating a sufficiently large number of elementary processes. In the theory, this condition permits computation of average probabilities for scattering through the replacement of sums by integrals over contributions from elementary probabilities for collisions on individual scattering centers.

• (iv) Detectors must be sufficiently far from the target, so that it can be said with certainty that they are located in the asymptotic region. This would guarantee that the differential cross section $\mathrm{d}Q/\mathrm{d}\Omega_i$ is correctly related to the absolute value of the scattering amplitude $f(\Omega_i)$ which is the coefficient of the outgoing spherical wave for the stationary scattering states from (1.11). Simultaneously, the distance r between the detector and the target must be much larger that the radius d_0 of the collimator ($r \gg d_0$). This would enable us to distinguish scattered particles from those reaching the detector without collisions with the target. As an example, taking e.g. that $d_0 \approx 10^{-1}$cm, it would be sufficient to choose $r \approx 10^2$cm to fulfill the condition (iv) and, of course, this is trivially feasible in the usual experimental circumstances. It is a typical empirical fact that in scattering experiments with short-range potentials, measured cross sections become independent of the distance $r' \approx r$ between the interaction domain and the detector for sufficiently large distances i.e. if the measuring instrument is situated in the asymptotic spatial region of scattering. In non-stationary theory, this fact is interpreted by saying that the corresponding observables must possess their limiting values when $t \to \mp\infty$. Thus, one of the tasks of the theory is to find the appropriate asymptotic constants of motion. In the experiment, the initial information related to such constants is obtained through the analysis of the incident beam by removing the target from its place. Namely, at an arbitrarily chosen initial time t, we would first measure the probability distribution of these constants of motion (impulse, spin, isospin, etc) that characterize the incident beam without the presence of the target. Then we assert that such a determined distribution corresponds to the incident beam in the remote past ($t \to -\infty$) i.e. a long time before collision takes place in the actual experiment when the target is placed at its location[1]. This is a realistic assertion, since a typical collision time T_0 is short (of the order of e.g. $\approx 10^{-15}$s or less). Such a small value of T_0 justifies the said interpretation even if the projectile is released from its source immediately before the collision. Similarly, and in a symmetric reasoning, due to the shortness of the time duration T_0 of a typical collision, every finite time $t \gg T_0$ chosen as the beginning of measuring the examined observable, which is associated with the stationary scattering state, could justifiably be considered as the distant future ($t \to +\infty$). Thus, if measurements of the final information (after the collision) on the distribution of the investigated constants of motion yield approximately the same result as the corresponding data from the initial configuration, we can be reasonably sure that indeed we are talking about the asymptotic constants of motion. We then say that such constants of motion are practically the same for the remote past and the distant future (of course, having in mind the relative meaning of the time

[1]The target place is usually the origin in the laboratory system of reference.

T_0). In the corresponding theoretical description, we must strictly apply the limits $t \to \mp\infty$, since there does not exist an alternative way which would guarantee that we obtained the asymptotic stationary state vectors with the outgoing/incoming spherical waves, respectively. From the given arguments, the importance of establishing symmetry between the past and the future in the scattering problem can be clearly seen. This is because such a symmetry can help in establishing a meaningful correspondence between experimentally measured and theoretically predicted quantities.

It should be emphasized that what is usually considered to be the main result of a scattering experiment is not the answer to the otherwise customary question: if we know the state $\Psi_i(t_0)$ of a given physical system at the initial time t_0, what is the probability to find this system in a different state $\Psi_f(t)$ at a later time $t > t_0$? The appropriate modification of this question relevant to the scattering experiment is: for a given initial state $\Psi_i(t_0)$ established in the considered system when the target is removed at an infinite separation from the projectile source, what is the probability that the system is found in the state $\Psi_f(t)$ when $t \to +\infty$ after returning the target to its original place?

• (v) The distance among the particles of the target must be much larger than the range of the perturbation interactions. This range is the characteristic radius after which the potential V can be ignored as e.g. in nuclear collisions. For example, if the target is in its solid state, then the typical distance r_0 among the constituent particles of the target is of the order of the Bohr radius a_0 i.e. $r_0 \approx a_0 \approx 10^{-8}$cm. Certainly, this distance is much greater from the range R of nuclear interactions, since R is of the order of one fermi ($R \approx 10^{-13}\text{cm} \equiv 1f$). The reason for imposing the condition (v) is in attempting to eliminate a situation where the incident particle on a given scattering center senses the fields of neighboring scattering centers. Of course, here it is understood that the scattering centers in the target are distributed in a random fashion. Moreover, such centers are assumed to be distributed in such a way that, on the average, they appear uniformly and without mutual overlap. This would eliminate phase correlations among waves scattered at different scattering centers. The theory of scattering examines single events at a time, such as a collision of one particle from the incident beam with one typical scattering center from the target. A failure in a measurement to secure that the coherent phase effects from different scattering centers are approximately eliminated would invalidate any comparison between the experimental and theoretical data.

• (vi) Each individual particle from the incident beam must be scattered only on one particle of the target (binary collisions). In other words, double, triple or multiple collisions of the projectile with the target must be eliminated along the way of the incident beam through a target of a given thickness. For a given target in the solid state, this condition can be approximately achieved by preparing the target in the form of thin slices (foils). If such a solid state target is imagined as a set of parallel layers separated by the distance $r_0 \approx 10^{-8}$cm, then the probability W of scattering of the projectile on each layer will be approximately equal to $Q/(\pi r_0^2)$. Here, Q is the total cross section and πr_0^2 is the classical geometric cross section. To get a rough estimate for W, we can take the value $r_0 \approx 10^{-8}$cm as the standard distance between the individual particles of the target and using $Q < 10^{-25}\text{cm}^2$ as a typical cross section measured in experiments on nuclear collisions, it follows $W \approx Q/(\pi r_0^2) < 10^{-9}$. Thus, in a sufficiently thin target e.g. of thickness of the order $\approx 10^{-2}$cm, the probability of multiple scattering is about $10^{-3} \ll 1$. Under such circumstances, the condition (vi) can be considered as fulfilled.

• (vii) The de Broglie wavelength λ of particles in an incident beam must be much smaller than the mutual separation r_0 between adjacent scattering centers in the target. As an illustration, consider a standard experiment on nuclear collisions for an incident beam of protons of kinetic energy 100 MeV. Let the particles in a metallic target be mutually separated by $r_0 \approx 10^{-8}$cm. This would guarantee the absence of nuclear interactions among the constituent particles of the target. The corresponding incident velocity v_i is: $v_i = [2 \cdot 10\text{MeV} \cdot (1.6 \times 10^{-19}\text{J/eV})/(1836.12 \times 9.1 \cdot 10^{-31}\text{kg})]^{1/2} \approx 1.38 \times 10^6\text{m/s}$, so that the associated impulse p becomes $p = m_p v_i \approx 2.31 \times 10^{-21}\text{kg} \cdot \text{m/s}$ where $m_p \approx 1836.12 m_e$ and $m_e \approx 9.1 \times 10^{-31}$kg are the masses of the proton and electron, respectively. With these data, the de Broglie wavelength reads $\lambda = h/p = 6.6 \times 10^{-34}\text{Js}/(2.31 \times 10^{-31}\text{kg} \cdot \text{m/s}) \approx 2.86 \times 10^{-13}$cm. This implies $\lambda \ll r_0$, so that the condition (vii) is satisfied. However, this condition is not fulfilled in diffraction phenomena in crystals, as in the case of diffraction of atoms on a crystal of nickel or in diffraction of thermal neutrons, diffraction of X-rays, etc. The condition (vii) imposed upon the scattering experiment eliminates interference effects, so that the total transition amplitude T_{if} becomes equal to the N fold value of one typical individual amplitude. In other words, individual collisions of particles from the projectile A on individual scattering centers from the target B are mutually unrelated and, as such, cannot lead to a global coherent effect. However, in principle, quantum mechanical descriptions of scattering of particle beams with targets are capable (at least in principle) of examining individual collisions of the projectile with each of the N particles of the target. This would allow us to write the total transition amplitude T_{if} for the global process as a sum of these individual transition amplitudes $\mathrm{T}_{i_n f_n}$ where $n \leq N$. In such a case, the corresponding differential cross section is proportional to the quantity $|\mathrm{T}_{if}|^2 = |\sum_{n=1}^{N} \mathrm{T}_{i_n f_n}|^2$ which obviously incorporates the interference effects.

• (viii) Experiments on atomic and subatomic levels must possess their internal tests for verifying the needed randomness and statistical character of elementary collisional events. Here, the term 'randomness' understands that after the given experiment has been repeated a number of times under identical conditions, the examined elementary events will have different probabilities (statistical fluctuations)[2]. In other words, a detailed distribution of individual observations cannot be reproduced from one measurement to the other. However, the usual statistical characteristics e.g. the average value of the probability, fluctuation indices, etc, are reproducible if measurements are repeated a large number of times under the identical conditions. For example, in the well-known Frank-Hertz experiment [32] on inelastic collisions of electrons with atoms, the detectors could register individually scattered electrons at certain selected scattering angles and for some values of the incident electron energy. The time interval between such successive detections and the sequence in which the electrons are scattered at different angles and energies is random. On the other hand, the statistical distribution of these time intervals, as well as scattering angles and energies or other related observables are reproducible within the limit of the experimental error bars. For example, radioactive decays are inherently unpredictable in the sense that it is impossible to say when any given particular radioactive nucleus will be disintegrated. Thus, one of the two identical radioactive nuclei can live longer or shorter than the other, before it fully decays, pointing to the randomness of the decay event. Thus far, the particular characteristics have not been identified for unequivocal discrimination of the conditions of short-lived from long-lived nuclei of the same kind. This statement is in apparent contradiction with the well-known fact that for nuclei of a given kind there exists a certain definite time called the lifetime τ. However, this contradiction is solved by resorting to the statistical character of the observable τ, via considerations of a large number of identical radioactive nuclei, rather than only two or a few at a time. Hence, here the only predictable quantity is the probability of decay of a large number of the same nuclei in a given time interval. Let the total number of nuclei that decay for an infinitesimally small time interval $\mathrm{d}t$ be $N\,\mathrm{d}W$. Then the corresponding decrease $\mathrm{d}N$ of the number of the radioactive nuclei within $\mathrm{d}t$ is given by the negative increment of the total number N via $\mathrm{d}N = -N\mathrm{d}t/\tau$. Integrating this latter differential equation over time with the initial condition $N(t=0) = N_0$, it follows $N = N_0 \exp(-t\tau)$. Recall that the randomness of radioactive decays of nuclei is so reliable that it is used via an algorithm for computer simulations as one of the best generators of random numbers. The condition (viii) is needed in order to accommodate the basic concept of probability, which can be applied only to reproducible experiments. All the experimental data are based upon probability and, therefore, their adequate verification can be obtained by repeating the same measurements a large number of times under the same conditions and observing the variations of the recorded findings. In such a case, each particular result will be a fraction of the total number of the repeated measurement. This means

[2]Of course, the notion of randomness has its significance also for a single elementary process, with the meaning that e.g. the place and the time of the occurrence of an individual event cannot be predicted.

that there exists a definite probability for obtaining such a result[3]. Finding such average values of the probability is the central task of scattering theory.

2.1 Elementary events versus multiple scatterings

After having listed the conditions (i)–(viii), the following important question emerges: How can we know with certainty that the events of a given kind indeed stem from elementary processes i.e. from one scattering at a time of one particle from the incident beam on a single particle (say, an atom) from the target? Could it be that the same result is due to a cumulative effect as a consequence of certain effective interactions of the given impact particle with two or more mutually unrelated i.e. independent atoms from the target B? The answer to this type of question can be given in the following way. Assume that a criterion for defining a given class of events is determined by verifying whether or not the event took place inside a certain fixed volume in the space. Then the event which includes the interaction of the incident particle with one atom at a time (an elementary process) must happen with the rate which is directly proportional to the number of atoms in the examined volume. The count of such elementary events by a detector will differ from a more complex event involving cumulative effects due to interactions of the impact particle with e.g. any two unrelated atoms from the target. This is because the rate or the frequency of occurrence of this latter event will be directly proportional to the pairs of the independent atoms i.e. to the square of the number of atoms in the given volume. Of course, more general higher-order events are also possible, involving e.g. interactions of one impact particle with n independent atoms. The rate of such events will be directly proportional to the nth power of the number of atoms in the investigated volume. Therefore, a clear answer to the posed question will be available by performing measurements that would identify those elementary processes whose rate of occurrence is directly proportional to the number of atoms in the target in the given volume, and simultaneously eliminate all other events as irrelevant. In this way, we can conclude that the rate of elementary processes must be directly proportional to the density of the material traversed by the given incident beam. If we do not use a counter as a detector, but rather the events are observed through the corresponding tracks that are left on an emulsion from the passage of the incident particles, then the rate of the events based on elementary processes must be proportional to the length of

[3]Of course, it could happen in some very special cases that the probability of one of these individual elements is equal to unity. Then we shall say that the corresponding measurement is deterministic.

the examined track and to the density of the material of the target. This means that the important quantity which must be known while observing elementary processes along the given length of the track is the rate of such processes per unit track length and per unit density of the traversed material. In such a case, the obtained rate will be independent of the track length and the density of the material of the target. Or, for events within a well-defined layer of the material from which the target is built, the event rate due to elementary processes must be proportional to the thickness of the examined layer. Thus, for example, in Rutherford's experiment [1] on the α – Au collisions, it was essential to examine large deflections of projectiles from their incident direction. Here, tests with gold foils of varying thickness lead to a large deflection of α particles by about $\sim 5°$ that can be attributed to elementary processes, provided that a typical foil of Au is sufficiently thin e.g. of the order of 1μ.

2.2 Average probabilities

When the conditions (i)–(viii) are fulfilled, as is indeed the case with most well-designed measurements, it is possible to carry out a direct comparison between theory and experiment on scattering, since each scattering center ν of the target acts as if it were isolated from the rest[4]. This allows us to focus our attention upon single collisional phenomena, since it is now justified to study one typical collision of one particle from the incident beam of particles on only one scattering center ν from the target. Firstly, theoretical predictions for the probability $W(\nu)$ are sufficient for only a single event i.e. for a collision of one incident particle with one scattering center ν from the target. Secondly, accounting for the conditions (i)–(viii) listed above[5] enables us to arrive at the theoretical average value of the probability $\bar{W}$ by summing over the contributions from the probabilities $W(\nu)$ for single scattering events stemming from each of the individual scattering centers ν. Here, it is pertinent to ask the following question: could it also be possible to experimentally obtain the probability $W_{\rm e}(\nu)$ for an elementary process involving only one particle from the projectile A and one scattering center from the target B considered as an 'ideal' ensemble (projectile+target)? The answer is: 'yes', in principle, but 'no', in practice. This is because we would first measure $W_{\rm e}(\nu)$ several times under the identical experimental conditions. If the findings differ, we would

[4]If the conditions (i)–(viii) are not satisfied, then a much more demanding analysis of the measured data is required, involving multiple scattering phenomena treated within particle transport physics [33].

[5]Especially neglecting multiple scattering within the target as well as taking advantage of the random distribution of scattering centers in the target.

conclude that we are dealing with a random event. However, if we repeat the measurement a large number of times under the identical conditions and find that a particular deflection of the projectile from its incident direction is more probable than scattering at other angles, then we would say that the examined collision possesses a statistical character. In this way, we could, in principle, carry out the measurement of the probability $W_{\rm e}(\nu)$ with a certain standard deviation. Then the corresponding theoretical result $W(\nu)$ should agree with the experimental finding $W_{\rm e}(\nu)$ to within the estimated standard deviation. Of course, such an idealized experiment is entirely unfeasible in practice. This is because the voltage impulse in the measuring detector from only one scattering particle would be negligibly small and, as such, below the threshold sensitivity of any realistic measuring instrument. Hence the impossibility to register experimentally a single elementary event. In other words, it is indispensable to have a sufficiently large number of particles in the incident beam in order to intensify the examined effect on the detector.

2.3 Total cross sections

Fulfillment of the conditions (i)–(viii) will enable us to introduce, in a simple way, the main observable in the scattering experiment i.e. the cross section. We shall do that first on a geometrical level for a similar, but macroscopic problem. Consider a certain plane b of the size of e.g. a school blackboard of the surface s_b on which we suspend in a random fashion and without overlapping some n_b small identical plates each of the individual surface Q_0. Let us now imagine that a hunter shoots randomly (e.g. from the dark) at the plane b hitting the surface s_b with some $n^a_{\rm inc}$ bullets whose fraction $n^a_{\rm sc}$ falls onto the plates. The question which is posed here is: how to determine the geometric surface Q_0 of one plate? The total surface covered by the plates is n_bQ_0, whereas the rate by which the bullets hit the plane b is equal to the number $n^a_{\rm inc}$ per unit time (this is the incident flux). From the total number $n^a_{\rm inc}$ of the bullets that fell onto the plane b, only the fraction $n^a_{\rm sc}$ hits the plates. Thus, the rate of hitting the plates is equal to the number $n^a_{\rm sc}$ per unit time (this is the scattered flux). Since we are dealing with random events, there exists the following self-evident proportionality $n^a_{\rm sc} : n^a_{\rm inc} = n_bQ_0 : s_b$. This gives at once the sought result for Q_0 as

$$Q_0 = \frac{n^a_{\rm sc}}{n^a_{\rm inc}} \frac{s_b}{n_b}. \tag{2.1}$$

In this way, using the known quantities $n^a_{\rm inc}, n^a_{\rm sc}, n_b$ and s_b, we arrive at the surface Q_0 of one individual plate. Such a calculated quantity Q_0 could be justifiably called the cross section for an event (an elementary process) which represents the hit of a single bullet onto one plate. Entirely similarly, in nuclear physics, experimentalists bombard a selected target by a chosen nuclear projectile beam. In such experiments, one measures the rate in which the events of a given type occur. Although the details of individual collisions are unknown in this microscopic realm, much information gained from these experiments remains independent of the number or the density of atoms in the target or the flux of the incident particles. For example, if we bombard a thin foil of e.g. gold ^{197}Au by deuterons d as the projectiles of the impact energy 30 MeV, we could have several different events, such as the following nuclear reactions

$$d + {}^{197}\text{Au} \longrightarrow p + {}^{198}\text{Au} \qquad d + {}^{197}\text{Au} \longrightarrow n + {}^{198}\text{Hg} \tag{2.2}$$

where n denotes the neutron. Each of these events, including many other processes that are not mentioned, possesses their cross sections Q. This cross section can be found in a way which is very similar to the outlined macroscopic geometrical model by considering the following general scattering problem. Let the entrance channel be comprized of the projectile A(i) and target B(i') with the set of well-defined quantum numbers $\{i, i'\}$. For convenience, let the target B be in the form of a metallic foil[6]. Further, let a sufficiently large number (per unit time) of structureless particles from A reach B in approximately parallel directions with a distribution of kinetic energy strongly peaked at a given fixed value. Let a part of the total surface of the target B exposed to the incident beam be denoted by S_{B} and let x be the thickness of the metallic foil. If there are N_{B} scattering centers in the given volume xS_{B}, then the corresponding density ρ_{B} i.e. the number of particles N_{B} in the volume xS_{B} is equal to $\rho_{\text{B}} = N_{\text{B}}/(xS_{\text{B}})$. If by Q we denote the elementary surface within the target i.e. the cross section for scattering of one incident particle on one scattering center, then the total surface, which is the total cross section for scattering on all the N_{B} scattering centers will be $N_{\text{B}}Q = \rho_{\text{B}}xS_{\text{B}}Q$. Let $N^{\text{A}}_{\text{inc}}$ be the rate by which the projectile hits the surface S_{B} and, moreover, let N^{A}_{sc} be the rate of the emergence of the scattered particles, signifying the events with deflections from the incident direction. Then, because of the random character of such events, the following evident proportionality will hold true $N^{\text{A}}_{\text{sc}} : N^{\text{A}}_{\text{inc}} = N_{\text{B}}Q : S_{\text{B}}$, which gives

$$Q = \frac{N^{\text{A}}_{\text{sc}}}{N^{\text{A}}_{\text{inc}}} \frac{S_{\text{B}}}{N_{\text{B}}}. \tag{2.3}$$

Although the expressions (2.1) and (2.3) are formally identical, it is clear that the analogy of the two considered problems is, in fact, complete. Of course, in a collision for which (2.3) is valid, the situation is much more

[6]This does not diminish the generality of the analysis whose results can be trivially extended to encompass the other aggregate states of the target.

complicated than the preceeding macroscopic model with the cross section (2.1). This is because in a scattering of the projectile A on the target B, there are more possibilities for the final result, as mentioned in the condition (viii) for scattering experiment. The result (2.3) can be rewritten in terms of the incident and scattered fluxes as is more customary in the literature. Namely, the cross section (differential or total) for a certain type of events in a given collision is defined by the quotient of the number of such events (per unit time and per scattering center) and the average flux of the incident particles $\jmath_{\rm inc}^{\rm A}$ relative to the target. The average incident flux $\bar{\jmath}_{\rm inc}^{\rm A}$ represents the number of particles $N_{\rm inc}^{\rm A}$ from the projectile A that pass per unit time through unit surface $S_{\rm A}$ perpendicular to the incident direction, so that

$$\bar{\jmath}_{\rm inc}^{\rm A} = \rho_{\rm A}\, v_i = \frac{\jmath_{\rm inc}^{\rm A}}{S_{\rm A}}. \tag{2.4}$$

Here, the quantity $\jmath_{\rm inc}^{\rm A}$ is the incident flux, defined as the number $N_{\rm A}$ of those particles from A that reach the target per unit time along nearly parallel directions having a pronounced peak around a given value. The dimension of the incident flux is given by $\rm s^{-1}$ i.e. $[\,\jmath_{\rm inc}^{\rm A}\,] = \rm s^{-1}$. In (2.4), the observable $\rho_{\rm A}$ is the projectile density i.e. the number of particles from the incident beam A per unit volume. Further, v_i is the relative average velocity (with respect to B) resulting from all the individual velocities of particles from A, whereas $S_{\rm A}$ is the cross-sectional surface of the incident beam. Under the assumption of the absence of dispersion of the incident beam, we have $S_{\rm A} \approx S_{\rm B}$, where $S_{\rm B}$ is that part of the whole target surface which is being hit by the projectiles. Dimensional analysis gives $[\,\bar{\jmath}_{\rm inc}^{\rm A}\,] = [\,\rho_{\rm A}\,]\cdot[\,v_i\,] = {\rm cm}^{-3}\cdot({\rm cm/s}) = {\rm cm}^{-2}s^{-1}$ or $[\,\bar{\jmath}_{\rm inc}^{\rm A}\,] = [\,\jmath_{\rm inc}^{\rm A}\,]\cdot[\,S^{-1}\,] = {\rm s}^{-1}\cdot{\rm cm}^{-2}$ and this is correct, as required by the definition (2.4) of the average incident flux. In the quoted definition of the cross section, under the notion ‘event’ we understand an occurrence in which the detector is hit, provided that the projectiles have previously collided with the target. Also, the detector is assumed to be in the asymptotic region of scattering i.e. at a macroscopic distance r from the target, as required by the condition (iv).

With the target assumed to be sufficiently thin, let us now introduce the three-dimensional quantity $N_{\rm B}$ representing the number of the scattering centers in the target B within the target effective volume $v_{\rm eff}$ which is actually exposed to interactions with the incident beam. If the target is taken as being comprized of thin layers of length x measured perpendicularly to the incident direction, we shall have

$$N_{\rm B} = S_{\rm B}\, x\, \rho_{\rm B} \equiv \rho_{\rm B} v_{\rm eff} \tag{2.5}$$

where $v_{\rm eff} = xS_{\rm B} \approx xS_{\rm A}$ with $\rho_{\rm B}$ being the effective target density i.e. the number $N_{\rm B}$ of particles in the volume $v_{\rm eff}$. Thus, we have $[\rho_{\rm B}\,] = {\rm cm}^{-3}$, so that according to (2.5) the quantity $N_{\rm B}$ is obviously a dimensionless observable. The earlier introduced dimensionless quantity $N_{\rm sc}^{\rm A}$ represents the number of those particles from the projectile A that interact with the scattering centers and, hence, undergo deflections from the incident direction. As such, the number $N_{\rm sc}^{\rm A}$ per unit time, in fact, represents the scattered radial flux $\jmath_{\rm sc}^{\rm A} = N_{\rm sc}^{\rm A}/t$, so that $[\jmath_{\rm sc}^{\rm A}] = {\rm s}^{-1}$. If the conditions (i)-(viii) in the examined experiment are fulfilled, the quantity $\jmath_{\rm sc}^{\rm A}$ will be directly proportional to the product of the incident flux $\bar{\jmath}_{\rm inc}^{\rm A}$ and the number $N_{\rm B}$ of the scattering centers from the target, so that

$$\jmath_{\rm sc}^{\rm A} \propto \bar{\jmath}_{\rm inc}^{\rm A} N_{\rm B}. \tag{2.6}$$

Calling upon the earlier defined general cross section for particle collisions and bearing in mind the performed dimensionality analysis, it follows from (2.6) that precisely the total cross section Q is the sought proportionality constant of the dimension of a surface

$$\jmath_{\rm sc}^{\rm A} = \bar{\jmath}_{\rm inc}^{\rm A} N_{\rm B} Q. \tag{2.7}$$

If here we exploit the relation $N_{\rm B}\,\bar{\jmath}_{\rm inc}^{\rm A} = \jmath_{\rm inc}^{\rm A} N_{\rm B}/S_{\rm B} = \jmath_{\rm inc}^{\rm A}\, x\, \rho_{\rm B}$, we shall obtain the following relationships

$$\begin{aligned} Q &= \frac{1}{N_{\rm B}} \frac{\jmath_{\rm sc}^{\rm A}}{\bar{\jmath}_{\rm inc}^{\rm A}} \\ &= \frac{1}{x\rho_{\rm B}} \frac{\jmath_{\rm sc}^{A}}{\jmath_{\rm inc}^{A}}. \end{aligned} \tag{2.8}$$

When we use wave packets for representing the incident beam, it is clear that their localization enables them to describe individual particles from the projectile A. Therefore, we can give an equivalent definition of the total cross section in terms of the flux (2.8) or the total number of particles viz

$$\begin{aligned} Q &= \frac{N_{\rm sc}^{\rm A}}{N_{\rm inc}^{\rm A}} \frac{1}{x\rho_{\rm B}} \\ &= \frac{N_{\rm sc}^{\rm A}}{N_{\rm inc}^{\rm A}} \frac{S_{\rm B}}{N_{\rm B}}. \end{aligned} \tag{2.9}$$

However, if we use the plane waves for representing the incident beam, their delocalization prevents them from describing any particle from the projectiles. In such a case, only the interpretation (2.8) of the cross section in terms of the flux has a meaning. Inspecting (2.9), it could be concluded at first glance that the cross section Q depends upon the parameters $N_{\rm inc}^{\rm A}$, x and $\rho_{\rm B}$. However, this is only apparent, since every change in these parameters is automatically compensated by the corresponding alteration of the measured number $N_{\rm sc}^{\rm A}$. In this way, the cross section Q remains only a function of the energy of the incident beam. This energy is not present explicitly in (2.9), but it is evidently present there in an indirect manner through the incident flux i.e. via the intensity of the incident beam.

In order to provide yet another adequate interpretation of the cross section for collision problems, we can also exploit an analogy between the expressions (2.4) and (2.9). Thus, e.g. in Rutherford's experiment [1] of measuring the cross sections Q for the α − Au scattering, one first determines the incident flux $\jmath_{\rm inc}^{\rm A}$ i.e. the number $N_{\rm inc}^{\rm A}$ of α particles per unit time in the projectile by removing the target gold foils and inserting a detector along the line of the incident direction. Afterwards, one puts back the target foils onto their place between the collimating diaphragms and the detector. This time the detector is placed along any chosen direction, which has an angle $\theta \geq 0°$ relative to the incident direction. Such an arrangement permits detection of the scattered flux $\jmath_{\rm sc}^{\rm A}$ i.e. the number $N_{\rm sc}^{\rm A}$ of the deflected α particles per unit time that were scattered by the target through the angle θ hitting the detector. Obviously, the rate of the elementary events registered in this way is given by the ratio $N_{\rm sc}^{\rm A}/N_{\rm inc}^{\rm A}$. This rate is directly proportional to the thickness of the target foil and the target density $\rho_{\rm B}$ expressed as the number of scattering centers $N_{\rm B}$ per unit of the effective volume $v_{\rm eff}$. Thus we see that the quotient $Q = N_{\rm sc}^{\rm A}/(N_{\rm inc}^{\rm A} x\rho_{\rm B})$ represents the rate of the deflection per unit time for those α particles that collided with the target Au.

An intuitive meaning of the statistical notion 'rate' for the definition of Q can be given transparently by the following quasi-geometric model. Let us suppose that each atom of e.g. the gold foil from Rutherford's experiment [1] contains an elementary target of surface Q which is much smaller than the surface of the atom itself. This supposition is entirely realistic since the deflection of one α−particle from the incident direction occurs only because of a single collision with one nucleus of the atom of gold (this nucleus being, of course, much smaller that the whole atom Au). Since the radius r_n of the nucleus is about $r_n \approx 1.2\, A^{1/3} f$ where A is the atomic mass number, it follows that in the case of $^{197}_{79}$Au we have $r_n \approx 7f$. Then, the corresponding surface of the nucleus of a gold atom is $Q \equiv \pi r_n^2 \approx 1.5b$. This elementary surface S_n is much smaller than the surface of the atom of gold and, therefore, it is clear that the total surface of all such elementary surfaces Q will still be only a fraction $Qx\rho_{\rm B}$ of the surface of the foil of the target B. If the foil is so thin that the mutual overlapping of the elementary target surfaces Q virtually does not occur, then it is obvious that the fraction $Qx\rho_{\rm B}$ will be much smaller than unity i.e. $Qx\rho_{\rm B} \ll 1$. Under such circumstances, from the given $N_{\rm inc}^{\rm A}$ incident particles of the projectile beam A sent to the target B, only a small fraction $N_{\rm inc}^{\rm A} Qx\rho_{\rm B}$ shall hit the detector. This is nothing other than counting the scattered particles $N_{\rm sc}^{\rm A}$ i.e. $N_{\rm inc}^{\rm A} Qx\rho_{\rm B} = N_{\rm sc}^{\rm A}$, in agreement with (2.9). For this reason, we customarily talk about a rate or frequency of collisions as if we were dealing with the rate of elementary targets of the effective surface Q. In this way, we justify the terminology after which the quotient Q from (2.8) is called the effective total surface of the target, or a more frequently employed term is the effective total cross section (or simply the total cross section) of one atom for occurrence of the given collisional process. In the French literature, the equivalent term is the effective total impact cross section ('la section efficace de choc totale') where the adjective 'effective' is a measure of efficiency of the target in removing a certain number of particles from the incident beam after having interacted with the projectile.

Nevertheless, in giving analogies one should bear in mind that they are not meant to be exceedingly far-reaching. Instead they should serve merely as auxiliary ways to better comprehend some phenomena from the micro-world using similar terms from the corresponding macroscopic phenomena. Indeed, only in certain exceptional circumstances, do cross sections in atomic or nuclear physics have a substantial connection with the real geometrical surface of the target. Instead, such microscopic cross sections relate to an effective surface as a measure of probability that certain collisional events will take place. For example, in e.g. mechanical collisions between two macroscopic bodies from the discussed geometrical model, we talk about a collision only if there is physical contact of the bodies. In contrast, however, there need not be any such contact at all in atomic or nuclear collisions, and yet collisions could still take place. This is because in e.g. Rutherford's α + Au scattering [1], a relatively strong interaction during a time interval T_0 which is small compared with the observation time $t\,(T_0 \ll t)$ i.e. the electrostatic repulsive Coulomb potential $2 \times 79/r$, exhibits a marked effect on the motion of the $\alpha-$particle. This is one of the most fundamental characteristics of a collision problem, since the relationship $T_0 \ll t$ causes the collisional system to be substantially different from any other non-collisional, dynamical system. Namely, since a large difference exists between the two times T_0 and $t\,(T_0 \ll t)$, it is possible that the movement of one particle or both colliding particles undergoes sudden changes that enable the separation of a relatively clean situation 'before' and 'after' the collision. Such a feature of scattering is of paramount importance since by being able to effectively discriminate the 'before from after' situations, we can acquire certain basic information about the collisional system despite the eventual lack of knowledge about the actual interactions (e.g. nuclear interactions) between the colliding particles.

The concept of scattering can also be used to study spontaneous decays of a given particle into two or more other particles. As an illustration, let us consider decay of an elementary particle called a sigma particle into a pion and a neutron

$$\sum\nolimits^{-} \longrightarrow \pi^{-} + n. \tag{2.10}$$

Here, we do not have the usual collisional phenomenon of two particles that approach each other in the entrance channel, except if we consider the inverse process relative to (2.10). Nevertheless, the decay of the $\sum^{-}$ particle still has a feature which is typical for collisions i.e. a clear distinction between the situation 'before' and 'after' the decay, so that we can gain certain very useful information about (2.10), despite a relatively poor knowledge of the interactions during a short-lasting disintegration event. This emphasis on the possibility of a clear difference between the asymptotic notions 'before' and 'after' the collision does not mean, at all, that scattering theory is incapable of describing many interesting physical effects that occur in the intermediate stage of the collision. Quite the contrary, scattering theory possesses very powerful methods that are capable of 'following' a scattering phenomenon as it develops in its intermediate steps, where we usually encounter short-lived metastable states of compound particles formed from the projectile and the target. This is the case e.g. with a collision of a proton of energy 25 MeV with the nucleus of the atom of silver ^{107}Ag. In this scattering, the two colliding particles can come so close to each other that their electrostatic long-range Coulomb repulsion becomes totally negligible relative to the corresponding strong short-range attractive nuclear interaction. In such a case, the incident proton can penetrate the nucleus of silver and thus form a compound nucleus. Such a compound nucleus is in a metastable state and, therefore, will decay to two different particles following the nuclear reaction called transmutation

$$p + {}^{107}\mathrm{Ag} \longrightarrow \alpha + {}^{104}\mathrm{Pb} \tag{2.11}$$

where ^{104}Pb is the atom of lead. Here, the decay time is about 10^{-18}s.

2.4 Differential cross sections

The outlined plausible definitions of the total cross section, in fact, contain more information that what is actually offered by (2.8) and (2.9). This is because in an experiment it is feasible to use the counters that register scattered particles in any possible fixed direction. Namely, it is obvious that there are $v_i \Delta Q'$ particles from A that, after being scattered on the target, pass through the elementary solid angle $\Delta\Omega_i'$ around the deflected direction towards the detector. This angle $\Delta\Omega_i' = (1/r^2)\Delta S_d$ is seen from the center of the surface S_{B} of the target under the arc $(1/r^2)\Delta S_d$. Here, ΔS_d is the so-called 'sensitivity surface' of the detector as a part of the surface of the detector lying perpendicular to the direction of the scattered particles at the distance r from the center of the target B. Let us first consider an elastic collision of the projectile A with the target B. Let $\Delta \jmath_{\mathrm{sc}}^{\mathrm{A}}$ be a finite increment of the scattered flux which is the number $\Delta N_{\mathrm{sc}}^{\mathrm{A}}$ of particles from A that are scattered elastically per unit time within the solid angle $\Delta\Omega_i'$ centered around $\{\theta', \phi'\}$. In such a case, for a sufficiently thin metallic foil of the target B of length x, it is justified to map (2.8) into a form of the finite differences

$$\begin{aligned} \Delta \jmath_{\mathrm{sc}}^{\mathrm{A}} &= N_{\mathrm{B}} \frac{\bar{\jmath}_{\mathrm{inc}}^{\mathrm{A}}}{\Delta Q'} = N_{\mathrm{B}} \bar{\jmath}_{\mathrm{inc}}^{\mathrm{A}} \frac{\Delta Q'}{\Delta\Omega_i'} \Delta\Omega_i' \\ &= N_{\mathrm{B}} \bar{\jmath}_{\mathrm{inc}}^{\mathrm{A}} \frac{\Delta Q'}{\Delta\Omega_i'} \frac{\Delta S_d}{r^2}. \end{aligned} \tag{2.12}$$

Here the primed quantities refer to the laboratory reference system in which the measurement is actually performed[7]. Applying the limit $\Delta\Omega_i' \longrightarrow 0$ to (2.12) i.e. passing from the difference to the differential calculus, we shall obtain the corresponding infinitesimally small change of the scattering flux $\mathrm{d}\jmath_{\mathrm{sc}}^{\mathrm{A}}$ in the form

$$\mathrm{d}\jmath_{\mathrm{sc}}^{\mathrm{A}} = N_{\mathrm{B}}\bar{\jmath}_{\mathrm{inc}}^{\mathrm{A}} F(\Omega_i')\mathrm{d}\Omega_i'. \tag{2.13}$$

This equation can also be expressed in an equivalent form via an infinitesimal change of the number of the scattered particles. Thus, if $\mathrm{d}N_{\mathrm{sc}}^{\mathrm{A}}$ is the number of particles from the projectile A scattered per unit time within the solid angle $\mathrm{d}\Omega_i'$, we shall have

$$\mathrm{d}N_{\mathrm{sc}}^{\mathrm{A}} = N_{\mathrm{B}}N_{\mathrm{inc}}^{\mathrm{A}} F(\Omega_i')\mathrm{d}\Omega_i'. \tag{2.14}$$

The proportionality constant $F(\Omega_i')$ between $\mathrm{d}\jmath_{\mathrm{sc}}^{\mathrm{A}}$ and $N_{\mathrm{B}}\bar{\jmath}_{\mathrm{inc}}^{\mathrm{A}}\mathrm{d}\Omega_i'$ from (2.13) is called the effective differential cross section (or simply differential cross section), which is defined by

$$F(\Omega_i') \equiv \frac{\mathrm{d}Q'}{\mathrm{d}\Omega_i'} = \frac{1}{N_{\mathrm{B}}\,\bar{\jmath}_{\mathrm{inc}}^{\mathrm{A}}}\frac{\mathrm{d}\jmath_{\mathrm{sc}}^{\mathrm{A}}}{\mathrm{d}\Omega_i'}. \tag{2.15}$$

It is also possible to immediately establish the following equivalent expression for the differential cross section

$$\begin{aligned}\frac{\mathrm{d}Q'}{\mathrm{d}\Omega_i'} &= \frac{S_{\mathrm{B}}}{\jmath_{\mathrm{inc}}^{\mathrm{A}}\,N_{\mathrm{B}}}\frac{\mathrm{d}\jmath_{\mathrm{sc}}^{\mathrm{A}}}{\mathrm{d}\Omega_i'}\\ &= \frac{1}{x\,\rho_{\mathrm{B}}\,\jmath_{\mathrm{inc}}^{\mathrm{A}}}\frac{\mathrm{d}\jmath_{\mathrm{sc}}^{\mathrm{A}}}{\mathrm{d}\Omega_i'}\\ &= \frac{r^2}{x\,\rho_{\mathrm{B}}\,\jmath_{\mathrm{inc}}^{\mathrm{A}}}\frac{\mathrm{d}\jmath_{\mathrm{sc}}^{\mathrm{A}}}{\mathrm{d}S_d}.\end{aligned} \tag{2.16}$$

Despite the explicit relations $N_{\mathrm{B}} = S_{\mathrm{B}}\,x\,\rho_{\mathrm{B}}$ and $\bar{\jmath}_{\mathrm{inc}}^{\mathrm{A}} = \rho_{\mathrm{A}}v_i = N_{\mathrm{A}}v_i/(yS_{\mathrm{A}})$, where y is the longitudinal length of the incident wave packet, the cross section $F(\Omega_i')$ from (2.15) is nevertheless independent of the quantities $r, N_{\mathrm{A}}, y, S_{\mathrm{A}}$ or $x, \rho_{\mathrm{B}}, S_{\mathrm{B}}$. This is because any change in these quantities will be compensated by the corresponding alteration of the detected number $N_{\mathrm{sc}}^{\mathrm{A}}$. However, crucially, the cross section $F(\Omega_i')$ depends on the solid angle Ω_i' and the incident velocity v_i. In the research literature on scattering experiments, the most frequently employed is the middle line in the string of the equations in (2.16) abbreviated by

$$\frac{\mathrm{d}Q'}{\mathrm{d}\Omega_i'} = \frac{1}{x\,\rho_{\mathrm{B}}}\frac{I(\Omega_i')}{I_0} \tag{2.17}$$

---

[7]Most experiments are carried out in the laboratory coordinate system of reference, except an important class of measurements involving crossed beams.

$$I(\Omega_i') \equiv \frac{\mathrm{d}\jmath_{\mathrm{sc}}^{\mathrm{A}}}{\mathrm{d}\Omega_i'} \qquad I_0 \equiv \jmath_{\mathrm{inc}}^{\mathrm{A}}. \tag{2.18}$$

From the basic working formula (2.17), it follows that, with the exception of the determination of x and ρ_{B} as trivial, the measurement of the differential cross section $\mathrm{d}Q'/\mathrm{d}\Omega_i'$ consists essentially of detecting the intensities $I(\Omega_i')$ and I_0 of the scattered and incident flux, respectively[8]. In the case of e.g. ionic beams, these intensities can be measured simply as the corresponding currents. The intensity I_0 is measured without the presence of the target and with the detector located along the incident direction ($\theta' = 0°$). In this way, we register the transmitted beam i.e. the incident beam alone. Afterwards, the target is returned to its place and the detector is located in a selected fixed direction under a certain angle θ' relative to the direction of incidence. This permits the measurement of $I(\Omega_i')$ at the same impact energy as in the case of I_0.

It is easy to show that the same result (2.10) for the total cross section can be obtained in any other system of reference as in the laboratory frame. However, this is not the case for the differential cross section (2.16) or (2.17). The passage to the system of the center of mass (c.m.), which is the easiest for theoretical analysis and calculations, can be achieved by using the following simple relationship: $\mathrm{d}Q'/\mathrm{d}\Omega_i' = (\mathrm{d}Q/\mathrm{d}\Omega_i)(\mathrm{d}\Omega_i/\mathrm{d}\Omega_i')$. Thus, the ratio of the two differential cross sections in the two different reference systems is of a purely kinematical nature. Specifically, it can be readily demonstrated that

$$\frac{\mathrm{d}Q'}{\mathrm{d}\Omega_i'} = \frac{(1 + 2\mu_0 \cos\theta + \mu_0^2)^{3/2}}{|1 + \mu_0 \cos\theta|} \frac{\mathrm{d}Q}{\mathrm{d}\Omega_i} \tag{2.19}$$

where $\theta_{\mathrm{cm}} \equiv \theta\,, \phi_{\mathrm{cm}} \equiv \phi$ and $\mu_0 = M_{\mathrm{A}}/M_{\mathrm{B}}$. The total cross section Q does not depend upon the reference system and can be obtained either from (2.17) or (2.19) by integrating over the angles

$$Q = \int_0^\pi \mathrm{d}\theta' \int_0^{2\pi} \mathrm{d}\phi' \frac{\mathrm{d}Q'}{\mathrm{d}\Omega_i'} = \int_0^\pi \mathrm{d}\theta \int_0^{2\pi} \mathrm{d}\phi \frac{\mathrm{d}Q}{\mathrm{d}\Omega_i}. \tag{2.20}$$

2.5 Total probabilities

Of course, we can also define the total probability W for having an incident particle deflected by a non-zero angle after interacting with a scattering center

[8]Of course, in a realistic scattering experiment e.g. between two atomic systems there is a whole sequence of subsidiary and supplementary measurements that is needed to ensure the correctness of the entire procedure e.g. eliminating the influence of possible cascade effects, repopulation of the states of interest, etc. Such details are beyond the scope of this book and can be found in specialized literature dealing with measurements on scattering.

from the target. This is equivalent to saying that a given particle has been removed from the projectile beam[9]. The said probability W for such an event is given by

$$W = \frac{N_{\mathrm{sc}}^{\mathrm{A}}}{N_{\mathrm{A}}}. \tag{2.21}$$

By inserting (2.5), (2.6) and (2.8) into (2.21) we find

$$W = \rho_{\mathrm{B}}\, x\, Q = \frac{N_{\mathrm{B}}}{S_{\mathrm{B}}} Q. \tag{2.22}$$

Here, according to (2.6), the product $\rho_{\mathrm{B}}\, x = N_{\mathrm{B}}/S_{\mathrm{B}}$ represents the average surface density i.e. the number N_{B} of particles from the target B in the given surface S_{B}. The probability W from (2.16) and (2.17) depends upon various experimentally measurable parameters e.g. $N_{\mathrm{A}}, \rho_{\mathrm{B}}$ and x, whereas the cross section Q is a function only of the energy $E > 0$ of the system and is related strictly to the internal microscopic structure of the ensemble (A+B). In this sense, Q can be considered as a measure of the tendency of the projectile A and target B to mutually interact at a given fixed energy $E > 0$. Equivalently, the observable Q can be viewed as a certain 'effective surface' which captures a part of the incident beam.

2.6 Transmission phenomena

The definition (2.22) is valid only if $W \ll 1$. In other words, if the target is not thin, we must introduce an appropriate correction. The same remark also applies to (2.21). Further, the earlier mentioned attempt to extrapolate the differential cross sections towards the forward angle $\theta \approx 0°$ can be substituted by a measurement of the cross section $Q(\theta \approx 0°, \phi)$. Namely, instead of detecting the deflected particles, we could register only the transmitted particles. This is an alternative approach for a scattering experiment via registering the remaining particles from the incident beam A i.e. detecting those particles that were not removed from the ensemble of the projectiles. Such a concept of measurement differs radically from the usual strategy in scattering experiments. Recall that e.g. Rutherford's measurement [1] on the α – Au scattering is based upon detecting scattered projectiles i.e. those α–particles

[9] A given particle from an ensemble of the projectiles is conceived as persisting to belong to that ensemble if it continues to move along the same incident direction ($\theta = 0°$) throughout the collision. Scattering takes place at any $\theta > 0°$ in which case this particle is viewed as being removed from the ensemble of the projectiles, since it has ceased to move in the direction of incidence.

that collide with the target and are deflected by an angle $\theta > 0°$. The mentioned alternative measurement would consist of inserting a foil between two collimating diaphragms. The front and back diaphragms would represent the obstacle to the incident and scattered beam, respectively, throughout the two surfaces except, of course, a small hole in the center of each of the collimating diaphragms. These two collimating holes should be of identical form and size. The source of the projectiles would preceed the front collimating diaphragm, whereas the detector would be placed behind the second diaphragm along the incident direction. Thus, we would have the following constellation of devices aligned in the direction of the incident beam: the projectile source, the first collimating hole, the target, the second collimating hole and the detector. As such, the incident beam on its way to the target passes through the front collimator, and then collides with the target, as in the case of the standard scattering experiments. However, this time the back collimator behind the target effectively stops all those particles that scatter through angles $\theta > 0°$ because they are deflected from the direction of incidence. Simultaneously, through its hole, this second diaphragm filters out all those particles from the incident beam that do not scatter at all ($\theta = 0°$). Such non-deflected i.e. transmitted particles from the incident beam are registered by the detector which is located behind the second collimating diaphragm. If we now increase the density of the foil of the target, there will be an increased number of deflections of the projectiles, meaning that a smaller number of particles will hit the detector. Specifically, adding new foils in the target in the form of successive layers of the same thickness reduces the particle count precisely for the amount equal to the counting rate of the particle deflections. This is because the rate of particle deflections within a given layer does not depend on the other layers. In such a circumstance, the number $N^{\rm A}_{\rm inc}$ of the non-deflected particles from the incident beam decreases in each layer of the infinitesimally small thickness $\mathrm{d}x$. This decrease $\mathrm{d}N^{\rm A}_{\rm inc}$ is directly proportional to the quantities $\mathrm{d}x, \rho_{\rm B}$ and $N^{\rm A}_{\rm inc}$. In order to emphasize this fact more transparently, we shall single out the functional dependence upon the thickness x of the target foil and write $N^{\rm A}_{\rm inc}(x)$. Hence, the function $N^{\rm A}_{\rm inc}(x)$ satisfies the following relationship: $\mathrm{d}N^{\rm A}_{\rm inc}(x) \propto -N^{\rm A}_{\rm inc}(x)\rho_{\rm B}\mathrm{d}x$ where the minus sign indicates that we are dealing with a negative increment signalling a decrease of $N^{\rm A}_{\rm inc}(x)$ with augmentation of x. Introducing the quantity $Q_{\rm tr}$ as a proportionality constant, we arrive at the corresponding differential equation for the increment of the transmitted particles as

$$\mathrm{d}N^{\rm A}_{\rm inc}(x) = -Q_{\rm tr}\rho_{\rm B}N^{\rm A}_{\rm inc}(x)\mathrm{d}x. \tag{2.23}$$

The solution of this first-order differential equation with the constant coefficient $Q_{\rm tr}\rho_{\rm B}$ is $N^{\rm A}_{\rm inc}(x) = C\exp(-Q_{\rm tr}\rho_{\rm B}x)$, where C is the integration constant. The constant C can be determined from the boundary condition at $x = 0$ requiring that $N^{\rm A}_{\rm inc}(x)$ for a very thin single foil of the target ($x \approx 0$) coincides with the initial number $N^{\rm A}_{\rm inc}$ i.e. $N^{\rm A}_{\rm inc}(x \approx 0) \approx N^{\rm A}_{\rm inc}$, so that $C \approx N^{\rm A}_{\rm inc}$

and, therefore

$$N_{\text{inc}}^{\text{A}}(x) = N_{\text{inc}}^{\text{A}}\mathrm{e}^{-Q_{\text{tr}}\rho_{\text{B}}x} \qquad N_{\text{inc}}^{\text{A}} = N_{\text{inc}}^{\text{A}}(0). \tag{2.24}$$

This is a very important law of exponential attenuation of the transmitted beam. An experimental verification of the law (2.24) would enable us to have a sensitive test determining whether the ensuing deflections of the scattered particles are indeed due to elementary processes. The corresponding probability acquires the value of the following quotient

$$W_{\text{tr}} = \frac{N_{\text{inc}}^{\text{A}}(x)}{N_{\text{inc}}^{\text{A}}} = \mathrm{e}^{-Q_{\text{tr}}\rho_{\text{B}}x}. \tag{2.25}$$

The damping constant Q_{tr} has the dimension of surface and is called the transmission cross section for removing one particle from the incident beam as a consequence of a single interaction of that particle with only one scattering center from the target B. If a deflection for $\theta > 0°$ is the only possible result of the said interaction, then Q_{tr} must coincide with the expression (2.15) for Q which is obtained by integrating the differential cross section $\mathrm{d}Q/\mathrm{d}\Omega_i$ over all the directions through which the deflections took place. In order to have a precise measurement of the observable Q_{tr}, it is necessary that the exponential weakening of the quantity $N_{\text{inc}}^{\text{A}}(x)/N_{\text{inc}}^{\text{A}}$ from (2.20) can be detected all the way up to a small number of the remaining particles such as $N_{\text{inc}}^{\text{A}}(x)/N_{\text{inc}}^{\text{A}} \approx 10^{-3}$. Of course, this kind of experiment is most convenient for light incident particles (electrons, positrons, etc), but it becomes impractical for heavy projectiles (protons, α−particles, neutrons, etc). Namely, in collisions of e.g. an α−particle with a gold foil as the target [1], most projectiles, due to their heavy mass, deflect only slightly from their incident direction by small fractions of 1°. Thus, the second collimating diaphragm behind the target is not able to stop the majority of scattered projectiles that will pass through the hole and hit the detector as if they were transmitted and not scattered. This implies that the quotient $N_{\text{inc}}^{\text{A}}(x)/N_{\text{inc}}^{\text{A}}$ will not be small and this would lead to insufficiently accurate results for the measured cross sections of transmission Q_{tr}. Since the probability W_{tr} is a measure of the intensity of the beam transmitted through the target, we see from (2.20) that this intensity decays exponentially with the increase of the target thickness. Thus, the outlined alternative experiment of scattering determines the attenuation of the intensity of the incident beam as a function of the target thickness x. Based upon such measured data and using (2.20), it is easy to deduce the transmission cross sections Q_{tr} viz

$$Q_{\text{tr}} = \frac{\ln \mathrm{W}_{\text{tr}}^{-1}}{\rho_{\text{B}}x}. \tag{2.26}$$

The described principle is the essence of the measurement called the attenuation experiment in which we basically compare the intensities of the scattered and transmitted beams of particles[10]. The probabilities W and $W_{\rm tr}$ from (2.16) and (2.20) for the two discussed types of scattering measurements cannot be compared directly, since we are dealing with two mutually exclusive events. In the first kind of experiment (e.g. Rutherford's experiment [1]) only the scattered particles are detected. By contrast, in the second kind of scattering experiment involving measurement of attenuation of the incident beam, only non-scattered particles are detected. However, if the examined collision is such that there are only two possibilities for the incident particle i.e. to be deflected ($\theta > 0°$) or not ($\theta = 0°$), as is the case for an elastic scattering, then the probabilities W and $W'_{\rm tr} \equiv 1 - W_{\rm tr}$ relate to the same event. This is a consequence of the law of conservation of probability which gives

$$\begin{aligned} W'_{\rm tr} \equiv 1 - W_{\rm tr} &= 1 - \frac{N^{\rm A}_{\rm inc}(x)}{N^{\rm A}_{\rm inc}} \\ &= 1 - {\rm e}^{-Q_{\rm tr}\rho_{\rm B}x} . \end{aligned} \tag{2.27}$$

If the cross section Q from (2.16) and (2.17) is also obtained for the same target thickness x as for $Q_{\rm tr}$ from (2.21), we will be able to obtain the sought relation between W and $W_{\rm tr}$, which is simultaneously a measure of the error in probability W

$$W_{\rm tr} = 1 - {\rm e}^{-W} . \tag{2.28}$$

In the limiting case of a thin target, we have that for $W = \rho_{\rm B}\, x\, Q \ll 1$ the relation (2.22) is reduced to $W_{\rm tr} \approx W$ i.e. to the previous case (2.17).

2.7 Quantum mechanical currents and cross sections

Finally, let us establish a relationship between certain fundamental quantum mechanical predictions and the corresponding measured data using merely the basic notions from the concept of the scattering experiment as outlined in this chapter. With this goal, let us first show that the 'alpha and omega' of scattering theory i.e. the equations (1.10) and (1.11) can be brought into accord with the associated expressions from the present chapter. Here, in the first place, we have in mind equation (2.10) whose version in the center of mass system reads as ${\rm d}\jmath^{\rm A}_{\rm sc} = N_{\rm B}\bar{\jmath}^{\rm A}_{\rm inc} F(\Omega_i){\rm d}\Omega_i$, where $F(\Omega_i) = {\rm d}Q/d\Omega_i$ is the differential cross section ${\rm d}Q/{\rm d}\Omega_i \equiv ({\rm d}Q/{\rm d}\Omega_i)_{\rm c.m.}$. We saw that $F(\Omega_i)$ is independent of $S_{\rm A}$ or $S_{\rm B}$, so that both of these two latter quantities can be

[10] As stated by the condition (i) in Section 2, such measurements will be statistically significant provided that a large number of collisional events could be recorded, which is possible if the incident beam is not of weak intensity.

replaced by the unit surface. In such a case, the average relative incident flux $\bar{\jmath}^{\rm A}_{\rm inc}$ will represent the flux $\jmath^{\rm A}_{\rm inc}$ per unit surface and per unit time. Then, under the assumption that the coherent effects among different scattering centers in the target can be ignored as negligible, it follows that the quantity $\mathrm{d}\jmath^{\rm A}_{\rm sc}$ represents the infinitesimal scattering outgoing flux. This is the number of outgoing particles that, while leaving the $N_{\rm B}$ scattering centers from the target, fall onto an infinitesimal solid angle $\mathrm{d}\Omega_i$ per unit time

$$\frac{\mathrm{d}\jmath^{\rm A}_{\rm sc}}{N_{\rm B}} = \jmath^{\rm A}_{\rm inc} F(\Omega_i)\mathrm{d}\Omega_i. \tag{2.29}$$

The corresponding quantum mechanical probability flux can be found from the well-known formula for the particle current [7]

$$\boldsymbol{\jmath} = \frac{1}{2i\mu}\left[\psi^*(\boldsymbol{r})\boldsymbol{\nabla}_{\boldsymbol{r}}\psi(\boldsymbol{r}) - \psi^*(\boldsymbol{r})\boldsymbol{\nabla}_{\boldsymbol{r}}\psi(\boldsymbol{r})\right] \tag{2.30}$$

where μ is the reduced mass of the involved particles. The incident plane wave $\phi_{\boldsymbol{k}_i}(\boldsymbol{r}) = (2\pi)^{-3/2}\exp(i\boldsymbol{k}_i\cdot\boldsymbol{r})$ generates the flux $\boldsymbol{\jmath}^{\rm A}_{\rm inc}$ per unit time and per unit surface which is perpendicular to the incident wave vector $\boldsymbol{k}_i$. Thus using (2.24), it easily follows that $(2\pi)^3\boldsymbol{\jmath}^{\rm A}_{\rm inc} = \boldsymbol{v}_i = \boldsymbol{k}_i/\mu$ i.e.

$$\jmath^{\rm A}_{\rm inc} = (2\pi)^{-3}v_i. \tag{2.31}$$

Likewise, the scattered probability flux is obtained by means of the outgoing wave for the appropriate scattering state taken at an asymptotically large distance r from the target i.e. using the asymptote

$$\Psi^+_{\boldsymbol{k}_i}(\boldsymbol{r}) \underset{r\to\infty}{\longrightarrow} \phi_{\boldsymbol{k}_i}(\boldsymbol{r}) + \Psi^{\rm sc}_{\boldsymbol{k}_i}(\boldsymbol{r}) \tag{2.32}$$

where $\Psi^{\rm sc}_{\boldsymbol{k}_i}(\boldsymbol{r}) = (2\pi)^{-3/2}f(\Omega_i)r^{-1}\exp(ik_ir)$ as in (1.10). It can be shown that the two additive terms in (2.32) do not interfere with each other while evaluating the flux $\jmath^{\rm A}_{\rm sc}$ by using (2.29), except for $\theta = 0°$ (forward scattering). This particular case of collision at zero scattering angle is treated separately with the result expressed through the so-called optical theorem

$$Q = \frac{4}{\pi k_i}\mathrm{Im}\, f(0°) \tag{2.33}$$

which is a consequence of conservation of the probability flux[11]. Thus, the optical theorem is a convenient way of computing the total cross section Q when one knows the value of the imaginary part of the transition amplitude for forward scattering. The relation (2.33) stems from a destructive interference between the two terms $\Phi_{\boldsymbol{k}_i}(\boldsymbol{r})$ and $\Psi^{\rm sc}_{\boldsymbol{k}_i}(\boldsymbol{r})$ of the asymptote (2.32) of

[11]Hereafter the symbols $\mathrm{Re}(z)$ and $\mathrm{Im}(z)$ denote the real and imaginary part of the complex number z, respectively.

the scattering wave function $\Psi^{+}_{\boldsymbol{k}_i}(\boldsymbol{r})$ behind the collisional region ($\theta \approx 0°$). Stated equivalently, the 'shadow' which the target casts in the direction of forward scattering diminishes the intensity of the incident beam, so that the scattered particles are removed from the beam by an amount proportional to the value of the total cross section Q. Thus, except for the case $\theta = 0°$, the interference term which stems from the two constituents $\Phi_{\boldsymbol{k}_i}(\boldsymbol{r})$ and $\Psi^{\rm sc}_{\boldsymbol{k}_i}(\boldsymbol{r})$ of the asymptote $\Psi^{\rm sc}_{\boldsymbol{k}_i}(\boldsymbol{r})$ from (2.32) can be left out. This justifies the procedure of computing separately the incident and outgoing flux. Hence, using (2.29) for the scattering wave $\Psi^{\rm sc}_{\boldsymbol{k}_i}(\boldsymbol{r})$, it is readily found that at large distances r from the target, this latter scattering wave produces the outgoing radial flux in the direction $\Omega_i = \{\theta, \phi\}$ per unit time and per unit surface perpendicular to the direction of the outgoing flux according to

$$\jmath^{\rm A}_{\rm sc} = (2\pi)^{-3}\, v_i |f(\Omega_i)|^2. \tag{2.34}$$

However, what we need is not the outgoing flux per unit time and per unit surface, but rather the infinitesimal number $\mathrm{d}\jmath^{\rm A}_{\rm sc}$ of outgoing particles per unit time that, after being scattered on the $N_{\rm B}$ scattering centers of the target, fall onto the infinitesimal surface $\mathrm{d}S_d = r^2 \mathrm{d}\Omega_i$ perpendicular to the radial direction. This infinitesimal number is deduced from (2.34) as

$$\begin{aligned}(2\pi)^3 \frac{\mathrm{d}\jmath^{\rm A}_{\rm sc}}{N_{\rm B}} &= v_i \frac{|f(\Omega_i)|^2}{r^2} \mathrm{d}S_d = v_i \frac{|f(\Omega_i)|^2}{r^2} (r^2 \mathrm{d}\Omega_i) \\ &= v_i |f(\Omega_i)|^2 \mathrm{d}\Omega_i \end{aligned} \tag{2.35}$$

so that

$$\frac{\mathrm{d}\jmath^{\rm A}_{\rm sc}}{N_{\rm B}} = \jmath^{\rm A}_{\rm inc}\, |f(\Omega_i)|^2 \mathrm{d}\Omega_i \tag{2.36}$$

where (2.31) and (2.34) are used. A direct comparison of (2.29) with (2.36) gives

$$\jmath^{\rm A}_{\rm inc} F(\Omega_i) \mathrm{d}\Omega_i = \jmath^{\rm A}_{\rm inc}\, |f(\Omega_i)|^2 \mathrm{d}\Omega_i. \tag{2.37}$$

Here, all the quantities are positive and this permits the identification of the differential cross section $F(\Omega_i) \equiv \mathrm{d}Q/\mathrm{d}\Omega_i$ in the form

$$\frac{\mathrm{d}Q}{\mathrm{d}\Omega_i} = |f(\Omega_i)|^2 \tag{2.38}$$

in agreement with (1.11) (QED). This analysis simultaneously shows under which circumstances the one-particle scattering theory can be used to describe a realistic scattering experiment.

3

Continuous spectrum and eigen-problems of resolvents

3.1 Completeness and separability of the Hilbert spaces

In chapters 3–10, we shall give a theoretical description of scattering by using spectral operator analysis and theory of infinite-dimensional Hilbert vector spaces. The first issue to be addressed here is the introduction of the notion of a spectrum of an operator in a general way, which will be applicable to both discrete and continuous variables. As is well-known [7], in an examination of stationary states of a given physical system, the main task is reduced to searching for the solutions of the time-independent Schrödinger equation

$$\hat{\mathrm{H}}\Psi = E\Psi. \tag{3.1}$$

Since this is a second-order differential equation, it is clear that the boundary conditions imposed to (3.1) will play a central role in selection of the proper, physical solutions. In the case of closed i.e. bound states, the eigen-values E are negative and discrete, whereas for scattering states, energy E becomes positive and takes its continuous values. In order to correctly formulate the eigen-value problem of type (3.1), we will resort to the theory of Hilbert vector spaces. In so doing, we assume that all the dynamical properties of quantum mechanical systems are contained in the structure of the self-adjoint linear Hamilton operator $\hat{\mathrm{H}}$ from the separable Hilbert space $\mathcal{H}$, defined in the field of complex numbers $\mathbb{C}$. Here, an abstract Hilbert vector space $\mathcal{H}$ is understood as a unitary, complete and separable complex-valued vector space i.e. a linear space of normalizable solutions of the Schrödinger equation (3.1), with positive-definite scalar products $\langle\psi|\psi\rangle \geq 0$. Unitarity means that the space $\mathcal{H}$ is equipped with a definition of the scalar (inner) product as a conjugate bilinear, Hermitean symmetrical[1] and scalar function in $\mathcal{H} \times \mathcal{H}$ i.e. with its values (images) in field $\mathbb{C}$. Space $\mathcal{H}$ is complete, since it contains all the limiting values of its convergent Cauchy series/sequences. Fundamental or the

[1] A mapping which is simultaneously anti-linear and linear in regard to the first and second factors (arguments), respectively, is said to be conjugate bilinear. Hermitean symmetry symbolizes the property $\langle\phi|\psi\rangle = \langle\psi|\phi\rangle^*$.

Cauchy series are those sequences that possess strong convergence i.e. convergence in the norm $|| \circ || \equiv \langle \circ | \circ \rangle^{1/2}$. Separability signifies that $\mathcal{H}$ is a countable infinite-dimensional space. Stated more precisely, space $\mathcal{H}$ is separable if it contains a countable set $\mathcal{S} \equiv \{\psi_k\}_{k=1}^{\infty}$, which is everywhere dense in $\mathcal{H}$ i.e. for each $\psi \in \mathcal{H}$ and an arbitrary positive number $\epsilon > 0$, there exists at least one element ψ_k such that $||\psi_k - \psi|| \leq \epsilon$. In other words, any vector $\psi \in \mathcal{H}$ can be approximated with an element from $\mathcal{S}$ to an arbitrary accuracy. The separability property of $\mathcal{H}$ always holds in quantum mechanical systems of particles with non-zero masses. Contrary to this, however, in field theory where one studies massless particles, systems with infinitely many degrees of freedom are routinely encountered, and the separability condition for $\mathcal{H}$ is not fulfilled. In such a case, one deals with non-separable Hilbert spaces $\mathcal{H}$ and this represents one of the severest obstacles to formulating scattering theory. We stress that the axiom of completeness and separability of $\mathcal{H}$ is of decisive importance for quantum scattering theory, not only in regard to its conceptual foundation of the basic principles, but also with respect to numerical applications that enable comparisons between theory and experiment. Namely, completeness of $\mathcal{H}$ assures the existence of at least one complete ortho-normal set $\{\phi_k\}$, which can be used as a basis for a so-called expansion theorem

$$\psi = \sum_{k=1}^{\infty} c_k \phi_k \qquad \langle \phi_m | \phi_n \rangle = \delta_{nm} \tag{3.2}$$

where δ_{nm} is the Kronecker δ−symbol: $\delta_{nn} = 1$ and $\delta_{nm} = 0\,(n \neq m)$. Expansion coefficients c_k are given by the scalar product

$$c_k = \langle \phi_k | \psi \rangle \tag{3.3}$$

and the infinite sum in (3.2) is convergent in the norm as a consequence of completeness and separability of $\mathcal{H}$. Set $\{\phi_k\}_{k=1}^{n}\,(n < \infty)$ can be easily found by applying e.g. the Gramm-Schmidt successive orthogonalization to a certain non-orthogonal basis $\{\psi_k\}_{k=1}^{n<\infty}$ according to [34]

$$\left.\begin{array}{l} \chi_{k+1} = \psi_{k+1} - \sum_{j=1}^{k} \langle \phi_j | \psi_k \rangle \phi_j \\[2ex] \phi_1 = \dfrac{\psi_1}{||\psi_1||} \qquad \phi_{k>1} = \dfrac{\chi_k}{||\chi_k||} \end{array}\right\}. \tag{3.4}$$

An entirely analogous procedure of the Gramm-Schmidt othogonalization can also be formulated in the case when our starting point $\{\psi_k\}_{k=1}^{\infty}$ contains countable infinite number of elements. Further, separability of $\mathcal{H}$ guarantees that, even if the infinite sum (3.2) is truncated at a given finite $n < \infty$ $(n \in \mathbb{N})$ and the coefficients c_k $(1 \leq k \leq n)$ are approximated by rational complex numbers, the norm of each element $\psi \in \mathcal{H}$ can be numerically computed with arbitrary accuracy ϵ through some powerful algorithms. Hence, separability of the Hilbert space is the basis of all the numerical applications of quantum mechanics. Except for a few publications (see e.g. Ref. [7]), this statement is virtually non-existent in the literature, but obviously deserves full attention. The conditions of completeness and separability are automatically fulfilled for finite-dimensional vector spaces.

3.2 The key realizations of abstract vector spaces

In quantum mechanics there exist two realizations of the abstract Hilbert space $\mathcal{H}$ and they are of particular importance for scattering theory. These are the Hilbert concrete spaces L^2 and $\mathcal{H}_0$ filled with the square-integrable elements and the vectors as matrix columns, respectively. More precisely, the Hilbert space of states whose elements are the solutions of the Schrödinger eigen-value problem, is a concrete example of an abstract vector space. Elements of an abstract vector space are not specified at all, but instead one is dealing with a set of axioms satisfied by these elements. Nevertheless, for the sake of semantic convenience, vector spaces containing the solutions of the Schrödinger equation will be termed the Hilbert abstract spaces $\mathcal{H}$ as in Ref. [34], whereas the realizations L^2 and $\mathcal{H}_0$ will be called concrete spaces. The Hilbert space L^2 is comprized of the wave functions $\langle x|\psi\rangle = \psi(x)$, where x symbolizes the coordinates of all the particles of the considered quantum system. These state vectors $\psi(x)$ are absolutely square-integrable in the sense of the Lebesgue integral $\int_0^\infty \mathrm{d}x\,|\psi(x)|^2 < \infty$, where the dimension of the integral is equal to the number of the attendant particles multiplied by three (for each of the degrees of freedom of every particle, not counting the spin). In the Hilbert space L^2, the scalar product is defined by the relation

$$\langle\psi_1|\psi_2\rangle = \int_0^\infty \mathrm{d}x\,\psi_1^*(x)\psi_2(x). \tag{3.5}$$

The sequential Hilbert space $\mathcal{H}_0$ contains infinite-dimensional vector columns $\mathbf{c} = \{c_k\}$ with the following properties $\sum_{k=1}^\infty |c_k|^2 < \infty$. In this vector space, the scalar product is introduced as

$$(\mathbf{a}, \mathbf{b}) = \sum_{k=1}^{\infty} a_k^* b_k \tag{3.6}$$

where $\{a_k\}$ and $\{b_k\}$ are the components of $\mathbf{a}$ and $\mathbf{b}$, respectively. Further, it is possible to establish an isomorph mapping between L^2 and $\mathcal{H}_0$ symbolized as $L^2 \cong \mathcal{H}_0$, as well as the one-to-one correspondence via the expansion theorem (3.2)

$$\psi \longleftrightarrow \mathbf{c}. \tag{3.7}$$

Isomorph mapping between two spaces is a correspondence which conserves the algebraic structure when passing from one space to another. In other words, a linear combination of elements with arbitrary scalars from one space is transformed to an analogous linear combination of elements in the other space. Among the most important consequences of isomorphism $L^2 \cong \mathcal{H}_0$ is conservation of structure of the scalar products when switching from L^2 to $\mathcal{H}_0$ as transcribed by

$$\langle\psi_1|\psi_2\rangle = (\mathbf{c}_1, \mathbf{c}_2) \tag{3.8}$$

which corresponds to the Parseval relation for complete systems.

3.3 Isomorphism of vector spaces

The relation of isomorphism ($\cong$) between two spaces is an equivalence relation for which the properties of reflexivity, symmetry and transitivity are valid. It can be shown that all finite-dimensional spaces (abstract and/or concrete) are mutually isomorph. Here, it is of interest to consider the following essential question: what are the implications of isomorphism between abstract and concrete spaces encountered in practice? Namely, if any abstract vector space $\mathrm{X}^n_{\mathcal{F}} \equiv \mathrm{X}^n$ defined on the field $\mathcal{F}$ is isomorph with e.g. a concrete vector space $\mathcal{X}^n_{\mathcal{F}} \equiv \mathcal{X}^n$ containing sequences of numbers of the type of algebraic real or complex vectors $\boldsymbol{x} = (x_1, x_2, ..., x_n)$, where $\{x_k\}_{k=1}^n$ represent the coordinates (components) of the $n-$dimensional arithmetic vector $\boldsymbol{x}$

$$\mathrm{X}^n_{\mathcal{F}} \cong \mathcal{X}^n_{\mathcal{F}} \tag{3.9}$$

then the work with abstract, general spaces would appear as an unnecessary burden. Here, however, the relation between general and particular has a special connotation, since isomorphism (3.9) is established for a particular basis $b_\psi \equiv \{\psi_k\}_{k=1}^n$. This, nevertheless, does not inconvenience the abstract space X^n, since the basis b_ψ is arbitrarily chosen, meaning that isomorphism (3.9) will also hold for any other selection of the basis. Hence, there exists invariance of all the statements and proofs of lemmas and theorems in the abstract space X^n with respect to the basis choice. Contrary to this, if the prototype space $\mathcal{X}^n$ is to be considered as general regarding all the conclusions drawn for a concrete basis, then the analogous statements in $\mathcal{X}^n$ must always be completed with proofs of independence of conclusions on the basis choice. This is certainly a tedious work which would additionally complicate the analysis. Generally speaking, when one examines certain general features, then abstract elements appear to provide the most concise and transparent framework, while concreteness frequently represents a burden. This also means that it is easier to work with abstract than with concrete spaces. The reason is in the fact that the notion introduced into the theory of abstract vector spaces, as well as statements and theorems proved in these spaces, is based solely on the underlying algebraic structure i.e. on axioms defining these abstract spaces. In so doing, one ignores the nature of the elements of abstract vector spaces as inessential and this represents a significant advantage. By contrast, in concrete spaces one is unable to determine, without additional analysis, how much a drawn conclusion depends upon the particular nature of the elements of a vector space and of the algebraic structure of the space. Nevertheless, in most practical cases, it appears as optimal to combine general with particular concepts, via partitioning the analysis into two stages. For instance, the general features of the studied system could be first examined in the appropriate abstract space X^n. Subsequently, in the second operative stage, one could choose a suitable basis in X^n and pass to $\mathcal{X}^n$ by means of a convenient specification of coordinates/components. Such a specification presupposes a given one-to-one correspondence between the vector ψ and the expansion scalar coefficients $\{c_k\}_{k=1}^{\infty}$ from (3.2). Here the elements $\{c_k\}_{k=1}^{\infty}$ are called coordinates of the state vector ψ. This also clearly transcends one of the most important properties of a given basis in the Hilbert space $\mathcal{H}$, namely the fact that with each vector $\psi \in \mathcal{H}$ one associates a finite or infinite sequence of scalars $\{c_1, c_2, ..., c_n, ...\} \in \mathbb{C}$. The usefulness of this concept is due to converting all the operations with abstract vectors from $\mathcal{H}$ to the corresponding computations with scalars from $\mathbb{C}$. Thus, the expression (3.2) *de facto* defines a mapping $\mathcal{H} \longrightarrow \mathbb{C}$, which can be shown to be linear. The particular space $\mathcal{X}^n$ is convenient for practical computations, as is obvious in quantum scattering theory. Likewise, the corresponding abstract space X^n is more flexible for the appropriate interpretations of geometrical and physical concepts. For a 'geometrical reasoning' i.e. for perceptions via the usual geometrical notions, it is certainly most natural to choose the space $\mathcal{X}^n$. This is because the latter space imposes itself as a logical extension of the usual three-dimensional space $\mathcal{X}^3$, or more simply, the Euclidean space $\mathbb{R}^3$ of geometrical vectors with familiar geometrical visualizations. If, on the other hand, the first stage of the analysis has been performed in the concrete representation $\mathcal{X}^n$, then subsequently in the abstract space X^n one could achieve the necessary 'algebraization' of geometrical perceptions. With this, one in fact 'translates' geometrical objects from $\mathcal{X}^n$ into the algebraic ones from X^n. We shall try as much as feasible to adhere to these useful methodological recommendations in the course of the present analysis.

It should be emphasized that finite-dimensional vector spaces are advantageous because their usage and/or application does not require any additional restrictions, except that a chosen basis should have a finite number of elements. A diametrically opposite situation is encountered in infinite-dimensional spaces for which nearly every statement must be supplemented by certain limitations/constraints or assumptions besides the primary request of having a basis with an infinite number of elements. For instance, in addition to algebraic structures that are sufficient for a finite-dimensional space, examinations in an infinite-dimensional space necessitate a topological structure. There is a special mathematical branch which deals with infinite-dimensional spaces called functional analysis. In physics, both finite and infinite-dimensional spaces are in use, but the latter spaces are requested from the outset for the majority of problems, especially in quantum scattering theory because of the continuous part of the spectrum of Hamiltonians.

The eigen-value problem (3.1) of Hamilton operator $\hat{\mathrm{H}}$ is determined by the boundary conditions imposed to state vectors Ψ. For closed or bound systems ($E < 0$), it is required that the vector Ψ belongs to the Hilbert space L^2 of square-integrable elements. Specifically, in the case of scattering of a spinless particle on a given potential, the Hilbert space is $\mathcal{H} = L^2(\mathbb{R}^3)$, so that the condition for square integrability of the vector $\Psi \in L^2(\mathbb{R}^3)$ is given by

$$\int_{\mathbb{R}^3} \mathrm{d}\boldsymbol{r}\, |\Psi(\boldsymbol{r})|^2 < \infty. \tag{3.10}$$

Since here the integration extends over the whole space $\mathbb{R}^3$, the invoked three-dimensional integral will be convergent, provided that $\Psi(\boldsymbol{r}) \equiv \langle \boldsymbol{r} | \Psi \rangle$ decreases sufficiently fast for $r \to \infty$. In the case of bound states of e.g. atomic systems, the corresponding wave functions decline exponentially at large distances from the center of interaction which keeps together all the participating particles. On the other hand, the asymptotic form of a wave function for scattering states ($E > 0$) is given by (1.10). The expression (1.10) implies that the state vector $\Psi_{\boldsymbol{k}}^{+}(\boldsymbol{r})$ oscillates in the asymptotic region and, therefore, cannot be an element of the L^2–space. Nevertheless, it is convenient to assume that e.g. a one-dimensional scattering state $\Psi(E, x) \equiv \langle x | \Psi(E) \rangle$ can be expressed as the first-order derivative with respect to the energy E of an auxiliary vector $\Phi(E, x) \in L^2$. In this way, the procedures developed for bound states could automatically be adapted for scattering states. For example, the plane wave e^{ikx} does not belong to the space L^2, but an auxiliary function of the type $(\mathrm{e}^{ikx} - 1)/(ix)$ is quadratically integrable

$$\begin{aligned} \mathrm{e}^{ikx} &= \frac{\mathrm{d}}{\mathrm{d}k} \varphi_k(x) \qquad k^2 = 2mE \\ \varphi_k(x) &= \frac{\mathrm{e}^{ikx} - 1}{ix} \longrightarrow \begin{cases} k & x \to 0 \\ 0 & x \to \infty. \end{cases} \end{aligned} \tag{3.11}$$

Quadratic integrability of the function $(\mathrm{e}^{ikx} - 1)/(ix)$ from (3.11) is secured by the asymptotic behavior at infinity. Namely, the absolute value of the function e^{ikx} is situated between -1 and $+1$, so that the oscillations of the numerator e^{ikx} from (3.11) are attenuated by the term $1/x$ as $x \to \infty$. Thus, the representation (3.11) for the plane wave permits the usage of the standard formalism of bound states for the ansatz $(\mathrm{e}^{ikx} - 1)/(ix)$ with the understanding that the derivative $\mathrm{d}/\mathrm{d}k$ should be applied at the end of the calculations of the considered observable. This is one of the alternatives to the more customary procedure of forming a wave packet as in (1.8). In (1.8), instead of differentiation, one carries out an integration over the energy with a weight function $w(\boldsymbol{k})$ which is sharply peaked only in a limited range and negligible elsewhere. Such a weight function damps the plane wave at large distances. Hence quadratic integrability of the wave packet (1.8). To gain a further insight into the way by which the ansatz $\varphi_k(x)$ also possesses a cut-off function with an effect of localization of the plane wave, let us rewrite (3.11) as

$$\phi_k(x) = N\frac{\mathrm{e}^{ikx}-1}{ix} = Nk\mathrm{e}^{ikx/2}\mathrm{sinc}(kx/2) \qquad \mathrm{sinc}(y) = \frac{\sin y}{y} \tag{3.12}$$

where we introduced the normalization factor N, so that $\phi_k(x) = N\varphi_k(x)$ with $\varphi_k(x)$ being unnormalized. The constant N can be determined by normalizing $\phi_k(x)$ to unity according to

$$\int_0^\infty \mathrm{d}x\phi_k^*(x)\phi_k(x) = 1. \tag{3.13}$$

Inserting here $\phi_k(x)$ from (3.12) and using the result

$$\int_0^\infty \mathrm{d}x\,\mathrm{sinc}^2(kx/2) = \frac{2\pi}{k} \tag{3.14}$$

it follows

$$\varphi_k(x) = \sqrt{\frac{k}{2\pi}}\mathrm{e}^{ikx/2}\mathrm{sinc}(kx/2). \tag{3.15}$$

Thus the function $\phi_k(x)$ contains the plane wave $\exp(ikx/2)$ which is damped at large distances x by the cut-off function $\mathrm{sinc}(kx/2)$. This damping is due to the presence of the hyperbole $1/y$ multiplying $\sin y$ in the sinc-function $\mathrm{sinc}(y)$ where $y = kx/2$. The sinc-function has the main lobe (maximum) at $x = 0$ and decays rapidly in an oscillatory/undulatory manner as x increases. Such a behavior leads to quadratic integrability of the function $\phi_k(x)$. Hence, this procedure achieves the same end effect as that of forming a wave packet. However, in most cases it is more efficient to perform the parametric differentiation $\mathrm{d}/\mathrm{d}k$ of the final results obtained with $\phi_k(x)$ than to carry out numerical integration from the corresponding wave packet.

3.4 Eigen-problems for continuous spectra

The continuum spectrum is associated with scattering states describing the colliding aggregates that are departing from each other in the asymptotic region of infinitely large spatial distances. To such free 'states' of particles we customarily associate plane waves $\phi_{\boldsymbol{k}_i} \propto \exp(i\boldsymbol{k}_i \cdot \boldsymbol{r})$ that are characterized by the so-called wave vector or the propagation vector $\boldsymbol{k}_i$. However, the plane waves are extended throughout the space and, as such, cannot describe a particle which must be localized in a given limited part of the whole space. To remedy the situation, we form e.g. a time-dependent travelling wave packet $\phi(t)$. This is an integral over the momentum $\boldsymbol{k}$ given by (1.8). The integral (1.8) is, in fact, a linear superposition of plane waves $\phi_{\boldsymbol{k}} = (2\pi)^{-3/2} \exp(i\boldsymbol{k}\cdot\boldsymbol{r})$ with a weight function $w(\boldsymbol{k})$. The role of this auxiliary function $w(\boldsymbol{k})$ is to cut off the individual plane waves $\phi_{\boldsymbol{k}}$ outside given finite parts of the space. Such a weight function is said to be peaked in a limited region where it attains its maximum, but outside of which is practically zero. In this way, the whole wave packet $\phi(t)$ is concentrated around e.g. the initial value $\boldsymbol{k}_i$ of the momentum. In contrast to a plane wave $\phi_{\boldsymbol{k}_i}$, the wave packet $\phi(t)$ formed in the described way becomes normalizable and, as such, belongs to the Hilbert space $\mathcal{H}$ of physically interpretable state vectors. This is because their norms are finite despite the fact that such wave packets belong to the continuous part of the Hamiltonian spectrum.

Scattering states ($E > 0$) are of utmost importance for collision theory and, as such, they should be examined from the most relevant aspects of spectral analysis of operators. To this end, let $\Lambda(\hat{\mathrm{A}})$ be the spectrum of a self-adjoint linear operator $\hat{\mathrm{A}}$ from the Hilbert space $\mathcal{H}$. In the case of a finite-dimensional space ($\dim\mathcal{H} = n \in \mathbb{N}$), the spectrum $\Lambda(\hat{\mathrm{A}})$ is defined as the set of real numbers λ that satisfy the eigen-value problem

$$\hat{\mathrm{A}}\psi = \lambda\psi \qquad \psi \in \mathcal{D}_{\hat{\mathrm{A}}} \subseteq \mathcal{H} \qquad \lambda \in \mathbb{R} \tag{3.16}$$

where ψ is a non-zero eigen-vector ($\psi \neq \emptyset$). If $\{\phi_k\}_{k=1}^{n} \subset \mathcal{H}$ is an orthonormal basis, then the set of all the eigen-values λ is given by the solutions of the secular equation

$$|\mathrm{A}_{pq} - \lambda\delta_{pq}| = 0 \tag{3.17}$$

where A_{pq} is the expected value of the operator $\hat{\mathrm{A}}$ taken over the basis functions ϕ_p and ϕ_q

$$\mathrm{A}_{pq} = \langle\phi_p|\hat{\mathrm{A}}|\phi_q\rangle. \tag{3.18}$$

In such a case, the spectrum $\Lambda(\hat{\mathrm{A}})$ is defined by the set $\{\lambda_k\}_{k=1}^{n} \subseteq \mathbb{R}$ whose general element λ_k is the solution of the algebraic equation (3.17) containing a polynomial of degree n in the variable λ. This latter polynomial is called the eigen-polynomial or the secular polynomial or the characteristic polynomial [34]. Every algebraic equation always possesses at least one root in the field of complex numbers $\mathbb{C}$ which is algebraically closed. An attempt to transfer the eigen-value problem (3.16) directly to the infinite-dimensional Hilbert space $\mathcal{H}$ would not meet with success because of the presence of the continuous part of the spectrum which is infinitely many times degenerate (infinite multiplicity). Namely, when the eigen-value λ lies in the continuum, the corresponding eigen-vector has an infinite norm and, therefore, as a non-normalizable vector, cannot belong to $\mathcal{H}$. Thus the standard eigen-value problem (3.16) has a meaning only for a discrete spectrum of the Hamiltonian $\hat{\mathrm{H}}$. Scattering states are not the elements of the Hilbert space $\mathcal{H}$ of normalizable state vectors. Finiteness of the norm in e.g. the coordinate representation is expressed via the standard condition $||\psi||^2 = \int \mathrm{d}\boldsymbol{r}\,|\psi(\boldsymbol{r})|^2 < \infty$ as in (3.10). Such a quadratic integrability of the wave function $\psi(\boldsymbol{r}) \in L^2(\mathbb{R}^3)$ can be fulfilled only if $\psi(\boldsymbol{r})$ tends to zero as $r \to \infty$. However, as mentioned, this is not the case with scattering states. Indeed according to the asymptotic form (1.10) the scattering wave function $\Psi_{\boldsymbol{k}}^{+}(\boldsymbol{r})$ has an oscillatory behavior at infinitely large distances r and, as such, does not tend to zero at $r \to \infty$. In order to arrive at a valid definition of a continuous spectrum, we shall transform (3.16) into a more convenient form which is fully equivalent to (3.16) for finite-dimensional spaces and which carries over to infinite-dimensional spaces without any essential change. The exit from this situation is offered precisely by the starting expression (3.16) itself, which implies that the equation

$$\hat{\mathrm{B}}\psi \equiv (\lambda\hat{1} - \hat{\mathrm{A}})\psi = \emptyset \qquad \psi \in \mathcal{H} \qquad \psi \neq \emptyset \tag{3.19}$$

has the solutions. On the other hand, for finite-dimensional spaces, we say that the operator $\hat{\mathrm{B}}$ is invertible i.e. that it has its inverse operator $\hat{\mathrm{B}}^{-1}$ if and only if the equation

$$\hat{\mathrm{B}}\psi = \phi \tag{3.20}$$

possesses precisely one solution for any vector $\phi \in \mathcal{H}$. It is easy to check that this will indeed be the case if and only if the equation

$$\hat{\mathrm{B}}\psi = \emptyset \tag{3.21}$$

has only the trivial solution $\psi = \emptyset$. Since $\hat{\mathrm{B}} = \lambda\hat{1} - \hat{\mathrm{A}}$, it follows that λ is the eigen-value of $\hat{\mathrm{A}}$ if and only if there exists a non-zero vector $\psi \neq \emptyset$ which under the action of the operator $\lambda\hat{1} - \hat{\mathrm{A}}$ becomes the zero vector. In other words, the condition that λ belongs to the spectrum of $\hat{\mathrm{A}}$ is that the operator $\lambda\hat{1} - \hat{\mathrm{A}}$ is singular. Thus, the feature (3.19) gives us the possibility to define the spectrum $\Lambda(\hat{\mathrm{A}})$ of the operator $\hat{\mathrm{A}}$ in the following general way. Let us denote by $\sigma(\hat{\mathrm{A}})$ the resolvent set, which represents the collection of all the complex numbers $z \in \mathbb{C}$ for which the operator $z\hat{1} - \hat{\mathrm{A}}$ is invertible

$$\sigma(\hat{\mathrm{A}}) = \{z \in \mathbb{C} : \exists\,(z\hat{1} - \hat{\mathrm{A}})^{-1}\}. \tag{3.22}$$

For such values of z we shall define the resolvent operator $\hat{\mathrm{R}}(z)$, or equivalently, the Green operator via

$$\hat{\mathrm{R}}(z) = (z\hat{1} - \hat{\mathrm{A}})^{-1} \qquad z \in \sigma(\hat{\mathrm{A}}). \tag{3.23}$$

By definition, the resolvent $\hat{\mathrm{R}}(\lambda)$ is a linear operator, which according to (3.20) can act on all vectors $\phi \in \mathcal{H}$ i.e. $\hat{\mathrm{R}}(\lambda)\phi = \psi$. Namely, as is well-known [7], if we are given the eigen-value problem (3.16), then the operator analytic function $f(\hat{\mathrm{A}})$ will satisfy the like equation $f(\hat{\mathrm{A}})\psi = f(\lambda)\psi$. Hence, the resolvent eigen-value problem is given by

$$\hat{\mathrm{R}}(z)\psi = r(z)\psi \tag{3.24}$$

with the same eigen-functions ψ as in the case of the operator $\hat{\mathrm{A}}$ from (3.16) and the eigen-values $r(z)$ that are connected to λ viz

$$r(z) = (z - \lambda)^{-1} \qquad \lambda = z - \frac{1}{r(z)}. \tag{3.25}$$

Of course, the order of all the degenerate eigen-values (if any) of the operators $\hat{\mathrm{A}}$ and $\hat{\mathrm{R}}(z)$ is the same. In this way, we finally arrive at the following general definition of the spectrum. Spectrum $\Lambda(\hat{\mathrm{A}})$ of the operator $\hat{\mathrm{A}}$ is the set $\{z \in \mathbb{C} \, : \, z \notin \sigma(\hat{\mathrm{A}})\}$ of all the eigen-values $z \in \mathbb{C}$ for which $z\hat{1} - \hat{\mathrm{A}}$ is not an invertible operator, and that is the set which is precisely complementary to the resolvent set $\sigma(\hat{\mathrm{A}})$ in $\mathbb{C}$

$$\Lambda(\hat{\mathrm{A}}) = \sigma^{\perp}(\hat{\mathrm{A}}) = \{z \in \mathbb{C} \, : \, z \notin \sigma(\hat{\mathrm{A}})\}. \tag{3.26}$$

It can be readily verified that this general definition is entirely equivalent to the usual definition of the spectrum related to (3.16) and, as such, could also be used in the case of bound states ($E < 0$). However, the real advantage of the generalized notion of the spectrum is in its possibility of a direct extension to infinite-dimensional spaces without any further change. In quantum scattering theory [7], one of the central places is occupied by the free and bound resolvent Hamiltonian operators $\hat{\mathrm{H}}_0$ and $\hat{\mathrm{H}}$, respectively

$$\hat{\mathrm{G}}_0(z) = (z\hat{1} - \hat{\mathrm{H}}_0)^{-1} \qquad \hat{\mathrm{G}}(z) = (z\hat{1} - \hat{\mathrm{H}})^{-1}. \tag{3.27}$$

If E is the total energy of the studied system and $z = E + i\varepsilon$ with ε being an infinitesimally small positive quantity, then we can introduce the free and the total Green operators by

$$\hat{\mathrm{G}}_0^{\pm}(E) = \operatorname*{Lim}_{\varepsilon \to 0+} \hat{\mathrm{G}}_0(E \pm i\varepsilon) \qquad \hat{\mathrm{G}}^{\pm}(E) = \operatorname*{Lim}_{\varepsilon \to 0+} \hat{\mathrm{G}}(E \pm i\varepsilon). \tag{3.28}$$

Hereafter, the symbol Lim in (3.28) denotes the strong limit which signifies convergence in the norm [7]. The importance of these resolvents for all physics, and not only for the theory of scattering, can be best appreciated by observing that the spectrum $\Lambda(\hat{\mathrm{G}})$ directly yields the spectrum $\Lambda(\hat{\mathrm{H}})$. This is because the poles and branch points of the resolvent $\hat{\mathrm{G}}(E)$ yield the discrete and continuous parts of the spectrum of $\hat{\mathrm{H}}$, respectively. As a consequence, it is not necessary any longer to try to explicitly solve (3.1), since the complete information about the studied system is contained in the spectrum of the corresponding resolvent operators. Thus, the Green operators allow examination of the complete spectrum of the given Hamiltonian on the same footing for both the discrete and continuous parts. Moreover, these resolvent operators can provide the eigen-functions in two different ways, either through the resolvent eigen-problem (3.24) or by means of the appropriate transition T-matrix. One of the most illustrative examples along these lines is the Coulomb Green function associated with the Coulomb potential.

The discrete spectrum $\Lambda_{\rm d}$ of the Hamiltonian $\hat{\rm H}$ related to the eigen-problem (3.16) is such that in the general case the following relationship is valid

$$\Lambda_{\rm d} \subseteq \Lambda. \tag{3.29}$$

Thus, generally, the set of all the eigen-values $\{\lambda_k\}_{k=1}^{n} \equiv \Lambda_{\rm d}$ of $\hat{\rm H}$ is a subset of the whole spectrum $\Lambda \equiv \Lambda(\hat{\rm H})$. The equality sign in (3.29) signifies that the discrete spectrum $\Lambda_{\rm d}$ could represent the whole spectrum Λ. Such a situation is encountered in the well-known Sturmian eigen-problem for the Coulomb potential [35]. The set of all the eigen-states ψ from (3.16) spans the vector subspace $\mathcal{H}_{\rm d}$, so that $\psi \in \mathcal{D}_{\hat{\rm A}} \equiv \mathcal{H}_{\rm d} \subseteq \mathcal{H}$. If $\hat{\rm P}_{\rm d}$ is the projection operator to the subspace $\mathcal{H}_{\rm d}$, then we say that $\hat{\rm P}_{\rm d}$ reduces the operator $\hat{\rm A}$. This circumstance, in the case of a bounded operator $\hat{\rm A}$, is expressed via the following commutator

$$[\hat{\rm P}_{\rm d}, \hat{\rm A}] = \hat{0}. \tag{3.30}$$

An operator $\hat{\rm A}$ is bounded if its norm is finite i.e.

$$||\hat{\rm A}\psi|| \le M||\psi|| < \infty \tag{3.31}$$

where M is a finite positive constant. The orthogonal complement of $\mathcal{H}_{\rm d}$ is another subspace denoted hereafter by $\mathcal{H}_{\rm c} = \mathcal{H}_{\rm d}^{\perp}$ such that the corresponding projection operator $\hat{\rm P}_{\rm c} = \hat{1} - \hat{\rm P}_{\rm d}$ also reduces $\hat{\rm A}$. Thus, the operator $\hat{\rm A}$ is reduced to two parts, each of which acts separately in their respective disjoint subspaces $\mathcal{H}_{\rm d}$ and $\mathcal{H}_{\rm c}$. In the usual, standard approach to the scattering problem, the theory is developed exclusively within the eigen-problem (3.16), as if there were only the operators from the subspace $\mathcal{H}_{\rm d}$. The same formalism is also used for operators from $\mathcal{H}_{\rm c}$ for which, however, the theory is not applicable. Strictly speaking, such a hybrid approach is mathematically unjustified and, moreover, it presents one of the main causes for the appearance of many heuristic formulae in the standard scattering theory. As stated, the solution to this undesirable situation is in resorting to the resolvent formalism which from the outset is equally valid for the discrete and continuous spectrum. We have already given the starting point in the proper definition of the spectrum. As an illustration in this direction, let us show how, by means of the resolvent operator, the spectrum of a self-adjoint operator is comprized of real numbers alone. This well-known fact can now be formulated as a stringent statement given by the following theorem:

<u>*Theorem*</u> 1: The spectrum of a self-adjoint operator $\hat{\rm A} : \mathcal{D}_{\hat{\rm A}} \longrightarrow \mathcal{R}_{\hat{\rm A}}$ represents a closed subset of the real axis $\mathbb{R}$. Consequently, the resolvent set contains half-planes $\mathbb{C}^+$ and $\mathbb{C}^-$. If $z \in \mathbb{C}^{\pm}$, then it follows that $\hat{\rm R}^{\dagger}(z) = \hat{\rm R}(z^*)$

where z^* is the complex conjugate of the complex number z. In such a case the following assessment of the resolvent norm is valid

$$||\hat{\mathrm{R}}(z)|| \le \frac{1}{|\mathrm{Im(z)}|}. \tag{3.32}$$

Proof: One of the ways of defining the set of complex numbers $\mathbb{C}$ is through $\mathbb{C} = \mathbb{R} \cup \mathbb{C}^+ \cup \mathbb{C}^-$, where $\mathbb{C}^\pm = \{z \in \mathbb{C} : z = x + iy, y \equiv \mathrm{Im}(z) \gtrless 0\}$ and $\cup$ is the usual symbol for the set relation of union. Let z be a complex number with a non-zero imaginary part ($z \in \mathbb{C}$, $y \neq 0$) where $\mathbb{C} = \{z : z = x+iy;\, x, y \in \mathbb{R}\}$. Then we have $(z\hat{1}-\hat{\mathrm{A}})^\dagger = z^*\hat{1}-\hat{\mathrm{A}}^\dagger = z^*\hat{1}-\hat{\mathrm{A}}$ where we used the condition that $\hat{\mathrm{A}}$ is a self-adjoint operator ($\hat{\mathrm{A}}^\dagger = \hat{\mathrm{A}}$). From here it follows that after taking the adjoint of the resolvent operator $\hat{\mathrm{R}}(z) \equiv (z\hat{1} - \hat{\mathrm{A}})^{-1}$, we shall have

$$\hat{\mathrm{R}}^\dagger(z) = \hat{\mathrm{R}}(z^*) \tag{3.33}$$

where $(\hat{\mathrm{B}}^{-1})^\dagger = (\hat{\mathrm{B}}^\dagger)^{-1}$. Then (3.33) together with the resolvent $\hat{\mathrm{R}}(i)$ can be used as the proof of self-adjointness of a given operator. Namely,

$$\begin{aligned} \hat{\mathrm{A}}^\dagger + i\hat{1} &= (\hat{\mathrm{A}} - i\hat{1})^\dagger = -[\hat{\mathrm{R}}^{-1}(i)]^\dagger \\ &= -[\hat{\mathrm{R}}^\dagger(i)]^{-1} = -[\hat{\mathrm{R}}(-i)]^{-1} = \hat{\mathrm{A}} + i\hat{1} \end{aligned}$$

$$\therefore \qquad \hat{\mathrm{A}}^\dagger + i\hat{1} = \hat{\mathrm{A}} + i\hat{1} \tag{3.34}$$

from which it follows $\hat{\mathrm{A}}^\dagger = \hat{\mathrm{A}}$. It is obvious from (3.33) that for real values of z i.e. $z\,(= z^* \equiv x)$, the resolvent $\hat{\mathrm{R}}(x)$ becomes a Hermitean operator

$$\hat{\mathrm{R}}^\dagger(x) = \hat{\mathrm{R}}(x) \qquad x \in \mathbb{R}. \tag{3.35}$$

Let us now evaluate the squared norm $||(z\hat{1} - \hat{\mathrm{A}})\psi||^2$ where $\psi \in \mathcal{D}_{\hat{\mathrm{A}}}$. Given that $z = x + iy\,(x, y \in \mathbb{R}$, $y \neq 0)$ we obtain

$$\begin{aligned} ||(z\hat{1} - \hat{\mathrm{A}})\psi||^2 &= \langle (z\hat{1} - \hat{\mathrm{A}})\psi | (z\hat{1} - \hat{\mathrm{A}})\psi \rangle \\ &= \langle (x\hat{1} - \hat{\mathrm{A}})\psi + iy\psi | (x\hat{1} - \hat{\mathrm{A}})\psi + iy\psi \rangle \\ &= ||(x\hat{1} - \hat{\mathrm{A}})\psi||^2 + y^2||\psi||^2 \ge y^2||\psi||^2 \end{aligned}$$

$$\therefore \qquad ||(z\hat{1} - \hat{\mathrm{A}})\psi||^2 \ge y^2||\psi||^2. \tag{3.36}$$

The inequality means that the operator $z\hat{1}-\hat{\mathrm{A}}$ has a bounded inverse operator. In other words, the resolvent $\hat{\mathrm{R}}(z) = (z\hat{1} - \hat{\mathrm{A}})^{-1}$ exists as a bounded operator, since according to (3.36)

$$||\hat{\mathrm{R}}(z)|| \le \frac{1}{|y|} \tag{3.37}$$

which implies $z \in \sigma(\hat{\mathrm{A}})$. Thus, if z is chosen in such a way to lie outside the circle of radius ρ around the eigen-value E i.e. $|z - \rho| \geq \rho$, then it follows

$$||\hat{\mathrm{R}}(z)\chi|| \leq \frac{1}{|\rho|}||\chi|| \qquad \forall \chi \in \mathcal{H}. \tag{3.38}$$

When defined in this way, the resolvent $\hat{\mathrm{R}}(z)$ appears to be a bounded operator. Therefore, the work with $\hat{\mathrm{R}}(z)$ is much easier than with the Hamiltonian $\hat{\mathrm{H}}$ which is an unbounded operator. For further discussion regarding theorem 1, we mention that a linear operator $\hat{\mathrm{A}} : \mathcal{H}_1 \longrightarrow \mathcal{H}_2$ is called closed, if and only if from $\psi_n \in \mathcal{D}_{\hat{\mathrm{A}}}$ $(n \in \mathbb{N})$, $\psi_n \underset{n\to\infty}{\longrightarrow} \psi \in \mathcal{H}_1$ and $\hat{\mathrm{A}}\psi_n \underset{n\to\infty}{\longrightarrow} \phi \in \mathcal{H}_2$, it follows $\psi \in \mathcal{D}_{\hat{\mathrm{A}}}$ and $\hat{\mathrm{A}}\psi = \phi$. If the operator $\hat{\mathrm{A}}$ is closed, then the set

$$\Theta_{\hat{\mathrm{A}}} \equiv \{\psi \in \mathcal{D}_{\hat{\mathrm{A}}} : \hat{\mathrm{A}}\psi = \emptyset\} \tag{3.39}$$

coincides with the closed subspace $\mathcal{H}_1$. Further, if $\hat{\mathrm{A}}$ is closed and moreover $\Theta_{\hat{\mathrm{A}}} = \oslash$, where $\oslash \equiv \{0\}$ is the empty set, then the inverse $\hat{\mathrm{A}}^{-1}$ is also a closed operator. On the basis of this finding, we assert that the resolvent operator $\hat{\mathrm{R}}(z)$ is defined in the whole Hilbert space $\mathcal{H}$. This is because the orthogonal complement of $\mathcal{R}_{z\hat{1}-\hat{\mathrm{A}}}$ is given by

$$\mathcal{R}^{\perp}_{z\hat{1}-\hat{\mathrm{A}}} = \Theta_{z\hat{1}-\hat{\mathrm{A}}} = \oslash \tag{3.40}$$

where $\mathcal{R}_{z\hat{1}-\hat{\mathrm{A}}}$ is the image region (range) of the operator $z\hat{1} - \hat{\mathrm{A}}$. Hence, the resolvent $\hat{\mathrm{R}}(z)$ is a closed operator and the set $\mathcal{R}_{z\hat{1}-\hat{\mathrm{A}}}$ is everywhere dense in $\mathcal{H}$. This together with the fact that the resolvent operator $\hat{\mathrm{R}}(z)$ is bounded and closed leads to the relation

$$\mathcal{R}_{z\hat{1}-\hat{\mathrm{A}}} = \mathcal{H}. \tag{3.41}$$

A linear operator $\hat{\mathrm{A}} : \mathcal{D}_{\hat{\mathrm{A}}} \longrightarrow \mathcal{R}_{\hat{\mathrm{A}}}$ is closed if its graph $\mathcal{G}_{\hat{\mathrm{A}}}$ represents a closed vector subspace. Here, the graph $\mathcal{G}_{\hat{\mathrm{A}}}$ of a linear operator $\hat{\mathrm{A}}$ is the direct sum $\mathcal{D}_{\hat{\mathrm{A}}} \oplus \mathcal{R}_{\hat{\mathrm{A}}}$, which is a set of the pairs of the form $\{\psi, \hat{\mathrm{A}}\psi\}$ i.e. $\mathcal{G}_{\hat{\mathrm{A}}} \equiv \{\psi, \hat{\mathrm{A}}\psi : \psi \in \mathcal{D}_{\hat{\mathrm{A}}}\}$. Finally, in order to convince ourselves that the spectrum $\Lambda(\hat{\mathrm{A}})$ is closed, it is sufficient to show that the set $\sigma(\hat{\mathrm{A}})$ is open. This step is carried out by recalling that the set $\sigma(\hat{\mathrm{A}})$ will represent an open set, if around each $z \in \sigma(\hat{\mathrm{A}})$ one could construct a small disc which is entirely contained in $\sigma(\hat{\mathrm{A}})$. We shall return to this important proof in chapter 5.

3.5 Normal and Hermitean operators

It is clear from (3.33) that in the general case of a complex value z, the resolvent $\hat{\mathrm{R}}(z)$ is not a self-adjoint operator. However, the resolvent $\hat{\mathrm{R}}(z)$ is a normal operator, since it commutes with its adjoint operator $\hat{\mathrm{R}}^\dagger(z)$

$$[\hat{\mathrm{R}}(z), \hat{\mathrm{R}}^\dagger(z)] = \hat{0}. \tag{3.42}$$

Here, we can make a digression and discuss the problem of possible complex observables in the context of normal operators. In that regard, it would be interesting to see whether is possible to have a complex number as a result of a measurement. In other words, it is instructive to see whether one could measure complex dynamic variables. Since every complex number $z = x + iy \in \mathbb{C}$ is determined by its pair of real numbers $\{x, y\}$ i.e. $x = \mathrm{Re}(z) \in \mathbb{R}$ and $y = \mathrm{Im}(z) \in \mathbb{R}$, an experiment could be envisaged to measure separately x and y. Then from the measured values for x and y, we would form the combination $x + iy = z$ and in this way we might legitimately state that the found complex number z *is* indeed the experimental result. In order to experimentally record a complex dynamic observable, it is necessary to perform two simultaneous, separate measurements. Such two measurements could be interpreted without any difficulty within classical physics. However, difficulties arise when dealing with quantum measurements. Namely, as is well-known [7] from the standpoint of quantum physics, two measurements cannot be considered as being exactly simultaneous, since they generally interfere with each other. Even if two given measurements have been performed in immediate succession, the first experiment would customarily disturb the state of the examined system and, therefore, lead to an uncertainty which could, in turn, influence the subsequent measurement. Because of this occurrence, it is usually said, as in the book of Dirac [36] that, in principle, quantum measurements yield only real-valued results for dynamic variables. Such a view has repeatedly been promoted in the literature to justify the choice of exclusively Hermitean operators ($\hat{\mathrm{A}}^\dagger = \hat{\mathrm{A}}$) as the representatives of observables since such operators possess only real eigen-values. As is well-known, the operator $\hat{\mathrm{A}}$ is Hermitean if it satisfies the relation

$$\langle \psi | \hat{\mathrm{A}} \phi \rangle = \langle \hat{\mathrm{A}} \psi | \phi \rangle \qquad \forall \psi \qquad \phi \in \mathcal{D}_{\hat{\mathrm{A}}} \tag{3.43}$$

under the assumption that the definition domain $\mathcal{D}_{\hat{\mathrm{A}}}$ is everywhere dense in the Hilbert space $\mathcal{H}$. If additionally the following relation is valid

$$\mathcal{D}_{\hat{\mathrm{A}}} = \mathcal{D}_{\hat{\mathrm{A}}^\dagger} \tag{3.44}$$

then the operator $\hat{\mathrm{A}}$ is said to be self-adjoint. From here it is clear that self-adjointness imposes a stricter requirement that hermiticity. In other words, self-adjointness implies hermiticity, but the opposite is not true. For a larger class of operators used in physics, self-adjointness is equivalent to hermiticity. By selecting self-adjoint operators for the representatives of observables, one implicitly emphasizes that measurements of a complex dynamic variable z are quantum mechanically inadmissible due to the lack of commutation of the operators associated with the real (x) and imaginary (y) part of z. However, such a standpoint, which could be found in some textbooks on quantum mechanics, is incorrect. This is because there exists an entire class of complex operators $\hat{\mathrm{Z}} = \mathrm{Re}(\hat{\mathrm{Z}}) + i\,\mathrm{Im}(\hat{\mathrm{Z}})$ whose real and imaginary parts commute

$$[\mathrm{Re}(\hat{\mathrm{Z}}), \mathrm{Im}(\hat{\mathrm{Z}})] = \hat{0}. \tag{3.45}$$

Thus, if by $\mathcal{Z} = \mathrm{Re}(\mathcal{Z}) + i\,\mathrm{Im}(\mathcal{Z})$ we denote a complex dynamic variable associated with the operator $\hat{\mathrm{Z}}$, it follows that the real $\mathrm{Re}(\mathcal{Z})$ and imaginary $\mathrm{Im}(\mathcal{Z})$ part can be measured simultaneously, such that one experimental recording does not disturb the other. Operators with the feature (3.42) are called normal operators. An operator $\hat{\mathrm{A}}$ is normal if and only if it commutes with its own adjoint operator i.e.

$$[\hat{\mathrm{A}}, \hat{\mathrm{A}}^{\dagger}] = \hat{0}. \tag{3.46}$$

A self-adjoint operator is normal, but the opposite need not be true.

3.6 Strong and weak topology

The discrete part of the spectrum of the operator $\hat{\mathrm{A}}$ i.e. its reduction to the subspace $\mathcal{H}_{\mathrm{d}}$ behaves in many aspects as if it were due to a self-adjoint operator in a finite-dimensional space. Indeed, normalized vectors ϕ_n $(n \in \mathbb{N})$ form a complete and ortho-normalized basis in $\mathcal{H}_{\mathrm{d}}$, so that every vector $\psi \in \mathcal{H}_{\mathrm{d}}$ could be expanded according to $\psi = \sum_{n=1}^{\infty} c_n \phi_n$ where $c_n = (\psi, \phi_n)$ as in (3.2). In quantum mechanics, the expression (3.2) is known as the expansion theorem. The same expression, however, from the mathematical viewpoint represents a special case of the spectral theorem. Although the operator $\hat{\mathrm{A}}$ in the discrete subspace $\mathcal{H}_{\mathrm{d}}$ shares a number of properties of operators in finite-dimensional spaces, nevertheless there are several important differences that must always be kept in mind. In the first place, the infinite sum in (3.2) can be divergent. In other words, there is a convergence problem for which it is necessary to fix the topology. Topology in which the series from (3.2) converges is termed strong topology [7].

Let us recall the two extremely important axioms known as completeness and separability of the Hilbert space $\mathcal{H}$. In particular, these axioms are crucial for the concept of the strong limit which plays a central role in scattering theory [7]. A certain sequence $\{\psi_n\}_{n=1}^{\infty}$ comprized of vectors ψ_n from the Hilbert space $\mathcal{H}$ i.e. $\psi_n \in \mathcal{H}$ $(n \in \mathbb{N})$, is said to be the Cauchy sequence if for each infinitesimally small number $\epsilon > 0$ there exists a number N_ϵ, such that the following inequality hold

$$||\psi_n - \psi_m|| < \epsilon \qquad n, m > N_\epsilon \tag{3.47}$$

where $n, m, N_\epsilon \in \mathbb{N}$. The Cauchy sequence is equivalently called the fundamental sequence. Not every Cauchy sequence is convergent. The axiom on completeness of the space $\mathcal{H}$ requires that the Cauchy sequences are convergent in the norm i.e. that they converge strongly [7]. A given sequence $\{\psi_n\}_{n=1}^{\infty}$ of vectors ψ_n from the space $\mathcal{H}$ i.e. $\psi_n \in \mathcal{H}\,(n \in \mathbb{N})$ converges strongly to the unique vector $\psi \in \mathcal{H}$, as symbolized by

$$\psi_n \underset{n\to\infty}{\Longrightarrow} \psi \tag{3.48}$$

if the sequence of numbers $||\psi_n - \psi||$ tends to zero

$$||\psi_n - \psi|| \underset{n\to\infty}{\longrightarrow} 0. \tag{3.49}$$

The state ψ is indeed the unique limiting vector, as can be shown by using the Schwartz inequality

$$||\psi + \phi|| \le ||\psi|| + ||\phi||. \tag{3.50}$$

To this end, let us suppose that there are two limiting vectors $\psi, \psi' \in \mathcal{H}$ of a strongly convergent sequence $\{\psi_n\}_{n=1}^{\infty}$ i.e. in addition to (3.48), we assume that the following relation is valid

$$\psi_n \underset{n\to\infty}{\Longrightarrow} \psi'. \tag{3.51}$$

Then it follows

$$\begin{aligned} ||\psi' - \psi|| &= ||(\psi' - \psi_n) + (\psi_n - \psi)|| \\ &\le ||\psi' - \psi_n|| + ||\psi_n - \psi|| \underset{n\to\infty}{\longrightarrow} 0 \end{aligned}$$

$$\therefore \quad ||\psi' - \psi|| \underset{n\to\infty}{\longrightarrow} 0 \tag{3.52}$$

and this together with (3.48) and (3.51) unequivocally implies that $\psi' = \psi$. This latter relation stems from $||\psi - \psi'|| = 0$ which holds true if and only if $\psi' = \psi$ (QED). Let us now show that each convergent sequence is a Cauchy sequence. Strong convergence of the sequence $\{\psi_n\}$ to the vector ψ i.e. $||\psi_n - \psi|| \underset{n\to\infty}{\longrightarrow} 0$, means that for every $\epsilon' > 0$ there exists a natural number $N_{\epsilon'}$ such that $||\psi_n - \psi|| < \epsilon'$ for $\forall\, n \ge N_{\epsilon'}$. On the other hand, we have

$$\begin{aligned} ||\psi_n - \psi_m|| &= ||(\psi_n - \psi) + (\psi - \psi_m)|| \\ &\le ||\psi_n - \psi|| + ||\psi - \psi_m)|| \\ &< \epsilon' + \epsilon' = \frac{\epsilon}{2} + \frac{\epsilon}{2} = \epsilon \end{aligned}$$

$$\therefore \quad ||\psi_n - \psi_m|| < \epsilon \tag{3.53}$$

where we set $\epsilon' = \epsilon/2$. Thus, for each $\epsilon' > 0$ there exists $N_{\epsilon'} \in \mathbb{N}$ for which the inequality $||\psi_n - \psi_m|| < \epsilon \equiv \epsilon'/2$ is valid whenever $\forall n, m \geq N_{\epsilon'}$ for $n, m \in \mathbb{N}$. Therefore, every strong convergent sequence is also a fundamental sequence. The opposite does not need to hold true i.e. general vector spaces might have Cauchy sequences that are not strongly convergent. In addition to the strong limit, there is also the weak limit which defines weak topology operating with the scalar products instead of the norm in the following way. A sequence $\{\psi_n\}$ converges weakly to its limiting value ψ i.e.

$$\psi_n \underset{n\to\infty}{\longrightarrow} \psi \tag{3.54}$$

if it holds

$$\langle\phi|\psi_n\rangle \underset{n\to\infty}{\longrightarrow} \langle\phi|\psi\rangle \qquad \forall\,\phi \in \mathcal{H} \tag{3.55}$$

in the sense

$$\lim_{n\to\infty} |\langle\phi|\psi_n\rangle - \langle\phi|\psi\rangle| = 0. \tag{3.56}$$

Referring to these definitions, it is possible to formulate the following lemma:

<u>*Lemma*</u> 1: If $\psi_n \underset{n\to\infty}{\longrightarrow} \psi$ and $||\psi_n||^2 \underset{n\to\infty}{\longrightarrow} ||\psi||^2$, then it follows that $\underset{n\to\infty}{\mathrm{Lim}}\ \psi_n = \psi$. The proof is simple if we use the definition of strong convergence, according to which the relation $\psi_n \underset{n\to\infty}{\Longrightarrow} \psi$, in turn, implies validity of the relation (3.49). Therefore, we have

$$\begin{aligned}
||\psi_n - \psi||^2 &= \langle\psi_n - \psi|\psi_n - \psi\rangle \\
&= ||\psi_n||^2 + ||\psi||^2 - \{\langle\psi_n|\psi\rangle + \langle\psi|\psi_n\rangle\} \\
&\underset{n\to\infty}{\longrightarrow} ||\psi_n||^2 + ||\psi||^2 - \{||\psi||^2 + ||\psi||^2\} \\
&\underset{n\to\infty}{\longrightarrow} ||\psi_n||^2 - ||\psi||^2 \underset{n\to\infty}{\longrightarrow} 0
\end{aligned}$$

$$\therefore \qquad ||\psi_n - \psi||^2 \underset{n\to\infty}{\longrightarrow} 0 \qquad \text{(QED)}. \tag{3.57}$$

The obvious corollary of lemma 1 reads: if $||\psi_n||^2 = 1$ and $\psi_n \underset{n\to\infty}{\longrightarrow} \psi$, then we shall have $\psi_n \underset{n\to\infty}{\Longrightarrow} \psi$ *if and only if* $||\psi||^2 = 1$. Hence, strong convergence $\psi_n \underset{n\to\infty}{\Longrightarrow} \psi$ means according to $||\psi_n - \psi|| \underset{n\to\infty}{\longrightarrow} 0$ that ψ_n tends to the vector ψ in the sense of the 'length' $d(\psi_n - \psi)$ of $\psi_n - \psi$ tending to zero as $n \to \infty$ via the relation

$$d(\psi_n - \psi) \equiv ||\psi_n - \psi|| \underset{n\to\infty}{\longrightarrow} 0. \tag{3.58}$$

Designed in this way, the strong limit has a very clear physical interpretation [7, 37]. Namely, the relation $\psi_n \underset{n\to\infty}{\longrightarrow} \psi$ means that in the limit $n \to \infty$ the state ψ_n is experimentally indistinguishable from the state ψ of the considered system, as follows from the expressions

$$\begin{aligned}
|\langle\phi|\psi_n\rangle - \langle\phi|\psi\rangle| &= |\langle\phi|\,\{|\psi_n\rangle - |\psi\rangle\}\,| \\
&\leq ||\phi|| \cdot ||\psi_n - \psi|| \\
&= ||\psi_n - \psi|| = 1
\end{aligned}$$

$$\therefore \qquad |\langle\phi|\psi_n\rangle - \langle\phi|\psi\rangle| \le ||\psi_n - \psi|| \qquad ||\phi|| = 1. \tag{3.59}$$

If a sequence $\{\psi_k\}_{k=1}^{\infty}$ of abstract elements converges strongly, then this will also be the case for any concrete representation of the abstract elements. Namely, the relation (3.59) obviously indicates that from the condition $\psi_n \underset{n\to\infty}{\Longrightarrow} \psi$, it obligatorily follows $\langle\phi|\psi_n\rangle \underset{n\to\infty}{\longrightarrow} \langle\phi|\psi\rangle$ for an arbitrary fixed vector $\phi \in \mathcal{H}$. Thus, if ψ_n converges strongly, then this will also be the case for all its components in an arbitrary representation. It is easy to show that the opposite is true as well, but only in the case of a finite-dimensional Hilbert space. This is intuitively understood in the case of the finite-dimensional space $\mathbb{R}^3$, where the very notion and the basic features of convergence along a direction and on a plane remain the same by the passage to the three-dimensional space. This is because in this case we are using the customary terminology of the type 'sufficiently close', irrespective of whether or not we are concerned with a set of points lying on a line, or on a plane or in a three-dimensional space. A similar consideration can be equally well extended to the spaces of higher dimensions e.g. from $\mathbb{R}^3$ to $\mathbb{R}^n$ and, therefore, to an arbitrary finite-dimensional space X^n due to isomorphism (3.9) of all finite-dimensional spaces. However, the situation is much more complicated for infinite-dimensional spaces for which convergence of the components of ψ_n need not imply convergence of the state vectors ψ_n themselves for $n \longrightarrow \infty$. This could be most conveniently shown by considering the set $\{\psi_n\}$ from the abstract Hilbert space $\mathcal{H}$ as an infinite ortho-normalized set. Then the well-known Riemann-Lebesgue lemma [7] asserts that

$$\langle\phi|\psi_n\rangle \underset{n\to\infty}{\longrightarrow} 0 \qquad \forall\,\phi \in \mathcal{H}. \tag{3.60}$$

This, in the case of the zero limiting value, is equivalent to the existence of weak convergence. Nevertheless, despite the existence of the weak limit, the set $\{\psi_n\}$ does not represent the Cauchy sequence, since we have

$$\begin{aligned}\lim_{n,m\to\infty} ||\psi_n - \psi_m||^2 &= \lim_{n,m\to\infty} \langle\psi_n - \psi_m|\psi_n - \psi_m)\rangle \\ &= \lim_{n,m\to\infty} \{\langle\psi_n|\psi_n\rangle + \langle\psi_m|\psi_m\rangle - 2\mathrm{Re}(\langle\psi_n|\psi_m\rangle)\} = 2\end{aligned}$$

$$\therefore \qquad \lim_{n,m\to\infty} ||\psi_n - \psi_m||^2 = 2. \tag{3.61}$$

In other words, the strong limit does not exist due to the relation

$$||\psi_n - \psi_m|| = \sqrt{2} \qquad \text{(QED)} \tag{3.62}$$

for $n, m \in \mathbb{N}$. In this proof, we used the abstract elements of the Hilbert space $\mathcal{H}$ and applied the Riemann-Lebesgue lemma [7]. The fact that these abstract elements ψ_n are chosen in a special form of an ortho-normal infinite set does not diminish the generality of the conclusion, since for the breakdown of a theorem it is sufficient to find a single counter-example, as we just did. This is sufficient for a demonstration that weak convergence does not generally imply strong convergence.

These illustrations indicate the real reason for insisting on the existence of the strong limits in scattering theory, as also analyzed in Ref. [7]. In examining convergence of state vectors via the strong limit, we could single out e.g. the square-integrable functions $\psi(t,x) \in \mathcal{H} = L^2(\mathbb{R}^n)$

$$\psi(t,x) = \psi(x)\mathrm{e}^{itx} \tag{3.63}$$

where $x = \{x_1, x_2, ..., x_n\}$ is a point in the $n-$dimensional space. Due to the oscillatory character of the rhs of (3.63), it is certain that the function $\psi(t,x) \equiv \langle x|\psi(t)\rangle$ does not tend to zero when $t \longrightarrow \infty$

$$\psi(t,x) \underset{t\longrightarrow\infty}{\not\Longrightarrow} \emptyset. \tag{3.64}$$

This means that the strong limit does not exist and the same conclusion could also be reached via

$$\begin{aligned}
\lim_{t\longrightarrow\infty} ||\psi(t)||^2 &= \lim_{t\longrightarrow\infty} \int \mathrm{d}x\, |\psi(t,x)|^2 \\
&= \int \mathrm{d}x\, |\psi(0,x)|^2 = ||\psi(0)||^2
\end{aligned}$$

$$\therefore \qquad \lim_{t\longrightarrow\infty} ||\psi(t)||^2 = ||\psi(0)||^2 \tag{3.65}$$

where $0 < ||\psi(x)|| < \infty$. Let us now take another square-integrable element $\phi(x)$ from $L^2(\mathbb{R}^n)$. Then the Riemann-Lebesgue lemma (3.60) yields directly

$$\lim_{t\longrightarrow\infty} \langle\phi|\psi'\rangle = 0 \qquad \psi' \equiv \psi(t,x). \tag{3.66}$$

This implies the existence of the weak limit i.e. the fulfillment of the condition (3.60) for any fixed square-integrable function $\phi(x) \in L^2(\mathbb{R}^n)$ although according to (3.65) the strong limit does not exist. Notice that the link between the weak and strong limits could easily be understood on an intuitive basis via a Gaussian wave packet without any recourse to the Riemann-Lebesgue lemma [7]. With this goal, let $\psi_0(t) = \hat{\mathrm{U}}_0(t)\psi_0$ be a Gaussian wave packet. This wave packet expands and spreads with the passage of time by following the customary $|t|^{-3/2}-$behavior at large values of time t [7]. Simultaneously, the center of gravitation of this wave packet strictly obeys classical mechanics through the Hamilton-Jacobi equations of motions. Because of the said spreading, the corresponding wave function in the configuration space $\psi_0(t,\boldsymbol{r}) = \langle \boldsymbol{r}\,|\psi_0\rangle$ tends to zero at large $|t|$ at any fixed point $\boldsymbol{r}$. Therefore, the overlap integral $\langle\phi|\psi_0(t)\rangle$ between $|\psi_0(t)\rangle$ and any fixed vector $|\phi\rangle$ will tend to zero when $t \longrightarrow \infty$, since $\langle\phi|\psi_0(t)\rangle \propto t^{-3/2}$. However, from the other side, we have $||\psi_0(t)|| = ||\psi_0|| = 1$ which certainly does not tend to zero when $t \longrightarrow \infty$. This illustrates in a plausible way that, although in the considered case, weak convergence exists via $\lim_{t\longrightarrow\infty}\langle\boldsymbol{r}\,|\psi_0(t)\rangle \longrightarrow 0$, the associated strong limit does not exist i.e. $\lim_{t\longrightarrow\infty} ||\psi_0(t)|| \not\longrightarrow 0$.

3.7 Compact operators for mapping of the weak to strong limits

The key practical question in the above analysis is the following: is there a way to transform a weakly convergent sequence of vectors into the corresponding sequence for which the strong limit exist? The answer is in the affirmative. Namely, one among the several equivalent definitions of the so-called compact operators can be helpful in answering the posed question. This useful definition is: an operator $\hat{\mathrm{C}}$ is compact if and only if it maps every weakly convergent vector sequence into a strongly convergent one. For a compact operator $\hat{\mathrm{C}}$, we then assert that the sequence $\{\phi_n\} \equiv \{\hat{\mathrm{C}}\psi_n\}$ possesses the strong limit whenever the corresponding sequence $\{\psi_n\}$ converges weakly. The feature of compactedness of operators, or equivalently, complete continuity, is very important in studying spectra of Hamiltonians by means of resolvents. Additionally, compact operators are important in view of the fact that their properties manifest most directly the analogy with operators in finite-dimensional matrices. Notice that a compact sequence of vectors represents a set in which every element has one strongly convergent subsequence. A sequence is said to be weakly compact if it possesses weakly convergent subsequences. Thus, we can say that a certain operator $\hat{\mathrm{C}}$ will be compact if for any bound vector sequence ψ_n, we have a sequence $\{\hat{\mathrm{C}}\psi_n\}$ which contains a strongly convergent subsequence. Such an operator $\hat{\mathrm{C}}$ is necessarily continuous. An operator $\hat{\mathrm{A}}$ is continuous if and only if it is bounded i.e. if (3.31) is valid. Obviously, every continuous operator is closed. Similarly, a vector ψ is bounded if it has a finite norm

$$||\psi|| < M' < \infty \tag{3.67}$$

where M' is a finite positive constant. If we provisionally assume that a given continuous operator $\hat{\mathrm{C}}$ is not bounded, then a sequence of normalized vectors ψ_n must exist such that $||\hat{\mathrm{C}}\psi_n|| > \mathrm{c}$ where c is a positive constant. Thus, such a sequence could not have any convergent subsequence. Consequently, the provisional assumption is false and, therefore, the opposite is true i.e. continuity of an operator necessarily implies its boundedness. In this regard, let us prove the following lemma:

Lemma 2: Every compact linear operator is bounded.

Proof: Assume that the opposite is true i.e. that a compact, linear operator $\hat{\mathrm{A}} : \mathrm{X} \longrightarrow \mathrm{Y}$, which maps the space X to Y, is not bounded. In such a case there would exist a sequence $\{\phi_n\}_{n=1}^{\leq\infty}$ of a finite or infinite number of terms $\phi_1, \phi_2, ...$, such that we have $||\phi_n|| = 1$ and $||\hat{\mathrm{A}}\phi_n|| \geq n$ for every number $n \in \mathbb{N}$. However, since the operator $\hat{\mathrm{A}}$ is compact, then in the space X there must exist a subsequence comprized of the elements $\phi_{n(k)}$ such that $\hat{\mathrm{A}}\phi_{n(k)}$

represents a vector $\psi \in \mathrm{Y}$ when $k \to \infty$

$$||\hat{\mathrm{A}}\phi_{n(k)}|| \underset{k\to\infty}{\longrightarrow} ||\psi||. \tag{3.68}$$

However, this contradicts the condition $||\hat{\mathrm{A}}\phi_{n(k)}|| \geq n(k) \underset{k\to\infty}{\longrightarrow} \infty$. Therefore, the assumption is false i.e. its negation is correct stating that every compact operator is bounded (QED).

An operator $\hat{\mathrm{C}}$ could also be said to be compact if and only if it maps every bounded vector sequence into a compact vectors sequence. Operators are fundamental entities in quantum mechanics but, nevertheless, for their physical interpretation we need their matrix elements. The feature of compactness can also be introduced on the level of matrix elements. This leads to yet another definition of compact operators. Namely, an operator $\hat{\mathrm{C}}$ is said to be compact if and only if $\langle\phi_n|\hat{\mathrm{C}}\psi_n\rangle \underset{n\to\infty}{\longrightarrow} \langle\phi|\hat{\mathrm{C}}\psi\rangle$ whenever there exists weak convergence $\phi_n \underset{n\to\infty}{\longrightarrow} \phi$ and $\psi_n \underset{n\to\infty}{\longrightarrow} \psi$.

3.8 Strong differentiability and strong analyticity

Here, we introduce strong differentiability and analyticity of abstract vectors[2]. In the general case, the weak and strong derivatives exist and differ from each other by the use of the weak and strong limits, respectively. A linear operator $\hat{\mathrm{L}}_x \equiv f'(x)$ is called the strong derivative of the function $f(x)$ at the point x if it holds

$$f(x+h) - f(x) - \hat{\mathrm{L}}_x h = \mathcal{O}(h) \qquad \hat{\mathrm{L}}_x \equiv f'(x) = \frac{\mathrm{d}f(x)}{\mathrm{d}x} \tag{3.69}$$

where $\hat{\mathrm{L}}_x h \in \mathcal{R}_{\hat{\mathrm{L}}_x}$ and $h \in \mathcal{D}_{\hat{\mathrm{L}}_x}$. If, for an abstract function of its abstract argument, there exist the weak and strong derivatives, then they are identical to each other. Moreover, if the abstract function is a function of a real or complex variable, with respect to which the derivatives are taken, then it can be shown that the weak and strong derivatives are the same. Thus, there exists only one kind of derivative of vectors depending upon a numerical variable such as time t in non-stationary scattering theory. Consequently, without any ambiguity one can talk about analyticity of functions-vectors of one complex variable. These notions become essential in scattering theory when studying analytic properties of wave functions or operators that have complex energy as their independent variable e.g. Green functions, the scattering S- and transition T-matrix elements [7].

[2]Under the term 'vector' we understand an element of an abstract vector space. In a special case, such an abstract vector can be reduced to a vector in the ordinary sense of the word as a geometrical vector i.e. an oriented segment of a line.

4

Linear and bilinear functionals

4.1 Linear functionals for mapping between vector spaces and scalar fields

Functionals, as a special form of mapping between vector spaces and scalar fields, play a very important role in scattering theory. For this reason they deserve a concise analysis which we shall present here. For a given finite or infinite-dimensional abstract space $\mathrm{X}_{\mathcal{F}}$ defined on the scalar field $\mathcal{F}$ we shall introduce a linear functional. This is a linear form $f : \mathrm{X}_{\mathcal{F}} \longrightarrow \mathcal{F}$ of the type of a linear operator, which performs correspondence between vectors from $\mathrm{X}_{\mathcal{F}}$ and scalars from $\mathcal{F}$ viz

$$f(\psi_1 + \psi_2) = f(\psi_1) + f(\psi_2) \qquad \forall\,\psi_1, \psi_2 \in \mathrm{X}_{\mathcal{F}} \tag{4.1}$$

$$f(\lambda\psi) = \lambda f(\psi) \qquad \forall\,\psi \in \mathrm{X}_{\mathcal{F}} \qquad \forall\,\lambda \in \mathcal{F} \tag{4.2}$$

or in a more compact form

$$\left.\begin{aligned} f(\lambda\psi_1 + \mu\psi_2) = \lambda f(\psi_1) + \mu f(\psi_2) \\ \forall\,\psi_1, \psi_2 \in \mathrm{X}_{\mathcal{F}} \qquad \forall\,\lambda, \mu \in \mathcal{F} \end{aligned}\right\}. \tag{4.3}$$

Of course, the definition of the linear functional (4.3) can be automatically extended to an arbitrary number of independent variables ($n < \infty$) in the evident form

$$\begin{aligned} f(\lambda_1\psi_1 + \lambda_2\psi_2 + \cdots + \lambda_n\psi_n) = \lambda_1 f(\psi_1) + \lambda_2 f(\psi_2) \\ + \cdots + \lambda_n f(\psi_n) \end{aligned} \tag{4.4}$$

for $\forall\,\psi_k \in \mathrm{X}_{\mathcal{F}}$ and $\forall\,\lambda_k \in \mathcal{F}\,(1 \le k \le n)$. The functional $f(\psi)$ is bounded if it holds

$$|f(\psi)| \le c||\psi|| \qquad \forall\,\psi \in \mathrm{X}_{\mathcal{F}} \tag{4.5}$$

where c is a finite positive real number ($c < \infty$). Here, it is understood that the norm $||\psi||$ is introduced via the following axioms that are related to the scalar product

$$\left.\begin{array}{c} ||\psi|| \geq 0 \qquad ||\psi|| = 0 \qquad \Longleftrightarrow \quad \psi = \emptyset \\ \\ ||\mu\psi|| = |\mu| \cdot ||\psi|| \qquad ||\psi + \phi|| \leq ||\psi|| + ||\phi|| \\ \\ \forall\, \psi, \phi \in \mathrm{X}_{\mathcal{F}} \qquad \mu \in \mathcal{F} \end{array}\right\}. \tag{4.6}$$

An obvious example of a functional is the usual scalar or inner asymmetric product (or Hermitean-symmetric) $\langle\phi|\psi\rangle$ defined by means of the standard five axioms

- Axiom 1: $\langle\phi|\alpha\psi_1 + \beta\psi_2\rangle = \alpha\langle\phi|\psi_1\rangle + \beta\langle\phi|\psi_2\rangle$
- Axiom 2: $\langle\alpha\phi_1 + \beta\phi_2, \psi\rangle = \alpha^*\langle\phi_1|\psi\rangle + \beta^*\langle\phi_2|\psi\rangle$
- Axiom 3: $\langle\phi|\psi\rangle = \langle\psi|\phi\rangle^*$
- Axiom 4: $\langle\psi|\psi\rangle \geq 0$
- Axiom 5: $\langle\psi|\psi\rangle = 0 \qquad \Longleftrightarrow \qquad \psi = \emptyset$ (4.7)

where $\phi, \psi, \psi_1, \psi_2 \in \mathcal{H}$, $\alpha, \beta \in \mathbb{C}$ and, as usual, the star superscript denotes complex conjugation. It then follows that the scalar product $\langle\phi|\psi\rangle$ is anti-linear and linear functional relative to the first (ϕ) and the second (ψ) term, respectively. On the other hand, a linear functional f is, by definition, in the class of linear operators. Therefore, it is clear that the set of all the images of such an operator i.e. the range $\mathcal{R}_f$ will be conveniently determined if we know how f acts on the elements of a given basis $b_\beta \equiv \{\beta_1, \beta_2, ..., \beta_n\}$. This is indeed the case, as can be shown by referring to the expansion theorem in a finite-dimensional space

$$\psi = \sum_{k=1}^{n} \psi_k^\beta \beta_k \qquad \psi_k^\beta \in \mathcal{F} \tag{4.8}$$

where $\psi_k^\beta \equiv \psi(\beta_k)$ are the expansion coefficients i.e. the coordinates of the vector ψ with respect to the basis b_β. Employing (4.4) we arrive at the following formula

$$[f]\,\psi = [f]\sum_{k=1}^{n} \psi_k^\beta \beta_k = \sum_{k=1}^{n} [f]\,\psi_k^\beta \beta_k = \sum_{k=1}^{n} \psi_k^\beta\,[f]\,\beta_k$$

$$\therefore \qquad [f]\,\psi = \sum_{k=1}^{n} \psi_k^\beta\,[f]\,\beta_k \tag{4.9}$$

where the square brackets around f as indicated by $[f]$ symbolize the action of the functional f in the analogy with the standard notion of a function

$$[f]\,\psi \equiv f(\psi). \tag{4.10}$$

In this way, the expression (4.8) becomes

$$f(\psi) = \sum_{k=1}^{n} \psi_k^\beta f(\beta_k). \tag{4.11}$$

This shows that if the set $b_\beta = \{\beta_k\}_{k=1}^n$ is a basis in the space $X_\mathcal{F}^n$, then the general and explicit functional f with the feature (4.4) is given by the linear combination

$$\begin{aligned} f(\psi) &= f(\psi_1^\beta \beta_1 + \psi_2^\beta \beta_2 + \cdots + \psi_n^\beta \beta_n) \\ &= \psi_1^\beta f(\beta_1) + \psi_2^\beta f(\beta_2) + \cdots + \psi_n^\beta f(\beta_n). \end{aligned} \tag{4.12}$$

Notice the concreteness of the expression (4.12) with respect to the general property (4.4). In the expression (4.12) the quantities $\beta_1, \beta_2, ..., \beta_n$ are the elements of the basis $b_\beta \subseteq X_\mathcal{F}$ and $\psi_1^\beta, \psi_2^\beta, ..., \psi_n^\beta$ are the expansion coefficients, whereas $\psi_1, \psi_2, ..., \psi_n$ are arbitrary vectors from $X_\mathcal{F}^n$ and $\lambda_1, \lambda_2, ..., \lambda_n \in \mathcal{F}$. Further, it clearly follows from (4.12) that the image $f(\psi)$ of the functional of any vector $\psi \in \mathcal{D}_f$ is fully determined by the knowledge of the elements that are obtained from the action of the same functional $f(\beta_k)$ $(1 \leq k \leq n)$ on the basis functions b_β. The obtained general form (4.12) of the linear functional formally resembles the usual way of presenting the scalar product

$$\langle\phi|\psi\rangle = \sum_k \phi_k^{e*} \psi_k^e. \tag{4.13}$$

Nevertheless, it must be stressed that the derivation of (4.12) is carried out for a general vector space $X_\mathcal{F}^n$ which does not need to be unitary. However, in the special case of unitary spaces $Y_\mathcal{F}^n$ that are of great importance for scattering theory, we shall show that (4.12) indeed coincides with the standard expression (4.13) of the scalar product $\langle\phi|\psi\rangle$ which is, as mentioned before, a linear functional with respect to its second term ψ. This constitutes the content of the Ries-Freshe theorem.

4.2 The Ries-Freshe theorem

Theorem 2 (Ries-Freshe): In order that a mapping $f : Y_\mathcal{F}^n \longrightarrow \mathcal{F}$ $(n \in \mathbb{N})$ could represent a linear functional defined in a unitary finite-dimensional vector space $Y_\mathcal{F}^n$ on the scalar field $\mathcal{F}$, it is necessary and sufficient that there exists the unique vector $\phi \in Y_\mathcal{F}^n$ with the feature

$$f(\psi) \equiv f_\phi(\psi) = \langle\phi|\psi\rangle \qquad \forall\,\psi \in Y_\mathcal{F}^n \tag{4.14}$$

where $\langle\phi|\psi\rangle$ is the scalar product of two abstract vectors defined by Axioms $1-5$ from (4.7).

Proof: Let us first prove sufficiency. Here, 'sufficiency' means that if f maps (4.14) into an unitary space $\mathrm{Y}_{\mathcal{F}}^{n}$, then f is a linear functional i.e. it has the property (4.12). To show this, we shall use the fact that every finite-dimensional space has a basis which can be ortho-normalized by e.g. the Gramm-Schmidt procedure [34]. Thus, let us assume that in a space $\mathrm{Y}_{\mathcal{F}}^{n}$ we are given an ortho-normalized basis $b_e \equiv \{e_1, e_2, ..., e_n\}$ where $\langle e_j|e_k\rangle = \delta_{jk}$. Then according to the expansion theorem (4.8), an arbitrary element $\psi \in \mathrm{Y}_{\mathcal{F}}^{n}$ can be developed in the basis b_e as

$$\phi = \sum_{k=1}^{n} \phi_k^e e_k. \tag{4.15}$$

All the expansion coefficients $\{\phi_k^e\}_{k=1}^{n}$ i.e. the coordinates of the vector ϕ in the basis b_e, can be found by forming the scalar product $\langle e_j|\phi\rangle$ via (4.15) and using the condition of ortho-normalizability of the basis b_e

$$\langle e_j|\phi\rangle = \langle e_j|\sum_{k=1}^{n}\phi_k^e e_k\rangle = \sum_{k=1}^{n}\phi_k^e\langle e_j|e_k\rangle = \sum_{k=1}^{n}\phi_k^e\delta_{jk} = \phi_j^e$$

$$\therefore \qquad \langle e_j|\phi\rangle = \phi_j^e \tag{4.16}$$

or equivalently

$$\phi_k^e = \langle e_k|\phi\rangle \qquad 1 \le k \le n. \tag{4.17}$$

We assumed that (4.14) is valid, but when attempting to directly identify the term $\langle e_k|\phi\rangle$ as a linear functional f, an apparent difficulty arises due to the fact that the scalar product $\langle e_k|\phi\rangle$ is anti-linear in its first term e_k. However, according to Axiom 3 from (4.7), the scalar product is Hermitean-symmetric $\langle\phi|\psi\rangle = \langle\psi|\phi\rangle^*$, so that

$$\phi_k^e = \langle e_k|\phi\rangle = \langle\phi|e_k\rangle^* = f_\phi^*(e_k). \tag{4.18}$$

This means that $f_\phi(e_k) = \langle\phi|e_k\rangle$ is a linear functional in which case the expansion (4.16) acquires the form

$$\phi = \sum_{k=1}^{n} f_\phi^*(e_k)e_k \qquad \phi \in \mathrm{Y}_{\mathcal{F}}^{n}. \tag{4.19}$$

This development is equivalent to the general starting expression (4.15). Concreteness of the formula (4.19) is transparent in that it specifies the coordinates ϕ_k^e of the vector ϕ in the basis b_e via the linear functional $f_\phi^{e*}(e_k)$, as prescribed by (4.17). Nevertheless, due to the arbitrariness of ϕ, the relation (4.15) is valid for any other $\psi \in \mathrm{Y}_{\mathcal{F}}^{n}$ i.e.

$$\psi = \sum_{k=1}^{n} \psi_k^e e_k \qquad \psi \in \mathrm{X}_{\mathcal{F}} \tag{4.20}$$

where $\{\psi_k^e\}_{k=1}^n$ are the coordinates of the vector ψ in the basis b_e. Therefore, we shall have

$$f_\phi(\psi) = f_\phi(\sum_{k=1}^n \psi_k^e e_k) = \langle\phi|\sum_{k=1}^n \psi_k^e e_k\rangle$$
$$= \sum_{k=1}^n \psi_k^e \langle\phi|e_k\rangle = \sum_{k=1}^n \psi_k^e f_\phi(e_k)$$

$$\therefore \qquad f_\phi(\psi) = \sum_{k=1}^n \psi_k^e f_\phi(e_k). \tag{4.21}$$

This, according to the definition (4.12) means that the mapping $f_\phi(\psi)$, which is of the form $\langle\phi|\psi\rangle$, represents a linear functional. Clearly, such a finding is in accordance with Axiom 1 from the definition (4.7) of the scalar product i.e. $\langle\phi|\lambda_1\psi_1 + \lambda_2\psi_2 + \cdots + \lambda_n\psi_n\rangle = \lambda_1\langle\phi|\psi_1\rangle + \lambda_2\langle\phi|\psi_2\rangle + \cdots + \lambda_n\langle\phi|\psi_n\rangle$. To prove the necessity of the Ries-Freshe theorem, we must invert the order of the proof from the sufficiency. In other words, now 'necessity' means that from the supposition of f being a linear functional i.e. a mapping with the feature (4.12), it follows that f must have the form of the scalar product (4.14). The supposition that $f_\phi(\psi)$ is a linear functional implies the validity of (4.12) which, by means of (4.21), can be rewritten in terms of the ortho-normalized basis b_e as

$$f_\phi(\psi) = \sum_{k=1}^n \psi_k^e f_\phi(e_k). \tag{4.22}$$

This expression must hold true for an element ϕ from the given unitary space $\mathrm{Y}_{\mathcal{F}}^n$. We are searching a vector $\phi \in \mathrm{X}_{\mathcal{F}}^n$ which determines the functional $f_\phi(\psi)$ precisely in the form $\langle\phi|\psi\rangle$. Since the element $\phi \in \mathrm{Y}_{\mathcal{F}}^n$ can be expanded in the basis b_e via (4.15) i.e. $\phi = \sum_{k=1}^n \phi_k^e e_k = \sum_{k=1}^n \langle e_k|\phi\rangle e_k$, we conclude that the kth expansion coefficient ϕ_k^e already possesses the sought form of the 'elementary' linear functional $f_\phi(e_k) = \langle\phi|e_k\rangle = \phi_k^{e*}$. However, we need the proof that the 'total' linear functional $f_\phi(\psi)$ indeed also has the required form $\langle\phi|\psi\rangle$. This can be done if the obtained 'elementary' i.e. the kth functional $f_\phi(e_k)$ is rewritten in its equivalent form ϕ_k^{e*} which after substitution into (4.22) leads to the final conclusion

$$f_\phi(\psi) = \sum_{k=1}^n \phi_k^{e*}\psi_k^e = \langle\phi|\psi\rangle \qquad \text{(QED)}. \tag{4.23}$$

Here, we used the standard form (4.13) of the scalar product $\langle\phi|\psi\rangle$. This proves the condition of necessity. In the course of the demonstration, the existence of the 'elementary' linear functional $f_\phi(e_k)$ is also proven in the form $\langle\phi|e_k\rangle^*$

for a vector $\phi \in X_{\mathcal{F}}$. By the same procedure, the expansion (4.20) for ψ can be written as

$$\psi = \sum_{k=1}^{n} f_{\psi}^{*}(e_k)e_k \qquad \psi \in X_{\mathcal{F}}. \tag{4.24}$$

In this way, the intermediate step $f_{\phi}(\psi) = \sum_{k=1}^{n} \phi_k^{e*}\psi_k^e$ from the derivation of the result (4.23) can be conveniently expressed by

$$f_{\phi}(\psi) = \sum_{k=1}^{n} f_{\phi}(e_k)f_{\psi}^{*}(e_k)$$
$$= \sum_{k=1}^{n} \langle\phi|e_k\rangle\langle\psi|e_k\rangle^{*}. \tag{4.25}$$

Again, this is nothing but the well-known Parseval representation of the scalar product in finite-dimensional spaces

$$\langle\phi|\psi\rangle = \sum_{k=1}^{n} \langle\phi|e_k\rangle\langle e_k|\psi\rangle. \tag{4.26}$$

Finally, we need to prove the uniqueness of the element $\phi \in Y_{\mathcal{F}}$. To this end, we shall assume the opposite i.e. that there exist two such elements $\phi, \chi \in Y_{\mathcal{F}}^{n}$ for which $f(\psi)$ is a linear functional. This implies that in addition to (4.14) in the form $f(\psi) = \langle\phi|\psi\rangle$, we also have $f(\psi) = \langle\chi|\psi\rangle$ for $\forall\,\psi \in Y_{\mathcal{F}}^{n}$. In such a case, it is obvious that $\langle\phi - \chi|\psi\rangle = 0$. Since this latter relation is valid for any $\psi \in Y_{\mathcal{F}}^{n}$, it also must hold true for a particular ψ selected as $\psi = \phi - \chi$ so that $\phi - \chi \in Y_{\mathcal{F}}^{n}$ where $Y_{\mathcal{F}}^{n}$ is a linear space. For such a choice of the vector ψ, we also have $\langle\phi - \chi|\phi - \chi\rangle \equiv ||\phi - \chi||^2 = 0$. This, according to the second of the features listed in (4.6) can be fulfilled if and only if $\chi = \phi$, which proves the uniqueness of the element ϕ. This completes the proof of the whole Ries-Freshe theorem (QED).

4.3 Bilinear functionals

In the case of an infinite-dimensional unitary space $Y_{\mathcal{F}}^{\infty}$, the Reis-Freshe theorem is valid only for bounded linear functional of the type (4.5). Notice that the linear functional (4.14) possesses the following feature

$$f_{\phi}(\psi) = f_{\psi}^{*}(\phi) \tag{4.27}$$

which follows from Axiom 3 of the scalar product (4.7) i.e. $\langle\phi|\psi\rangle = \langle\psi|\phi\rangle^{*}$. Additionally, Axiom 2 of the scalar product (4.7) i.e. its property of anti-linearity regarding the first term ψ i.e. $\langle\lambda_1\phi_1 + \lambda_2\phi_2|\psi\rangle = \lambda_1^{*}\langle\phi_1|\psi\rangle + \lambda_2^{*}\langle\phi_2|\psi\rangle$, implies that the expression (4.14) satisfies the relation

$$f_{\lambda\phi}(\psi) = \lambda^* f_\phi(\psi). \tag{4.28}$$

Also the following feature of the linear functional holds true

$$f_{\lambda\phi+\mu\chi}(\psi) = \lambda^* f_\phi(\psi) + \mu^* f_\chi(\psi). \tag{4.29}$$

In addition to linear functionals, there are also bilinear forms of mappings that are very important for scattering theory. The function $g(\phi, \psi)$ whose domain is given by the tensor product of $\mathrm{X}^n_{\mathcal{F}}$ with itself i.e. $\mathcal{D}_g = \mathrm{X}^n_{\mathcal{F}} \times \mathrm{X}^n_{\mathcal{F}}$, and whose range $\mathcal{R}_g$ represents the scalar field $\mathcal{F}$ on which the abstract vector space $\mathrm{X}^n_{\mathcal{F}}$ is defined, is called a bilinear form if it fulfills the following conditions

$$g(\alpha_1\phi_1 + \alpha_2\phi_2, \psi) = \alpha_1^* g(\phi_1, \psi) + \alpha_2^* g(\phi_2, \psi) \tag{4.30}$$

$$g(\phi, \beta_1\psi_1 + \beta_2\psi_2) = \beta_1 g(\phi, \psi_1) + \beta_2 g(\phi, \psi_2). \tag{4.31}$$

By definition, a bilinear form is bounded if we have

$$|g(\phi, \psi)| \le c\,||\phi|| \cdot ||\psi|| \tag{4.32}$$

where c is a finite real and positive constant. A bilinear functional satisfies a very important theorem which reads as:

Theorem 3: If $\mathrm{X}^n_{\mathcal{F}}$ is a unitary space $\mathrm{Y}^n_{\mathcal{F}}$, then for each bilinear functional $g(\phi, \psi)$, there exists a bounded linear operator $\hat{\mathrm{A}}$ with the characteristics

$$g(\phi, \psi) = \langle\phi|\hat{\mathrm{A}}\psi\rangle \tag{4.33}$$

such that

$$||A|| = \sup\left\{\frac{g(\phi, \psi)}{||\phi|| \cdot ||\psi||}\right\}. \tag{4.34}$$

Proof: The relation (4.34) follows from the Schwartz inequality [7], which in this case is given by

$$|\langle\phi|A\psi\rangle| \le ||\phi|| \cdot ||\hat{\mathrm{A}}\psi||. \tag{4.35}$$

Here, the equality sign holds for $\phi = \hat{\mathrm{A}}\psi$ in which case we immediately find

$$||\hat{\mathrm{A}}|| = \sup\left\{\frac{|\langle\phi|\hat{\mathrm{A}}\psi\rangle|}{||\phi|| \cdot ||\psi||}\right\}. \tag{4.36}$$

This is valid for any choice of normalizable vectors ϕ and ψ in accordance with (4.34). The proof of theorem 3 on bilinear functionals from unitary spaces is based upon the following obvious observation: if ϕ is a fixed vector then $g(\phi, \psi)$ is reduced to a linear functional in ψ

$$g(\phi, \psi) = \langle \phi | \hat{\mathrm{A}} \psi \rangle = f_\phi(\hat{\mathrm{A}}\psi). \tag{4.37}$$

Similarly, if the vector ψ is fixed, then the bilinear functional (4.33) is an anti-linear functional in ϕ. Now sufficiency of theorem 3 states that if the mapping g has the form (4.33), then it is a bilinear functional. This can be proven by starting from

$$g(\phi, \psi) = f_\phi(\hat{\mathrm{A}}\psi). \tag{4.38}$$

According to the Ries-Freshe theorem, the mapping $f_\phi(\chi) = \langle \phi | \chi \rangle$ is a linear functional in χ. This also holds for $f_\phi(\hat{\mathrm{A}}\psi)$, since the operator $\hat{\mathrm{A}}$ is linear

$$\begin{aligned} g(\phi, \sum_{k=1}^{n} \beta_k \psi_k^\beta) &= f_\phi(\hat{\mathrm{A}} \sum_{k=1}^{n} \beta_k \psi_k^\beta) = f_\phi(\sum_{k=1}^{n} \beta_k \hat{\mathrm{A}} \psi_k^\beta) \\ &= \sum_{k=1}^{n} \beta_k f_\phi(\hat{\mathrm{A}} \psi_k^\beta) = \sum_{k=1}^{n} \beta_k g(\phi, \psi_k^\beta) \end{aligned}$$

$$\therefore \qquad g(\phi, \sum_{k=1}^{n} \beta_k \psi_k^\beta) = \sum_{k=1}^{n} \beta_k g(\phi, \psi_k^\beta) \tag{4.39}$$

so that the mapping (4.33) indeed satisfies the condition (4.31). Similarly, by fixing the vector ψ and using (4.28) and (4.29) for a linear functional, the following relations emerge

$$\begin{aligned} g(\alpha_1\phi_1 + \alpha_2\phi_2, \psi) &= f_{\alpha_1\phi_1+\alpha_2\phi_2}(\hat{\mathrm{A}}\psi) \\ &= \alpha_1^* f_{\phi_1}(\hat{\mathrm{A}}\psi) + \alpha_2^* f_{\phi_2}(\hat{\mathrm{A}}\psi) \\ &= \alpha_1^* g(\phi_1, \psi) + \alpha_2^* g(\phi_2, \psi) \end{aligned}$$

$$\therefore \qquad g(\alpha_1\phi_1 + \alpha_2\phi_2, \psi) = \alpha_1^* g(\phi_1, \psi) + \alpha_2^* g(\phi_2, \psi) \tag{4.40}$$

in agreement with the requirement (4.30). Therefore, the mapping $g(\phi, \psi)$ of the form (4.33) represents a bilinear functional. This proves sufficiency of theorem 3. On the other hand, if we assume that $g(\phi, \psi)$ is a bilinear functional, then for a unitary space $\mathrm{Y}_{\mathcal{F}}^n$ this functional will obligatorily have the form (4.33). This is recognized as the necessity condition of theorem 3. The necessity will now be proven by referring to the Ries-Freshe theorem. Thus, if (4.31) holds, then the mapping $g(\phi, \psi)$ is a linear functional in ψ, as stated in (4.38). According to the Ries-Freshe theorem, such a linear functional $f_\phi(\hat{\mathrm{A}}\psi)$ is of the form

$$f_\phi(\hat{\mathrm{A}}\psi) = \langle \phi | \hat{\mathrm{A}}\psi \rangle. \tag{4.41}$$

Since (4.38) is valid, it immediately follows

$$g(\phi, \psi) = \langle \phi | \hat{\mathrm{A}} \psi \rangle. \tag{4.42}$$

This completes the proof on bilinear functionals. Its variant from (4.30) and (4.31) is also valid for an infinite-dimensional space $\mathrm{Y}_{\mathcal{F}}^{\infty}$, provided that we are restricted to a bounded bilinear functional (4.32).

5

Definition of a quantum scattering event

5.1 Hamiltonian operators and boundedness

Infinite-dimensional Hilbert spaces are of utmost importance for scattering theory and this leads to the following critical question: is the central operator i.e. the Hamiltonian $\hat{\mathrm{H}}$, bounded? As seen in chapter 3, boundedness is the most essential feature of operators defined in infinite-dimensional vector spaces. The answer to this key question is in the negative. The Hamilton operator belongs to a class of so-called positive operators that are also called semi-bounded operators. Stated more precisely, due to the existence of the minimum of the total energy of the investigated system, the Hamiltonian $\hat{\mathrm{H}}$ is an operator which is bounded from below. This is mathematically expressed by the following requirement for non-negativity of the expected values of the Hamiltonian operator

$$E = \langle\Psi|\hat{\mathrm{H}}\Psi\rangle \geq 0 \qquad \forall\,\Psi \in \mathcal{D}_{\hat{\mathrm{H}}} \tag{5.1}$$

where $\mathcal{D}_{\hat{\mathrm{H}}} \subset \mathcal{H}$ is the domain i.e. the definition region of the operator $\hat{\mathrm{H}}$ which represents a closed linear manifold in $\mathcal{H}$. Here, we emphasize the especially important fact that $\mathcal{D}_{\hat{\mathrm{H}}}$ is not equal to the whole space $\mathcal{H}$. Rather, $\mathcal{D}_{\hat{\mathrm{H}}}$ coincides only with a subspace of $\mathcal{H}$. However, this latter subspace is everywhere dense in $\mathcal{H}$. The notion of a linear manifold is in close connection with the customary notion of a subspace of a given vector space. Let us denote by ψ the elements of the vector space $\mathcal{H}$ and by ψ' the elements of the corresponding subspace $\mathcal{H}' \subset \mathcal{H}$. Then the set of all the vectors $\psi + \psi'$ $(\psi \in \mathcal{H}, \psi' \in \mathcal{H}')$ is called a linear manifold and it is denoted as $\psi + \mathcal{H}'$

$$\mathcal{H}' + \psi'' = \{\psi \in \mathcal{H} : \psi = \psi' + \psi'' \,,\, \psi' \in \mathcal{H}' \,,\, \exists\,\psi'' \in \mathcal{H}\}\,. \tag{5.2}$$

From the obvious geometric associations, a linear manifold $\psi + \mathcal{H}'$ is also called the translation of the subspace $\mathcal{H}'$ for the vector ψ. For each element $\psi \in \mathcal{H}$, the linear manifold $\psi + \mathcal{H}'$ is determined entirely. Moreover, the element $\psi \in \mathcal{H}$ simultaneously belongs to the linear manifold $\psi + \mathcal{H}'$. Namely, since $\mathcal{H}'$ is a subspace, it follows that $\emptyset \in \mathcal{H}$, so that $\psi = (\psi + \emptyset) \in \psi + \mathcal{H}'$. The relation (5.1) expresses the most important feature of all physical systems and that is the fact the total energy E possesses a lower limit which corresponds to the lowest ($E \equiv E_{\rm min}$) i.e. bound state of the system. Nevertheless, the Hamilton operator $\hat{\rm H}$ is not bounded from above. This implies that, from the mathematical viewpoint, the Hamiltonian $\hat{\rm H}$ is an unbounded operator. Notice that from the physical viewpoint, it might appear at first glance that it could be more acceptable to write $E = \langle\Psi|\hat{\rm H}\Psi\rangle \geq E_{\rm min}$ instead of (5.1). This is because in the inequality (5.1) we arbitrarily assigned the zero value as the lower limit of the energy E. But this does not represent any essential restriction, since the appearance of any additive constant in the Hamiltonian $\hat{\rm H}$ would have no physical implication. Namely, an extra constant term in $\hat{\rm H}$ could only lead to a scaling on the energy axis where the starting value of the physical energy is chosen arbitrarily anyway. In other words, the expression (5.1) is justified, since the lower limit as the zero limit for the energy can always be obtained by a convenient choice of a constant term in $\hat{\rm H}_0$ and $\hat{\rm H}$.

5.2 Evolution operators and the Møller wave operators

For a mathematical formulation of both the free ($\hat{\rm V} = \hat{0}$) and the bound ($\hat{\rm V} \neq 0$) dynamics, the unperturbed and the total (free + bound) Hamilton operators $\hat{\rm H}_0$ and $\hat{\rm H}$, respectively, do not represent the convenient starting points for the analysis. This is because the domains of $\hat{\rm H}_0$ and $\hat{\rm H}$ are only subspaces of the whole Hilbert space $\mathcal{H}$ and, as such, do not represent closed linear manifolds of normalizable state vectors (wave packets). Besides, both $\hat{\rm H}_0$ and $\hat{\rm H}$ are unbounded operators. Needless to say, the work with subspaces is much more difficult than with the whole Hilbert space $\mathcal{H}$. Much more fruitful mathematical entities for the description of the time evolution of quantum systems are the unperturbed (free) and total evolution unitary and bounded operators $\hat{\rm U}_0(t)$ and $\hat{\rm U}(t)$. The definition domains of the evolution operators $\hat{\rm U}_0(t)$ and $\hat{\rm U}(t)$ coincide with those of $\mathcal{H}$, respectively. Furthermore, for the time variable t in the interval from $-\infty$ to $+\infty$ the operators $\hat{\rm U}_0(t)$ and $\hat{\rm U}(t)$ represent the one-parameter infinitesimal group of the time translation with the generators $\hat{\rm H}_0$ and $\hat{\rm H}$, respectively. The most convenient consequence of the switch from $\{\hat{\rm H}_0, \hat{\rm H}\}$ to $\{\hat{\rm U}_0(t), \hat{\rm U}(t)\}$ is the fact that all the theorems and lemmas from scattering theory given in Ref. [7] and in the present book will remain valid for both finite and infinite-dimensional vector space. A great practical advantage in the work with the evolution operators $\hat{\rm U}_0(t)$ and $\hat{\rm U}(t)$ is especially clear in the case of conservative systems for which we have the following simple exponentials

$$\hat{\rm U}_0(t) = {\rm e}^{-i\hat{\rm H}_0 t} \qquad \hat{\rm U}(t) = {\rm e}^{-i\hat{\rm H}t} \qquad t \in \mathbb{R}. \tag{5.3}$$

Here, the first obvious gain is that multiplication of these exponential operators could be carried out without any limitation. This is because the range of one such operator is always contained in the domain of the other evolution operator. One such product of utmost importance for scattering theory is the product of the evolution operators $\hat{\mathrm{U}}^\dagger(t)$ and $\hat{\mathrm{U}}_0(t)$. The limiting values (if they exist) of the product $\hat{\mathrm{U}}^\dagger(t)\hat{\mathrm{U}}_0(t)$ when $t \to \mp\infty$ are called the Møller wave operators $\hat{\Omega}^\pm$ that are defined as

$$\hat{\Omega}^\pm = \operatorname*{Lim}_{t\to\mp\infty} \{\hat{\mathrm{U}}^\dagger(t)\hat{\mathrm{U}}_0(t)\}. \tag{5.4}$$

Incidently, it may not seem logical that the order of the appearance of the signs $\pm$ is opposite on the lhs and rhs of (5.4). In other words, it would appear as more natural that $\hat{\Omega}^\pm$ are the limiting values of the operators $\hat{\mathrm{U}}^\dagger(t)\hat{\mathrm{U}}_0(t)$ when $t \to \pm\infty$. However, we shall see shortly that the convention for the signs $\mp$ appearing in (5.4) corresponds exactly to the usual notions of the incoming/outgoing waves, respectively.

5.3 The Cauchy strong limit in non-stationary scattering theory

Whenever the symbol Lim for the strong limit i.e. convergence in the norm is associated with time variable $t \to \mp\infty$, as in (5.4), we are talking about the Cauchy strong limit in the non-stationary scattering theory [7, 37]. We shall see in chapter 6 that there also exists the Abel strong limit in stationary scattering theory. The special limiting passage Lim in (5.4) relates to the so-called q−numbers that obey the non-commutative algebra. The connection of the Lim procedure with the usual limit (lim) used for the so-called c−numbers obeying the commutative algebra is given in the considered case by

$$\lim_{t\to\mp\infty} ||\{\hat{\mathrm{U}}^\dagger(t)\hat{\mathrm{U}}_0(t) - \hat{\Omega}^\pm\}\Psi|| = 0 \qquad \forall\,\Psi \in \mathcal{H}. \tag{5.5}$$

This latter relation is written concisely viz

$$\hat{\mathrm{U}}^\dagger(t)\hat{\mathrm{U}}_0(t)\Psi \underset{t\to\mp\infty}{\Longrightarrow} \hat{\Omega}^\pm\Psi \tag{5.6}$$

where again the special symbol $\Longrightarrow$ denotes the strong limit [7]. A real physical meaning of these relations is best perceived in the way in which the wave operators $\hat{\Omega}^\pm$ act upon the unperturbed state Ψ_0 from the spectrum of the Hamiltonian $\hat{\mathrm{H}}_0$

$$\Psi^\pm = \hat{\Omega}^\pm\Psi_0 \qquad \Psi_0 \in \mathcal{H} \tag{5.7}$$

where the vectors $\Psi^\pm$ are called the scattering states. In Ref. [7] the key role of the strong limit is illuminated from many basic aspects and especially in the context of asymptotic convergence in the coordinate representation within the Schrödinger picture of quantum mechanics. Additionally, the goal of the present chapter is to highlight the possibility of formulating general and *complete* scattering theory which could be entirely generated from a single equation such as (5.4). For example, we shall show that (5.4) and (5.7) lead directly the Lippmann-Schwinger integral equations

$$\Psi^{\pm} = \Psi_0 + [(E \pm i\varepsilon)\hat{1} - \hat{\mathrm{H}}]^{-1}\Psi^{\pm} \qquad \varepsilon \to 0^{+}. \tag{5.8}$$

Here, the symbol $\varepsilon \to 0^{+}$ denotes the approach to zero through the positive numbers and this determines the selection of the so-called physical Riemann sheet in the complex energy plane. By this procedure, we impose the boundary conditions of the outgoing (e^{ikr}/r) or incoming (e^{-ikr}/r) wave. The integral equations from (5.8) permit a systematic examination of collisions e.g. by means of their iterative solutions via the method of successive substitutions or via variational principles.

5.4 Three criteria for a quantum collisional system

The relations (5.4)–(5.7) constitute the foundation of the three basic criteria for the definition of a quantum mechanical one-channel collisional system:

(i) The first condition which is necessary for a two-body quantum system to be a collisional system is that there exist the following two strong limits

$$\operatorname*{Lim}_{t\to\mp\infty} \hat{\mathrm{U}}^{\dagger}(t)\hat{\mathrm{U}}_0(t)\Psi_0 = \Psi^{\pm} \qquad \exists\,\Psi^{\pm} \in \mathcal{H} \qquad \forall\,\Psi_0 \in \mathcal{H}. \tag{5.9}$$

The strong limit (5.6) is assumed to be applicable to an arbitrary vector $\Psi \in \mathcal{H}$. Therefore, it will also be valid for the unperturbed state $\Psi_0 \in \mathcal{H}$, so that

$$\hat{\mathrm{U}}^{\dagger}(t)\hat{\mathrm{U}}_0(t)\Psi_0 \underset{t\to\mp\infty}{\Longrightarrow} \hat{\Omega}^{\pm}\Psi_0 = \Psi^{\pm} \tag{5.10}$$

and this is precisely the condition (5.9). If the examined system is invariant with respect to the anti-unitary transformation of time inversion[1], then it is not necessary to carry out both limits $t \to \mp\infty$. Rather, it is sufficient to prove the existence of the limit for e.g. $t \to -\infty$ from which the existence of the other limit would automatically follow for $t \to +\infty$. Otherwise, in this part of the general proof of the existence of the wave operators $\hat{\Omega}^{\pm}$, we must

[1]This is the case for collisions of two spinless particles (one-channel scattering) as analyzed in Ref. [7].

demonstrate that the domains i.e. the definition regions of the operators $\hat{\Omega}^{\pm}$ are the same and equal to the whole Hilbert space $\mathcal{H}$

$$\mathcal{D}_{\hat{\Omega}^+} = \mathcal{D}_{\hat{\Omega}^-} \equiv \mathcal{H}. \tag{5.11}$$

This could be rationalized by noticing that the rhs of (5.7) has a meaning if the state vector Ψ_0 belongs to the sets $\mathcal{D}_{\hat{\Omega}^+}$ and $\mathcal{D}_{\hat{\Omega}^-}$. Simultaneously, Ψ_0 is an arbitrary element from $\mathcal{H}$ and, therefore, it is clear that the sets $\mathcal{D}_{\hat{\Omega}^+}$ and $\mathcal{D}_{\hat{\Omega}^-}$ must coincide with the whole Hilbert space $\mathcal{H}$, as stated in the condition (5.11).

(ii) The second condition for the definition of the quantum collision system is the requirement that the ranges i.e. the images of the operators $\hat{\Omega}^{\pm}$ coincide with the subspace $\mathcal{H}_{\rm ac}$ of state vectors of the absolutely continuous part of the spectrum of the Hamiltonian $\hat{\rm H}$

$$\mathcal{R}_{\hat{\Omega}^+} = \mathcal{R}_{\hat{\Omega}^-} = \mathcal{H}_{\rm ac} \subset \mathcal{H}. \tag{5.12}$$

Since $\hat{\rm H}$ is a self-adjoint operator in the separable Hilbert space $\mathcal{H}$, it is generally possible to carry out the unique decomposition of $\mathcal{H}$ into three orthogonal non-degenerate spectra[2] of the Hamilton operator $\hat{\rm H}$

$$\mathcal{H} = \mathcal{H}_{\rm d} \oplus \mathcal{H}_{\rm ac} \oplus \mathcal{H}_{\rm sc}. \tag{5.13}$$

Here, the indices $\{{\rm d}, {\rm ac}, {\rm sc}\}$ correspond to the discrete, absolutely continuous and singularly continuous part of the spectrum of $\hat{\rm H}$, respectively. Specifically, $\mathcal{H}_{\rm ac}$ is contained in the subspace $\mathcal{H}_{\rm c} \equiv \mathcal{H}_{\rm d}^{\perp}$ of continuous states, whereas $\mathcal{H}_{\rm ac} \subset \mathcal{H}_{\rm c}$. Singularly continuous subspace $\mathcal{H}_{\rm sc} = \mathcal{H}_{\rm ac}^{\perp}$ is the orthogonal complement of $\mathcal{H}_{\rm ac}$. One can contemplate that no observable exists with a singularly continuous spectrum. The reason for this would be important to identify especially in view of Dirac's [36] conviction that every self-adjoint operator should be associated with a physically measurable quantity i.e. an observable.

(iii) The third condition for the definition of a quantum collision system is obtained by imposing the requirement that there are no states in the continuum part of the spectrum of the total Hamiltonian $\hat{\rm H}$ that are not scattering states associated with the operator $\hat{\rm H}_0$. If such states should exist, then the description of the collision problem by means of the operators $\hat{\rm H}$ and $\hat{\rm H}_0$ would

---

[2]Recall that a non-degenerate spectrum has all distinct eigen-values each of which is associated with only one eigen-state. In such a case, the result of the corresponding measurement is also said to be non-degenerate. By contrast, in a degenerate spectrum, one or more (or even all) eigen-values might coincide with each other. In such a case, one eigen-value can correspond to more than one eigen-state. One of the well-known examples of a degenerate spectrum is the set of bound states of atomic hydrogen with the energy $E_n = -1/(2n^2)$, which is the eigen-value of the Hamiltonian $-\nabla^2/2 - 1/r$, where n is the principal quantum number. This energy E_n is the same for all the eigen-states $\psi_{nlm}(\boldsymbol{r})$ characterized by n irrespective of the value of the orbital (l) and magnetic (m) quantum numbers.

be complete and this is precisely the situation encountered in many-channel scatterings (processes, reactions, etc). Clearly, this third condition demands that the continuous spectra of the operators $\hat{\mathrm{H}}$ and $\hat{\mathrm{H}}_0$ coincide with each other. To formulate this precisely in mathematical terms, we shall explicitly introduce the already employed subspace of bound states $\mathcal{H}_{\mathrm{d}} \subset \mathcal{H}$ which is comprized of closed linear manifolds of normalizable vectors of the operator $\hat{\mathrm{H}}$ and these are all the solutions ψ_n of the Schrödinger eigen-problem $\hat{\mathrm{H}}\psi_n = E_n\psi_n$ with real eigen-values E_n

$$\mathcal{H}_{\mathrm{d}} \equiv \mathcal{B} = \left\{\psi : \hat{\mathrm{H}}\psi_n = E_n\psi_n \ , \ E_n \in \mathbb{R}\right\}. \tag{5.14}$$

The orthogonal complement $\mathcal{H}_{\mathrm{d}}^{\perp} \equiv \mathcal{H}_{\mathrm{c}}$ of the subspace $\mathcal{H}_{\mathrm{d}}$ is $\mathcal{H}_{\mathrm{c}}$ which contains the elements from the continuum spectrum of the operator $\hat{\mathrm{H}}$

$$\mathcal{H}_{\mathrm{c}} \equiv \mathcal{R} = \{\phi : \langle\phi|\psi\rangle = 0 \ , \ \forall\, \psi \in \mathcal{H}_{\mathrm{d}}\}. \tag{5.15}$$

It then follows that the third condition imposes the following restriction

$$\mathcal{H}_{\mathrm{c}} = \mathcal{H}_{\mathrm{ac}}. \tag{5.16}$$

In the case when $\hat{\mathrm{H}}_0$ is the operator of the kinetic energy, it is easily shown that the spectrum of $\hat{\mathrm{H}}_0$ is absolutely continuous.

We shall show that the listed three conditions (i)–(iii) constitute the essential basic elements of the standard scattering theory. Thus, for example, we shall see that for a special choice of plane waves $\Psi_0 \propto \exp(i\boldsymbol{k}_i \cdot \boldsymbol{r})$, the wave functions of the type $\Psi^{\pm} = \operatorname*{Lim}_{t\to\mp\infty} \hat{\mathrm{U}}^{\dagger}(t)\hat{\mathrm{U}}_0(t)\Psi_0$ represent those customary states that in the standard scattering theory behave asymptotically at $r \to \infty$ as prescribed by (1.10). Further, if the limit in (5.9) exists, then this would imply also the existence of the following limit

$$\operatorname*{Lim}_{t\to\mp\infty} \hat{\mathrm{U}}_0^{\dagger}(t)\hat{\mathrm{U}}(t)\Psi^{\pm} = \Psi_0 \qquad \forall\, \Psi_0 \in \mathcal{H}. \tag{5.17}$$

This is indeed true, as can be shown by recalling that the strong limit (5.9) is based upon convergence in terms of the norm. This means that for an arbitrary $\epsilon > 0$ there exists a certain time T and

$$||\hat{\mathrm{U}}^{\dagger}(t)\hat{\mathrm{U}}_0(t)\Psi_0 - \Psi^{+}|| < \epsilon \tag{5.18}$$

for every $t > T$. However, since the evolution operators are unitary ($\hat{\mathrm{A}}^{\dagger}\hat{\mathrm{A}} = \hat{\mathrm{A}}\hat{\mathrm{A}}^{\dagger} = \hat{1}$, $\hat{\mathrm{A}} = \hat{\mathrm{U}}_0, \hat{\mathrm{U}}$), it immediately follows from (5.18) that

$$\begin{aligned}
||\hat{\mathrm{U}}^{\dagger}(t)\hat{\mathrm{U}}_0(t)\Psi_0 - \Psi^{+}|| &= ||\hat{\mathrm{U}}^{\dagger}(t)\{\hat{\mathrm{U}}_0(t)\Psi_0 - \hat{\mathrm{U}}(t)\Psi^{+}\}|| \\
&= ||\hat{\mathrm{U}}_0(t)\Psi_0 - \hat{\mathrm{U}}(t)\Psi^{+}|| \\
&= ||\hat{\mathrm{U}}_0(t)\{\Psi_0 - \hat{\mathrm{U}}_0^{\dagger}(t)\hat{\mathrm{U}}(t)\Psi^{+}\}|| \\
&= ||\Psi_0 - \hat{\mathrm{U}}_0^{\dagger}(t)\hat{\mathrm{U}}(t)\Psi^{+}|| < \epsilon
\end{aligned}$$

$$\therefore \qquad ||\hat{\mathrm{U}}^\dagger(t)\hat{\mathrm{U}}_0(t)\Psi_0 - \Psi^+|| < \epsilon \qquad \text{(QED)} \tag{5.19}$$

where we used the relations of isometry $||\hat{\mathrm{U}}_0|| = 1$ and $||\hat{\mathrm{U}}^\dagger|| = 1$. In (5.17) and (5.19), the vector $\Psi_I^+(t) \equiv \hat{\mathrm{U}}_0^\dagger(t)\hat{\mathrm{U}}(t)\Psi^+$ is recognized as a non-stationary state of the system in the Dirac or the interaction picture of quantum mechanics [7]. Further, the vector Ψ^+ is the corresponding stationary state in the Heisenberg picture of quantum mechanics [7]. In such a case, the existence of the limiting value (5.17) signifies that the time-dependent state vector $\Psi_I^+(t)$ in the Dirac picture tends to a stationary state Ψ_0 when $t \to -\infty$, namely $\mathrm{Lim}_{t\to-\infty} \Psi_I^+(t) \equiv \mathrm{Lim}_{t\to-\infty} \hat{\mathrm{U}}^\dagger(t)\hat{\mathrm{U}}_0(t)\Psi^+ = \Psi_0$. However, the kinematics and the dynamics that are governed by $\hat{\mathrm{H}}_0$ and $\hat{\mathrm{V}}$, respectively, are totally separated in the Dirac picture [7]. More precisely, in the Dirac picture, the time dependence of an operator $\hat{\mathrm{A}}$ satisfies the Tomanaga-Schwinger equation of motion of the operator $i(\mathrm{d}/\mathrm{d}t)\hat{\mathrm{A}}_I = [\hat{\mathrm{A}}_I, \hat{\mathrm{H}}_0] + i(\partial/\partial t)\hat{\mathrm{A}}_I$. Simultaneously, in the same picture, the state vectors vary with the passage of time, obeying the equation $i(\partial/\partial t)\Psi_I^+(t) = \hat{\mathrm{V}}_I(t)\Psi_I^+(t)$ with $\hat{\mathrm{V}}_I(t) = \hat{\mathrm{U}}_0^\dagger(t)\hat{\mathrm{V}}\hat{\mathrm{U}}_0(t)$, where $\hat{\mathrm{V}}$ is the interaction potential operator $\hat{\mathrm{V}}\,(= \hat{\mathrm{H}} - \hat{\mathrm{H}}_0)$. It is then clear that the state vector $\Psi_I^+(t)$ will be stationary, provided that $i(\partial/\partial t)\Psi_I^+(t) = 0$ which, in turn, requires that $\hat{\mathrm{V}}_I = \hat{0}$ i.e. $\hat{\mathrm{V}} = \hat{0}$. Therefore, the fact that the state vector $\Psi_I^+(t)$ tends to a stationary vector in the remote past ($t \to -\infty$) imposes that the interaction $\hat{\mathrm{V}}$ must be negligibly small in the same limit. A similar reasoning is not necessary to repeat for $t \to +\infty$, since this becomes automatically valid if the condition (5.12) is fulfilled. This remark illustrates how the two typical limits for the stationary ($r \to \infty$) and non-stationary ($|t| \to \infty$) scattering theory harmoniously cohere by complementing each other.

6

The adiabatic theorem and the Abel strong limit

6.1 Adiabatic theorem for scattering states

The analysis from the final paragraph of chapter 5 is based upon the view that the effective interaction potential $\hat{V}$ disappears at large negative and positive values of time ($t \rightarrow \pm\infty$). This is because in the Dirac picture of quantum mechanics, even when $\hat{V}$ does not depend explicitly upon time, the probability at $|t| \rightarrow \infty$ becomes predominantly high for finding the particle far away from the scattering centre where $\hat{V} \approx \hat{0}$. Therefore, the physical results should remain unaltered if we assume that the interaction $\hat{V}$ is 'adiabatically turned on and off'. This can be conveniently accomplished by the exponential screening of the interaction via the replacement of $\hat{V}$ by $\hat{V}e^{-\varepsilon|t|}$. Here, $\varepsilon > 0$ is an infinitesimally small positive number, such that the limit $\varepsilon \rightarrow 0^+$ should be taken at the end of the calculations. In fact, by using $|t|$ we can carry out the analysis for both Møller wave operators $\hat{\Omega}^+$ and $\hat{\Omega}^-$. This concept is known as the adiabatic theorem. The meaning of this exponential screening of the potential $\hat{V}$ via attenuation or damping is seen by observing that for every proper normalizable physical state, there exists a sufficiently small number ε such that the function $\exp(-\varepsilon|t|)$ is significantly different from unity only for the values of time t for which the interaction $\hat{V}$ vanishes. For small values of $|t|$ and infinitesimally small $\varepsilon > 0$, it is obvious that $\hat{V}$ and $\hat{V}\exp(-\varepsilon|t|)$ coincide with each other. In contrast to this, for large values of $|t|$ the difference between the operators $\hat{V}$ and $\hat{V}\exp(-\varepsilon|t|)$ becomes appreciable, since in this case the damping factor $\exp(-\varepsilon|t|)$ tends to zero, thus also causing $\hat{V}\exp(-\varepsilon|t|)$ to vanish. Thus, $\hat{V}$ and $\hat{V}\exp(-\varepsilon|t|)$ will again coincide with each other only if the potential $\hat{V}$ itself tends to zero for large values of $|t|$. This is precisely the case for short-range potentials that are, by definition, negligible at long distances at which the probability is high to find the colliding particle for large values of time $|t|$.

In this context, a very important question emerges: is the adiabatic theorem applicable to scattering states in view of the fact that they are improper i.e. non-normalizable? The answer is in the affirmative. Namely, as long as we keep $\varepsilon \neq 0$, the adiabatic theorem is also valid for improper states, such as plane waves, provided that they are averaged via an integration over a weight function such that (5.9) holds true. This additional procedure of averaging within improper delocalized state vectors by forming localized wave packets is a consequence of the fact that the improper states are normalized to the Dirac $\delta-$function. As is well-known, the $\delta-$function is not a function in the strict meaning of the word, but rather it is a distribution which can be interpreted physically only underneath an integral [24]. After these remarks, we are in a position to formulate the adiabatic theorem as follows[1]:

Theorem 4: If the standard time-dependent Schrödinger equation $i(\partial/\partial_t)\Psi(t,\boldsymbol{r}) = (\hat{\mathrm{H}}_0 + \hat{\mathrm{V}})\Psi(t,\boldsymbol{r})$ is solved for the auxiliary interaction $\hat{\mathrm{V}}(\boldsymbol{r}) \times \exp(-\varepsilon|t|)$ instead of the original potential operator $\hat{\mathrm{V}} \equiv \hat{\mathrm{V}}(\boldsymbol{r})$ with the boundary condition at large distances $\Psi(t,\boldsymbol{r}) \underset{t\to-\infty}{\Longrightarrow} \exp(i\boldsymbol{k}_i \cdot \boldsymbol{r} - iEt)$, then for any finite time t the limit $\varepsilon \to 0^+$ yields a stationary solution of the Schrödinger equation $\hat{\mathrm{H}}\Psi(\boldsymbol{r}) = E\Psi(\boldsymbol{r})$. A similar reasoning is also valid if we consider the other time asymptote, $t \to +\infty$.

The physical meaning of this theorem is that when the interaction $\hat{\mathrm{V}}\,(= \hat{\mathrm{H}} - \hat{\mathrm{H}}_0)$ is turned on adiabatically, then any eigen-vector of the operator $\hat{\mathrm{H}}_0$ is transformed into an eigen-vector of $\hat{\mathrm{H}}$ with the same eigen-energy for both limits $t \to -\infty$ and $t \to +\infty$. From here, the following two important consequences emerge:

(a) Only the perturbation potential $\hat{\mathrm{V}}$ is responsible for the transition of the system between its two unperturbed states that are the initial state i from the remote past ($t \to -\infty$) and the final state f from the distant feature ($t \to +\infty$).

(b) The transition $i \longrightarrow f$ is physically measurable, since it preserves the energy E of the whole system via the conservation law $E_i = E_f \equiv E$. We say that the transition takes place on the energy shell [7]. This is reflected in the fact that the scattering $\hat{\mathrm{S}}-$operator commutes with the Hamiltonian $\hat{\mathrm{H}}_0$ as symbolized by $[\hat{\mathrm{S}}, \hat{\mathrm{H}}_0] = \hat{0}$.

6.2 Adiabatic theorem and existence of wave operators

Let us now show how the adiabatic theorem is connected with the existence of the Møller wave operators $\hat{\Omega}^{\pm}$. According to (5.4), the operators $\hat{\Omega}^{\pm}$ are defined as the Cauchy strong limits of the product of the evolution operators $\hat{\mathrm{U}}^{\dagger}(t)\hat{\mathrm{U}}_0(t) = \exp(i\hat{\mathrm{H}}t)\exp(-i\hat{\mathrm{H}}_0 t)$ at $t \to \mp\infty$, so that

$$\Psi^{\pm} = \hat{\Omega}^{\pm}\Psi_0 = \underset{t\to\mp\infty}{\mathrm{Lim}} \{\hat{\mathrm{U}}^{\dagger}(t)\hat{\mathrm{U}}_0(t)\Psi_0\}$$

[1] The adiabatic theorem played an important role at a certain stage in the development of field theory.

$$= \Psi_0 + i \int_0^{\mp\infty} \mathrm{d}\tau \hat{\mathrm{U}}^\dagger(\tau)\hat{\mathrm{V}}\hat{\mathrm{U}}_0(\tau)\Psi_0$$

$$\therefore \qquad \Psi^\pm = \hat{\Omega}^\pm \Psi_0 = \Psi_0 + i \int_0^{\mp\infty} \mathrm{d}\tau \hat{\mathrm{U}}^\dagger(\tau)\hat{\mathrm{V}}\hat{\mathrm{U}}_0(\tau)\Psi_0. \tag{6.1}$$

6.3 The Abel strong limit in stationary scattering theory

As in Ref. [7], the integrals from (6.1) are not only convergent, but they are also absolutely convergent. This means that if the integrand $\hat{\mathrm{U}}^\dagger(\tau)\hat{\mathrm{V}}\hat{\mathrm{U}}_0(\tau)\Psi_0$ is replaced by the norm $||\hat{\mathrm{U}}^\dagger(\tau)\hat{\mathrm{V}}\hat{\mathrm{U}}_0(\tau)\Psi_0||$, then the new integral will still be convergent. Using this latter fact, let us now formulate a lemma which will relate the strong limit (5.17) with the adiabatic theorem.

Lemma 3: If the integral $\int_0^\infty \mathrm{d}\tau\psi(\tau)$ is absolutely convergent i.e.

$$\int_0^\infty \mathrm{d}\tau||\psi(\tau)|| < \infty \tag{6.2}$$

then, it follows

$$\int_0^\infty \mathrm{d}\tau\psi(\tau) = \operatorname*{Lim}_{\varepsilon\to 0+} \int_0^\infty \mathrm{d}\tau\, \mathrm{e}^{-\varepsilon\tau}\psi(\tau) = \operatorname*{Lim}_{\varepsilon\downarrow 0} \int_0^\infty \mathrm{d}\tau\, \mathrm{e}^{-\varepsilon\tau}\psi(\tau)$$

$$\therefore \qquad \int_0^\infty \mathrm{d}\tau\psi(\tau) = \operatorname*{Lim}_{\varepsilon\downarrow 0} \int_0^\infty \mathrm{d}\tau\, \mathrm{e}^{-\varepsilon\tau}\psi(\tau). \tag{6.3}$$

Here the symbol $\varepsilon \downarrow 0$, just as $\varepsilon \to 0^+$, serves to emphasize the way in which ε tends to zero through positive numbers with the understanding that $\varepsilon > 0$. The limiting procedure with such symbols is known as the Abel limit. Moreover, if these limits are associated with the symbol Lim as in (6.3), then again we are referring to the strong limits i.e. convergence in the norm when $\varepsilon \to 0^+$ or $\varepsilon \downarrow 0$. The relation (6.3) can be proven by writing the integral from 0 to ∞ as the sum of two integrals for the sub-intervals $t \in [0, T]$ and $t \in [T, \infty]$, where $T > 0$ is an arbitrary number. In such a case, we obtain

$$\lim_{\varepsilon\to 0+} \left|\left| \int_0^\infty \mathrm{d}\tau\psi(\tau) - \int_0^\infty \mathrm{d}\tau\, \mathrm{e}^{-\varepsilon\tau}\psi(\tau) \right|\right|$$

$$= \lim_{\varepsilon\to 0+} \left|\left| \int_0^T \mathrm{d}\tau \left(1 - \mathrm{e}^{-\varepsilon\tau}\right)\psi(\tau) + \int_T^\infty \mathrm{d}\tau \left(1 - \mathrm{e}^{-\varepsilon\tau}\right)\psi(\tau) \right|\right| = 0. \tag{6.4}$$

Here, each of the integrals from 0 to T and from T to ∞ is separately equal to zero when $\varepsilon \to 0^+$ (QED). Namely, if ε is infinitesimally small and T is chosen to be large, then the second integral from T to ∞ on the rhs of (6.4) represents the contribution from large values $\tau \gtrsim 1/\varepsilon$ to the total integral $\int_0^\infty \mathrm{d}\tau[1 - \exp(-\varepsilon\tau)]\psi(\tau)$. In such a case of large values of τ, the term $\exp(-\varepsilon\tau)$ tends to zero, so that the integral $\int_T^\infty \mathrm{d}\tau[1 - \exp(-\varepsilon\tau)]\psi(\tau)$ is reduced to $\int_T^\infty \mathrm{d}\tau\psi(\tau)$ which vanishes for large T. This is also clear from the assumption of lemma 3, since if the integral $\int_0^\infty \mathrm{d}\tau\psi(\tau)$ converges, then its contribution to the final result from the region of large values τ is certainly negligible. Simultaneously and, of course, for the same choice of an infinitesimally small $\varepsilon > 0$ and a very large T, the first integral from the rhs of (6.4) i.e. $\int_0^T \mathrm{d}\tau[1 - \exp(-\varepsilon\tau)]\psi(\tau)$, represents the contribution from small values of $\tau \ll 1/\varepsilon$ to the whole integral $\int_0^\infty \mathrm{d}\tau[1 - \exp(-\varepsilon\tau)]\psi(\tau)$. Since $\varepsilon \to 0^+$ and $\tau \ll 1/\varepsilon$ imply that $\exp(-\varepsilon\tau)$ is close to 1, we shall have $1 - \exp(-\varepsilon\tau) \underset{\varepsilon\to 0^+}{\longrightarrow} 0$, so that $\int_0^T \mathrm{d}\tau[1 - \exp(-\varepsilon\tau)]\psi(\tau) \underset{\varepsilon\to 0^+}{\Longrightarrow} \emptyset$. The manner in which we carry out the proof also highlights the meaning of the identity (6.3) via the analysis of the contribution of the integrand $\exp(-\varepsilon\,\tau)\psi(\tau)$ to the integral $\int_0^\infty \mathrm{d}\tau \exp(-\varepsilon\,\tau)\psi(\tau)$ from the region of small and large values of τ. This can be summarized in the following way. The exponential factor $\exp(-\varepsilon\,\tau)$ is approximately equal to unity for small τ i.e. for $\tau \ll 1/\varepsilon$ when $\varepsilon \to 0^+$. This means that the region of small values of τ is in accord with the lhs of (6.3). Further, the contribution from the same exponential factor to the integral on the rhs of (6.3) is certainly negligible for large τ i.e. for $\tau \gtrsim 1/\varepsilon$ when $\varepsilon \to 0^+$. This is because the factor $\exp(-\varepsilon\,\tau)$ now tends to zero, which attenuates the whole integrand $\exp(-\varepsilon\,\tau)\psi(\tau)$, so that the convergence of the integral from the rhs of (6.2) is enhanced. Lemma 3 asserts that the integral $\underset{\varepsilon\to 0^+}{\mathrm{Lim}} \int_0^\infty \mathrm{d}\tau \exp(-\varepsilon\,\tau)\psi(\tau)$ is convergent, which means that the contribution to the integrand from the region of large values of τ is negligible. Therefore, if we choose a sufficiently small $\varepsilon > 0$, as dictated explicitly by the limiting procedure $\varepsilon \to 0^+$, then the damping factor $\exp(-\varepsilon\,\tau)$ will not alter the value of the integral $\int_0^\infty \mathrm{d}\tau \exp(-\varepsilon\,\tau)\psi(\tau)$ for large τ. Now we can apply lemma 3 i.e. the identity (6.3) to (6.1), so that

$$\Psi^\pm = \hat{\Omega}^\pm\Psi_0 = \Psi_0 \mp i \underset{\varepsilon\to 0^+}{\mathrm{Lim}} \int_0^{\mp\infty} \mathrm{d}\tau\, \mathrm{e}^{\pm\varepsilon\,\tau}\hat{\mathrm{U}}^\dagger(\tau)\hat{\mathrm{V}}\hat{\mathrm{U}}_0(\tau)\Psi_0. \tag{6.5}$$

A few remarks are in order regarding the relationship which exists between (6.1) and (6.5). First, the expression (6.5) is meaningful i.e. the indicated limit $\varepsilon \to 0^+$ exists if and only if the integral from (6.1) is convergent when $t \to \mp\infty$. In such a case, (6.5) holds true rigorously and the Abel limit exists for every normalizable vector $\Psi_0 \in \mathcal{H}$. Second, convergence of the starting integral (6.1) is based upon relatively subtle oscillations of its integrand. Such oscillations are due to the fact that the wave packet Ψ_0 moves and spreads. As noted in chapter 5 e.g. a Gaussian wave packet spreads out following the $t^{-3/2}$–law when $t \to \infty$ [7]. Compared to such a behavior, the integral from (6.5) is exponentially convergent, since now the product of the damping factor and the Gaussian wave packet falls much faster, i.e. as $\exp(-\varepsilon\,\tau)\tau^{-3/2}$ where $\varepsilon > 0$ and $\tau > 0$. This advantage leads to convergence acceleration by the passage from the integral $\int_0^\infty \mathrm{d}\tau\,\psi(\tau)$ to its counterpart $\mp \lim_{\varepsilon\to 0^+} \int_0^{\mp\infty} \mathrm{d}\tau\, \exp(\mp\varepsilon\,\tau)\psi(\tau)$. The price for this achievement is additional effort, relative to (6.1), since we must take the limit $\varepsilon \to 0^+$ once the integration over τ has been carried out. However, this supplementary limit is not a significant problem, especially since that there is another benefit from the whole procedure. Namely, as shown in Ref. [7] through one of the essential steps in the formalism of stationary scattering states, precisely this additional limit $\varepsilon \to 0^+$ permits an adequate interpretation of improper plane waves $|\boldsymbol{p}\,\rangle$, when they are juxtaposed to the corresponding normalizable proper physical state (wave packet) $|\Psi_0\rangle$.

6.4 Exponential screening of potentials and adiabatic theorem

Given the outlined features, the connection between the concept of the exponential damping with the adiabatic theorem becomes evident. Namely, roughly speaking, the equations (6.1) and (6.5) differ from each other in that the potential $\hat{\mathrm{V}}$ from (6.1) is replaced by $\hat{\mathrm{V}}_\varepsilon(\tau) \equiv \hat{\mathrm{V}}\exp(-\varepsilon\,|\tau|\,)$ $(\varepsilon > 0)$ in (6.5), which necessitates the additional limit $\varepsilon \to 0^+$ at the end of the analysis. However, changing $\hat{\mathrm{V}}$ to $\hat{\mathrm{V}}_\varepsilon(\tau)$ automatically alters the evolution operator $\hat{\mathrm{U}}(t)$ due to the definition $\exp(-i\hat{\mathrm{H}}t) = \exp(-i[\hat{\mathrm{H}}_0 + \hat{\mathrm{V}}]t)$. Nevertheless, in the case of a short-range potential, we saw that $\hat{\mathrm{V}}$ and $\hat{\mathrm{V}}\exp(-\varepsilon|\tau|)$ did not differ appreciably for either small or large values τ. This obviously facilitates the interpretation of (6.5) in terms of the adiabatic theorem. Thus, equivalence of (6.1) and (6.5) can also be considered as a way of expressing the fact that the evolution of any scattering state will be unaltered if the interaction potential $\hat{\mathrm{V}}$ is substituted by its screened counterpart $\hat{\mathrm{V}}\exp(-\varepsilon\,|\tau|\,)$ for a sufficiently small number $\varepsilon > 0$. Precisely this conclusion is expected, since the projectile in an arbitrary scattering state will eventually get so far away from the collision center that the interaction $\hat{\mathrm{V}}$ shall cease to have any appreciable influence on the incident particle. Therefore, for any scattering state we can choose $\varepsilon > 0$ in such a way that the potentials $\hat{\mathrm{V}}$ and $\hat{\mathrm{V}}\exp(-\varepsilon\,|\tau|\,)$ will not significantly differ from each other. We saw that the exception to this takes place at large time $\tau \gtrsim 1/\varepsilon\,(\varepsilon \to 0^+\,)$, in which case the exponential screening $\exp(-\varepsilon\,|\tau|)$ disappears too fast, so that the whole term $\hat{\mathrm{V}}\exp(-\varepsilon\,|\tau|\,)$ becomes nearly zero. However, at large times t the projectile itself is far away from the scattering center, so that effectively the original potential $\hat{\mathrm{V}}$ is also close to zero for short-range interactions. Hence, again even at large times t, the difference between $\hat{\mathrm{V}}$ and $\hat{\mathrm{V}}\exp(-\varepsilon\,|\tau|\,)$ becomes irrelevant. Thus, the concept of the adiabatic theorem according to which in scattering theory the potential $\hat{\mathrm{V}}$ is replaced by $\hat{\mathrm{V}}\exp(-\varepsilon\,|\tau|\,)$ $(\varepsilon > 0)$ has a solid physical ground and a clear interpretation for short-range potentials.

Note that it has been shown by Dollard [38]–[40] that neither the adiabatic switching nor exponential screening (or any type of screening for that matter) is applicable to Coulomb potentials. This very important finding adds two more features to many other characteristics [7] that are otherwise standard in scattering theory on short-range interactions, but are invalid for long-range Coulomb potentials. Of course, the most well-known departure of Coulomb scattering from the usual concept valid for short-range interactions is the so-called asymptotic convergence problem of Dollard [41], or equivalently, the correct boundary conditions for scattering wave functions and perturbations [7, 42], as reviewed intensively over the years [44]–[50].

6.5 Adiabatic theorem and the Green operators

We shall now demonstrate how the adiabatic theorem helps to carry out the integration over τ. With this goal, we shall utilize the unperturbed eigen-value problem $\hat{\mathrm{H}}_0\Psi_0 = E\Psi_0$ which gives

$$\hat{\mathrm{U}}_0(t)\Psi_0 = \mathrm{e}^{-iEt}\Psi_0 \tag{6.6}$$

so that, it follows for $\varepsilon > 0$

$$\Psi^+ = \Psi_0 + i \lim_{\varepsilon\to 0+} \int_{-\infty}^{0} \mathrm{d}\tau\, \mathrm{e}^{-i(E+i\varepsilon)\tau}\hat{\mathrm{U}}^\dagger(t)\hat{\mathrm{V}}\Psi_0. \tag{6.7}$$

The remaining operator integral can be calculated by transforming it to the corresponding integral of a scalar function. To achieve this we shall employ the spectral decomposition of the unity operator in terms of the complete set $\{\Psi_\nu\}$ of ortho-normal stationary proper vectors of the Hamiltonian $\hat{\mathrm{H}}$ which is present in $\hat{\mathrm{U}}^\dagger(\tau) = \exp(i\hat{\mathrm{H}}\tau)$. Here the collective label ν encompasses both the discrete and the continuous spectrum

$$\hat{1} = \sum\!\!\!\!\!\!\int_\nu \mathrm{d}\nu\, |\Psi_\nu\rangle\langle\Psi_\nu| \tag{6.8}$$

where the composite symbol $\sum\!\!\!\!\!\!\int_\nu$ denotes the sum over the discrete and integration over the continuous part of the spectrum of the operator $\hat{\mathrm{H}}$. The discrete spectrum stems from the possibility that the potential $\hat{\mathrm{V}}$ sustains the bound states i.e. that in a given part of the space we have $\mathrm{V} < 0$. The continuous spectrum of the operator $\hat{\mathrm{H}}$ originates from $\hat{\mathrm{H}}_0$ according to the earlier introduced condition (iii) for the definition of the quantum collision system from chapter 5. The eigen-problem $\hat{\mathrm{H}}\Psi_\nu = E_\nu \Psi_\nu$ implies

$$\hat{\mathrm{U}}^\dagger(t)\Psi_\nu = \mathrm{e}^{iE_\nu t}\Psi_\nu \tag{6.9}$$

so that

$$\begin{aligned}
i\int_{-\infty}^{0} \mathrm{d}\tau\, \mathrm{e}^{i(\hat{\mathrm{H}}-[E+i\varepsilon]\hat{1})\tau}\hat{\mathrm{V}}|\Psi_0\rangle &= i\int_{-\infty}^{0}\mathrm{d}\tau \sum\!\!\!\!\!\!\!\int_\nu \mathrm{d}\nu\, \mathrm{e}^{i(\hat{\mathrm{H}}-[E+i\varepsilon]\hat{1})\tau}|\Psi_\nu\rangle\langle\Psi_\nu|\hat{\mathrm{V}}|\Psi_0\rangle \\
&= i\sum\!\!\!\!\!\!\!\int_\nu \mathrm{d}\nu \int_{-\infty}^{0}\mathrm{d}\tau\, \mathrm{e}^{i(E_\nu - E - i\varepsilon)\tau}|\Psi_\nu\rangle\langle\Psi_\nu|\hat{\mathrm{V}}|\Psi_0\rangle \\
&= \sum\!\!\!\!\!\!\!\int_\nu \mathrm{d}\nu\,(E - E_\nu + i\varepsilon)^{-1}|\Psi_\nu\rangle\langle\Psi_\nu|\hat{\mathrm{V}}|\Psi_0\rangle \\
&= [(E+i\varepsilon)\hat{1} - \hat{\mathrm{H}}]^{-1}\hat{\mathrm{V}}|\Psi_0\rangle
\end{aligned}$$

$$\therefore \qquad i\int_{-\infty}^{0}\mathrm{d}\tau\, \mathrm{e}^{i(\hat{\mathrm{H}}-[E+i\varepsilon]\hat{1})\tau}\hat{\mathrm{V}}|\Psi_0\rangle = [(E+i\varepsilon)\hat{1} - \hat{\mathrm{H}}]^{-1}\hat{\mathrm{V}}|\Psi_0\rangle. \tag{6.10}$$

In this derivation, we first replace the order integration over τ and summation over ν. This is justified due to the presence of the term $\varepsilon\tau$ ($\varepsilon > 0$) in the exponent. Such an exponential damping makes the whole integrand tend to zero when $\tau \to -\infty$. The final step of the identification of the operator $[(E + i\varepsilon)\hat{1} - \hat{\mathrm{H}}]^{-1}$ from the expansion over ν comes from the following prescription of the spectral representation of the Green resolvent operator

$$\begin{aligned}
\hat{\mathrm{G}}(E+i\varepsilon) &= [(E+i\varepsilon)\hat{1} - \hat{\mathrm{H}}]^{-1} \\
&= \sum\!\!\!\!\!\!\!\int_\nu \mathrm{d}\nu\;\; (E - E_\nu + i\varepsilon)^{-1}|\Psi_\nu\rangle\langle\Psi_\nu|.
\end{aligned} \tag{6.11}$$

This expression is obtained when multiplying $\hat{\mathrm{G}}(E+i\varepsilon)$ by the unity operator and then subsequently exploiting (6.8) and (3.24).

6.6 Adiabatic theorem and the Lippmann-Schwinger equations

It is also seen in (6.10) that, because of the exponential damping, there is no contribution to the integral over τ from its lower bound at $\tau = -\infty$. Inserting the result (6.9) into (6.7), we arrive at the Lippmann-Schwinger equations

$$\begin{aligned}
\Psi^+ &= \Psi_0 + \operatorname*{Lim}_{\varepsilon\to 0+}\;\; \hat{\mathrm{G}}(E+i\varepsilon)\,\hat{\mathrm{V}}\,\Psi_0 \\
&\equiv \Psi_0 + \hat{\mathrm{G}}(E + i\,0)\,\hat{\mathrm{V}}\,\Psi_0 \\
&\equiv \Psi_0 + \hat{\mathrm{G}}^+(E)\,\hat{\mathrm{V}}\,\Psi_0.
\end{aligned} \tag{6.12}$$

Here, we employ the label $\hat{\mathrm{A}}(E + i0)$ to symbolize the limiting value of the Abel limit $\varepsilon \to 0^+ \equiv \varepsilon \downarrow 0$ of an arbitrary linear operator $\hat{\mathrm{A}}(z)$ when the variable $z = E + i\varepsilon$ tends to the real axis E from above for $\varepsilon > 0$

$$\hat{\mathrm{A}}(E + i0) = \operatorname*{Lim}_{\varepsilon \to 0^+} \hat{\mathrm{A}}(E + i\varepsilon) = \operatorname*{Lim}_{\varepsilon \downarrow 0} \hat{\mathrm{A}}(E + i\varepsilon) = \hat{\mathrm{A}}^+(E). \tag{6.13}$$

With this convention, the free and total Green operators $\hat{\mathrm{G}}_0^{\pm}(E)$ and $\hat{\mathrm{G}}^{\pm}(E)$ acquire the form

$$\hat{\mathrm{G}}_0^{\pm}(E) = \operatorname*{Lim}_{\varepsilon \to 0^+} \hat{\mathrm{G}}_0(E \pm i\varepsilon) = \hat{\mathrm{G}}_0(E \pm i0) = [(E \pm i0)\hat{1} - \hat{\mathrm{H}}_0]^{-1} \tag{6.14}$$

$$\hat{\mathrm{G}}^{\pm}(E) = \operatorname*{Lim}_{\varepsilon \to 0^+} \hat{\mathrm{G}}(E \pm i\varepsilon) = \hat{\mathrm{G}}(E \pm i0) = [(E \pm i0)\hat{1} - \hat{\mathrm{H}}]^{-1}. \tag{6.15}$$

The result similar to (6.12) can also be obtained for Ψ^- in a symmetric form

$$\begin{aligned} \Psi^- &= \Psi_0 + \operatorname*{Lim}_{\varepsilon \to 0^+} \hat{\mathrm{G}}(E - i\varepsilon)\,\hat{\mathrm{V}}\,\Psi_0 \\ &\equiv \Psi_0 + \hat{\mathrm{G}}(E - i\,0)\,\hat{\mathrm{V}}\,\Psi_0 \\ &\equiv \Psi_0 + \hat{\mathrm{G}}^-(E)\,\hat{\mathrm{V}}\,\Psi_0. \end{aligned} \tag{6.16}$$

Let us emphasize that in a rigorous work on the Lippmann-Schwinger equation, one must first carry out the analysis on $\hat{\mathrm{G}}(z)$ for a complex variable $z \in \mathbb{C}$ and subsequently perform the corresponding limit to the physical real axis, as we just did.

7

Non-stationary and stationary scattering via the strong limits

7.1 The Abel limit and Lippmann-Schwinger equations

The strong limits from (5.17) of the Møller operators $\hat{\Omega}^{\pm}$ are of central importance to scattering theory. One of the key illustrations in this direction can be provided by showing that the Lippmann-Schwinger equations for scattering states could be derived from these limits. To demonstrate this fact, we shall use the method of exponential screening of functions, as already explored in the adiabatic theorem for a special case of the operator potential interaction $\hat{\mathrm{V}}$. In general, the method of exponential attenuation of functions or operators can be formulated in the form of the following lemma:

Lemma 4: Let $\psi(t) \in \mathcal{H}$ be a state vector for which the Laplace transform $\int_0^\infty \mathrm{d}t\,\mathrm{e}^{-\varepsilon t}\psi(t)$ for $\varepsilon > 0$ exists and is given by

$$\operatorname*{Lim}_{t\to\mp\infty} \psi(t) = \psi^{\pm}. \tag{7.1}$$

Then the following equality is valid

$$\mp\operatorname*{Lim}_{\varepsilon\to 0+} \varepsilon \int_0^{\mp\infty} \mathrm{d}t\,\mathrm{e}^{\pm\varepsilon t}\psi(t) = \psi^{\pm}. \tag{7.2}$$

Recalling (6.3), we see that the strong limit in (7.2) is the Abel limit, so that we can use the appropriate notation

$$\mp\operatorname*{Lim}_{\varepsilon\downarrow 0} \varepsilon \int_0^{\mp\infty} \mathrm{d}t\,\mathrm{e}^{\pm\varepsilon t}\psi(t) = \psi^{\pm} \qquad (\varepsilon > 0). \tag{7.3}$$

The procedure of the Abel limiting process can be best understood by an example of an ordinary scalar function $f(t)$ i.e. $f(t) \in \mathcal{F}$, where $\mathcal{F}$ is a given scalar field. Then, we will assume that the limiting value of $f(t)$ can be found when e.g. $t \to -\infty$ and, as such, shall be denoted by f^+. The statement (7.2) applies equally well to ordinary scalar functions as well as to abstract vectors of linear spaces. Therefore, we can say that the same value f^+ should be attainable through the Abel limit

$$f^+ = \lim_{\varepsilon\to 0+} \varepsilon \int_{-\infty}^{0} \mathrm{d}t\,\mathrm{e}^{\varepsilon t} f(t). \tag{7.4}$$

Notice first that for the special case of the scalar function $f(t)$ defined as a constant for every value of the independent variable t e.g. $f(t) = f^+$, the elementary integration immediately yields the result

$$\varepsilon \int_{-\infty}^{0} \mathrm{d}t\, \mathrm{e}^{\varepsilon\, t} f^+ = f^+ \varepsilon \int_{-\infty}^{0} \mathrm{d}t\, \mathrm{e}^{\varepsilon\, t} = f^+. \tag{7.5}$$

However, if here the lower infinite bound ($t = -\infty$) is replaced by a finite value T ($-\infty < T < 0$), then it easily follows that the finite integration region $t \in [-T, 0]$ gives the zero contribution to the integral on the lhs of (7.5) when $\varepsilon \to 0^+$, so that

$$\begin{aligned} \lim_{\varepsilon \to 0^+} \varepsilon \int_{-|T|}^{0} \mathrm{d}t\, \mathrm{e}^{\varepsilon\, t} f^+ &= f^+ \lim_{\varepsilon \to 0^+} \varepsilon \int_{-|T|}^{0} \mathrm{d}t\, \mathrm{e}^{\varepsilon\, t} = \lim_{\varepsilon \to 0^+} \mathrm{e}^{\varepsilon\, t}\Big|_{t=-|T|}^{t=0} \\ &= f^+ \lim_{\varepsilon \to 0^+} \left(1 - \mathrm{e}^{-\varepsilon |T|}\right) = 0 \end{aligned}$$

$$\therefore \qquad \lim_{\varepsilon \to 0^+} \varepsilon \int_{-|T|}^{0} \mathrm{d}t\, \mathrm{e}^{\varepsilon\, t} f^+ = 0. \tag{7.6}$$

Next we consider a more general case where $f(t)$ is not a constant function and has the limiting value f^+ when $t \to -\infty$. Then in the limit $\varepsilon \to 0^+$ the whole contribution to the integral $\int \mathrm{d}t\, f(t) \exp(\varepsilon t)$ from the semi-infinite interval $t \in [-\infty, 0]$ is still reduced to the term f^+ alone, as can be shown by a simple change of variable

$$\varepsilon t = \tau. \tag{7.7}$$

This leads to

$$\begin{aligned} \lim_{\varepsilon \to 0^+} \varepsilon \int_{-\infty}^{0} \mathrm{d}t\, \mathrm{e}^{\varepsilon\, t} f(t) &= \lim_{\varepsilon \to 0^+} \int_{-\infty}^{0} \mathrm{d}\tau\, \mathrm{e}^{\tau} f(\tau/\varepsilon) \\ &= \int_{-\infty}^{0} \mathrm{d}\tau\, \mathrm{e}^{\tau} \lim_{\varepsilon \to 0^+} f(\tau/\varepsilon) \\ &= \int_{-\infty}^{0} \mathrm{d}\tau\, \mathrm{e}^{\tau} f^+ = f^+ \int_{-\infty}^{0} \mathrm{d}\tau\, \mathrm{e}^{\tau} = f^+ \end{aligned}$$

$$\therefore \qquad \lim_{\varepsilon \to 0^+} \varepsilon \int_{-\infty}^{0} \mathrm{d}t\, \mathrm{e}^{\varepsilon\, t} f(t) = f^+ \qquad \text{(QED)} \tag{7.8}$$

where we exchanged the order of the limiting process and integration. This is justified by the existence of underlying uniform convergence. From here we see that the concept of the strong limit of abstract vectors of elements of a general linear space is an entirely natural extension of the usual uniform convergence from scalar functions. This is a very useful analogy.

Let us now pass to a more general proof of lemma 4 for abstract elements of vector spaces. For stationary vectors such as $\psi(t) \equiv \psi^-$ or $\psi(t) \equiv \psi^+$ the equality (7.3) is reduced to the identity similarly to the corresponding case of a constant scalar function $f(t) \equiv f^+$ or $f(t) \equiv f^-$. In the case when $\psi(t)$ is not a constant vector for any t, it is illustrative to carry out the analysis for e.g. $\psi^- = \emptyset$. In such a case, strong convergence of $\psi(t)$ means that for an arbitrary number $\epsilon' > 0$ we can choose T such that $||\psi(t) - \psi^-|| = ||\psi(t)|| < \epsilon'$ whenever $t > T$. Thus, setting $\epsilon' \equiv \epsilon/2$ $(\epsilon > 0)$, we have for $T > 0$

$$\begin{aligned}
\left\|\varepsilon\int_0^\infty \mathrm{d}t\,\mathrm{e}^{-\varepsilon t}\psi(t)\right\| &= \varepsilon\left\|\int_0^T \mathrm{d}t\,\mathrm{e}^{-\varepsilon t}\psi(t) + \int_T^\infty \mathrm{d}t\,\mathrm{e}^{-\varepsilon t}\psi(t)\right\| \\
&\le \varepsilon\left\|\int_0^T \mathrm{d}t\,\mathrm{e}^{-\varepsilon t}\psi(t)\right\| + \varepsilon\left\|\int_T^\infty \mathrm{d}t\,\mathrm{e}^{-\varepsilon t}\psi(t)\right\| \\
&\le \varepsilon\int_0^T \mathrm{d}t\,\mathrm{e}^{-\varepsilon t}\,||\psi(t)|| + \varepsilon\int_T^\infty \mathrm{d}t\,\mathrm{e}^{-\varepsilon t}||\psi(t)|| \\
&\le \varepsilon\int_0^T \mathrm{d}t\,\mathrm{e}^{-\varepsilon t}||\psi(t)|| + \varepsilon\frac{\epsilon}{2}\left\|\int_T^\infty \mathrm{d}t\,\mathrm{e}^{-\varepsilon t}\right\| \\
&\le \varepsilon\int_0^T \mathrm{d}t\,\mathrm{e}^{-\varepsilon t}||\psi(t)|| + \frac{\epsilon}{2}\mathrm{e}^{-\varepsilon T} \\
&\le \varepsilon\int_0^T \mathrm{d}t\,||\psi(t)|| + \frac{\epsilon}{2} < \frac{\epsilon}{2} + \frac{\epsilon}{2} = \epsilon
\end{aligned}$$

$$\therefore \qquad \left\|\varepsilon\int_0^\infty \mathrm{d}t\,\mathrm{e}^{-\varepsilon t}\psi(t)\right\| < \epsilon \qquad \text{(QED)}. \tag{7.9}$$

This proves strong convergence of the vector $\psi(t)$ to $\psi^- = \emptyset$, since the norm $||\varepsilon\int_0^\infty \mathrm{d}t\,\mathrm{e}^{-\varepsilon t}\{\psi(t) - \psi^-\}||$ is shown to be vanishingly small in the limit $\varepsilon \longrightarrow 0^+$. In the demonstration (7.9), we select $\varepsilon < \epsilon/(2\int_0^T \mathrm{d}t\,||\psi(t)||)$ in the integral from T to ∞ and use the inequality $||\psi(t)|| \le \epsilon' \equiv \epsilon/2$ which is justified by the condition of lemma 4. We also use the Schwartz inequality for $\exp(-\varepsilon t)$ and $\exp(-\varepsilon T)$ via $|\exp(-\varepsilon\tau)| = |1 - \varepsilon\tau + \varepsilon^2\tau^2/2 - \cdots| \le 1 - \varepsilon\tau + \varepsilon^2\tau^2/2 - \cdots \le 1$. Let us now apply lemma 4 to scattering states

$$\Psi(t) \equiv \hat{\Omega}(t)\Psi_0 \tag{7.10}$$

$$\hat{\Omega}(t) = \hat{\mathrm{U}}^\dagger(t)\hat{\mathrm{U}}_0(t). \tag{7.11}$$

Due to the presence of the evolution operators $\hat{\mathrm{U}}$ and $\hat{\mathrm{U}}_0$ and their underlying feature of isometry, it is possible to prove lemma 4 in a different way without considering two special cases with ψ^+ being a non-zero vector and $\psi^+ = \emptyset$. To this end, we shall first assume that the following strong limits exist

$$\begin{aligned}
\Psi^{\pm} &= \underset{t\to\mp\infty}{\mathrm{Lim}} \quad \Psi(t) \\
&= \underset{t\to\mp\infty}{\mathrm{Lim}} \quad \hat{\Omega}(t)\Psi_0 \\
&= \underset{t\to\mp\infty}{\mathrm{Lim}} \quad \hat{\mathrm{U}}^{\dagger}(t)\hat{\mathrm{U}}_0(t)\Psi_0 \\
&= \underset{t\to\mp\infty}{\mathrm{Lim}} \quad \mathrm{e}^{i\hat{\mathrm{H}}t}\,\mathrm{e}^{-i\hat{\mathrm{H}}_0 t}\Psi_0 \quad = \quad \hat{\Omega}^{\pm}\Psi_0
\end{aligned}$$

$$\therefore \qquad \Psi^{\pm} = \underset{t\to\mp\infty}{\mathrm{Lim}} \quad \Psi(t) = \hat{\Omega}^{\pm}\Psi_0 \tag{7.12}$$

where (5.3), (7.10) and (7.11) are used. Then lemma 4 asserts that there must also exist the Abel limit with the same limiting value as follows

$$\Psi^{\pm} = \mp \underset{\varepsilon\to 0^+}{\mathrm{Lim}} \quad \varepsilon \int_0^{\mp\infty} \mathrm{d}\tau\, \mathrm{e}^{\pm\varepsilon\tau}\hat{\mathrm{U}}^{\dagger}(\tau)\hat{\mathrm{U}}_0(\tau)\Psi_0. \tag{7.13}$$

Proof of lemma 4: Before proving that (7.13) follows from (7.12), we shall interpret the corresponding strong limits. For instance, the expression (7.12) means that

$$\left\|\hat{\mathrm{U}}^{\dagger}(t)\hat{\mathrm{U}}_0(t)\Psi_0 - \Psi^-\right\| < \epsilon' \tag{7.14}$$

for an arbitrary $\epsilon' > 0$ starting from $t > T$. A similar interpretation is valid for (7.13), since the Abel limit also represents strong convergence i.e. convergence in the norm. Namely, choosing an arbitrary number $\epsilon > 0$, the norm of the difference between the state vectors from the lhs and rhs of (7.13) can be made as small as we desire i.e.

$$\left\|\varepsilon \int_0^{\infty} \mathrm{d}\tau\, \mathrm{e}^{-\varepsilon\tau}\hat{\mathrm{U}}^{\dagger}(\tau)\hat{\mathrm{U}}_0(\tau)\Psi_0 - \Psi^-\right\| < \epsilon \tag{7.15}$$

when t becomes greater that certain $T > 0$. Splitting the $\tau-$interval $[0, \infty]$ into two sub-intervals according to $[0, T] + [T, \infty]$, we shall have

$$\begin{aligned}
&\left\|\varepsilon \int_0^{\infty} \mathrm{d}\tau\, \mathrm{e}^{-\varepsilon\tau}\hat{\mathrm{U}}^{\dagger}(\tau)\hat{\mathrm{U}}_0(\tau)\Psi_0 - \Psi^-\right\| = \\
&\qquad\qquad \varepsilon\left\|\int_0^{\infty} \mathrm{d}\tau\, \mathrm{e}^{-\varepsilon\tau}\{\hat{\mathrm{U}}^{\dagger}(\tau)\hat{\mathrm{U}}_0(\tau)\Psi_0 - \Psi^-\}\right\| \\
&\le \varepsilon \int_0^{\infty} \mathrm{d}\tau\, \mathrm{e}^{-\varepsilon\tau}||\hat{\mathrm{U}}^{\dagger}(\tau)\hat{\mathrm{U}}_0(\tau)\Psi_0 - \Psi^-|| \\
&= \int_0^{\infty} \mathrm{d}\tau\, \mathrm{e}^{-\tau}||\hat{\mathrm{U}}^{\dagger}(\tau/\varepsilon)\hat{\mathrm{U}}_0(\tau/\varepsilon)\Psi_0 - \Psi^-||
\end{aligned}$$

$$
\begin{aligned}
&= \left\{\int_0^T + \int_T^\infty\right\} \mathrm{d}\tau\, \mathrm{e}^{-\tau} ||\hat{\mathrm{U}}^\dagger(\tau/\varepsilon)\hat{\mathrm{U}}_0(\tau/\varepsilon)\Psi_0 - \Psi^-|| \\
&\leq \int_0^T \mathrm{d}\tau\, \mathrm{e}^{-\tau}\{||\hat{\mathrm{U}}^\dagger(\tau/\varepsilon)\hat{\mathrm{U}}_0(\tau/\varepsilon)\Psi_0|| + ||\Psi^-||\} + \frac{\epsilon}{2}\int_T^\infty \mathrm{d}\tau\, \mathrm{e}^{-\tau} \\
&= \int_0^T \mathrm{d}\tau\, \mathrm{e}^{-\tau}(||\Psi_0|| + ||\Psi^-||) + \frac{\epsilon}{2}\int_T^\infty \mathrm{d}\tau\, \mathrm{e}^{-\tau} \\
&= (1-\mathrm{e}^{-T})(||\Psi_0|| + ||\Psi^-||) + \frac{\epsilon}{2}\mathrm{e}^{-T} \\
&\leq T(||\Psi_0|| + ||\Psi^-||) + \frac{\epsilon}{2}\mathrm{e}^{-T} \\
&\leq T(||\Psi_0|| + ||\Psi^-||) + \frac{\epsilon}{2} \\
&\leq \frac{\epsilon}{2} + \frac{\epsilon}{2} = \epsilon
\end{aligned}
$$

$$
\therefore \qquad \left\|\varepsilon\int_0^\infty \mathrm{d}\tau\, \mathrm{e}^{-\varepsilon\tau}\hat{\mathrm{U}}^\dagger(\tau)\hat{\mathrm{U}}_0(\tau)\Psi_0 - \Psi^-\right\| \leq \epsilon \qquad \text{(QED)}. \tag{7.16}
$$

Here, we used the expansion $1-\exp(-T) = 1-(1-T+\cdots) = T+\cdots$ as well as the majorization $\exp(-\xi) \leq 1\,(\xi \geq 0)$. Also in the integral over $\tau \in [T,\infty]$ we employed the assumption (7.14) of lemma 4 for $\varepsilon' = \epsilon/2$. Additionally, we made the following choice of the otherwise arbitrary number T

$$
T \leq \frac{1}{2}\frac{\epsilon}{||\Psi_0|| + ||\Psi^-||}. \tag{7.17}
$$

Applying lemma 4 to the state vector $\Psi(t)$ from (7.10), we find

$$
\begin{aligned}
\mp\varepsilon\int_0^{\mp\infty} \mathrm{d}t\, \mathrm{e}^{\pm\varepsilon t}\Psi(t) &= \mp\varepsilon\int_0^{\mp\infty} \mathrm{d}t\, \mathrm{e}^{\pm\varepsilon t}\{\mathrm{e}^{i\hat{\mathrm{H}}t}\,\mathrm{e}^{-i\hat{\mathrm{H}}_0 t}\Psi_0\} \\
&= \mp\varepsilon\int_0^{\mp\infty} \mathrm{d}t\, \mathrm{e}^{\pm\varepsilon t}\{\mathrm{e}^{i\hat{\mathrm{H}}t}\,\mathrm{e}^{-iEt}\Psi_0\} \\
&= \mp\varepsilon\int_0^{\mp\infty} \mathrm{d}t\, \mathrm{e}^{it[\hat{\mathrm{H}}-(E\mp i\varepsilon)\hat{1}]}\Psi_0 \\
&= \pm i\varepsilon[(E\pm i\varepsilon)\hat{1} - \hat{\mathrm{H}}]^{-1}\Psi_0
\end{aligned}
$$

$$
\therefore \qquad \mp\varepsilon\int_0^{\mp\infty} \mathrm{d}t\, \mathrm{e}^{\pm\varepsilon t}\Psi(t) = \pm i\varepsilon[(E\pm i\varepsilon)\hat{1} - \hat{\mathrm{H}}]^{-1}\Psi_0 \tag{7.18}
$$

where the integration is carried out by means of the spectral expansion of the unity operator in terms of the eigen-vectors of the Hamiltonian $\hat{\mathrm{H}}$, as in (6.10). Thus employing (7.13) and (7.18) we have

$$
\Psi^\pm = \pm i \lim_{\varepsilon\to 0^+} \varepsilon[(E\pm i\varepsilon)\hat{1} - \hat{\mathrm{H}}]^{-1}\Psi_0. \tag{7.19}
$$

Further, the usage of the assumed separable form of the total Hamiltonian via $\hat{\mathrm{H}} = \hat{\mathrm{H}}_0 + \hat{\mathrm{V}}$ permits the arrival at the following transformation of the term $\pm i\varepsilon[(E \pm i\varepsilon)\hat{1} - \hat{\mathrm{H}}]^{-1}\Psi_0$ from the rhs of (7.19)

$$\begin{aligned}
\pm i\varepsilon[(E \pm i\varepsilon)\hat{1} - \hat{\mathrm{H}}]^{-1}\Psi_0 &= \pm i\varepsilon[(E \pm i\varepsilon)\hat{1} - \hat{\mathrm{H}}_0 - \hat{\mathrm{V}}]^{-1}\Psi_0 \\
&= \pm i\varepsilon\{[(E \pm i\varepsilon)\hat{1} - \hat{\mathrm{H}}_0][\hat{1} - (E\hat{1} - \hat{\mathrm{H}}_0 \pm i\varepsilon\hat{1})^{-1}\hat{\mathrm{V}}]\}^{-1}\Psi_0 \\
&= \pm i\varepsilon\{\hat{1} - [(E \pm i\varepsilon)\hat{1} - \hat{\mathrm{H}}_0]^{-1}\hat{\mathrm{V}}\}^{-1}[(E \pm i\varepsilon)\hat{1} - \hat{\mathrm{H}}_0]^{-1}\Psi_0 \\
&= \pm i\varepsilon\{\hat{1} - [(E \pm i\varepsilon)\hat{1} - \hat{\mathrm{H}}_0]^{-1}\hat{\mathrm{V}}\}^{-1}(\pm i\varepsilon)^{-1}\Psi_0 \\
&= \{\hat{1} - [(E \pm i\varepsilon)\hat{1} - \hat{\mathrm{H}}_0]^{-1}\hat{\mathrm{V}}\}^{-1}\Psi_0
\end{aligned}$$

$$\therefore \qquad \pm i\varepsilon[(E \pm i\varepsilon)\hat{1} - \hat{\mathrm{H}}]^{-1}\Psi_0 = \{\hat{1} - [(E \pm i\varepsilon)\hat{1} - \hat{\mathrm{H}}_0]^{-1}\hat{\mathrm{V}}\}^{-1}\Psi_0. \tag{7.20}$$

Here, we made use of the following relation

$$\pm i\varepsilon\,[(E \pm i\varepsilon)\hat{1} - \hat{\mathrm{H}}_0]^{-1}\Psi_0 = \Psi_0 \tag{7.21}$$

which is evident from the unperturbed eigen-problem $(\hat{\mathrm{H}}_0 - E\hat{1})\Psi_0 = 0$. Letting $\varepsilon \to 0^+$ in (7.20) according to the requirement in (7.19), it follows

$$\begin{aligned}
\Psi^\pm &= \lim_{\varepsilon \to 0^+} \{\hat{1} - [(E \pm i\varepsilon)\hat{1} - \hat{\mathrm{H}}_0]^{-1}\hat{\mathrm{V}}\}^{-1}\Psi_0 \\
&= [\hat{1} - \hat{\mathrm{G}}_0^\pm(E)\hat{\mathrm{V}}]^{-1}\Psi_0
\end{aligned}$$

$$\therefore \qquad \Psi^\pm = [\hat{1} - \hat{\mathrm{G}}_0^\pm(E)\hat{\mathrm{V}}]^{-1}\Psi_0. \tag{7.22}$$

Since $\Psi^\pm = \hat{\Omega}^\pm\Psi_0$, as per (5.7), it is clear that by means of (7.22) we can readily identify the wave operators $\hat{\Omega}^\pm$ in the forms

$$\begin{aligned}
\hat{\Omega}^\pm &= \lim_{\varepsilon \to 0^+} \{\hat{1} - [(E \pm i\varepsilon)\hat{1} - \hat{\mathrm{H}}_0]^{-1}\hat{\mathrm{V}}\}^{-1} \\
&= [\hat{1} - \hat{\mathrm{G}}_0^\pm(E)\hat{\mathrm{V}}]^{-1}.
\end{aligned} \tag{7.23}$$

If both sides of (7.23) are multiplied by the operator $\hat{1} - \hat{\mathrm{G}}_0^\pm(E)\hat{\mathrm{V}}$ from the left, we will obtain the following two Lippmann-Schwinger equations for $\hat{\Omega}^\pm$

$$\hat{\Omega}^\pm = \hat{1} + \hat{\mathrm{G}}_0^\pm(E)\hat{\mathrm{V}}\hat{\Omega}^\pm. \tag{7.24}$$

When these Møller operators are applied to Ψ_0 through (5.7), the Lippmann-Schwinger equations are obtained directly for scattering states $\Psi^\pm$ as

$$\begin{aligned}
\Psi^\pm &= \hat{\Omega}^\pm\Psi_0 \\
&= [\hat{1} + \hat{\mathrm{G}}_0^\pm(E)\hat{\mathrm{V}}\hat{\Omega}^\pm]\Psi_0 \\
&= \Psi_0 + \hat{\mathrm{G}}_0^\pm(E)\hat{\mathrm{V}}\{\hat{\Omega}^\pm\Psi_0\} \\
&= \Psi_0 + \hat{\mathrm{G}}_0^\pm(E)\hat{\mathrm{V}}\Psi^\pm
\end{aligned}$$

$$\therefore \qquad \Psi^{\pm} = \Psi_0 + \hat{G}_0^{\pm}(E)\hat{V}\Psi^{\pm}. \tag{7.25}$$

The same expression as in (7.25) can be obtained directly by applying the operator $\hat{1} - \hat{G}_0^{\pm}(E)\hat{V}$ from the right to both sides of (7.22). The Abel strong limit (7.13), which enabled the derivation of the Lippmann-Schwinger equation (7.25), is also capable of yielding the formal solution of these integral equations. This solution is contained in the intermediate step (7.19), since using the unperturbed eigen-problem $(\hat{H}_0 - E\hat{1})\Psi_0 = \emptyset$, we have

$$\begin{aligned}
\Psi^{\pm} &= \pm i\varepsilon\hat{G}^{\pm}(E)\Psi_0 = \hat{G}^{\pm}(E)i\varepsilon\Psi_0 \\
&= \hat{G}^{\pm}(E)[(E \pm i\varepsilon)\hat{1} - \hat{H}_0 - \hat{V} + \hat{V}]\Psi_0 \\
&= \hat{G}^{\pm}(E)[(E \pm i\varepsilon)\hat{1} - \hat{H} + \hat{V}]\Psi_0 \\
&= [\hat{1} + \hat{G}^{\pm}(E)\hat{V}]\Psi_0 \\
&= \Psi_0 + \hat{G}^{\pm}(E)\hat{V}\Psi_0
\end{aligned}$$

$$\therefore \qquad \Psi^{\pm} = \Psi_0 + \hat{G}^{\pm}(E)\hat{V}\Psi_0 \tag{7.26}$$

in agreement with (6.12) and (6.13). This is the solution of (7.25), since the vector $\Psi^{\pm}$ is eliminated from the rhs of (7.26), unlike the situation in the corresponding integral equation (7.25). Nevertheless, the solution (7.26) is still of a purely formal nature since it is expressed through the total Green operator $\hat{G}^{\pm}(E)$, which is just as difficult to handle as the original equation for the scattering states $\Psi^{\pm}$. By combining the expressions (5.7) and (7.26), we can arrive at yet another useful representation of the Møller operators $\hat{\Omega}^{\pm}$ via the total Green resolvents $\hat{G}^{\pm}(E)$

$$\begin{aligned}
\hat{\Omega}^{\pm} &= \hat{1} + \operatorname*{Lim}_{\varepsilon\to 0+} [\hat{H} - (E \mp i\varepsilon)\hat{1}]^{-1}\hat{V} \\
&= \hat{1} + \hat{G}^{\pm}(E)\hat{V}.
\end{aligned} \tag{7.27}$$

If the identities $\hat{G}^{\pm}(E) = \hat{G}_0^{\pm}(E)[\hat{1} - \hat{V}\hat{G}_0^{\pm}(E)]^{-1} = [\hat{1} - \hat{G}_0^{\pm}(E)\hat{V}]^{-1}\hat{G}_0^{\pm}(E)$ are multiplied by $\hat{V}$, it will follow

$$\hat{G}_0^{\pm}(E)\hat{T}^{\pm}(E) = \hat{G}^{\pm}(E)\hat{V} \tag{7.28}$$

$$\hat{T}^{\pm}(E)\hat{G}_0^{\pm}(E) = \hat{V}\hat{G}^{\pm}(E) \tag{7.29}$$

where $\hat{T}^{\pm}(E)$ are the transition $\hat{T}$–operators

$$\hat{T}^{\pm}(E) = [\hat{1} - \hat{V}\hat{G}_0^{\pm}(E)]^{-1}\hat{V} = \hat{V}[\hat{1} - \hat{G}_0^{\pm}(E)\hat{V}]^{-1}. \tag{7.30}$$

Using (7.28), it is possible to express $\hat{\Omega}^{\pm}$ via the transition $\hat{T}$–operators according to

$$\hat{\Omega}^{\pm} = \hat{1} + \hat{G}_0^{\pm}(E)\hat{T}^{\pm}(E). \tag{7.31}$$

The accomplished analysis completes the proof that the strong limit (5.9) leads to the time-independent i.e. stationary Lippmann-Schwinger integral equations that represent the cornerstone of the quantum scattering theory.

From the performed analysis it is clear that all the results are independent of the manner of averaging in the integral representations of the Møller wave operators. The same expression (7.25), as well as the existence of the wave operator $\hat{\Omega}^+$ can be proven by employing the following averaging

$$\hat{\Omega}^+ = \lim_{T\to+\infty} \frac{1}{T}\int_0^T dt\, \hat{U}^\dagger(t)\hat{U}_0(t). \tag{7.32}$$

With this formula, the results for $\hat{\Omega}^\pm$ appear as a generalization of the well-known Neumann mean ergodic theorem.

The results from (7.25) can also be obtained from the same starting point as we just did, but proceeding afterwards in a different way. Namely, let us take the intermediate result (7.19), which can be rewritten via the total Green resolvent $\hat{G}(z)$ from (6.15) as

$$\Psi^\pm = \pm i \lim_{\varepsilon\to 0+} \varepsilon\,\hat{G}(E\pm i\varepsilon)\Psi_0. \tag{7.33}$$

If the obvious identity

$$(z-z')\hat{1} = (z\hat{1}-\hat{H}) - (z'\hat{1}-\hat{H}) = \hat{G}^{-1}(z) - \hat{G}^{-1}(z') \tag{7.34}$$

is multiplied from the left by $\hat{G}(z')$, it follows

$$\begin{aligned}\hat{G}(z') - \hat{G}(z) &= (z-z')\hat{G}(z)\hat{G}(z')\\ &= (z-z')\hat{G}(z')\hat{G}(z).\end{aligned} \tag{7.35}$$

On the other hand, the relation $\hat{H} = \hat{H}_0 + \hat{V}$ can be expressed via

$$\hat{V} = \hat{H} - \hat{H}_0 = (z\hat{1}-\hat{H}_0) - (z\hat{1}-\hat{H}) = \hat{G}_0^{-1}(z) - \hat{G}^{-1}(z) \tag{7.36}$$

where $\hat{G}_0(z)$ is the free Green resolvent (6.15). Multiplications of (7.36) from the left by $\hat{G}_0(z)$ and from the right by $\hat{G}(z)$ or in the opposite order, lead to the Lippmann-Schwinger equation for the resolvent operator $\hat{G}(z)$

$$\begin{aligned}\hat{G}(z) &= \hat{G}_0(z) + \hat{G}_0(z)\hat{V}\hat{G}(z)\\ &= \hat{G}_0(z) + \hat{G}(z)\hat{V}\hat{G}_0(z).\end{aligned} \tag{7.37}$$

With the help of (7.37), the relation (7.33) now becomes

$$\begin{aligned}\Psi^\pm = &\pm i \lim_{\varepsilon\to 0+} \varepsilon\,\hat{G}_0(E\pm i\varepsilon)\Psi_0\\ &\pm i \lim_{\varepsilon\to 0+} \varepsilon\,\hat{G}_0(E\pm i\varepsilon)\,\hat{V}\,\hat{G}(E\pm i\varepsilon)\Psi_0.\end{aligned} \tag{7.38}$$

Further, in terms of $\hat{G}_0(z)$, the expression (7.21) acquires the form

$$\pm i\varepsilon\,\hat{G}_0(E\pm i\varepsilon)\Psi_0 = \Psi_0 \tag{7.39}$$

so that by means of (7.38), we have

$$\begin{aligned}\Psi^{\pm} &= \pm i \lim_{\varepsilon\to 0+} \varepsilon \hat{\mathrm{G}}_0(E \pm i\varepsilon)\Psi_0 \\ &\pm i \lim_{\varepsilon\to 0+} \hat{\mathrm{G}}_0(E \pm i\varepsilon)\hat{\mathrm{V}}\varepsilon \lim_{\varepsilon\to 0+} \hat{\mathrm{G}}(E \pm i\varepsilon)\Psi_0 \\ &= \Psi_0 + \lim_{\varepsilon\to 0+} \hat{\mathrm{G}}_0(E \pm i\varepsilon)\Psi^{\pm} \\ &\equiv \Psi_0 + \hat{\mathrm{G}}_0(E \pm i0)\hat{\mathrm{V}}\Psi^{\pm} = \Psi_0 + \hat{\mathrm{G}}_0^{\pm}(E)\hat{\mathrm{V}}\Psi^{\pm}\end{aligned}$$

$$\therefore \quad \Psi^{\pm} = \Psi_0 + \hat{\mathrm{G}}_0^{\pm}(E)\hat{\mathrm{V}}\Psi^{\pm} \tag{7.40}$$

where (7.33) is used. The obtained expression (7.40) is in full agreement with the previously derived Lippmann-Schwinger equations for scattering states $\Psi^{\pm}$ from (7.25).

Notice that the signs of the terms $\pm\varepsilon$ in (7.39) coincide with the order of the superscripts $\pm$ in the Møller wave operators $\hat{\Omega}^{\pm}$ and the scattering states $\Psi^{\pm}$ from the same equations. Such an ordering of the signs $\pm$ in $\hat{\Omega}^{\pm}$ is in accord with the superscripts of the Green operators $\hat{\mathrm{G}}^{\pm}(E) = \lim_{\varepsilon\to 0+} \hat{\mathrm{G}}(E \pm i\varepsilon)$ in (6.15) where the signs $\pm$ have a physical interpretation relative to the corresponding boundary conditions for the outgoing/incoming waves, respectively. Now such an interpretation is via (7.38) directly transferred into the Green $\hat{\mathrm{G}}^{\pm}(z)$ and Møller $\hat{\Omega}^{\pm}$ operators. Their superscripts $\pm$ are introduced in (5.4) so as to correspond to the limits $t \to \mp\infty$ of the operator $\hat{\Omega}(t)$ from (7.11), and this also explains an apparently illogical convention (5.4).

7.2 The Abel limit and Fourier integrals

It is important to emphasize that the performed derivations of the Lippmann-Schwinger equation (7.25) are based upon the assumption that the Møller wave operators $\hat{\Omega}^{\pm}$ exist as the Cauchy strong limits of the product of the evolution operators $\hat{\mathrm{U}}^{\dagger}(t)\hat{\mathrm{U}}_0(t)$ when $t \to \mp\infty$. This assumption further implies the existence of the Abel strong limits (7.13) when $\varepsilon \to 0^+$ and this, in turn, secures the existence of scattering states $\Psi^{\pm} \equiv \Psi^{\pm}(E)$ as the solutions of the time-independent Lippmann-Schwinger equations (7.25). Thus, the introduction of the Abel strong limit in the analysis is precisely the place where time t disappears from the presented formalism of scattering theory. In so doing, the newly derived stationary Lippmann-Schwinger integral equations from (7.25) satisfy precisely the same boundary conditions when $\varepsilon \to 0^+$, as in the case of the corresponding time-dependent equations from the standard Fourier transform [7]

$$\Psi_{\alpha}^{+}(t) = \Psi_{i\alpha} - i \int_{-\infty}^{t} \mathrm{d}t' \, \mathrm{e}^{-\hat{\mathrm{H}}_0(t-t')}\hat{\mathrm{V}}\Psi_{\alpha}^{+}(t') \tag{7.41}$$

$$\Psi_\alpha^-(t) = \Psi_{f\alpha} - i\int_t^\infty \mathrm{d}t' \, \mathrm{e}^{-\hat{\mathrm{H}}_0(t-t')}\hat{\mathrm{V}}\Psi_\alpha^-(t'). \tag{7.42}$$

In other words, the passage from the Cauchy strong limit with $|t| \to \infty$ to the Abel strong limit with $\varepsilon \to 0^+$ actually carries out the transformation from the non-stationary state vectors $\Psi^\pm(t)$ to the stationary ones $\Psi^\pm(E)$. This accomplishes the sought link between the time-dependent and time-independent formalisms of scattering theory without recourse to the more customary Fourier integral. The final results e.g. (6.13) and (7.25) are identical with the corresponding expressions that are obtained by the Fourier integrals connecting the non-stationary and stationary state vectors and Green operators [7]

$$\begin{aligned}\Psi_\alpha^+(E) &= \Psi_{0\alpha}(E) + \hat{\mathrm{G}}_0^+(E)\hat{\mathrm{V}}\Psi_\alpha^+(E) \\ &= \Psi_{0\alpha}(E) + \hat{\mathrm{G}}^+(E)\hat{\mathrm{V}}\Psi_{0\alpha}(E)\end{aligned} \tag{7.43}$$

and

$$\begin{aligned}\Psi_\alpha^-(E) &= \Psi_{0\alpha}(E) + \hat{\mathrm{G}}_0^-(E)\hat{\mathrm{V}}\Psi_\alpha^-(E) \\ &= \Psi_{0\alpha}(E) + \hat{\mathrm{G}}^-(E)\hat{\mathrm{V}}\Psi_{0\alpha}(E).\end{aligned} \tag{7.44}$$

From the expounded subject, it is clear that the two formalisms i.e. 1) the Fourier transform between non-stationary and stationary $q-$numbers and 2) the Cauchy-Abel transforms between the strong limits of the Cauchy and Abel types of the same $q-$numbers are equivalent to each other. We established not only the global equivalence between the concepts 1) and 2), but also the local equivalence of the two strategies (stationary $\longleftrightarrow$ non-stationary) within each of the formalisms 1) and 2). This is accomplished on a totally general level by operating with abstract elements of linear spaces i.e. with vectors and operators in their arbitrary representations. The equivalence established in this way is valid in any of the special representations of the abstract linear spaces. This is reminiscent of the familiar setting in which the existence of the Cauchy and Abel strong limits obligatorily implies the existence of the corresponding weak limits.

8

Scattering matrix and transition matrix

8.1 The Abel limit and scattering operators

In this chapter, we shall focus on deriving the connection between the scattering matrix (or the S-matrix) and the transition matrix (or the T-matrix) by means of the formalism of the Abel strong limits from chapter 7. As is well-known [7], the $\hat{\mathrm{S}}$−operator is defined as the product of the two Møller wave operators according to

$$\begin{aligned}
\hat{\mathrm{S}} &= \hat{\Omega}^{-\dagger}\hat{\Omega}^{+} \\
&= \operatorname*{Lim}_{t\to+\infty}\ \operatorname*{Lim}_{t'\to-\infty}\ \hat{\Omega}^{-\dagger}(t)\hat{\Omega}^{+}(t') \\
&= \operatorname*{Lim}_{t\to+\infty}\ \operatorname*{Lim}_{t'\to-\infty}\ \{\hat{\mathrm{U}}_0^{\dagger}(t)\hat{\mathrm{U}}(t)\}\{\hat{\mathrm{U}}^{\dagger}(t')\hat{\mathrm{U}}_0(t')\} \\
&= \operatorname*{Lim}_{t\to+\infty}\ \operatorname*{Lim}_{t'\to-\infty}\ \{\mathrm{e}^{i\hat{\mathrm{H}}_0 t}\,\mathrm{e}^{-i\hat{\mathrm{H}}t}\}\{\mathrm{e}^{i\hat{\mathrm{H}}t'}\,\mathrm{e}^{-i\hat{\mathrm{H}}_0 t'}\}
\end{aligned}$$

$$\therefore\qquad \hat{\mathrm{S}} = \operatorname*{Lim}_{t\to+\infty}\ \operatorname*{Lim}_{t'\to-\infty}\ \{\mathrm{e}^{i\hat{\mathrm{H}}_0 t}\,\mathrm{e}^{-i\hat{\mathrm{H}}t}\}\{\mathrm{e}^{i\hat{\mathrm{H}}t'}\,\mathrm{e}^{-i\hat{\mathrm{H}}_0 t'}\}. \tag{8.1}$$

The order of the appearance of the two limits in (8.1) is irrelevant. This occurs because in the definition (8.1) of the stationary Møller wave operators $\hat{\Omega}^{\pm} = \operatorname*{Lim}_{\tau\to\mp\infty}\{\hat{\mathrm{U}}^{\dagger}(\tau)\hat{\mathrm{U}}_0(\tau)\}$, we set $\tau = t$ and $\tau = t'$ for $\hat{\Omega}^{-}$ and $\hat{\Omega}^{+}$, respectively. However, time τ appears as a dummy variable and, as such, associating τ with t or t' is arbitrary. In other words, we could have made the opposite choice such that $\tau = t'$ and $\tau = t$ correspond to $\hat{\Omega}^{-}$ and $\hat{\Omega}^{+}$, respectively

$$\hat{\mathrm{S}}' = \operatorname*{Lim}_{t'\to+\infty}\ \operatorname*{Lim}_{t\to-\infty}\ \{\mathrm{e}^{i\hat{\mathrm{H}}_0 t'}\,\mathrm{e}^{-i\hat{\mathrm{H}}t'}\}\{\mathrm{e}^{i\hat{\mathrm{H}}t}\,\mathrm{e}^{-i\hat{\mathrm{H}}_0 t}\}. \tag{8.2}$$

Here, by making the change $t \longleftrightarrow t'$, it follows

$$\hat{\mathrm{S}}' = \hat{\mathrm{S}} \qquad \text{(QED)}. \tag{8.3}$$

However, this proof indicates a benefit which will facilitate the further analysis. Namely, carrying out explicitly two limiting procedures in (8.1) is more difficult as well as cumbersome, and one could inquire about the possibility of considering only one limit. This is indeed feasible in the following manner. First, let us notice that the curly brackets in (8.2) can be left out. This is possible due to unitarity of the evolution operators, so that in all their combinations of the type

$$\{e^{i\hat{H}_0 t}e^{-i\hat{H}t}\}\{e^{i\hat{H}t'}e^{-i\hat{H}_0 t'}\} \quad , \quad e^{i\hat{H}_0 t}\{e^{-i\hat{H}t}e^{i\hat{H}t'}\}e^{-i\hat{H}_0 t'}$$
$$\{e^{i\hat{H}_0 t}e^{-i\hat{H}t}e^{i\hat{H}t'}\}e^{-i\hat{H}_0 t'} \quad , \quad e^{i\hat{H}_0 t}\{e^{-i\hat{H}t}e^{i\hat{H}t'}e^{-i\hat{H}_0 t'}\} \tag{8.4}$$

the domains and ranges remain always equal to the whole Hilbert space $\mathcal{H}$. The same conclusion about omitting the mentioned curly brackets in (8.1) can be reached by applying the Becker-Housdorf lemma on the product of two exponential operators

$$e^{\hat{A}}\, e^{\hat{B}} = e^{\hat{A}+\hat{B}+[\hat{A},\hat{B}]/2}. \tag{8.5}$$

In our case, the Hamilton operators $\hat{H}_0$ and $\hat{H}$ do not commute with each other, so that

$$e^{i\hat{H}_0 t}\, e^{-i\hat{H}t} \neq e^{i\hat{H}_0 t - i\hat{H}t}. \tag{8.6}$$

Here, the inequality will become the equality if, according to (8.5), we add the commutator $(it/2)[\hat{H}_0, -\hat{H}]$ to the operators $i\hat{H}_0 t - i\hat{H}t$ in the exponent on the rhs of (8.6). Thus, using Becker-Housdorf lemma in the product of the Møller operators $\hat{\Omega}^{-\dagger}$ and $\hat{\Omega}^{+}$, it follows

$$\hat{\Omega}^{-\dagger}(t)\hat{\Omega}^{+}(t') = e^{\hat{A}(t)}\, e^{\hat{A}^{\dagger}(t')} = e^{\hat{A}(t)+\hat{A}^{\dagger}(t')} \tag{8.7}$$

$$\hat{A}(\tau) = \frac{i}{\tau}(\hat{H}_0 - \hat{H}) + \frac{i}{2\tau}[\hat{H}_0, -\hat{H}]. \tag{8.8}$$

From the definition of the operator $\hat{A}(\tau)$, it is easily verified that the following relation holds

$$[\hat{A}(t), \hat{A}^{\dagger}(t')] = \hat{0}. \tag{8.9}$$

Thus, the expression (8.7) is indeed one more confirmation that the curly brackets can be dropped from (8.1) or redistributed in any other desired order with no influence whatsoever on the final result. For convenience, we shall set

$$\begin{aligned}\hat{\Omega}^{-\dagger}(t)\hat{\Omega}^{+}(t') &= e^{i\hat{H}_0 t}\,\{e^{-i\hat{H}t}e^{i\hat{H}t'}\}\,e^{-i\hat{H}_0 t'} \\ &= e^{i\hat{H}_0 t}\, e^{-i\hat{H}(t-t')}\, e^{-i\hat{H}_0 t'}\end{aligned} \tag{8.10}$$

where Becker-Housdorf lemma is used for the operator product $e^{-i\hat{H}t}e^{i\hat{H}t'}$. This leads to the following result for the $\hat{S}-$operator

$$\hat{S} = \lim_{t\to+\infty}\ \lim_{t'\to-\infty}\ e^{i\hat{H}_0 t}\, e^{-i\hat{H}(t-t')}\, e^{-i\hat{H}_0 t'}. \tag{8.11}$$

Here, we can freely set $t' = -t$ and thus arrive to a new expression for the $\hat{S}-$operator in terms of a single limiting process via

$$\begin{aligned}\hat{S} &= \lim_{t\to+\infty}\ e^{i\hat{H}_0 t}\, e^{-2i\hat{H}t}\, e^{i\hat{H}_0 t} \\ &\equiv \lim_{t\to+\infty}\ \hat{S}(t).\end{aligned} \tag{8.12}$$

8.2 Matrix elements of scattering operators

Of central importance in scattering theory are the S-matrix elements S_{if} of the $\hat{S}$–operator taken between the initial and final eigen-states Ψ_{0i} and Ψ_{0f} of the unperturbed Hamiltonian $\hat{H}_0$

$$\begin{aligned} S_{if} &= \langle\Psi_{0f}|\hat{S}|\Psi_{0i}\rangle \\ &= \lim_{t\to+\infty} \langle\Psi_{0f}|e^{i\hat{H}_0 t}\, e^{-2i\hat{H}t}\, e^{i\hat{H}_0 t}|\Psi_0\rangle \\ &\equiv \lim_{t\to+\infty} \langle\Psi_{0f}|\hat{S}(t)|\Psi_{0i}\rangle. \end{aligned} \tag{8.13}$$

From the limit over time in (8.13), we can pass to a more convenient integration over t if we differentiate the operator $\hat{S}(t)$ and utilize the relation $\hat{H} - \hat{H}_0 = \hat{V}$. Then, for a conservative system[1], we have

$$\frac{d}{d\tau}\hat{S}(\tau) = -i\{e^{i\hat{H}_0\tau}\hat{V}e^{-2i\hat{H}\tau}e^{i\hat{H}_0\tau} + e^{i\hat{H}_0\tau}\, e^{-2i\hat{H}\tau}V e^{i\hat{H}_0\tau}\}. \tag{8.14}$$

Let us now integrate both sides of (8.14) over τ in the interval $[0, t]$ and then find the limiting value when $t \to \infty$. In this way, the lhs of (8.14) becomes

$$\begin{aligned} \lim_{t\to+\infty} \int_0^t d\tau \frac{d}{d\tau}\hat{S}(\tau) &= \lim_{t\to+\infty} \hat{S}(\tau)\Big|_{\tau=0}^{\tau=t} \\ &= \lim_{t\to+\infty} \hat{S}(t) - \hat{1} = \hat{S} - \hat{1} \end{aligned}$$

$$\therefore \quad \lim_{t\to+\infty} \int_0^t d\tau \frac{d}{d\tau}\hat{S}(\tau) = \hat{S} - \hat{1} \tag{8.15}$$

where the relation $\hat{S}(0) = \hat{1}$ is used from (8.12). Thus, with the help of (8.15), it follows

$$\begin{aligned} S_{if} = \langle\Psi_{0f}|\Psi_{0i}\rangle - i\int_0^\infty d\tau \\ \times \langle\Psi_{0f}|e^{i\hat{H}_0\tau}\hat{V}e^{-2i\hat{H}\tau}e^{i\hat{H}_0\tau} + e^{i\hat{H}_0\tau}e^{-2i\hat{H}\tau}\hat{V}e^{i\hat{H}_0\tau}|\Psi_{0i}\rangle. \end{aligned} \tag{8.16}$$

Now applying lemma 3 from (6.3), we have

$$\begin{aligned} S_{if} = \langle\Psi_{0f}|\Psi_{0i}\rangle - i \lim_{\varepsilon\to 0^+} \int_0^\infty d\tau\, e^{-\varepsilon\tau} \\ \times \langle\Psi_{0f}|e^{i\hat{H}_0\tau}\hat{V}e^{-2i\hat{H}\tau}e^{i\hat{H}_0\tau} + e^{i\hat{H}_0\tau}\, e^{-2i\hat{H}\tau}\hat{V}e^{i\hat{H}_0\tau}|\Psi_{0i}\rangle. \end{aligned} \tag{8.17}$$

[1]A physical system is said to be conservative if the operators $\hat{H}, \hat{H}_0$ and $\hat{V}$ are time-independent.

The operator-valued integral over τ can be carried out similarly to (6.10) and thus reduced to an ordinary integral of a scalar function. To this end, we first exploit the relation

$$\hat{\mathrm{U}}_0^\dagger(\tau)|\Psi_{0i}\rangle = \mathrm{e}^{iE_i\tau}|\Psi_{0i}\rangle \qquad \langle\Psi_{0f}|\hat{\mathrm{U}}_0^\dagger(\tau) = \langle\Psi_{0f}|\mathrm{e}^{iE_f\tau} \tag{8.18}$$

where $E_i = p_i^2/(2m)$ and $E_f = p_f^2/(2m)$. Thus, we have

$$\begin{aligned} \mathrm{S}_{if} = \; & \langle\Psi_{0f}|\Psi_{0i}\rangle - i \lim_{\varepsilon\to 0+} \int_0^\infty \mathrm{d}\tau \, \mathrm{e}^{-\varepsilon\tau} \\ & \times \langle\Psi_{0f}|\mathrm{e}^{iE_f\tau}\hat{\mathrm{V}}\mathrm{e}^{-2i\hat{\mathrm{H}}\tau}\mathrm{e}^{iE_i\tau} + \mathrm{e}^{iE_f\tau}\,\mathrm{e}^{-2i\hat{\mathrm{H}}\tau}\hat{\mathrm{V}}\mathrm{e}^{iE_i\tau}|\Psi_{0i}\rangle. \end{aligned} \tag{8.19}$$

Finally, if in (8.19) we write $\hat{\mathrm{V}}\mathrm{e}^{-2i\hat{\mathrm{H}}\tau} = \hat{\mathrm{V}}\hat{1}\mathrm{e}^{-2i\hat{\mathrm{H}}\tau}$ and $\mathrm{e}^{-2i\hat{\mathrm{H}}\tau}\hat{\mathrm{V}} = \mathrm{e}^{-2i\hat{\mathrm{H}}\tau}\hat{1}\hat{\mathrm{V}}$ together with substituting the spectral representation (6.8) of the unity operator and apply (6.9), we shall have

$$\begin{aligned} \mathrm{S}_{if} = \; & \langle\Psi_{0f}|\Psi_{0i}\rangle \\ & - i \lim_{\varepsilon\to 0+} \sum\!\!\!\!\!\!\int_\nu \mathrm{d}\nu \, \langle\Psi_{0f}|\{|\Psi_\nu\rangle\langle\Psi_\nu|\hat{\mathrm{V}} \int_0^\infty \mathrm{d}\tau \, \mathrm{e}^{-2i(E_\nu - \bar{E}_{if} - i\varepsilon/2)\tau} \\ & + \int_0^\infty \mathrm{d}\tau \, \mathrm{e}^{-2i(E_\nu - \bar{E}_{if} - i\varepsilon/2)\tau}\hat{\mathrm{V}}|\Psi_\nu\rangle\langle\Psi_\nu|\}\Psi_{0i}\rangle \end{aligned} \tag{8.20}$$

$$\bar{E}_{if} = \frac{E_i + E_f}{2}. \tag{8.21}$$

By this device, the original operator-valued integral over τ is now reduced to an ordinary integral of a scalar function with the evident result

$$-i \int_0^\infty \mathrm{d}\tau \, \mathrm{e}^{-2i(E_\nu - \bar{E}_{if} - i\varepsilon/2)\tau} = \frac{1}{2}(\bar{E}_{if} - E_\nu + i\varepsilon/2)^{-1}. \tag{8.22}$$

In this way, equation (8.20) becomes

$$\begin{aligned} \mathrm{S}_{if} = \; & \langle\Psi_{0f}|\Psi_{0i}\rangle \\ & + \frac{1}{2} \lim_{\varepsilon\to 0+} \sum\!\!\!\!\!\!\int_\nu \mathrm{d}\nu \, \langle\Psi_{0f}|\{|\Psi_\nu\rangle\langle\Psi_\nu|\hat{\mathrm{V}}(\bar{E}_{if} - E_\nu + i\varepsilon/2)^{-1} \\ & + (\bar{E}_{if} - E_\nu + i\varepsilon/2)^{-1}\hat{\mathrm{V}}|\Psi_\nu\rangle\langle\Psi_\nu|\}\Psi_{0i}\rangle. \end{aligned} \tag{8.23}$$

Here, the sum over ν can be performed with the help of the spectral representation of the Green resolvent

$$\hat{\mathrm{G}}(z) = (z\hat{1} - \hat{\mathrm{H}})^{-1} = \sum\!\!\!\!\!\!\int_\nu \mathrm{d}\nu \, \frac{|\Psi_\nu\rangle\langle\Psi_\nu|}{z - E_\nu} \tag{8.24}$$

so that the expression (8.23) acquires the form

$$\begin{aligned} \mathrm{S}_{if} = \; & \langle\Psi_{0f}|\Psi_{0i}\rangle \\ & + \frac{1}{2} \lim_{\varepsilon\to 0+} \langle\Psi_{0f}|\hat{\mathrm{V}}\hat{\mathrm{G}}\left(\bar{E}_{if} + i\varepsilon/2\right) + \hat{\mathrm{G}}\left(\bar{E}_{if} + i\varepsilon/2\right)\hat{\mathrm{V}}|\Psi_{0i}\rangle. \end{aligned} \tag{8.25}$$

8.3 Transition operators

Due to the ortho-normality relation (see e.g. [7])

$$\langle\Psi_\alpha^\pm(E)|\Psi_{\alpha'}^\pm(E')\rangle = \delta(E-E')\delta_{\alpha,\alpha'} \tag{8.26}$$

the first term $\langle\Psi_{0f}|\Psi_{0i}\rangle$ in (8.25) is reduced to $\delta(E_i - E_f)\delta_{if}$. Thus, in order for the rhs of (8.25) to have a meaning, the matrix element of the operator $\hat{\mathrm{V}}\hat{\mathrm{G}} + \hat{\mathrm{G}}\hat{\mathrm{V}}$ must also have a factorized Dirac δ−function, which is infinite at $E_i = E_f$. Inspecting the argument of the Green resolvent in (8.25), we can notice that the closest combination which could indeed yield the δ−function $\delta(E_i - E_f)$ is given by one of several equivalent definitions of the Dirac function [7]. This stems from the ubiquitous symbolic relation

$$\frac{1}{x-x_0 \pm i\varepsilon} = \mathcal{P}\frac{1}{x-x_0} \mp i\pi\delta(x-x_0) \tag{8.27}$$

where $\mathcal{P}$ denotes the Cauchy principal value, so that

$$\begin{aligned} -2\pi\, i\, \delta(E_i - E_f) = &\ \operatorname*{Lim}_{\varepsilon\to 0^+} \frac{2i\varepsilon}{(E_i-E_f)^2 - \varepsilon^2} \\ = &\ \operatorname*{Lim}_{\varepsilon\to 0^+} \left(\frac{1}{E_f - E_i + i\varepsilon} + \frac{1}{E_i - E_f + i\varepsilon}\right). \end{aligned} \tag{8.28}$$

This suggests that the operator $\hat{\mathrm{V}}\hat{\mathrm{G}} + \hat{\mathrm{G}}\hat{\mathrm{V}}$ should be expressed via a product of another operator and a scalar $(E_f - E_i + i\varepsilon)^{-1} + (E_i - E_f + i\varepsilon)^{-1}$. Such a combination can be obtained by using (7.28) and (7.29)

$$\hat{\mathrm{G}}_0(z)\hat{\mathrm{T}}(z) = \hat{\mathrm{G}}(z)\hat{\mathrm{V}} \qquad \hat{\mathrm{T}}(z)\hat{\mathrm{G}}_0(z) = \hat{\mathrm{V}}\hat{\mathrm{G}}(z) \tag{8.29}$$

so that

$$\hat{\mathrm{V}}\hat{\mathrm{G}}(z) + \hat{\mathrm{G}}(z)\hat{\mathrm{V}} = \hat{\mathrm{T}}(z)\hat{\mathrm{G}}_0(z) + \hat{\mathrm{G}}_0(z)\hat{\mathrm{T}}(z) \tag{8.30}$$

where the $\hat{\mathrm{T}}$−operator is given by any of the following equivalent expressions

$$\begin{aligned} \hat{\mathrm{T}}(z) &= [\hat{1} - \hat{\mathrm{V}}\hat{\mathrm{G}}_0(z)]^{-1}\hat{\mathrm{V}} \\ &= \hat{\mathrm{V}}[\hat{1} - \hat{\mathrm{G}}_0(z)\hat{\mathrm{V}}]^{-1} \\ &= \hat{\mathrm{V}} + \hat{\mathrm{V}}\hat{\mathrm{G}}_0(z)\hat{\mathrm{T}}(z) \\ &= \hat{\mathrm{V}} + \hat{\mathrm{T}}(z)\hat{\mathrm{G}}_0(z)\hat{\mathrm{V}} \\ &= \hat{\mathrm{V}} + \hat{\mathrm{V}}\hat{\mathrm{G}}(z)\hat{\mathrm{V}} \end{aligned}$$

$$\therefore \qquad \hat{\mathrm{T}}(z) = \hat{\mathrm{V}} + \hat{\mathrm{V}}\hat{\mathrm{G}}(z)\hat{\mathrm{V}}. \tag{8.31}$$

Then, in the special case $z = \bar{E}_{if} + i\varepsilon/2$, it follows

$$\begin{aligned}\langle\Psi_{0f}|\hat{\mathrm{V}}\hat{\mathrm{G}}(z) &= \langle\Psi_{0f}|\hat{\mathrm{G}}_0(z)\hat{\mathrm{T}}(z) = \langle\Psi_{0f}|(z-E_f)^{-1}\hat{\mathrm{T}}(z)\\ &= 2\langle\Psi_{0f}|(E_i - E_f + i\varepsilon)^{-1}\hat{\mathrm{T}}(z)\left(\bar{E}_{if} + i\varepsilon/2\right)\end{aligned} \qquad (8.32)$$

$$\begin{aligned}\hat{\mathrm{G}}(z)\hat{\mathrm{V}}|\Psi_{0i}\rangle &= \hat{\mathrm{T}}(z)\hat{\mathrm{G}}_0(z)|\Psi_{0i}\rangle = \hat{\mathrm{T}}(z)(z-E_i)^{-1}|\Psi_{0i}\rangle\\ &= 2\hat{\mathrm{T}}(z)\left(\bar{E}_{if} + i\varepsilon/2\right)(E_f - E_i + i\varepsilon)^{-1}|\Psi_{0i}\rangle.\end{aligned} \qquad (8.33)$$

In this way, the second term in the S-matrix element (8.25) becomes

$$\begin{aligned}&\frac{1}{2}\lim_{\varepsilon\to 0+}\langle\Psi_{0f}|\hat{\mathrm{V}}\hat{\mathrm{G}}\left(\bar{E}_{if} + i\varepsilon/2\right) + \hat{\mathrm{G}}\left(\bar{E}_{if} + i\varepsilon/2\right)\hat{\mathrm{V}}|\Psi_{0i}\rangle\\ =&\lim_{\varepsilon\to 0+}\left(\frac{1}{E_f - E_i + i\varepsilon} + \frac{1}{E_i - E_f + i\varepsilon}\right)\langle\Psi_{0f}|\hat{\mathrm{T}}\left(\bar{E}_{if} + i\varepsilon/2\right)|\Psi_{0i}\rangle\\ =& -2\pi i\delta(E_i - E_f)\langle\Psi_{0f}|\hat{\mathrm{T}}\left(E + i\varepsilon\right)|\Psi_{0i}\rangle\end{aligned} \qquad (8.34)$$

where $\varepsilon' = \varepsilon/2 \equiv \varepsilon$. Here, due to presence of the overall multiplicative function $\delta(E_i - E_f)$, we have $\bar{E}_{if} = (E_i + E_f)/2 = E_i = E_f = E$. Substituting (8.34) into (8.25), together with the ortho-normalization $\langle\Phi_{0f}|\Phi_{0i}\rangle = \delta(E_i - E_f)\delta_{if}$, finally yields

$$\mathrm{S}_{if} = \delta(E_i - E_f)\left(\delta_{if} - 2\pi\, i\,\mathrm{T}_{if}\right) \qquad (8.35)$$

where T_{if} is the transition T-matrix element

$$\begin{aligned}\mathrm{T}_{if} =& \lim_{\varepsilon\to 0+}\langle\Psi_{0f}|\hat{\mathrm{T}}(E + i\varepsilon)|\Psi_{0i}\rangle\\ &= \langle\Psi_{0f}|\hat{\mathrm{T}}(E + i0)|\Psi_{0i}\rangle\\ &= \langle\Psi_{0f}|\hat{\mathrm{T}}^{+}(E)|\Psi_{0i}\rangle \equiv \mathrm{T}_{if}^{+} \qquad (E_f = E_i).\end{aligned} \qquad (8.36)$$

The expression (8.36) is the 'on-shell' T-matrix which describes only those transitions $i \longrightarrow f$ that conserve the total energy of the examined system, so that $E_i = E_f = E$. It is clear, however, that in a general case, the operator $\hat{\mathrm{T}}(z)$ can be defined 'off-shell' i.e. for the values of z which do not conserve the energy ($E_f \neq E_i$). Nevertheless, the physical meaning of the corresponding T-matrix element $\langle\Psi_{0f}|\hat{\mathrm{T}}(z)|\Psi_{0i}\rangle$ is only obtained by the passage $z \longrightarrow E$ which leads to the energy shell $E_i \longrightarrow E_f = E$. At first glance, one could doubt the usefulness of the matrix element $\langle\Psi_{0f}|\hat{\mathrm{T}}(z)|\Psi_{0i}\rangle$, since it is apparently indifferent to the energy conservation law. Here, it is pertinent to ask the following question: is the T-matrix really required to conserve the total energy? The answer is in the negative. This can be understood from the argument which runs as follows. The theory, via the commutator $[\hat{\mathrm{S}}, \hat{\mathrm{H}}_0] = \hat{0}$, demands only that the $\hat{\mathrm{S}}-$operator conserves the total energy of the system, and (8.35) confirms that this is indeed the case. Namely, the fact that the

$\delta(E_i - E_f)$ function multiplies the matrix $\mathrm{T}_{if}(E)$ in (8.35) guarantees that the product $\delta(E_i - E_f)\mathrm{T}_{if}$ vanishes outside the energy shell i.e. whenever $E_i \neq E_f$. However, there is no reason whatsoever that the T-matrix itself within the product $\delta(E_i - E_f)\mathrm{T}_{if}$ should also vanish off the energy shell, and this generally does not happen. In other words, the S-matrix as a whole is required to conserve the total energy of the system, but not necessarily the T-matrix. However, if the T-matrix does not need to preserve the total energy, one could inquire about the genuine meaning of the off-shell T-matrix, especially in view of the fact that only its on-shell limiting case can be physically interpreted. The following two reasons lend support to using the off-shell T-matrices: (i) an explicit solution of the Lippmann-Schwinger equation for the T-matrix, and (ii) the Faddeev equations for the exact solution of the three-body problem. In the first case, starting from (8.31), we have

$$\begin{aligned}
\langle\Psi_{0f}|\hat{\mathrm{T}}(z)|\Psi_{0i}\rangle &= \mathrm{V}_{if} + \langle\Psi_{0f}|\hat{\mathrm{V}}\hat{\mathrm{G}}_0(z)\hat{\mathrm{T}}(z)|\Psi_{0i}\rangle \\
&= \mathrm{V}_{if} + \int \mathrm{d}\boldsymbol{q}\,\langle\Psi_{0f}\hat{\mathrm{V}}|\Psi_{0q}\rangle\langle\Psi_{0q}|\hat{\mathrm{G}}_0(z)\hat{\mathrm{T}}(z)|\Psi_{0i}\rangle \\
&= \mathrm{V}_{if} + \int \mathrm{d}\boldsymbol{q}\,\mathrm{V}_{qf}\frac{1}{z - E_{\boldsymbol{q}}}\mathrm{T}_{iq}(z)
\end{aligned}$$

$$\therefore \qquad \langle\Psi_{0f}|\hat{\mathrm{T}}(z)|\Psi_{0i}\rangle = \mathrm{V}_{if} + \int \mathrm{d}\boldsymbol{q}\,\mathrm{V}_{qf}\frac{1}{z - E_{\boldsymbol{q}}}\mathrm{T}_{iq}(z) \tag{8.37}$$

$$\mathrm{X}_{qq'} = \langle\Psi_{0q'}|\hat{\mathrm{X}}|\Psi_{0q}\rangle \qquad \hat{\mathrm{X}} = \hat{\mathrm{V}}, \hat{\mathrm{T}} \tag{8.38}$$

where $E_{\boldsymbol{q}} = q^2/(2m)$. Without the explicit reference to the off-shell $\mathrm{T}_{iq}-$matrix elements, $\mathrm{T}_{qq'} = \langle\Psi_{0q'}|\hat{\mathrm{T}}|\Psi_{0q}\rangle$, it would be impossible to either numerically solve this equation or to analyze it theoretically. Namely, the lhs of (8.37) will be on-shell for $z = E_i + i0 = E_f + i0$, if the energy conservation law is imposed on the T-matrix elements themselves. Of course, in such a case, the rhs of (8.37) will also be on-shell, but only as a whole. However, simultaneously a part of the rhs of (8.37), namely T_{iq} is defined outside the energy shell. Such a hybrid situation is said to be one half-off-energy-shell, since $z = E_i + i0 \neq E_{\boldsymbol{q}} + i0$. The reason for this is the fact that the integrand $\mathrm{T}_{iq}(z)$ from the integral (8.37) varies as a function of the intermediate momentum $\boldsymbol{q} = (q, \theta_q, \phi_q)$, which belongs to the intervals $q \in [0, \infty]$, $\theta_q \in [0, \pi]$, $\phi_q \in [0, 2\pi]$. Therefore, there is no reason that, in general, the relation $E_i = E_f = E = p^2/(2m)$ should be fulfilled.

Finally, to comment on the mentioned reason (ii), it suffices to recall that the Faddeev system of two coupled integral equations for a problem of three bodies gives the unique solution expressed through two-particle off-shell T-matrices [51]–[54].

9

Spectral analysis of operators

9.1 The Abel limit with no recourse to the Cauchy limit

Thus far, we analyzed the non-stationary and stationary scattering theory on the same footing by means of the strong limits of the Cauchy and Abel types, respectively. The way in which we expounded on this concept was based upon the assumption that the Cauchy strong limits exist in the time-dependent scattering theory. Such an assumption helped us prove the existence of the strong Abel limit as one of the building blocks of the stationary scattering theory[1]. However, it would be important to see whether the stationary scattering theory within the strong limits could be built without needing recourse to its non-stationary counterpart. Moreover, it would be of utmost importance to prove the existence of the strong limit in the general case for an arbitrary representation of quantum scattering theory. These goals are especially reinforced in view of the three conditions (i)-(iii) from chapter 5 that are capable of identifying the sub-class of quantum collision systems from a broad class of quantum systems. To this end, we shall prove the existence of the wave Møller operators $\hat{\Omega}^{\pm}$ in a general case of an arbitrary representation, as required in the first of the three defining conditions of quantum mechanical one-channel scattering systems. Furthermore, the proof will be given for the strong Abel limit, but without any assumption of the existence of the corresponding strong Cauchy limit.

We introduce the auxiliary operators $\hat{\Omega}^{\pm}(\varepsilon)$ by means of the following Abel integral representation

$$\begin{aligned} \hat{\Omega}^{\pm}(\varepsilon) &= \mp\varepsilon \int_0^{\mp\infty} \mathrm{d}t\, \mathrm{e}^{\pm\varepsilon t}\hat{\Omega}(t) \\ &= \mp\varepsilon \int_0^{\mp\infty} \mathrm{d}t\, \mathrm{e}^{\pm\varepsilon t}\hat{\mathrm{U}}^{\dagger}(t)\hat{\mathrm{U}}_0(t) \qquad (\varepsilon > 0). \end{aligned} \tag{9.1}$$

Then, the wave Møller operators $\hat{\Omega}^{\pm}$ can be obtained as the strong Abel limits

$$\hat{\Omega}^{\pm} = \underset{\varepsilon\to 0^+}{\mathrm{Lim}}\, \hat{\Omega}^{\pm}(\varepsilon). \tag{9.2}$$

[1]The existence of the strong Cauchy limit can readily be proven within e.g. the coordinate representation of the Schrödinger picture [7].

To establish the existence of the operator $\hat{\Omega}^{\pm}$ without any reference to the existence of the corresponding strong Cauchy limit $\hat{\Omega}^{\pm} = \operatorname*{Lim}_{t\to\mp\infty} \hat{\Omega}(t)$, it is necessary to first examine the properties of the integral operators $\hat{\Omega}^{\pm}(\varepsilon)$ by formulating and proving several fundamental lemmas and theorems. In this context, the central role is played by the so-called spectral theorem of self-adjoint operators in the Hilbert space $\mathcal{H}$. This theorem will help us establish a number of useful transformations of the features from finite- to infinite-dimensional spaces. In the case of finite dimensions, a set $\{\psi_k\}_{k=1}^{n}$ of normalized vectors ψ_k which satisfies the eigen-problem (3.16) represents a complete ortho-normal set and, hence, a basis. The corresponding generalization of this latter fact to the case of an infinite-dimensional space represents precisely the spectral theorem. This can be demonstrated if, similarly to the previously performed procedure related to the resolvent eigen-problem (3.24), we first reformulate the results from a finite-dimensional case in such a way that the new statement is entirely equivalent to the old one, but with the advantage that further analysis could be extended without any change to an infinite-dimensional case.

9.2 The spectral theorem

Following the preceeding remarks, we are led to the precise formulation of the spectral theorem as follows:

Theorem 5: For each self-adjoint operator $\hat{\mathrm{A}}$ there exists a certain non-decreasing family $\mathcal{P}_\lambda$ of the projection operators $\hat{\mathrm{P}}_\lambda$ that depend upon a real variable $\lambda\,(-\infty < \lambda < +\infty)$ with the features

$$\operatorname*{Lim}_{\lambda\to-\infty} \hat{\mathrm{P}}_\lambda = \hat{0} \qquad \operatorname*{Lim}_{\lambda\to+\infty} P_\lambda = \hat{1} \tag{9.3}$$

$$\langle\psi|\hat{\mathrm{A}}\psi\rangle = \int_{-\infty}^{+\infty} \lambda du(\lambda) \qquad du(\lambda) = \langle\psi|\mathrm{d}\hat{\mathrm{P}}_\lambda\psi\rangle \qquad \forall\,\psi\in\mathcal{H}. \tag{9.4}$$

The integral in (9.4) is the standard Stieltjes integral. For the family $\mathcal{P}_\lambda$ of the projection operators $\hat{\mathrm{P}}_\lambda$, we say that it is non-decreasing with the augmentation of the scalar variable λ if for each vector $\psi \in \mathcal{H}$ the following inequality holds

$$\langle\psi|\hat{\mathrm{P}}_{\lambda_1}\psi\rangle \le \langle\psi|\hat{\mathrm{P}}_{\lambda_2}\psi\rangle \qquad \lambda_1 \le \lambda_2. \tag{9.5}$$

The relations (9.4) and (9.5) can be written in an abridged form without referring to a particular state ψ as

$$\hat{\mathrm{P}}_{\lambda_1} \le \hat{\mathrm{P}}_{\lambda_1} \qquad \lambda_1 \le \lambda_2 \tag{9.6}$$

$$\hat{A} = \int_{-\infty}^{+\infty} \lambda \, d\hat{P}_\lambda. \tag{9.7}$$

Of course, whenever (9.6) and (9.7) need to be interpreted, we must invariably resort to (9.4) and (9.5), respectively. Thus, e.g. (9.4) would be said to denote the vector $\psi \in \mathcal{H}$ from the domain of the operator $\hat{A}$ i.e. $\psi \in \mathcal{D}_{\hat{A}}$, if and only if the norm $||\hat{A}\psi||$ is finite

$$\begin{aligned} ||\hat{A}\psi|| &= \int_{-\infty}^{+\infty} \lambda^2 \, d\langle\psi|\hat{P}_\lambda\psi\rangle = \int_{-\infty}^{+\infty} \lambda^2 \, d\langle\psi|\hat{P}_\lambda^2\psi\rangle \\ &= \int_{-\infty}^{+\infty} \lambda^2 \, d\langle\hat{P}_\lambda\psi|\hat{P}_\lambda\psi\rangle = \int_{-\infty}^{+\infty} \lambda^2 d \, ||\hat{P}_\lambda\psi||^2 < \infty \end{aligned}$$

$$\therefore \qquad ||\hat{A}\psi|| < \infty. \tag{9.8}$$

Here, we used the following properties of the projection operator

$$\hat{P}_\lambda^\dagger = \hat{P}_\lambda \qquad \hat{P}_\lambda\hat{P}_\mu = \hat{P}_\mu\hat{P}_\lambda = \hat{P}_\lambda \qquad \lambda \le \mu. \tag{9.9}$$

According to the spectral representation (9.7), the application of the operator $\hat{A}$ to $\psi \in \mathcal{D}_{\hat{A}}$ is understood in the following sense

$$\hat{A}\psi = \int_{-\infty}^{+\infty} \lambda \, d(\hat{P}_\lambda\psi). \tag{9.10}$$

On the rhs of (9.10) we have an improper integral, which represents the twofold strong limit via the corresponding convergence in the norm within the Hilbert space $\mathcal{H}$

$$\int_{-\infty}^{+\infty} \lambda \, d(\hat{P}_\lambda\psi) = \operatorname*{Lim}_{\alpha\to-\infty} \; \operatorname*{Lim}_{\beta\to+\infty} \int_{\alpha}^{\beta} \lambda \, d(\hat{P}_\lambda\psi). \tag{9.11}$$

Of course, the integral representation (9.7) of the operator $\hat{A}$ can be interpreted as the Stieltjes integral valid for any pair of the elements $\phi, \psi \in \mathcal{H}$

$$\langle\phi|\hat{A}\psi\rangle = \int_{-\infty}^{+\infty} \lambda \, d\langle\phi|\hat{P}_\lambda\psi\rangle. \tag{9.12}$$

Regarding (9.8), we can say that the domain of the operator $\hat{A}$ is given by the following set

$$\mathcal{D}_{\hat{A}} = \{\psi \in \mathcal{H} : \int_{-\infty}^{+\infty} \lambda^2 d \, \langle\psi|\hat{P}_\lambda\psi\rangle < \infty\}. \tag{9.13}$$

The advantage in the introduction of the spectral representation (9.7) is especially manifested for operator functions e.g. $f(\hat{A})$ of a self-adjoint operator $\hat{A}$.

In the case when $f(\lambda)$ is a function of a real variable λ, the operator function $f(\hat{\mathrm{A}})$ is defined by

$$f(\hat{\mathrm{A}}) = \int_{-\infty}^{+\infty} f(\lambda)\,\mathrm{d}\hat{\mathrm{P}}_\lambda. \tag{9.14}$$

In a special case where f is a power function of the operator $\hat{\mathrm{A}}$ e.g. $f(\hat{\mathrm{A}}) = \hat{\mathrm{A}}^n$ $(n = 2, 3, 4, ...)$, the relation (9.14) is cast into the form

$$\hat{\mathrm{A}}^m = \int_{-\infty}^{+\infty} \lambda^m\,\mathrm{d}\hat{\mathrm{P}}_\lambda. \tag{9.15}$$

The domain $\mathcal{D}_{f(\hat{\mathrm{A}})}$ of the operator function $f(\hat{\mathrm{A}})$ is easily seen from (9.13) to be given by

$$\mathcal{D}_{f(\hat{\mathrm{A}})} = \{\psi \in \mathcal{H} : \int_{-\infty}^{+\infty} |f(\lambda)|^2 \mathrm{d}\,\langle\psi|\hat{\mathrm{P}}_\lambda\psi\rangle < \infty\}. \tag{9.16}$$

Moreover, it can be readily shown that the set $\mathcal{D}_{f(\hat{\mathrm{A}})}$ represents a linear manifold of all the vectors $\psi \in \mathcal{H}$ for which the convergence condition (9.16) holds true according to

$$\int_{-\infty}^{+\infty} |f(\lambda)|^2 \mathrm{d}\,\langle\psi|\hat{\mathrm{P}}_\lambda\psi\rangle < \infty. \tag{9.17}$$

In such a case, for each element $\phi \in \mathcal{H}$ we can form the scalar product $\langle\phi|f(\hat{\mathrm{A}})\psi\rangle$ with the following representation

$$\langle\phi|f(\hat{\mathrm{A}})\psi\rangle = \int_{-\infty}^{+\infty} f(\lambda)\,\langle\phi|\mathrm{d}\hat{\mathrm{P}}_\lambda\psi\rangle \tag{9.18}$$

where the function $\langle\phi|f(\hat{\mathrm{A}})\psi\rangle$ is a function of bounded variation in accordance with (9.5). Also, in the spirit of (9.7), expression (9.18) could be written in the abridged form (9.14). From here it is obvious that in a special case $f(\lambda) = \lambda$, we have $f(\hat{\mathrm{A}}) = \hat{\mathrm{A}}$, so that the relation (9.14) is reduced to (9.7). Notice that the performed analysis is general, as it deals with the abstract form of the operator function $f(\hat{\mathrm{A}})$, which does not depend upon the manner of description in a given concrete representation of the Hilbert space $\mathcal{H}$.

9.3 Unitary operators and strong topology

In this chapter, we primarily focus on the family of unitary operators $\hat{\mathrm{W}}(t)$ depending on a real parameter $t\,(-\infty < t < +\infty)$. These operators are given as exponentials of a self-adjoint operator $\hat{\mathrm{A}}$

$$\hat{\mathrm{W}}(t) = \mathrm{e}^{i\hat{\mathrm{A}}t} \qquad t \in \mathbb{R}. \tag{9.19}$$

Further, from the definition (9.14) of the spectral representation of the operator function $f(\hat{\mathrm{A}})$, we have

$$\hat{\mathrm{W}}(t) = \int_{-\infty}^{+\infty} \mathrm{e}^{i\lambda t} \mathrm{d}\hat{\mathrm{P}}_\lambda. \tag{9.20}$$

Moreover, the domain of the operator $\hat{\mathrm{W}}(t)$ is the whole Hilbert space $\mathcal{H}$

$$\mathcal{D}_{\hat{\mathrm{W}}(t)} = \mathcal{H}. \tag{9.21}$$

The operator family satisfies the 'group relation' viz

$$\hat{\mathrm{W}}(t)\hat{\mathrm{W}}(s) = \hat{\mathrm{W}}(t+s) = \hat{\mathrm{W}}(s)\hat{\mathrm{W}}(t) \tag{9.22}$$

and it is a representation of the additive group of real numbers. It is also possible to easily check from the spectral representation (9.20) itself that the operator $\hat{\mathrm{W}}(t)$ is defined in the whole Hilbert space $\mathcal{H}$ and that $\hat{\mathrm{W}}(t)$ is unitary, as expected according to (9.19). Namely, if $\hat{\mathrm{A}}$ is a self-adjoint operator, then $\exp(i\hat{\mathrm{A}}t)$ is a unitary operator for any $t \in \mathbb{R}$. One of the fundamental features of the operator $\hat{\mathrm{W}}(t)$ is the validity of the following equation in the sense of strong topology

$$\frac{\mathrm{d}\hat{\mathrm{W}}(t)}{\mathrm{d}t} = i\hat{\mathrm{A}}\,\hat{\mathrm{W}}(t). \tag{9.23}$$

This operator-valued differential equation is interpreted by saying that if $\psi \in \mathcal{D}_{\hat{\mathrm{A}}}$, then we also have that $\hat{\mathrm{W}}(t)\psi \in \mathcal{D}_{\hat{\mathrm{A}}}$ for each $t \in \mathbb{R}$ and

$$\frac{\mathrm{d}\hat{\mathrm{W}}(t)\psi}{\mathrm{d}t} = i\hat{\mathrm{A}}\,\hat{\mathrm{W}}(t)\psi. \tag{9.24}$$

This equality of the two operators must be established in terms of strong topology in which case we talk about strong differentiability. In other words, here the derivative $\mathrm{d}/\mathrm{d}t$ plays the role of the strong derivative. Therefore, the expression (9.23) will be valid provided that the following condition is fulfilled

$$\lim_{\tau \to 0} \left\| \frac{\hat{\mathrm{W}}(t+\tau) - \hat{\mathrm{W}}(t)}{\tau}\psi - i\hat{\mathrm{A}}\,\hat{\mathrm{W}}(t)\psi \right\| = 0. \tag{9.25}$$

We can prove this important relation as follows

$$\begin{aligned}
\frac{\hat{\mathrm{W}}(t+\tau) - \hat{\mathrm{W}}(t)}{\tau}\psi &= \int_{-\infty}^{+\infty} \frac{\mathrm{e}^{i(t+\tau)\lambda} - \mathrm{e}^{it\lambda}}{\tau} \mathrm{d}\,(\hat{\mathrm{P}}_\lambda\psi) \\
&= \int_{-\infty}^{+\infty} \frac{\lambda\left(\mathrm{e}^{i\lambda\tau/2} - \mathrm{e}^{-i\lambda\tau/2}\right)}{\tau\lambda} \mathrm{e}^{i\tau\lambda/2} \mathrm{d}\,(\hat{\mathrm{P}}_\lambda\psi) \\
&= i\int_{-\infty}^{+\infty} \frac{\lambda\,\sin(\lambda\tau/2)}{\lambda\tau/2} \mathrm{e}^{i(t+\tau/2)\lambda}\, \mathrm{d}(\hat{\mathrm{P}}_\lambda\psi)
\end{aligned}$$

$$\therefore \qquad \frac{\hat{\mathrm{W}}(t+\tau)-\hat{\mathrm{W}}(t)}{\tau}\psi = i\int_{-\infty}^{+\infty}\frac{\lambda\,\sin(\lambda\tau/2)}{\lambda\tau/2}\mathrm{e}^{i(t+\tau/2)\lambda}\,\mathrm{d}(\hat{\mathrm{P}}_\lambda\psi) \tag{9.26}$$

where the relation $\mathrm{e}^{i\lambda\tau/2}-\mathrm{e}^{-i\lambda\tau/2}=(i/2)\sin(\lambda\tau/2)$ is used. Taking the limit $\tau\to 0$ in (9.26) for any vector $\psi\in\mathcal{D}_{\hat{\mathrm{A}}}$, we obtain

$$\frac{\mathrm{d}\hat{\mathrm{W}}(t)}{\mathrm{d}t} = i\,\lim_{\tau\to 0}\int_{-\infty}^{+\infty}\frac{\lambda\,\sin(\lambda\tau/2)}{\lambda\tau/2}\mathrm{e}^{i(t+\tau/2)\lambda}\,\mathrm{d}\hat{\mathrm{P}}_\lambda. \tag{9.27}$$

This expression is convenient, since $x^{-1}\sin(x)$ tends to zero when $x\to 0$. However, the exchange of the order of the limit and the integral can be done only in the case of uniform convergence. The equality in (9.26) is also understood in the sense of strong topology i.e. convergence in the norm. Thus, treating the improper integral (9.26) according to (9.10) we shall have

$$\left\|\int_{|\lambda|\geq N}\frac{\lambda\sin(\lambda\tau/2)}{\lambda\tau/2}\mathrm{e}^{i(t+\tau/2)\lambda}\mathrm{d}\,(\hat{\mathrm{P}}_\lambda\psi)\right\|^2$$

$$\leq\int_{|\lambda|\geq N}\left|\frac{\lambda\sin(\lambda\tau/2)}{\lambda\tau/2}\mathrm{e}^{i(t+\tau/2)\lambda}\right|^2\mathrm{d}||\hat{\mathrm{P}}_\lambda\psi||^2\leq\int_{|\lambda|\geq N}|\lambda|^2\mathrm{d}\,||\hat{\mathrm{P}}_\lambda\psi||^2\longrightarrow 0$$

$$\therefore \qquad \left\|\int_{|\lambda|\geq N}\frac{\lambda\sin(\lambda\tau/2)}{\lambda\tau/2}\mathrm{e}^{i(t+\tau/2)\lambda}\right\|^2\leq\int_{|\lambda|\geq N}|\lambda|^2\mathrm{d}\,||\hat{\mathrm{P}}_\lambda\psi||^2\longrightarrow 0 \tag{9.28}$$

where $\psi\in\mathcal{D}_{\hat{\mathrm{A}}}$. Here, the region of large values of λ does not give any contribution to the integral. In (9.28) we use the majorization $|x^{-1}\sin(x)|\leq 1$. Further, the result (9.28), together with (9.26), implies that we have for any wave function $\psi\in\mathcal{D}_{\hat{\mathrm{A}}}$

$$\left\|\frac{\hat{\mathrm{W}}(t+\tau)-\hat{\mathrm{W}}(t)}{\tau}\psi\right\|^2\leq\int_{-\infty}^{+\infty}\lambda^2\mathrm{d}\,||\hat{\mathrm{P}}_\lambda\psi||^2<\infty \tag{9.29}$$

which means that $\hat{\mathrm{W}}(t)\psi$ is an element of the sub-space $\mathcal{D}_{\hat{\mathrm{A}}}$ i.e. $\hat{\mathrm{W}}(t)\psi\in\mathcal{D}_{\hat{\mathrm{A}}}$. In deriving (9.28), we exploit the fact that convergence to zero is uniform relative to τ when $N\to\infty$. In (9.29) we used also the relation (9.8) by which we select the vectors that belong to the set $\mathcal{D}_{\hat{\mathrm{A}}}$ out of all other vectors $\psi\in\mathcal{H}$. In this way, it follows that the integral $\int_{-\infty}^{+\infty}\tau^{-1}\left(\mathrm{e}^{i\tau\lambda}-1\right)\mathrm{e}^{it\lambda}\mathrm{d}(\hat{\mathrm{P}}_\lambda\psi)$ converges uniformly in τ. Therefore, the order of the limit and integration in (9.9) can be exchanged, so that

$$\begin{aligned}\frac{\mathrm{d}}{\mathrm{d}t}\hat{\mathrm{W}}(t)\psi &= \lim_{\tau\to 0}\frac{\hat{\mathrm{W}}(t+\tau)-\hat{\mathrm{W}}(t)}{\tau}\psi\\ &= i\,\lim_{\tau\to 0}\int_{-\infty}^{+\infty}\frac{\lambda\,\sin(\lambda\tau/2)}{\lambda\tau/2}\mathrm{e}^{i(t+\tau/2)\lambda}\,\mathrm{d}\,(\hat{\mathrm{P}}_\lambda\psi)\end{aligned}$$

$$= i \int_{-\infty}^{+\infty} \lambda \left\{ \lim_{\tau \to 0} \frac{\sin(\lambda\tau/2)}{\lambda\tau/2} e^{i(t+\tau/2)\lambda} \right\} d(\hat{P}_\lambda \psi)$$

$$= i \int_{-\infty}^{+\infty} \lambda e^{it\lambda} d(\hat{P}_\lambda \psi) = i \hat{A} \hat{W}(t)$$

$$\therefore \qquad \frac{d}{dt}\hat{W}(t)\psi = i \hat{A} \hat{W}(t) \qquad \text{(QED)}. \tag{9.30}$$

In (9.30) we used the expression

$$\hat{A} \hat{W}(t)\psi = \int_{-\infty}^{+\infty} \lambda e^{it\lambda} d(\hat{P}_\lambda \psi) \tag{9.31}$$

which follows from the definition (9.14) of the operator function $f(\hat{A}) = \hat{A}e^{i\hat{A}t}$.

Self-adjoint operators $\hat{A}$ and $\hat{B}$ are called unitarily equivalent if there exists a unitary operator $\hat{C}$ such that the domain of $\hat{B}$ and $\hat{C}\hat{A}\hat{C}^{-1}$ are equal and, moreover, for every vector ψ from that domain, it follows

$$\hat{B}\psi = \hat{C}\hat{A}\hat{C}^{-1}\psi. \tag{9.32}$$

If $\mathcal{P}_\lambda$ and $\mathcal{P}'_\lambda$ are the spectral families of the operators $\hat{A}$ and $\hat{B}$, respectively, then it holds

$$\mathcal{P}'_\lambda = \hat{C}\mathcal{P}_\lambda\hat{C}^{-1} \qquad \lambda \in \mathbb{R}. \tag{9.33}$$

From here it follows that the operators $\hat{W}_A(t) = e^{i\hat{A}t}$ and $\hat{W}_B(t) = e^{i\hat{B}t}$ are also unitarily equivalent i.e.

$$\hat{W}_B(t) = \hat{C}\hat{W}_A(t)\hat{C}^{-1} \qquad t \in \mathbb{R}. \tag{9.34}$$

This is symbolically written as

$$\hat{W}_B(t) \propto \hat{W}_A(t). \tag{9.35}$$

Lemma 5: The vectors $\hat{U}(t)\psi \equiv e^{-i\hat{H}t}\psi$ and $\hat{U}_0(t)\psi \equiv e^{-i\hat{H}_0 t}\psi$ are strongly continuous functions for every real variable $t \in \mathbb{R}$ and for all elements $\psi \in \mathcal{H}$ from the Hilbert space $\mathcal{H}$.

Proof: Similarly to the strong limit, strong continuity is also a mathematical notion connected to the norm of abstract elements of vector spaces. Thus, strong continuity of vectors $\hat{U}(t)\psi$ at the point $t = t_0$ means that for $\epsilon > 0$ there exists certain $\delta > 0$ such that $||\hat{U}(t)\psi - \hat{U}(t_0)\psi|| < \epsilon$ whenever $|t - t_0| < \delta$. Let us now write the norm $||\hat{U}(t)\psi - \hat{U}(t_0)\psi||$ in the form which is more convenient for our further analysis as

$$\begin{aligned}
||\hat{U}(t)\psi - \hat{U}(t_0)\psi)|| &= ||\hat{U}(t_0)\{\hat{U}^\dagger(t_0)\hat{U}(t)\psi - \psi\}|| \\
&= ||\hat{U}(t - t_0)\psi - \psi|| \\
&= ||\{\hat{1} - \hat{U}(t - t_0)\}\psi||
\end{aligned}$$

$$\therefore \qquad ||\hat{\mathrm{U}}(t)\psi - \hat{\mathrm{U}}(t_0)\psi)|| = ||\{\hat{1} - \hat{\mathrm{U}}(t - t_0)\}\psi|| \tag{9.36}$$

where we used isometry $||\hat{\mathrm{U}}(t_0)|| = 1$ and the group property of the evolution operator $\hat{\mathrm{U}}^\dagger(t_0)\hat{\mathrm{U}}(t) = \hat{\mathrm{U}}(t - t_0)$. Now the change of variable $t' = t - t_0$ clearly shows from (9.36) that continuity at the point $t = t_0$ is equivalent to continuity at $t = 0$ so that

$$||\{\hat{1} - \hat{\mathrm{U}}(t)\}\psi|| < \epsilon \qquad |t| < \delta. \tag{9.37}$$

This expression can also be written in a form which employs the scalar product as follows

$$\begin{aligned} ||\{\hat{1} - \hat{\mathrm{U}}(t)\}\psi|| &= \{\langle[\hat{1} - \hat{\mathrm{U}}(t)]\psi|[\hat{1} - \hat{\mathrm{U}}(t)]\psi\rangle\}^{1/2} \\ &= \{\langle\psi|\,\hat{1} - \hat{\mathrm{U}}(t) - \hat{\mathrm{U}}^\dagger(t) + \hat{\mathrm{U}}^\dagger(t)\hat{\mathrm{U}}(t)|\psi\rangle\}^{1/2} \\ &= 2\{\langle\psi|\,\hat{1} - [\hat{\mathrm{U}}(t) + \hat{\mathrm{U}}^\dagger(t)]/2|\psi\rangle\}^{1/2} \end{aligned}$$

$$\therefore \qquad ||\{\hat{1} - \hat{\mathrm{U}}(t)\}\psi|| = 2\{\langle\psi|\,\hat{1} - [\hat{\mathrm{U}}(t) + \hat{\mathrm{U}}^\dagger(t)]/2|\psi\rangle\}^{1/2} \tag{9.38}$$

where we used unitarity $\hat{\mathrm{U}}^\dagger(t)\hat{\mathrm{U}}(t) = \hat{1}$. Given a self-adjoint Hamiltonian operator $\hat{\mathrm{H}}$ and the evolution operator $\hat{\mathrm{U}}(t) = \mathrm{e}^{-i\hat{\mathrm{H}}t}$, it follows that there exists a sequence of non-decreasing projection operators $\hat{\mathrm{P}}_\lambda$ with the features (9.3), (9.6) and (9.9) where $\lambda \in \mathbb{R}$ is a real parameter $(-\infty < \lambda < +\infty)$. In accordance with (9.14), the spectral equation can be stated via

$$\langle\psi|\hat{\mathrm{U}}(t)\psi\rangle = \int_{-\infty}^{+\infty} \mathrm{e}^{-i\lambda t}\mathrm{d}\langle\psi|\hat{\mathrm{P}}_\lambda\psi\rangle. \tag{9.39}$$

Inserting (9.39) into (9.38) we obtain

$$||\{\hat{1} - \hat{\mathrm{U}}(t)\}\psi|| = \left\{2\int_{-\infty}^{+\infty} [\hat{1} - (\mathrm{e}^{-i\lambda t} + \mathrm{e}^{i\lambda t})/2]\mathrm{d}\langle\psi|\hat{\mathrm{P}}_\lambda\psi\rangle\right\}^{1/2}$$

$$= \left\{2\int_{-\infty}^{+\infty} [1 - \cos(\lambda t)]\mathrm{d}\langle\psi|\hat{\mathrm{P}}_\lambda\psi\rangle\right\}^{1/2} = 2\left\{\int_{-\infty}^{+\infty} \sin^2\left(\frac{\lambda t}{2}\right)\mathrm{d}\langle\psi|\hat{\mathrm{P}}_\lambda\psi\rangle\right\}^{1/2}$$

$$\therefore \qquad ||\{\hat{1} - \hat{\mathrm{U}}(t)\}\psi|| = 2\left\{\int_{-\infty}^{+\infty} \sin^2\left(\frac{\lambda t}{2}\right)\mathrm{d}\langle\psi|\hat{\mathrm{P}}_\lambda\psi\rangle\right\}^{1/2}. \tag{9.40}$$

Here, we can now conveniently perform majorization, since according to (9.5), the quantity $\langle\psi|\hat{\mathrm{P}}_\lambda\psi\rangle$ is a monotonically increasing function of λ. Because of this latter fact and the inequality $\sin^2 x \le x^2$, we can carry out majorization in the whole integrand in (9.40) via $(\lambda t/2)^2\mathrm{d}\langle\psi|\hat{\mathrm{P}}_\lambda\psi\rangle$, so that

$$||\{\hat{1} - \hat{\mathrm{U}}(t)\}\psi|| \le t\left\{\int_{-\infty}^{+\infty} \lambda^2\mathrm{d}\langle\psi|\hat{\mathrm{P}}_\lambda\psi\rangle\right\}^{1/2} \equiv t\,C. \tag{9.41}$$

Here, the quantity C is a positive finite constant, as per (9.8)

$$C = ||\hat{\mathrm{H}}\psi|| = \left\{\int_{-\infty}^{+\infty} \lambda^2 \mathrm{d}\langle\psi|\hat{\mathrm{P}}_\lambda\psi\rangle\right\}^{1/2} < \infty \tag{9.42}$$

where $\psi \in \mathcal{D}_{\hat{\mathrm{H}}}$ and $\hat{\mathrm{H}}\psi = \lambda\psi$. Let further $\phi \in \mathcal{H}$ be an arbitrary vector from the Hilbert space $\mathcal{H}$. Since the sub-space $\mathcal{D}_{\hat{\mathrm{H}}}$ is everywhere dense in $\mathcal{H}$, we shall have the inequality

$$||\phi - \psi|| < \frac{\epsilon}{4}. \tag{9.43}$$

For a fixed element $\psi \in \mathcal{D}_{\hat{\mathrm{H}}}$, we shall choose the parameter t such that $t < \epsilon/(2C)$ and this implies

$$\begin{aligned} ||\{\hat{1} - \hat{\mathrm{U}}(t)\}\phi|| &= ||\{\hat{1} - \hat{\mathrm{U}}(t)\}(\phi - \psi) + \{\hat{1} - \hat{\mathrm{U}}(t)\}\psi|| \\ &\le ||\{\hat{1} - \hat{\mathrm{U}}(t)\}(\phi - \psi)|| + ||\{\hat{1} - \hat{\mathrm{U}}(t)\}\psi|| \\ &\le ||\hat{1} - \hat{\mathrm{U}}(t)|| \cdot ||\phi - \psi|| + tC \\ &< \{\hat{1} + ||\hat{\mathrm{U}}(t)||\}||\phi - \psi|| + \frac{\epsilon}{2C}C \\ &\le 2\frac{\epsilon}{4} + \frac{\epsilon}{2} = \epsilon \end{aligned}$$

$$\therefore \qquad ||\{\hat{1} - \hat{\mathrm{U}}(t)\}\phi|| < \epsilon \qquad \text{(QED)} \tag{9.44}$$

where isometry $||\hat{\mathrm{U}}(t)|| = 1$ is utilized. Likewise, the proof can also be repeated for the free evolution operator $\hat{\mathrm{U}}_0(t)$. With this, the proof of the Theorem 5 is completed.

Lemma 6: Let an auxiliary operator $\hat{\Omega}(t)$ be defined by

$$\hat{\Omega}(t) \equiv \hat{\mathrm{U}}^\dagger(t)\hat{\mathrm{U}}_0(t). \tag{9.45}$$

Then, the state vector $\hat{\Omega}(t)\psi$ is strongly continuous function of time for $\forall\, t \in \mathbb{R}$ and $\forall\, \psi \in \mathcal{H}$.

Proof: Let us write the difference of the operators $\hat{\Omega}(t) - \hat{\Omega}(t_0)$ in a way which permits a direct application of lemma 5. This is achieved by adding and subtracting the operator $\hat{\mathrm{U}}^\dagger(t)\hat{\mathrm{U}}_0(t_0)$ from the difference $\hat{\Omega}(t) - \hat{\Omega}(t_0)$ viz

$$\begin{aligned} \hat{\Omega}(t) - \hat{\Omega}(t_0) &= \hat{\mathrm{U}}^\dagger(t)\hat{\mathrm{U}}_0(t) - \hat{\mathrm{U}}^\dagger(t_0)\hat{\mathrm{U}}_0(t_0) \\ &= \hat{\mathrm{U}}^\dagger(t)[\hat{\mathrm{U}}_0(t) - \hat{\mathrm{U}}_0(t_0)] + [\hat{\mathrm{U}}^\dagger(t) - \hat{\mathrm{U}}^\dagger(t_0)]\hat{\mathrm{U}}_0(t_0). \end{aligned} \tag{9.46}$$

Applying now the Schwartz inequality and lemma 5, we find

$$\begin{aligned} ||\{\hat{\Omega}(t) - \hat{\Omega}(t_0)\}\psi|| &= ||\hat{\mathrm{U}}^\dagger(t)[\hat{\mathrm{U}}_0(t) - \hat{\mathrm{U}}_0(t_0)]\psi + [\hat{\mathrm{U}}^\dagger(t) - \hat{\mathrm{U}}^\dagger(t_0)]\hat{\mathrm{U}}_0(t_0)\psi|| \\ &\le ||\hat{\mathrm{U}}^\dagger(t)\{\hat{\mathrm{U}}_0(t) - \hat{\mathrm{U}}_0(t_0)\}\psi|| \\ &+ ||\{\hat{\mathrm{U}}^\dagger(t) - \hat{\mathrm{U}}^\dagger(t_0)\}\hat{\mathrm{U}}_0(t_0)\psi|| \\ &= ||\{\hat{\mathrm{U}}_0(t) - \hat{\mathrm{U}}_0(t_0)\}\psi|| + ||\{\hat{\mathrm{U}}^\dagger(t) - \hat{\mathrm{U}}^\dagger(t_0)\}\chi|| \\ &\le \frac{\epsilon}{2} + \frac{\epsilon}{2} = \epsilon \end{aligned}$$

$$\therefore \qquad ||\{\hat{\Omega}(t) - \hat{\Omega}(t_0)\}\psi|| \leq \epsilon \qquad \text{(QED)} \tag{9.47}$$

where $\chi \equiv \hat{\mathrm{U}}_0(t_0)\psi$. Thus, strong continuity of the vector $\Omega(t)\psi$ at the point $t = t_0$ follows from lemma 5. Notice that the relation $||\hat{\mathrm{U}}(t)\chi - \hat{\mathrm{U}}(t_0)\chi|| < \epsilon$ holds for every $t, t_0 \in \mathbb{R}$. If in this latter expression, we replace t and t_0 by $-t$ and $-t_0$, respectively, we have that the following inequality is also valid $||\hat{\mathrm{U}}(-t)\chi - \hat{\mathrm{U}}(t_0)\chi|| = ||\hat{\mathrm{U}}^\dagger(t)\chi - \hat{\mathrm{U}}^\dagger(t_0)\chi|| < \epsilon$, and this is precisely the content of the statement (9.47).

Lemma 7: The given scalar function

$$f(t) \equiv \langle\phi|\hat{\Omega}(t)\psi\rangle \tag{9.48}$$

is weakly continuous for every $t \in \mathbb{R}$ and $\forall\, \phi, \psi \in \mathcal{H}$ and, moreover, the following inequality holds true

$$|f(t)| \leq ||\phi|| \cdot ||\psi||. \tag{9.49}$$

Proof: Weak continuity is always guaranteed by the existence of the corresponding strong continuity, since the strong limit implies the weak limit. This proves the first statement in lemma 7. On the other hand, boundedness of the function $f(t)$ for $\forall\, t \in \mathbb{R}$ and $\forall\, \phi, \psi \in \mathcal{H}$ follows from the Schwartz inequality (4.35) and unitarity of the operator $\Omega(t)$, so that

$$\begin{aligned} |f(t)| &= |\langle\phi|\hat{\Omega}(t)\psi\rangle| \leq ||\phi|| \cdot ||\hat{\Omega}(t)\psi|| \\ &= ||\phi|| \cdot ||\hat{\Omega}(t)|| \cdot ||\psi|| \leq ||\phi|| \cdot ||\psi|| \end{aligned}$$

$$\therefore \qquad |f(t)| \leq ||\phi|| \cdot ||\psi|| \qquad \text{(QED)}. \tag{9.50}$$

9.4 The Abel limit for the Møller wave operators

Lemma 8: We are given the function $F_\varepsilon(\phi, \psi)$ via the appropriate Abel integral representation

$$F_\varepsilon(\phi, \psi) = \varepsilon \int_0^\infty \mathrm{d}t\, \mathrm{e}^{-\varepsilon t} f(t) \qquad \varepsilon > 0 \tag{9.51}$$

where $f(t)$ is defined by (9.48). Then, $F_\varepsilon(\phi, \psi)$ represents a bounded function

$$|F_\varepsilon(\phi, \psi)| \leq ||\phi|| \cdot ||\psi|| \tag{9.52}$$

such that the integral on the rhs of (9.51) is absolutely convergent for $\forall\, \phi, \psi \in \mathcal{H}$ and $\varepsilon > 0$.

<u>*Proof*</u>: Using lemma 5, it follows that the integrand $\mathrm{e}^{-\varepsilon t}f(t)$ is bounded for $\forall t \in \mathbb{R}$, since both functions $\mathrm{e}^{-\varepsilon|t|}$ and $f(t)$ are bounded

$$\left|\mathrm{e}^{-\varepsilon t}f(t)\right| = \mathrm{e}^{-\varepsilon|t|}\,|f(t)| \leq |f(t)|. \tag{9.53}$$

Further, strong continuity of the function $\mathrm{e}^{-\varepsilon t}f(t)$ is proven in the same manner as in lemma 5 through the replacement of the operators $\hat{\mathrm{U}}(t)$ and $\hat{\mathrm{U}}_0(t)$ by $\mathrm{e}^{-\varepsilon|t|/2}\hat{\mathrm{U}}(t)$ and $\mathrm{e}^{-\varepsilon|t|/2}\hat{\mathrm{U}}_0(t)$, respectively. The integrand $\mathrm{e}^{-\varepsilon|t|}f(t)$ from (9.51) possesses the feature of strong continuity and boundedness. This guarantees integrability of the rhs of (9.51). Absolute integrability of the rhs of (9.51) means that if the integrand $\mathrm{e}^{-\varepsilon t}f(t)$ is replaced by its absolute value, then the function $F_\varepsilon(\phi,\psi)$ will still be integrable. This is due to boundedness of the integrand (9.53), so that

$$\begin{aligned}|F_\varepsilon(\phi,\psi)| &= \left|\varepsilon\int_0^\infty \mathrm{d}t\,\mathrm{e}^{-\varepsilon t}f(t)\right| \leq \varepsilon\int_0^\infty \mathrm{d}t\,\mathrm{e}^{-\varepsilon t}|f(t)| \\ &\leq ||\phi||\cdot||\psi||\,\varepsilon\int_0^\infty \mathrm{d}t\,\mathrm{e}^{-\varepsilon t} = ||\phi||\cdot||\psi||\end{aligned}$$

$$|F_\varepsilon(\phi,\psi)| \leq ||\phi||\cdot||\psi|| \qquad \text{(QED)}. \tag{9.54}$$

<u>*Lemma*</u> 9: The function $F_\varepsilon(\phi,\psi)$ is a bilinear functional whose upper bound is equal to unity.

<u>*Proof*</u>: Using the Axioms 1 and 2 from the definition (4.7) of the scalar product as well as its anti-linearity in the first and linearity in the second term, we shall have

$$\begin{aligned}F_\varepsilon(\alpha_1\phi_1+\alpha_2\phi_2,\psi) &= \varepsilon\int_0^\infty \mathrm{d}t\,\mathrm{e}^{-\varepsilon t}\langle\alpha_1\phi_1+\alpha_2\phi_2|\hat{\Omega}(t)\psi\rangle \\ &= \varepsilon\int_0^\infty \mathrm{d}t\,\mathrm{e}^{-\varepsilon t}\{\alpha_1^*\langle\phi_1|\hat{\Omega}(t)\psi\rangle+\alpha_2^*\langle\phi_2|\hat{\Omega}(t)\psi\rangle\} \\ &= \alpha_1^*\varepsilon\int_0^\infty \mathrm{d}t\,\mathrm{e}^{-\varepsilon t}\langle\phi_1|\hat{\Omega}(t)\psi\rangle+\alpha_2^*\varepsilon\int_0^\infty \mathrm{d}t\,\mathrm{e}^{-\varepsilon t}\langle\phi_2|\hat{\Omega}(t)\psi\rangle \\ &= \alpha_1^*F_\varepsilon(\phi_1,\psi)+\alpha_2^*F_\varepsilon(\phi_2,\psi)\end{aligned}$$

$$\therefore \qquad F_\varepsilon(\alpha_1\phi_1+\alpha_2\phi_2,\psi) = \alpha_1^*F_\varepsilon(\phi_1,\psi)+\alpha_2^*F_\varepsilon(\phi_2,\psi) \tag{9.55}$$

$$\begin{aligned}F_\varepsilon(\phi,\beta_1\psi_1+\beta_2\psi_2) &= \varepsilon\int_0^\infty \mathrm{d}t\,\mathrm{e}^{-\varepsilon t}\langle\phi,\hat{\Omega}(t)\{\beta_1\psi_1+\beta_2\psi_2\}\rangle \\ &= \varepsilon\int_0^\infty \mathrm{d}t\,\mathrm{e}^{-\varepsilon t}\{\beta_1\langle\phi|\hat{\Omega}(t)\psi_1\rangle+\beta_2\langle\phi|\hat{\Omega}(t)\psi_2\rangle\} \\ &= \beta_1\varepsilon\int_0^\infty \mathrm{d}t\,\mathrm{e}^{-\varepsilon t}\langle\phi_1|\hat{\Omega}(t)\psi\rangle+\beta_2\varepsilon\int_0^\infty \mathrm{d}t\,\mathrm{e}^{-\varepsilon t}\langle\phi_2|\hat{\Omega}(t)\psi\rangle \\ &= \beta_1F_\varepsilon(\phi,\psi_1)+\beta_2F_\varepsilon(\phi,\psi_2)\end{aligned}$$

$$\therefore \qquad F_\varepsilon(\phi, \beta_1\psi_1 + \beta_2\psi_2) = \beta_1 F_\varepsilon(\phi, \psi_1) + \beta_2 F_\varepsilon(\phi, \psi_2). \tag{9.56}$$

Here, we also utilized linearity of the operator $\hat{\Omega}(t)$. Now, from lemma 8, it follows that $|F_\varepsilon(\phi, \psi)| \leq ||\psi|| \cdot ||\psi||$. By comparing this latter relation with $g(\phi, \psi) \leq C||\phi|| \cdot ||\psi|| \, (C > 0)$ from (4.32), it follows that the upper bound of the function $F_\varepsilon(\phi, \psi)$ is indeed equal to unity i.e. $C = 1$. In this way, according to (4.30)–(4.33), the function $F_\varepsilon(\phi, \psi)$ can immediately be identified as a bilinear functional for $\forall \phi, \psi \in \mathcal{H}$ and $\varepsilon > 0 \, , t \in \mathbb{R}$. This completes the proof of lemma 9.

Hence, we can conclude that there exists a linear operator $\hat{\Omega}^+(\varepsilon)$ with the following features

$$F_\varepsilon(\phi, \psi) = \langle \phi | \hat{\Omega}^+(\varepsilon) \psi \rangle \tag{9.57}$$

$$||\hat{\Omega}^+(\varepsilon)|| \leq 1 \tag{9.58}$$

where $\hat{\Omega}^\pm(\varepsilon)$ are given in (9.1). Whenever there is no possibility for confusion, we shall leave out the superscripts $\pm$ from the operators in (9.1), so that $\hat{\Omega}(\varepsilon)$ is a collective label for the operators $\hat{\Omega}^+(\varepsilon)$ and $\hat{\Omega}^-(\varepsilon)$. This means that all the expressions in which the operator $\hat{\Omega}(\varepsilon)$ appears will automatically be valid for both $\hat{\Omega}^+(\varepsilon)$ and $\hat{\Omega}^-(\varepsilon)$.

It is clear from chapter 6 that for the purpose of quantum scattering theory, it is most important to examine the limit $\varepsilon \to 0$ of the operator $\hat{\Omega}(\varepsilon)$. More precisely, there are two basic kinds of limiting processes known as the strong and weak Abel limits. Namely, for the operator $\hat{\Omega}(\varepsilon)$, we say that there is the strong Abel limit $\hat{\Omega}$

$$\hat{\Omega} = \operatorname*{Lim}_{\varepsilon \to 0^+} \hat{\Omega}(\varepsilon) \tag{9.59}$$

if and only if

$$||\{\hat{\Omega}(\varepsilon) - \hat{\Omega}\}\psi|| \underset{\varepsilon \to 0^+}{\longrightarrow} 0. \tag{9.60}$$

Similarly, the operator $\hat{\Omega}(\varepsilon)$ converges weakly in the Abel sense i.e.

$$\hat{\Omega}(\varepsilon) \underset{\varepsilon \to 0^+}{\rightharpoonup} \hat{\Omega} \tag{9.61}$$

provided that

$$|\langle \phi | \hat{\Omega}(\varepsilon) \psi \rangle - \langle \phi | \hat{\Omega} \psi \rangle| \underset{\varepsilon \to 0^+}{\longrightarrow} 0. \tag{9.62}$$

A connection between the weak and strong Abel limits can be established by applying the Schwartz inequality (4.35) to the relation (9.62) yielding

$$\begin{aligned} \lim_{\varepsilon \to 0^+} \left| \langle \phi | \hat{\Omega}(\varepsilon) \psi \rangle - \langle \phi | \hat{\Omega} \psi \rangle \right| &= \lim_{\varepsilon \to 0^+} ||\langle \phi | \{\hat{\Omega}(\varepsilon) - \hat{\Omega}\} \psi \rangle|| \\ &\leq ||\phi|| \cdot \lim_{\varepsilon \to 0^+} ||\{\hat{\Omega}(\varepsilon) - \hat{\Omega}\}\psi|| = 0 \end{aligned}$$

$$\therefore \qquad \lim_{\varepsilon\to 0^+}\left|\langle\phi|\hat{\Omega}(\varepsilon)\psi\rangle - \langle\phi|\hat{\Omega}\psi\rangle\right| = 0. \tag{9.63}$$

Here, the zero limiting value is obtained by assuming that the strong Abel limit (9.59) exists. This proves that the Abel strong limit implies the weak limit, but the opposite does not hold. Such a conclusion is expected from the corresponding fact that the Cauchy strong limit leads to the weak limit.

Lemma 10: The set $\mathcal{L}$, introduced as a collection of limiting values $\hat{\Omega}\psi$ of the sequence $\{\hat{\Omega}(\varepsilon)\}$

$$\mathcal{L} = \{\psi \in \mathcal{H} : \exists \ \operatorname*{Lim}_{\varepsilon\to 0^+} \hat{\Omega}(\varepsilon)\psi\} \tag{9.64}$$

represents a closed linear manifold such that $\operatorname*{Lim}_{\varepsilon\to 0^+} \hat{\Omega}(\varepsilon)$ is a bounded linear operator on $\mathcal{B}$ with the bound[2] $||\hat{\Omega}|| \leq 1$.

Proof: Linearity of the operator $\hat{\Omega}(\varepsilon)$ follows from linearity of the integral itself. Namely, for any two complex numbers $\alpha, \beta \in \mathbb{C}$ and $\phi, \psi \in \mathcal{D}_{\hat{\Omega}(\varepsilon)}$, we have these properties

$$\begin{aligned}
\hat{\Omega}^{\pm}(\varepsilon)(\alpha\phi + \beta\psi) &= \mp\varepsilon \int_0^{\mp\infty} \mathrm{d}t\, \mathrm{e}^{\mp\varepsilon t}\hat{\mathrm{U}}^{\dagger}(t)\hat{\mathrm{U}}_0(t)(\alpha\phi + \beta\psi) \\
&= \alpha\{\mp\varepsilon \int_0^{\mp\infty} \mathrm{d}t\, \mathrm{e}^{\mp\varepsilon t} U^{\dagger}(t)U_0(t)\phi\} \\
&+ \beta\{\mp\varepsilon \int_0^{\mp\infty} \mathrm{d}t\, \mathrm{e}^{\mp\varepsilon t}\hat{\mathrm{U}}^{\dagger}(t)\hat{\mathrm{U}}_0(t)\psi\} \\
&= \alpha\hat{\Omega}^{\pm}(\varepsilon)\phi + \beta\hat{\Omega}^{\pm}(\varepsilon)\psi
\end{aligned}$$

$$\therefore \qquad \hat{\Omega}^{\pm}(\varepsilon)(\alpha\phi + \beta\psi) = \alpha\hat{\Omega}^{\pm}(\varepsilon)\phi + \beta\hat{\Omega}^{\pm}(\varepsilon)\psi \qquad \text{(QED)}. \tag{9.65}$$

The set $\mathcal{L}$ is indeed a linear manifold as seen from the following two facts: 1) $\hat{\Omega}(\varepsilon)$ is a linear operator and 2) the limit of the sum is equal to the sum of the limits. This second property of the operator $\hat{\Omega}(\varepsilon)$ can be proven as follows

$$\begin{aligned}
\operatorname*{Lim}_{\varepsilon\to 0^+} [\hat{\Omega}(\varepsilon)\phi + \hat{\Omega}(\varepsilon)\psi] &= \operatorname*{Lim}_{\varepsilon\to 0^+} \hat{\Omega}(\varepsilon)(\phi + \psi) \\
&\equiv \operatorname*{Lim}_{\varepsilon\to 0^+} \hat{\Omega}(\varepsilon)\xi \\
&= \hat{\Omega}\xi = \hat{\Omega}(\phi + \psi) = \hat{\Omega}\phi + \hat{\Omega}\psi \\
&= \operatorname*{Lim}_{\varepsilon\to 0^+} \hat{\Omega}(\varepsilon)\phi + \operatorname*{Lim}_{\varepsilon\to 0^+} \hat{\Omega}(\varepsilon)\psi
\end{aligned}$$

$$\therefore \qquad \operatorname*{Lim}_{\varepsilon\to 0^+} [\hat{\Omega}(\varepsilon)\phi + \hat{\Omega}(\varepsilon)\psi] = \operatorname*{Lim}_{\varepsilon\to 0^+} \hat{\Omega}(\varepsilon)\phi + \operatorname*{Lim}_{\varepsilon\to 0^+} \hat{\Omega}(\varepsilon)\psi \quad \text{(QED)} \tag{9.66}$$

[2]In fact, the sub-space $\mathcal{B}$ is already present in our analysis through the condition (iii) of the definition of a quantum collision system as a set of closed linear manifolds containing proper, normalizable state vectors of the eigen-problem (3.1) of the Hamilton operator $\hat{\mathrm{H}}$.

where $\psi, \phi \in \mathcal{L}$. Finally, let us verify that the sub-space $\mathcal{L}$ is closed. To this end, we consider the sequence $\{\psi_k\}_{k=1}^{\infty} \subset \mathcal{L}$ with the feature $\operatorname{Lim}_{n\to\infty} \psi_n = \psi$. Closeness of the sub-space $\mathcal{B}$ presumes that the limiting value ψ also belongs to $\mathcal{B}$. As we know, every strong sequence $\{\psi_k\}_{k=1}^{\infty}$ is interpreted via the relation $||\psi_n - \psi|| \underset{n\to\infty}{\longrightarrow} 0$. With this at hand, we find

$$\begin{aligned}||\{\hat{\Omega}(\varepsilon_1) - \hat{\Omega}(\varepsilon_2)\}\psi|| &= ||\{\hat{\Omega}(\varepsilon_1) - \hat{\Omega}(\varepsilon_2)\}\psi_n + \hat{\Omega}(\varepsilon_1)(\psi - \psi_n) \\ &+ \hat{\Omega}(\varepsilon_2)(\psi_n - \psi)|| \le ||\{\hat{\Omega}(\varepsilon_1) - \hat{\Omega}(\varepsilon_2)\}\psi_n|| \\ &+ ||\hat{\Omega}(\varepsilon_1)(\psi - \psi_n)|| + ||\hat{\Omega}(\varepsilon_2)(\psi_n - \psi)||\end{aligned}$$

$$\begin{aligned}\therefore \qquad ||\{\hat{\Omega}(\varepsilon_1) - \hat{\Omega}(\varepsilon_2)\}\psi|| &\le ||\{\hat{\Omega}(\varepsilon_1) - \hat{\Omega}(\varepsilon_2)\}\psi_n|| \\ &+ ||\hat{\Omega}(\varepsilon_1)(\psi - \psi_n)|| + ||\hat{\Omega}(\varepsilon_2)(\psi_n - \psi)||. \qquad (9.67)\end{aligned}$$

Given the meaning of strong convergence of the sequence $\{\psi_k\}_{k=1}^{\infty}$, we can now choose $\epsilon > 0$ such that for a fixed $n \in \mathbb{N}$, it follows

$$||\psi - \psi_n|| < \frac{\epsilon}{3}. \qquad (9.68)$$

In a similar manner, strong convergence implies that for a pair of positive numbers $\{\varepsilon_1, \varepsilon_2\}$ and for the mentioned fixed n we can select

$$||\{\hat{\Omega}(\varepsilon_1) - \hat{\Omega}(\varepsilon_1)\}\psi_n|| < \frac{\epsilon}{3}. \qquad (9.69)$$

In this way, using boundedness of the operator $\hat{\Omega}(\varepsilon)$ i.e.

$$||\hat{\Omega}(\varepsilon)|| < \infty \qquad (9.70)$$

we obtain from (9.64)

$$\begin{aligned}||\{\hat{\Omega}(\varepsilon_1) - \hat{\Omega}(\varepsilon_2)\}\psi|| &\le ||\{\hat{\Omega}(\varepsilon_1) - \hat{\Omega}(\varepsilon_2)\}\psi_n|| + ||\hat{\Omega}(\varepsilon_1)(\psi - \psi_n)|| \\ &+ ||\hat{\Omega}(\varepsilon_2)(\psi_n - \psi)|| = ||\{\hat{\Omega}(\varepsilon_1) - \hat{\Omega}(\varepsilon_2)\}\psi_n|| \\ &+ ||\hat{\Omega}(\varepsilon_1)|| \cdot ||\psi - \psi_n|| + ||\hat{\Omega}(\varepsilon_2)|| \cdot ||\psi_n - \psi|| \\ &\le ||\{\hat{\Omega}(\varepsilon_1) - \hat{\Omega}(\varepsilon_2)\}\psi_n|| + ||\psi - \psi_n|| + ||\psi_n - \psi|| \\ &\le \frac{\epsilon}{3} + \frac{\epsilon}{3} + \frac{\epsilon}{3} = \epsilon\end{aligned}$$

$$\therefore \qquad ||\{\hat{\Omega}(\varepsilon_1) - \hat{\Omega}(\varepsilon_2)\}\psi|| \le \epsilon. \qquad (9.71)$$

In other words, the strong limit

$$\operatorname{Lim}_{\varepsilon\to 0^+} \hat{\Omega}(\varepsilon)\psi = \hat{\Omega}\psi \qquad (9.72)$$

exists and this implies that the sub-space $\mathcal{B}$ is closed (QED). Further, the minimal lower bound of the operator $\hat{\Omega}(\varepsilon)$ is given by

$$\begin{aligned}\sup_{\psi\in\mathcal{B}}\frac{||\operatorname*{Lim}_{\varepsilon\to 0^+}\hat{\Omega}^{\pm}(\varepsilon)\psi||}{||\psi||} &= \sup_{\psi\in\mathcal{B}}\frac{||\operatorname*{Lim}_{\varepsilon\to 0^+}\mp\varepsilon\int_0^{\mp\infty}\mathrm{e}^{\mp\varepsilon t}\hat{\mathrm{U}}^{\dagger}(t)\hat{\mathrm{U}}_0(t)\psi||}{||\psi||}\\ &\leq\mp\frac{\operatorname*{Lim}_{\varepsilon\to 0^+}\varepsilon\int_0^{\mp\infty}\mathrm{e}^{\mp\varepsilon t}||\hat{\mathrm{U}}^{\dagger}(t)\hat{\mathrm{U}}_0(t)||\cdot||\psi||}{||\psi||}\\ &=\mp\operatorname*{Lim}_{\varepsilon\to 0^+}\varepsilon\int_0^{\mp\infty}\mathrm{e}^{\mp\varepsilon t}=1\end{aligned}$$

$$\therefore\qquad\sup_{\psi\in\mathcal{B}}\frac{||\operatorname*{Lim}_{\varepsilon\to 0^+}\hat{\Omega}^{\pm}(\varepsilon)\psi||}{||\psi||}\leq 1. \tag{9.73}$$

Lemma 10 shows that the set of vectors ψ, for which the strong Abel limit (9.72) exists, represents the sub-space $\mathcal{L}$ of the Hilbert space $\mathcal{H}$. Thus, the operator $\hat{\Omega}$ from (9.72) which is defined on the set $\mathcal{L}\subset\mathcal{H}$ represents a bounded, linear operator with the lower bound equal to unity ($||\hat{\Omega}||\leq 1$). However, it would be desirable if we could extend the definition of the operator $\hat{\Omega}$ to the whole space $\mathcal{H}$. This can be accomplished by introducing a linear and bounded wave operator defined on the whole Hilbert space $\mathcal{H}$ such that this operator is reduced to $\hat{\Omega}$ on the sub-space $\mathcal{L}$ having the same limiting value as that of $\hat{\Omega}$. Let $\mathcal{L}^{\perp}$ be the orthogonal complement of $\mathcal{L}$ and let us write for any vector $\psi\in\mathcal{H}$ the unique decomposition

$$\psi=\phi+\chi\qquad\phi\in\mathcal{L}\qquad\chi\in\mathcal{L}^{\perp}\qquad\mathcal{H}=\mathcal{L}\oplus\mathcal{L}^{\perp} \tag{9.74}$$

where, from the definition of the orthogonal complement, we have

$$\langle\phi|\chi\rangle=0. \tag{9.75}$$

In this way, the correspondence $\psi\longleftrightarrow\hat{\Omega}\phi$ represents the sought extension of the operator $\hat{\Omega}$ with the same limiting value. Nevertheless, in scattering theory, such a generalization is not necessary, since for collision systems the relation (9.74) is valid for every $\psi\in\mathcal{H}$. This is a direct consequence of lemma 4, which states that if the Cauchy strong limit $\operatorname*{Lim}_{t\to+\infty}\hat{\mathrm{U}}^{\dagger}(t)\hat{\mathrm{U}}_0(t)\psi=\psi^-$ exists, then there also exists the strong Abel limit $\operatorname*{Lim}_{\varepsilon\to 0^+}\varepsilon\int_0^{\infty}\mathrm{d}t\,\mathrm{e}^{-\varepsilon t}\hat{\mathrm{U}}^{\dagger}(t)\hat{\mathrm{U}}_0(t)\psi=\psi^-$ with the same limiting value ψ^-.

Let us now prove that the decomposition (9.74) is possible. We shall do that first for a finite-dimensional sub-space $\mathcal{L}\subset\mathcal{H}$ i.e. for $\dim\mathcal{L}=n<\infty$. As we know, every finite-dimensional space or sub-space has a basis, which can be ortho-normalized by the Gramm-Schmidt procedure (3.4). Then let $b_e\equiv\{e_k\}_{k=1}^n$ be such a basis, where $\langle e_j|e_k\rangle=\delta_{jk}$. In this case, for every vector $\psi = \phi + \chi \in \mathcal{H}$, the vectors ϕ and χ are fully determined by

$$\phi = \sum_{j=1}^{n} \langle e_j|\psi\rangle e_j \qquad \chi = \psi - \phi = \psi - \sum_{j=1}^{n} \langle e_j|\psi\rangle e_j. \tag{9.76}$$

Since the vector ϕ from (9.76) is expressed as a linear combination of the basis elements $e_j \in \mathcal{L}$, it is clear that ϕ also belongs to the sub-space $\mathcal{L}$ i.e. $\psi \in \mathcal{L}$. On the other hand, the vector χ is orthogonal on every basis element e_j $(1 \leq j \leq n)$, since it follows from (9.76) that

$$\langle e_k|\chi\rangle = \langle e_k|\phi\rangle - \sum_{j=1}^{n} \langle e_j|\psi\rangle\langle e_k|e_j\rangle = \langle e_k|\psi\rangle - \langle e_k|\psi\rangle = 0$$

$$\therefore \qquad \langle e_k|\chi\rangle = 0 \qquad \forall k \in [1, n]. \tag{9.77}$$

Accounting for (9.77) together with the fact that each element from $\mathcal{L}$ can be expanded on the basis b_e, it can be concluded that the vector χ is orthogonal to the whole sub-space $\mathcal{L}$

$$\psi \perp \mathcal{L} \tag{9.78}$$

or equivalently

$$\chi \in \mathcal{L}^{\perp}. \tag{9.79}$$

Uniqueness of the development (9.74) can be proven by assuming the opposite. In other words, we assume that $\psi \in \mathcal{L}$ can be written in the twofold way

$$\begin{aligned} \psi &= \phi + \chi & \phi &\in \mathcal{L} & \chi &\in \mathcal{L}^{\perp} \\ &= \phi' + \chi' & \phi' &\in \mathcal{L} & \chi' &\in \mathcal{L}^{\perp}. \end{aligned} \tag{9.80}$$

Then, we shall have

$$(\phi - \phi') + (\chi - \chi') = \emptyset \tag{9.81}$$

and since the zero vector $\emptyset$ is unique, it follows that

$$\phi' = \phi \qquad \chi' = \chi. \tag{9.82}$$

This proves the uniqueness of the expansion (9.74) (QED).

9.5 The link between the Møller operators and Green resolvents

Although we just established that the Møller wave operators $\hat{\Omega}^{\pm}$ exist in the whole Hilbert space $\mathcal{H}$, it cannot be stated without any proof that

$$\operatorname*{Lim}_{\varepsilon \to 0^+} \hat{\Omega}^{\pm\dagger}(\varepsilon) = \hat{\Omega}^{\pm\dagger}. \tag{9.83}$$

At first glance, it seems that (9.76) is simply a consequence of the corresponding expression (9.59). However, this is misleading, since in order for (9.59) to be true, the strong Cauchy limit $\psi^{\pm *} = \underset{t\to\mp\infty}{\text{Lim}} \hat{\text{U}}_0^\dagger(t)\hat{\text{U}}(t)\psi^*$ must exist for each vector $\psi \in \mathcal{H}$. However, this latter limit does not exist for $\psi \in \mathcal{R}^\perp = \mathcal{B}$ and, therefore, lemma 10 cannot be applied. The relation (9.59) of the strong limit for $\hat{\Omega}(\varepsilon)$ implies only the existence of the weak limit for $\hat{\Omega}^\dagger(\varepsilon)$ due to the Riemann-Lebesgue lemma [7]. However, since (9.79) denotes the strong limit, the existence of such a limit must be demonstrated by a separate proof. The strong limit (9.79) exists for state vectors $\psi \in \mathcal{R} \subset \mathcal{H}$ due to the applicability of lemma 10. Thus, it is necessary to prove the validity of (9.79) only for $\psi \in \mathcal{R}^\perp$. From this, of course, the condition (iii) of the definition of a quantum scattering system implies that $\psi \in \mathcal{B}$ where $\mathcal{B}$ is a sub-space spanned by proper normalizable state vectors of the eigen-value problem (3.1) of the Hamiltonian H. Let then $\{\phi_k\}_{k=1}^n$ $(n \leq \infty)$ be a complete, finite- or infinite-dimensional ortho-normal set of elements from $\mathcal{B}$. Thus according to the expansion theorem (3.2), any vector $\psi \in \mathcal{H}$ can be written in the form of a standard linear combination

$$\psi = \sum_{k=1}^{n} c_k\phi_k \qquad \sum_{k=1}^{n} |c_k|^2 < \infty \qquad (n \leq \infty). \tag{9.84}$$

The elements $\{\phi_k\}$ can be taken to be the solutions of the eigen-value problem

$$\hat{\text{H}}\phi_k = E_k\phi_k \tag{9.85}$$

where $\{E_k\}$ are the eigen-values that can, in principle, be degenerate. Then, by using linearity of the evolution operator $\hat{\text{U}}(t)$, it follows

$$\hat{\text{U}}(t)\psi = \hat{\text{U}}(t)\sum_{k=1}^{n\leq\infty} c_k\phi_k = \sum_{k=1}^{n\leq\infty} c_k\hat{\text{U}}(t)\phi_k = \sum_{k=1}^{n\leq\infty} c_k \text{e}^{-iE_kt}\phi_k$$

$$\therefore \qquad \hat{\text{U}}(t)\psi = \sum_{k=1}^{n} c_k \text{e}^{-iE_kt}\phi_k \qquad (n \leq \infty). \tag{9.86}$$

This gives

$$\begin{aligned}
\hat{\Omega}^\dagger(\varepsilon)\psi &= \varepsilon\int_0^\infty \text{d}t\,\text{e}^{-\varepsilon t}\hat{\text{U}}_0^\dagger(t)\hat{\text{U}}(t)\psi \\
&= \varepsilon\int_0^\infty \text{d}t\,\hat{\text{U}}_0^\dagger(t)\sum_{k=1}^{n\leq\infty} c_k\text{e}^{-i(E_k-i\varepsilon)t}\phi_k \\
&= \varepsilon\sum_{k=1}^{n\leq\infty} c_k\int_0^\infty \text{d}t\,\hat{\text{U}}_0^\dagger(t)\,\text{e}^{-i(E_k-i\varepsilon)t}\phi_k = -i\varepsilon\sum_{k=1}^{n\leq\infty} c_k\hat{\text{R}}_0(z_k)\phi_k
\end{aligned}$$

$$\therefore \qquad \hat{\Omega}^{\dagger}(\varepsilon)\psi = -i\,\varepsilon\sum_{k=1}^{n} c_k\hat{\mathrm{R}}_0(z_k)\phi_k \qquad (n \le \infty) \tag{9.87}$$

where $z_k = E_k - i\varepsilon$ and $R_0(z)$ is the resolvent operator

$$\hat{\mathrm{R}}_0(E_k - i\varepsilon) = i\int_0^{\infty} \mathrm{d}t\,\mathrm{e}^{i[\hat{\mathrm{H}}_0-(E_k-i\varepsilon)\hat{1}]t}\phi_k = [(E_k - i\varepsilon)\hat{1} - \hat{\mathrm{H}}_0]^{-1}. \tag{9.88}$$

In (9.87), the integration over t is carried out by using the solutions of the eigen-problem of the operator $\hat{\mathrm{H}}_0$. As we know from (3.37), the resolvent $\hat{\mathrm{R}}_0(z)$ is a bounded linear operator for all the values $\{E_k\}$ that are not in the spectrum of the Hamiltonian $\hat{\mathrm{H}}_0$ i.e. for every $\varepsilon > 0$. After these preliminaries, we can formulate and prove the following lemma:

Lemma 11: If the operator $\hat{\mathrm{H}}_0$ does not possess a discrete spectrum, then we have

$$\operatorname*{Lim}_{\varepsilon\to 0^+} \ \varepsilon\hat{\mathrm{R}}_0(z_k)\psi_k = 0 \qquad \forall\, k \in \mathbb{N}. \tag{9.89}$$

Proof: Let us denote by $z = E - i\varepsilon$ one of the elements from the set $\{z_k\}_{k=1}^{n}$ $(n \le \infty)$ related to the eigen-value E and eigen-state ϕ of the operator $\hat{\mathrm{H}}$. Then, we will have

$$\varepsilon^2||\hat{\mathrm{R}}_0(z)\phi||^2 = \varepsilon^2\langle\hat{\mathrm{R}}_0(z)\phi|\hat{\mathrm{R}}_0(z)\phi\rangle = \varepsilon^2\langle\phi|\hat{\mathrm{R}}_0^{\dagger}(z)\hat{\mathrm{R}}_0(z)\phi\rangle. \tag{9.90}$$

Using the feature (3.33) of the resolvent i.e. $\hat{\mathrm{R}}_0^{\dagger}(z) = \hat{\mathrm{R}}_0(z^*)$, it follows

$$\hat{\mathrm{R}}_0^{\dagger}(z) - \hat{\mathrm{R}}_0(z) = \hat{\mathrm{R}}_0(z^*) - \hat{\mathrm{R}}_0(z) = (z - z^*)\hat{\mathrm{R}}_0^{\dagger}(z)\hat{\mathrm{R}}_0(z) \tag{9.91}$$

so that the expression (9.90) becomes

$$\varepsilon^2||\hat{\mathrm{R}}_0(z)\phi||^2 = \frac{\varepsilon^2}{z - z^*}\langle\phi|\{\hat{\mathrm{R}}_0^{\dagger}(z) - \hat{\mathrm{R}}_0(z)\}\phi\rangle. \tag{9.92}$$

Calling upon the eigen-value resolvent problem (3.24) as well as the spectral representation (9.14) of an operator function, we find

$$\begin{aligned}
\varepsilon^2||\hat{\mathrm{R}}_0(z)\phi||^2 &= \frac{\varepsilon^2}{z - z^*}\int_{-\infty}^{+\infty}\left(\frac{1}{z^* - \lambda} - \frac{1}{z - \lambda}\right)\mathrm{d}\langle\phi|\hat{\mathrm{P}}_{\lambda}\phi\rangle \\
&= \varepsilon^2\int_{-\infty}^{+\infty}\frac{1}{(z^* - \lambda)(z - \lambda)}\mathrm{d}\langle\phi|\hat{\mathrm{P}}_{\lambda}\phi\rangle \\
&= \varepsilon^2\int_{-\infty}^{+\infty}\frac{1}{(\lambda - E)^2 + \varepsilon^2}\mathrm{d}\langle\phi|\hat{\mathrm{P}}_{\lambda}\phi\rangle \\
&= \varepsilon^2\int_{-\infty}^{+\infty}\frac{1}{\lambda^2 + \varepsilon^2}\mathrm{d}\langle\phi|\hat{\mathrm{P}}_{\lambda+E}\phi\rangle
\end{aligned}$$

$$\therefore \qquad \varepsilon^2||\hat{\mathrm{R}}_0(z)\phi||^2 = \varepsilon^2\int_{-\infty}^{+\infty}\frac{1}{\lambda^2 + \varepsilon^2}\mathrm{d}\langle\phi|\hat{\mathrm{P}}_{\lambda+E}\phi\rangle. \tag{9.93}$$

In this Stieltjes integral, the integrand $\varepsilon^2/(\lambda^2+\varepsilon^2)$ is a positive function for every $\lambda \in \mathbb{R}$ and $\varepsilon > 0$. Furthermore, the scalar product $\langle\phi|\hat{\mathrm{P}}_{\lambda+E}\phi\rangle$ is a non-decreasing function. From here it follows that it is possible to obtain an upper bound of the integral (9.93). This can be done by splitting the integration interval as $[-\infty,+\infty]$ into three parts $[-\infty,+\infty] = [-\infty,-\delta] + [-\delta,+\delta] + [+\delta,+\infty]$ where δ is an arbitrary positive number

$$\begin{aligned}
\varepsilon^2||\hat{\mathrm{R}}_0(z)\phi||^2 &= \varepsilon^2 \int_{-\infty}^{+\infty} \frac{1}{\lambda^2+\varepsilon^2} \mathrm{d}\langle\phi|\hat{\mathrm{P}}_{\lambda+E}\phi\rangle \\
&= \varepsilon^2 \left\{ \int_{-\infty}^{-\delta} + \int_{-\delta}^{+\delta} + \int_{+\delta}^{+\infty} \right\} \frac{1}{\lambda^2+\varepsilon^2} \mathrm{d}\langle\phi|\hat{\mathrm{P}}_{\lambda+E}\phi\rangle \\
&\leq \left\{ \frac{\varepsilon^2}{\delta^2+\varepsilon^2} \int_{-\infty}^{-\delta} + \int_{-\delta}^{+\delta} + \frac{\varepsilon^2}{\delta^2+\varepsilon^2} \int_{+\delta}^{+\infty} \right\} \mathrm{d}\langle\phi|\hat{\mathrm{P}}_{\lambda+E}\phi\rangle \\
&= \left\{ \int_{-\delta}^{+\delta} + \frac{\varepsilon^2}{\delta^2+\varepsilon^2} \int_{-\infty}^{+\infty} \right\} \mathrm{d}\langle\phi|\hat{\mathrm{P}}_{\lambda+E}\phi\rangle \\
&= \langle\phi|\hat{\mathrm{P}}_{E+\delta}\phi\rangle - \langle\phi|\hat{\mathrm{P}}_{E-\delta}\phi\rangle + \frac{\varepsilon^2}{\delta^2+\varepsilon^2} \\
&= \langle\phi|\hat{\mathrm{P}}_{E+\delta} - \hat{\mathrm{P}}_{E-\delta}|\phi\rangle + \frac{\varepsilon^2}{\delta^2+\varepsilon^2}
\end{aligned}$$

$$\therefore \qquad \varepsilon^2||\hat{\mathrm{R}}_0(z)\phi||^2 \leq \langle\phi|\hat{\mathrm{P}}_{E+\delta} - \hat{\mathrm{P}}_{E-\delta}|\phi\rangle + \frac{\varepsilon^2}{\delta^2+\varepsilon^2} \qquad (\delta > 0) \tag{9.94}$$

where we carry out the majorization of the integrand $\varepsilon^2/(\lambda^2+\varepsilon^2) \leq 1$ for $\lambda \in [-\delta,+\delta]$ and $\varepsilon^2/(\lambda^2+\varepsilon^2) \leq \varepsilon^2/(\delta^2+\varepsilon^2)$ for $\lambda \in [-\infty,-\delta]$ and $\lambda \in [+\delta,+\infty]$. Applying the limiting process $\varepsilon \to 0^+$ to (9.94), we arrive at the result

$$\lim_{\varepsilon\to 0^+} \varepsilon^2 \,||\hat{\mathrm{R}}_0(z)\phi|| \leq \langle\phi|\hat{\mathrm{P}}_{E+0} - \hat{\mathrm{P}}_{E-0}|\phi\rangle = 0. \tag{9.95}$$

This is possible since the Hamiltonian $\hat{\mathrm{H}}_0$ does not possess the discrete spectrum i.e. the operator $\hat{\mathrm{U}}_0(t)$ has the spectral evolution without discontinuities

$$\hat{\mathrm{P}}_{E+0} = \hat{\mathrm{P}}_{E-0}. \tag{9.96}$$

In this way, it follows that

$$\lim_{\varepsilon\to 0^+} \varepsilon \,||\hat{\mathrm{R}}_0(z)\phi|| = 0 \tag{9.97}$$

$$\operatorname*{Lim}_{\varepsilon\to 0^+} \varepsilon \, \hat{\mathrm{R}}_0(z)\phi = \emptyset \tag{9.98}$$

$$\operatorname*{Lim}_{\varepsilon\to 0^+} \hat{\Omega}^\dagger \phi = \emptyset \qquad \forall \phi \in \mathcal{B} \qquad \text{(QED)}. \tag{9.99}$$

10

The existence and completeness of the Møller wave operators

10.1 Linearity and isometry of wave operators

From the properties (5.11), (5.12) and (5.16) of the definition of a quantum collision system, it can be concluded that the correspondence between $\psi \in \mathcal{H}$ and $\psi^+ = \hat{\Omega}^+\psi \in \mathcal{R}$ represents a linear, isometric mapping of the sub-space $\mathcal{H}$ to $\mathcal{R}$. Linearity is proven as follows

$$\begin{aligned}\operatorname*{Lim}_{t\to\mp\infty} \hat{\mathrm{U}}^\dagger(t)\hat{\mathrm{U}}_0(t)(\lambda_1\psi_1 + \lambda_2\psi_2) &= \lambda_1 \operatorname*{Lim}_{t\to\mp\infty} \hat{\mathrm{U}}^\dagger(t)\hat{\mathrm{U}}_0(t)\psi_1 \\ &\quad + \lambda_2 \operatorname*{Lim}_{t\to\mp\infty} \hat{\mathrm{U}}^\dagger(t)\hat{\mathrm{U}}_0(t)\psi_2 = \lambda_1\psi_1^\pm + \lambda_2\psi_2^\pm\end{aligned}$$

$$\therefore \quad \operatorname*{Lim}_{t\to\mp\infty} \hat{\mathrm{U}}^\dagger(t)\hat{\mathrm{U}}_0(t)(\lambda_1\psi_1 + \lambda_2\psi_2) = \lambda_1\psi_1^\pm + \lambda_2\psi_2^\pm \tag{10.1}$$

where $\lambda_1, \lambda_2 \in \mathbb{C}$ and $\psi_1, \psi_2 \in \mathcal{H}$. Isometry means that the mapping preserves the norm as stated via

$$||\psi|| = ||\psi^\pm|| = ||\hat{\Omega}^\pm\psi||. \tag{10.2}$$

Since $\hat{\mathrm{U}}_0(t)$ and $\hat{\mathrm{U}}(t)$ are isometric operators, we have

$$\begin{aligned}||\psi^\pm||^2 &= \langle\psi^\pm|\psi^\pm\rangle \\ &= \langle \operatorname*{Lim}_{t\to\mp\infty} \hat{\mathrm{U}}^\dagger(t)\hat{\mathrm{U}}_0(t)\psi| \operatorname*{Lim}_{t\to\mp\infty} \hat{\mathrm{U}}^\dagger(t)\hat{\mathrm{U}}_0(t)\psi\rangle \\ &= \operatorname*{Lim}_{t\to\mp\infty} \langle\hat{\mathrm{U}}^\dagger(t)\hat{\mathrm{U}}_0(t)\psi|\hat{\mathrm{U}}^\dagger(t)\hat{\mathrm{U}}_0(t)\psi\rangle \\ &= \operatorname*{Lim}_{t\to\mp\infty} \langle\psi|\hat{\mathrm{U}}_0^\dagger(t)\hat{\mathrm{U}}(t)\hat{\mathrm{U}}^\dagger(t)\hat{\mathrm{U}}_0(t)\psi\rangle \\ &= \operatorname*{Lim}_{t\to\mp\infty} \langle\psi|\hat{\mathrm{U}}_0^\dagger(t)\hat{\mathrm{U}}_0(t)\psi\rangle \\ &= \operatorname*{Lim}_{t\to\mp\infty} \langle\psi|\psi\rangle = ||\psi||^2\end{aligned}$$

$$\therefore \quad ||\psi^\pm||^2 = ||\psi||^2 \qquad \text{(QED)}. \tag{10.3}$$

This proof can also be accomplished by using isometry of the operator $\hat{\Omega}$ i.e. $\hat{\Omega}^\dagger\hat{\Omega} = \hat{1}$, so that

$$||\hat{\Omega}\psi||^2 = \langle\hat{\Omega}\psi|\hat{\Omega}\psi\rangle = \langle\psi|\hat{\Omega}^\dagger\hat{\Omega}\psi\rangle = \langle\psi|\psi\rangle = ||\psi|| \quad \text{(QED)} \tag{10.4}$$

where $\hat{\Omega}$ stands for both $\hat{\Omega}^\pm$.

10.2 Boundedness of wave operators in the whole Hilbert space

The two proven features of linearity and isometry permit the introduction of a pair of bounded operators $\hat{\Omega}^\pm$ defined on the whole space $\mathcal{H}$ with the range $\mathcal{R}$, such that

$$\hat{\Omega}^\pm\psi = \psi^\pm \qquad \hat{\Omega}^\pm\psi = \lim_{t\to\mp\infty} \hat{U}^\dagger(t)\hat{U}_0(t)\psi. \tag{10.5}$$

Boundedness of the evolution operators $\hat{U}^\dagger(t)$ and $\hat{U}_0(t)$ directly implies that the wave operators $\hat{\Omega}^\pm$ are also bounded i.e.

$$||\hat{\Omega}^\pm|| = ||\hat{U}^\dagger(t)\hat{U}_0(t)|| = ||\hat{U}^\dagger(t)|| \cdot ||\hat{U}_0(t)|| = 1. \tag{10.6}$$

Further, from the isometry property it follows that the set $\mathcal{R}$ represents a vector sub-space. This can be shown as follows. We first notice that boundedness of the operator $\hat{\Omega}$ in the whole Hilbert space $\mathcal{H}$ implies that the adjoint operator $\hat{\Omega}^\dagger$ can also be defined throughout $\mathcal{H}$ as a bounded, linear operator with the property

$$\langle\hat{\Omega}^\dagger\psi|\phi\rangle = \langle\psi|\hat{\Omega}\phi\rangle \qquad \forall\phi \in \mathcal{H}. \tag{10.7}$$

Here, it is pertinent to ask the following question: is it possible to conclude from (10.7) that the following relation is valid [7]

$$\hat{\Omega}^\dagger\hat{\Lambda} = \hat{0} \tag{10.8}$$

where $\hat{\Lambda}$ is the projection operator on the sub-space $\mathcal{B}$ of bound states? The answer is in the affirmative. The proof as well as the related detailed discussion can be found in chapter 7 of Ref. [7].

10.3 The Schur lemma on invariant sub-spaces for evolution operators

Let us now formulate and prove the Schur lemma which transcends one of the most important consequences of the intertwining relation of the type $\hat{H}\hat{\Omega}^\pm =$

$\hat{\Omega}^{\pm}\hat{\mathrm{H}}_0$, which finds many important applications in scattering theory [7].

Lemma 12 (Schur): Let $\Theta_{\hat{\Omega}} \subset \mathcal{H}$ be the null space of the operator $\hat{\Omega}$ i.e. a set of vectors $\psi \in \mathcal{H}$ for which $\hat{\Omega}\psi = \emptyset$

$$\Theta_{\hat{\Omega}} \equiv \Theta = \{\psi \in \mathcal{H} \ : \ \hat{\Omega}\psi = \emptyset\}. \tag{10.9}$$

Referring to the condition (5.11) of the definition of a quantum collision system, it is clear that the ortho-complement of the sub-space Θ represents the domain of the wave Møller operator $\hat{\Omega}$

$$\Theta^{\perp} = \mathcal{D}_{\hat{\Omega}} = \{\psi \in \mathcal{H} \ : \ \hat{\Omega}\psi = \chi \neq \emptyset\}. \tag{10.10}$$

Similarly, $\Theta_{\hat{\Omega}^{\dagger}} \equiv \Theta^{\dagger} \subset \mathcal{H}$ is the domain of the operators $\hat{\Omega}^{\dagger}$. Further, by $\mathcal{R}_{\hat{\Omega}} \equiv \mathcal{R} \subset \mathcal{H}$ and $\mathcal{R}_{\hat{\Omega}^{\dagger}} \equiv \mathcal{R}^{\dagger} \subset \mathcal{H}$ we denote the ranges of the operators $\hat{\Omega}$ and $\hat{\Omega}^{\dagger}$

$$\Theta_{\hat{\Omega}^{\dagger}} \equiv \Theta^{\dagger} = \{\psi \in \mathcal{H} \ : \ \hat{\Omega}^{\dagger}\psi = \emptyset\}. \tag{10.11}$$

On the account of (10.8), it follows

$$\mathcal{R}^{\perp} = \Theta^{\dagger} \qquad \Theta^{\perp} = \mathcal{R}^{\dagger}. \tag{10.12}$$

Then the Schur lemma asserts the following: the sets $\mathcal{R}$ and $\mathcal{R}^{\dagger}$ are invariant sub-space for the operators $\hat{\mathrm{U}}(t)$ and $\hat{\mathrm{U}}_0(t)$, respectively. Moreover, reduction of the operator $\hat{\mathrm{U}}(t)$ to $\mathcal{R}$ is unitarily equivalent to reduction of the operator $\hat{\mathrm{U}}_0(t)$ to $\mathcal{R}^{\dagger}$

$$\hat{\mathrm{U}}^{\mathcal{R}}(t) \propto \hat{\mathrm{U}}_0^{\mathcal{R}^{\dagger}}(t). \tag{10.13}$$

In our particular case, we have that $\mathcal{R}^{\dagger} = \Theta^{\perp}$ coincides with the whole space $\mathcal{H}$, since according to (10.1) the set $\mathcal{R}^{\dagger} = \Theta^{\perp} = \mathcal{D}_{\hat{\Omega}}$ represents the domain $\mathcal{D}_{\hat{\Omega}} = \mathcal{H}$ of the operator $\hat{\Omega}$. Hence, for collisional systems, it follows

$$\hat{\mathrm{U}}^{\mathcal{R}}(t) \propto \hat{\mathrm{U}}_0^{\mathcal{H}}(t) \equiv \hat{\mathrm{U}}_0(t). \tag{10.14}$$

On the other hand, the set $\mathcal{R}$ is the sub-space of continuum states of the Hamiltonian $\hat{\mathrm{H}}$

$$\mathcal{R} = \Theta^{\dagger\perp} = \mathcal{B}^{\perp}. \tag{10.15}$$

Therefore, the relation (10.14) can also be expressed via the statement: The operator $\hat{\mathrm{U}}_0(t)$ is unitarily equivalent to the reduction of the operator $\hat{\mathrm{U}}(t)$ of continuum states of $\mathcal{H}$. This brings us to a corollary of the Schur lemma: the continuous parts of the spectrum of the operators $\hat{\mathrm{H}}$ and $\hat{\mathrm{H}}_0$ are identical to each other. This is in accord with the condition (5.16) of the definition of a quantum collision system.

10.4 Intertwining relations for evolution operators and wave operators

Lemma 13: As before, we continue to use $\hat{\Omega}$ to denote both $\hat{\Omega}^+$ and $\hat{\Omega}^-$. Further, let $\hat{\Omega}_1$ represent an arbitrary bounded operator which satisfies the following intertwining relation

$$\hat{U}(t)\hat{\Omega}_1 = \hat{\Omega}_1\hat{U}_0(t). \tag{10.16}$$

Hereafter, the operator $\hat{\Omega}_1$ will be called the intertwining operator. Then between the corresponding ranges $\mathcal{R} \equiv \mathcal{R}_{\hat{\Omega}}$ and $\mathcal{R}_{\hat{\Omega}_1} \equiv \mathcal{R}_1$ the following relation exists

$$\mathcal{R} \subseteq \mathcal{R}_1. \tag{10.17}$$

Proof: Notice first that it is sufficient to carry out the proof for those intertwining operators that are isometric. Such a simplification is possible, since if $\hat{A}$ is any bounded intertwining operator, then there exists an isometric counterpart $\hat{B}$ whose range coincides exactly with the range of $\hat{A}$ i.e.

$$\mathcal{R}_{\hat{A}} = \mathcal{R}_{\hat{B}}. \tag{10.18}$$

This is a consequence of the well-known theorem on the so-called polar representation of linear bounded operators. According to this theorem, to every linear bounded operator $\hat{A}$, it is possible to associate a unique linear isometric operator $\hat{B}$ with the same left and right projection as those of $\hat{A}$

$$\hat{A} = \hat{B}\hat{C} \qquad \hat{C} = \left(\hat{A}^\dagger\hat{A}\right)^{1/2}. \tag{10.19}$$

The operator $\hat{B}$ is defined in such a way that for the elements ϕ of the form

$$\phi = \hat{C}\psi \tag{10.20}$$

the following relation holds true

$$\hat{B}\phi = \hat{A}\psi \tag{10.21}$$

i.e. $\hat{B}\phi = \hat{B}\hat{C}\psi = \hat{A}\psi$. Let $\mathcal{L}_C$ be a linear manifold of such elements ϕ. Then the operator $\hat{B}$ is homogeneous, additive and isometric in both $\mathcal{L}_C$ and in the closed linear manifold $\bar{\mathcal{L}}_C$. On the other hand, by definition, in the ortho-complement $\bar{\mathcal{L}}_C^\perp$, the operator $\hat{B}$ is identical to the zero operator

$$\hat{B}\chi = \hat{0} \qquad \forall\chi \in \bar{\mathcal{L}}_C^\perp. \tag{10.22}$$

Hence, the operator $\hat{B}$ is a partial isometry such that the left and the right projections of the operators $\hat{B}$ and $\hat{A}$ coincide. With the application of lemma

13 to the case under study, the operator $\hat{\mathrm{A}}$ is considered as an intertwined operator and, therefore, we can write

$$\hat{\mathrm{U}}(t)\hat{\mathrm{A}} = \hat{\mathrm{A}}\hat{\mathrm{U}}_0(t). \tag{10.23}$$

We shall prove that this feature implies

$$\hat{\mathrm{U}}(t)\hat{\mathrm{B}} = \hat{\mathrm{B}}\hat{\mathrm{U}}_0(t) \tag{10.24}$$

where the operator $\hat{\mathrm{B}}$ is defined by (10.22). In order to prove that (10.24) holds true, let us take that the vector ψ is given in an additive form

$$\psi = \phi + \chi \qquad \phi \in \mathcal{L}_{\mathrm{C}} \qquad \chi \in \mathcal{L}_{\mathrm{C}}^{\perp} \tag{10.25}$$

with the operator $\hat{\mathrm{C}}$ being given in (10.19). Employing (10.22) we obtain

$$\hat{\mathrm{B}}\psi = \hat{\mathrm{B}}\phi. \tag{10.26}$$

Further, the vector ϕ is of the form $\hat{\mathrm{C}}\xi$ for each $\xi \in \mathcal{H}$, so that

$$\hat{\mathrm{B}}\psi = \hat{\mathrm{A}}\xi \tag{10.27}$$

i.e. $\hat{\mathrm{B}}\psi = \hat{\mathrm{B}}\phi = \hat{\mathrm{B}}\hat{\mathrm{C}}\xi = \hat{\mathrm{A}}\xi$ where (10.19) is used. If we now apply the operator $\hat{\mathrm{U}}(t)$ to both sides of (10.27) and use (10.23), it follows

$$\hat{\mathrm{U}}(t)\hat{\mathrm{B}}\psi = \hat{\mathrm{A}}\hat{\mathrm{U}}_0(t)\xi \tag{10.28}$$

since $\hat{\mathrm{U}}(t)\hat{\mathrm{B}}\psi = \hat{\mathrm{U}}(t)\hat{\mathrm{A}}\xi = \hat{\mathrm{A}}\hat{\mathrm{U}}_0(t)\xi$. Therefore, in order to prove (10.24) we must first verify that we have

$$\hat{\mathrm{A}}\hat{\mathrm{U}}_0(t)\xi = \hat{\mathrm{B}}\hat{\mathrm{U}}(t)\psi. \tag{10.29}$$

Utilizing (10.25) together with the fact that the operators $\hat{\mathrm{U}}_0(t)$ and $\hat{\mathrm{C}}$ commute

$$[\hat{\mathrm{U}}_0(t), \hat{\mathrm{C}}] = \hat{0} \tag{10.30}$$

we deduce

$$\hat{\mathrm{U}}_0(t)\psi = \hat{\mathrm{U}}_0(t)\phi + \hat{\mathrm{U}}_0(t)\chi \qquad \hat{\mathrm{U}}_0(t)\phi \in \bar{\mathcal{L}}_{\mathrm{C}} \qquad \hat{\mathrm{U}}_0(t)\chi \in \mathcal{L}_{\mathrm{C}}^{\perp}. \tag{10.31}$$

If now the operator $\hat{\mathrm{B}}$ is applied to (10.31) and the expression (10.30) is used, it will follow

$$\begin{aligned} \hat{\mathrm{B}}\hat{\mathrm{U}}_0(t)\psi &= \hat{\mathrm{B}}\hat{\mathrm{U}}_0(t)\phi + \hat{\mathrm{B}}\hat{\mathrm{U}}_0(t)\chi = \hat{\mathrm{B}}\hat{\mathrm{U}}_0(t)\phi + \emptyset \\ &= \hat{\mathrm{B}}\hat{\mathrm{U}}_0(t)\hat{\mathrm{C}}\xi = \hat{\mathrm{B}}\hat{\mathrm{C}}\hat{\mathrm{U}}_0(t)\xi = \hat{\mathrm{A}}\hat{\mathrm{U}}_0(t)\xi \end{aligned}$$

$$\therefore \qquad \hat{\mathrm{B}}\hat{\mathrm{U}}_0(t)\psi = \hat{\mathrm{A}}\hat{\mathrm{U}}_0(t)\xi. \tag{10.32}$$

This proves the assumed equality (10.29) which, in turn, implies that the statement (10.24) holds true (QED).

10.5 The role of spectral projection operators

Let $\hat{\mathrm{B}}$ represent the operator $\hat{\Omega}_1$ introduced in lemma 12. Further, let $\hat{\mathrm{P}}_\lambda^{(0)}$ and $\hat{\mathrm{P}}_\lambda$ be the spectral projection operators associated with the representations of $\hat{\mathrm{U}}_0(t)$ and $\hat{\mathrm{U}}(t)$, respectively

$$\hat{\mathrm{U}}_0(t) = \int_{-\infty}^{+\infty} \mathrm{e}^{-i\lambda t}\, \mathrm{d}\hat{\mathrm{P}}_\lambda^{(0)} \qquad \hat{\mathrm{U}}(t) = \int_{-\infty}^{+\infty} \mathrm{e}^{-i\lambda t}\, \mathrm{d}\hat{\mathrm{P}}_\lambda. \tag{10.33}$$

Then, the following intertwining relation is also valid

$$\hat{\mathrm{P}}_\lambda \hat{\Omega}_1 = \hat{\Omega}_1 \hat{\mathrm{P}}_\lambda^{(0)}. \tag{10.34}$$

Operators $\hat{\mathrm{U}}_0(t)$ and $\hat{\mathrm{U}}(t)$ are unitarily equivalent. This is because there exists a unitary operator $\hat{\Omega}_1$ such that the domains of $\hat{\mathrm{U}}(t)$ and $\hat{\Omega}_1\hat{\mathrm{U}}_0(t)\hat{\Omega}_1^\dagger$ coincide and for each element ψ from these domains it follows $\hat{\mathrm{U}}(t)\psi = \hat{\Omega}_1\hat{\mathrm{U}}_0(t)\hat{\Omega}_1^\dagger\psi$. Since $\hat{\mathrm{P}}_\lambda$ and $\hat{\mathrm{P}}_\lambda^{(0)}$ are spectral operators associated with $\hat{\mathrm{U}}(t)$ and $\hat{\mathrm{U}}_0(t)$, respectively, then according to (10.1), we shall have

$$\hat{\mathrm{P}}_\lambda = \hat{\Omega}_1 \hat{\mathrm{P}}_\lambda^{(0)} \hat{\Omega}_1^\dagger. \tag{10.35}$$

If (10.34) is multiplied from the right by $\hat{\Omega}_1^\dagger$, it follows with the help of isometry and (10.35) that

$$\hat{\mathrm{P}}_\lambda \hat{\mathrm{P}}_1 = \hat{\Omega}_1 \hat{\mathrm{P}}_\lambda^{(0)} \hat{\Omega}_1^\dagger \hat{\mathrm{P}}_1 = \hat{\Omega}_1 \hat{\Omega}_1^\dagger \hat{\mathrm{P}}_\lambda^{(0)} \hat{\Omega}_1^\dagger = \hat{\mathrm{P}}_1 \hat{\Omega}_1 \hat{\mathrm{P}}_\lambda^{(0)} \hat{\Omega}_1^\dagger = \hat{\mathrm{P}}_1 \hat{\mathrm{P}}_\lambda \tag{10.36}$$

$$\hat{\mathrm{P}}_1 = \hat{\Omega}_1 \hat{\Omega}_1^\dagger. \tag{10.37}$$

In other words, $\hat{\mathrm{P}}_1$ is the left projection operator of $\hat{\Omega}_1$. Finally, we take the vector ψ from the range of the operator $\hat{\Omega}_1$ i.e. $\psi \in \mathcal{R}_1$ such that

$$\psi = \hat{\mathrm{P}}_1 \psi \tag{10.38}$$

which will be written in the form (10.25) i.e.

$$\psi = \phi + \chi \qquad \phi \in \mathcal{R} \qquad \chi \in \mathcal{R}^\perp \equiv \mathcal{B}. \tag{10.39}$$

Our task is to demonstrate that we have

$$\chi = \emptyset \qquad \forall \psi \in \mathcal{R}_1 \tag{10.40}$$

with the automatic consequence

$$\mathcal{R}_1 \subset \mathcal{R}. \tag{10.41}$$

This final step will be achieved by introducing the projection operator $\hat{\mathrm{P}}_k$ which corresponds to the eigen-value λ_k of the operator $\hat{\mathrm{U}}(t)$

$$\hat{\mathrm{P}}_k \equiv \hat{\mathrm{P}}_{\lambda_k+0} - \hat{\mathrm{P}}_{\lambda_k-0}. \tag{10.42}$$

Of course, the total projection $\hat{\mathrm{P}}$ associated with the sub-space $\mathcal{B}$ of bound states is given by the sum

$$\hat{\mathrm{P}} = \sum_k \hat{\mathrm{P}}_k. \tag{10.43}$$

The element χ belongs to the sub-space $\mathcal{B}$ meaning that we have

$$\hat{\mathrm{P}}\chi = \chi \tag{10.44}$$

and it also holds

$$\chi = \hat{\mathrm{P}}\psi \qquad [\hat{\mathrm{P}}, \hat{\mathrm{P}}_1] = \hat{0}. \tag{10.45}$$

Using (10.38) and (10.45) we easily find that the vector χ also belongs to the range $\mathcal{R}_1$ of the operator $\hat{\Omega}_1$ i.e.

$$\hat{\mathrm{P}}_1\chi = \hat{\mathrm{P}}_1\hat{\mathrm{P}}\psi = \hat{\mathrm{P}}\hat{\mathrm{P}}_1\psi = \hat{\mathrm{P}}\psi = \chi. \tag{10.46}$$

Next we shall calculate the norm of the vector χ by means of (10.44)

$$\begin{aligned}
||\chi||^2 &= \langle\chi|\chi\rangle = \langle\chi|\hat{\mathrm{P}}\chi\rangle \\
&= \langle\chi|\sum_k \hat{\mathrm{P}}_k\chi\rangle = \langle\chi|\sum_k \hat{\mathrm{P}}_k\hat{\mathrm{P}}_1\chi\rangle \\
&= \langle\chi|\sum_k (\hat{\mathrm{P}}_{\lambda_k+0} - \hat{\mathrm{P}}_{\lambda_k-0})\hat{\mathrm{P}}_1\chi\rangle \\
&= \langle\chi|\sum_k \hat{\Omega}_1[\hat{\mathrm{P}}^{(0)}_{\lambda_k+0} - \hat{\mathrm{P}}_{\lambda_k-0}]\hat{\Omega}_1^\dagger\chi\rangle \\
&= \langle\hat{\Omega}_1^\dagger\chi|\sum_k [\hat{\mathrm{P}}^{(0)}_{\lambda_k+0} - \hat{\mathrm{P}}_{\lambda_k-0}]\hat{\Omega}_1^\dagger\chi\rangle = 0.
\end{aligned}$$

$$\therefore \qquad ||\chi||^2 = 0. \tag{10.47}$$

Thus, the norm $||\chi||$ is equal to zero, since the operator $\hat{\mathrm{U}}_0(t)$ has the spectral representation without discontinuities i.e. (9.96) is valid for this operator. Due to (10.1), we have according to (10.47) that $\chi = \emptyset$ as required in (10.40). In this way, the relation (10.41) i.e. $\mathcal{R}_{\hat{\Omega}_1} \subseteq \mathcal{R}_{\hat{\Omega}}$, is proven (QED).

Let us now establish several additional properties of a linear bounded operator $\hat{\Omega}_1$ which satisfies an intertwining relation of the type

$$\hat{\Omega}_1\hat{\mathrm{U}}_0(t) = \hat{\mathrm{U}}(t)\hat{\Omega}_1 \qquad \hat{\Omega}_1^\dagger\hat{\mathrm{U}}(t) = \hat{\mathrm{U}}_0(t)\hat{\Omega}_1^\dagger. \tag{10.48}$$

We shall see that the examination of the operator $\hat{\Omega}_1$ will be very useful, since it will yield an alternative definition of the scattering $\hat{\mathrm{S}}-$operator. Let the sub-spaces $\mathcal{R}_1 \subset \mathcal{H}$ and $\mathcal{R}_1^\dagger \subset \mathcal{H}$ represent the ranges of the operators $\hat{\Omega}_1$ and $\hat{\Omega}_1^\dagger$, respectively. In such a case, according to the Schur lemma, it follows

$$\hat{\mathrm{U}}^{\mathcal{R}_1}(t) \propto \hat{\mathrm{U}}_0^{\mathcal{R}_1^\dagger}(t). \tag{10.49}$$

According to the feature (5.12) of the definition of a quantum scattering system, the set $\mathcal{R} \equiv \mathcal{R}_{\hat{\Omega}}$ denotes the common range of the operators $\hat{\Omega}^+$ and $\hat{\Omega}^-$. Bearing this in mind, let us introduce several lemmas.

Lemma 14: For any intertwining operator of a given scattering system, the operator

$$\hat{\Omega}_1(t) = \hat{\mathrm{U}}_0^\dagger(t)\hat{\Omega}_1\hat{\mathrm{U}}_0(t) \tag{10.50}$$

possesses its limit when $t \to \mp\infty$.

Proof: The usage of the intertwining relation (10.48) gives

$$\hat{\Omega}_1(t) = \hat{\mathrm{U}}_0^\dagger(t)\{\hat{\Omega}_1\hat{\mathrm{U}}_0(t)\} = \hat{\mathrm{U}}_0^\dagger(t)\hat{\mathrm{U}}(t)\hat{\Omega}_1. \tag{10.51}$$

According to the relation (5.17), we know that the vector $\hat{\mathrm{U}}_0^\dagger(t)\hat{\mathrm{U}}(t)\psi$ has the strong Cauchy limit when $t \to \mp\infty$ for every element $\psi \in \mathcal{R}$. Further, every vector ψ of the form

$$\psi = \hat{\Omega}_1\phi \tag{10.52}$$

obligatorily belongs to the sub-space $\mathcal{R}_1$, since according to (10.41) from lemma 13 we have that $\mathcal{R}_1 \subset \mathcal{R}$. Thus, we conclude that the sought limiting value $\operatorname{Lim}_{t\to\mp\infty} \hat{\Omega}_1\phi$ exists for every vector $\phi \in \mathcal{H}$ (QED).

Lemma 15: Any intertwining operator $\hat{\Omega}_1$ uniquely determines the two operators $\hat{\mathrm{X}}^+$ and $\hat{\mathrm{X}}^-$ with the following features

$$\hat{\Omega}_1 = \hat{\Omega}\hat{\mathrm{X}} \qquad [\hat{\mathrm{X}}, \hat{\mathrm{U}}_0(t)] = \hat{0} \tag{10.53}$$

where $\hat{\Omega} = \hat{\Omega}^+$ or $\hat{\Omega} = \hat{\Omega}^-$ and $\hat{\mathrm{X}} = \hat{\mathrm{X}}^+$ or $\hat{\mathrm{X}} = \hat{\mathrm{X}}^-$.

Proof: Operator $\hat{\mathrm{X}}$ can be obtained by multiplying both sides of (10.53) by $\hat{\Omega}^\dagger$ and using isometry $\hat{\Omega}^\dagger\hat{\Omega} = \hat{1}$ [7]

$$\hat{\Omega}^\dagger\hat{\Omega}_1 = \hat{\Omega}^\dagger\hat{\Omega}\hat{\mathrm{X}} = \hat{\mathrm{X}}. \tag{10.54}$$

Since on the other hand $\hat{\Omega}\hat{\Omega}^\dagger = \hat{1} - \hat{\mathrm{P}}_{\mathrm{Q}^\dagger}$ we can multiply (10.54) from the left by $\hat{\Omega}$ to arrive at

$$\hat{\Omega}\hat{\Omega}^\dagger\hat{\Omega}_1 = (\hat{1} - \hat{\mathrm{P}}_{\mathrm{Q}^\dagger})\hat{\Omega}_1 = \hat{\Omega}\hat{\mathrm{X}}. \tag{10.55}$$

From the relation $\mathrm{Q}^\dagger = \mathcal{R}^\perp$, it is seen that $\hat{\mathrm{P}}_{\mathrm{Q}^\dagger}$ is the projection in the sub-space $\mathcal{R}^\perp$. According to lemma 13, the range $\mathcal{R}_1$ is contained in $\mathcal{R}$, so that all the vectors $\psi \in \mathcal{R}_1$ are orthogonal to $\mathrm{Q}^\dagger = \mathcal{R}^\perp$. This implies

$$\hat{\mathrm{P}}_{\mathrm{Q}^\dagger}\hat{\Omega}_1 = \hat{0}. \tag{10.56}$$

In this way, the expression (10.55) is reduced to

$$\hat{\Omega}\hat{X} = \hat{\Omega}_1 \qquad \text{(QED)}. \tag{10.57}$$

Additionally, using the intertwining relation (10.48) as well as $\hat{U}_0(t)\hat{\Omega}^\dagger = \hat{\Omega}^\dagger\hat{U}(t)$, we find that the operators $\hat{X}$ and $\hat{U}_0(t)$ commute

$$\begin{aligned}[\hat{X}, \hat{U}_0(t)] &= \hat{X}\hat{U}_0(t) - \hat{U}_0(t)\hat{X} = \hat{\Omega}^\dagger\hat{\Omega}_1\hat{U}_0(t) - \hat{U}_0(t)\hat{\Omega}^\dagger\hat{\Omega}_1 \\ &= \hat{\Omega}^\dagger\hat{\Omega}_1\hat{U}_0(t) - \hat{\Omega}^\dagger\hat{U}(t)\hat{\Omega}_1 = \hat{\Omega}^\dagger\hat{\Omega}_1\hat{U}_0(t) - \hat{\Omega}^\dagger\hat{\Omega}_1\hat{U}_0(t) = \hat{0}\end{aligned}$$

$$\therefore \qquad [\hat{X}, \hat{U}_0(t)] = \hat{0} \qquad \text{(QED)}. \tag{10.58}$$

10.6 Completeness of the Møller wave operators

Definition 1: The intertwining operator $\hat{\Omega}_1$ is complete if it holds that

$$\mathcal{R}_1 = \mathcal{R}. \tag{10.59}$$

With respect to this definition, we have the following theorem.

Theorem 6: The intertwining operator $\hat{\Omega}_1$ is complete if and only if the operator $\hat{X}$ from lemma 14 has the corresponding inverse operator.

Proof: We shall first prove necessity. This means that if we assume that the operator $\hat{\Omega}_1$ is complete, then the relation (10.59) holds. In accord with the notation from lemma 12, let Θ_1 be the set of the elements $\psi \in \mathcal{H}$ for which we have $\hat{\Omega}_1\psi = \emptyset$ i.e.

$$\Theta_1 = \{\psi \in \mathcal{H} \ : \ \hat{\Omega}_1\psi = \emptyset\}. \tag{10.60}$$

Obviously, the sub-space Θ_1 is simultaneously the ortho-complement to $\mathcal{R}^\dagger$

$$\Theta_1 = \mathcal{R}_1^{\dagger\perp} = \mathcal{R}^{\dagger\perp} \tag{10.61}$$

where the feature (10.59) of completeness of the operator $\hat{\Omega}_1$ is used. We saw from lemma 14 that any vector ψ of the form $\psi \equiv \hat{\Omega}_1\phi$ belongs to $\mathcal{R}_1$ where $\phi \in \mathcal{H}$. In this way, we have from (10.54)

$$\hat{X}\phi = \hat{\Omega}^\dagger\psi \qquad \phi \in \mathcal{H} \tag{10.62}$$

due to $\hat{X}\phi = \hat{\Omega}^\dagger\hat{\Omega}_1\phi = \hat{\Omega}^\dagger\psi$. Let us now examine the range of the operator $\hat{X}$. According to (10.62) the range of the operator $\hat{X}$ is a set, say $\Omega^\dagger(\mathcal{R}_1)$, which is a sub-space of all the vectors $\hat{\Omega}^\dagger\psi$, where ψ is an element from $\mathcal{R}_1$

$$\Omega^\dagger(\mathcal{R}_1) = \{\psi_1 \ : \ \psi_1 = \hat{\Omega}^\dagger\psi \quad \psi \in \mathcal{R}_1\}. \tag{10.63}$$

In such a case, the operator $\hat{X}$ from lemma 15 maps the whole Hilbert space $\mathcal{H}$ to the sub-space $\hat{\Omega}^\dagger(\mathcal{R}_1) \subset \mathcal{H}$. Since $\hat{\Omega}_1$ is assumed to be complete, the relation (10.59) is valid with the understanding that $\mathcal{R}_1 = \mathcal{R}$ and $\mathcal{H}^\dagger(\mathcal{R}_1) = \mathcal{H}$. This implies that the range of the operator $\hat{X}$ is equal to the whole Hilbert space $\mathcal{H}$ as written viz

$$\mathcal{R}_{\hat{X}} \equiv \hat{\Omega}^\dagger(\mathcal{R}_1) = \mathcal{H}. \tag{10.64}$$

Further, we shall show that $\psi \neq \emptyset$ also implies $\phi \neq \emptyset$ where $\phi = \hat{X}\psi$

$$\phi \neq \emptyset \qquad \phi = \hat{X}\psi \neq \emptyset. \tag{10.65}$$

Let us suppose that the opposite is true i.e. that we have $\phi = \emptyset$ in which case, the following relations are valid

$$\hat{X}\psi = \hat{\Omega}^\dagger\hat{\Omega}_1\psi \equiv \hat{\Omega}^\dagger\chi = \emptyset. \tag{10.66}$$

According to (10.11) and (10.12), the relation $\hat{\Omega}^\dagger\chi = \emptyset$ from (10.66) means that $\chi \in \mathcal{R}^\perp$ i.e.

$$\chi \equiv \hat{\Omega}_1\psi \in \Theta^\dagger = \mathcal{R}^\perp. \tag{10.67}$$

On the other hand, we obviously have

$$\hat{\Omega}_1\psi \in \mathcal{R}_1 = \mathcal{R} \tag{10.68}$$

where we used completeness (10.59) of the operator $\hat{\Omega}_1$. Relations (10.67) and (10.68) i.e. $\hat{\Omega}_1\psi \in \mathcal{R}^\perp$ and $\hat{\Omega}_1\psi \in \mathcal{R}$, can be simultaneously satisfied only if $\hat{\Omega}_1\psi$ is the zero vector

$$\hat{\Omega}_1\psi = \emptyset. \tag{10.69}$$

Then, we conclude from (10.60), (10.61) and (10.67) that

$$\psi \in \Theta_1 = \mathcal{R}_1^{\dagger\perp} = \mathcal{R}^{\dagger\perp}. \tag{10.70}$$

Since for collisional systems we have

$$R^\dagger = \mathcal{H} \tag{10.71}$$

it follows that the ortho-complement to $\mathcal{R}^\dagger$ is an empty set $\oslash$

$$\mathcal{R}^{\dagger\perp} = \mathcal{R}_1^{\dagger\perp} = \Theta_1 = \oslash. \tag{10.72}$$

Thus, the relation (10.67) implies $\psi \in \oslash$ which means that $\psi = \emptyset$. In other words, we have $\hat{X}\psi = \emptyset$ if and only if $\psi = \emptyset$, meaning that the operator $\hat{X}$ is invertible i.e. there exists the inverse operator $\hat{X}^{-1}$ for each vector $\psi \in \mathcal{H}$. This completes the proof of necessity of the theorem 6.

For the proof of sufficiency of the theorem 6, we shall assume that the inverse operator $\hat{X}^{-1}$ exists and from this we shall show completeness of the operator $\hat{\Omega}_1$ i.e. the validity of the set relation (10.59). Thus, if the operator $\hat{X}^{-1}$ exists, the expression (10.54) can be rewritten in the form

$$\hat{\Omega} = \hat{\Omega}_1 \hat{X}^{-1}. \tag{10.73}$$

On the other hand, for any vector $\phi \in \mathcal{H}$, we know that the element $\hat{\Omega}\phi$ belongs to the sub-space $\mathcal{R}$, which is the range of the wave Møller operator $\hat{\Omega}$

$$\psi = \hat{\Omega}\phi \in \mathcal{R}. \tag{10.74}$$

However, by means of (10.73), the same vector ψ from (10.74) can also be expressed in the following equivalent way

$$\psi = \hat{\Omega}_1 \chi \qquad \chi = \hat{X}^{-1}\phi. \tag{10.75}$$

Of course, the vector $\hat{\Omega}_1\chi$ belongs to the range $\mathcal{R}_1$ of the operator $\hat{\Omega}_1$, so that

$$\psi = \hat{\Omega}_1 \chi \in \mathcal{R}_1. \tag{10.76}$$

The formulae (10.74) and (10.76) lead to the set relation of the form

$$\mathcal{R} \subset \mathcal{R}_1. \tag{10.77}$$

But, according to lemma 13, we have the relationship $\mathcal{R}_1 \subset \mathcal{R}$, which together with (10.77) finally gives $\mathcal{R} = \mathcal{R}_1$. This, according to the definition 1, means that the operator $\hat{\Omega}_1$ is complete thus proving sufficiency (QED). Hence the proof of the entire theorem 6 (QED).

10.7 Scattering operator derived from intertwining wave operators

Let us now give an exceptionally important corollary of the theorem 6:

Corollary 1: If $\hat{\Omega}_1$ and $\hat{\Omega}_2$ are two intertwining operators, then there exists a linear operator $\hat{Z}$ such that

$$\hat{\Omega}_1 = \hat{\Omega}_2 \hat{Z} \qquad [\hat{Z}, \hat{U}_0(t)] = \hat{0}. \tag{10.78}$$

Additionally, if

$$\hat{\Omega}_1 = \hat{\Omega}\,\hat{X}_1 \qquad \hat{\Omega}_2 = \hat{\Omega}\,\hat{X}_2 \tag{10.79}$$

then the following relation is valid for the operator $\hat{Z}$ from (10.78)

$$\hat{Z} = \hat{X}_2^{-1}\hat{X}_1. \tag{10.80}$$

Proof: Let us first extract $\hat{\mathrm{Z}}$ from the defining equation (10.78) via multiplication by $\hat{\Omega}_2^\dagger$ and with a subsequent usage of isometry

$$\hat{\Omega}_k^\dagger \hat{\Omega}_k = \hat{1} \qquad (k = 1, 2). \tag{10.81}$$

This procedure yields

$$\hat{\mathrm{Z}} = \hat{\Omega}_2^\dagger \hat{\Omega}_1 \qquad \hat{\Omega}_k \hat{\mathrm{U}}_0(t) = \hat{\mathrm{U}}(t)\hat{\Omega}_k \qquad \hat{\Omega}_k^\dagger \hat{\mathrm{U}}(t) = \hat{\mathrm{U}}_0(t)\hat{\Omega}_k^\dagger. \tag{10.82}$$

Now (10.48) can be applied to the operators $\hat{\Omega}_k$ $(k = 1, 2)$ that are assumed to be the intertwining operators. This latter feature together with the relation (10.82) combined with the property $\hat{\Omega}_1 = \hat{\Omega}_2 \hat{\mathrm{Z}}$ can be used to prove the commutation property from (10.78) as follows

$$\begin{aligned}
[\hat{\mathrm{Z}}, \hat{\mathrm{U}}_0(t)] &= \hat{\mathrm{Z}}\hat{\mathrm{U}}_0(t) - \hat{\mathrm{U}}_0(t)\hat{\mathrm{Z}} \\
&= \hat{\mathrm{Z}}\hat{\mathrm{U}}_0(t) - \hat{\mathrm{U}}_0(t)\hat{\Omega}_2^\dagger \hat{\Omega}_1 \\
&= \hat{\mathrm{Z}}\hat{\mathrm{U}}_0(t) - \hat{\Omega}_2^\dagger \hat{\mathrm{U}}(t)\hat{\Omega}_1 \\
&= \hat{\mathrm{Z}}\hat{\mathrm{U}}_0(t) - \hat{\Omega}_1^\dagger \hat{\Omega}_1 \hat{\mathrm{U}}_0(t) \\
&= \hat{\mathrm{Z}}\hat{\mathrm{U}}_0(t) - \hat{\mathrm{Z}}\hat{\mathrm{U}}_0(t) = \hat{0}
\end{aligned}$$

$$\therefore \qquad [\hat{\mathrm{Z}}, \hat{\mathrm{U}}_0(t)] = \hat{0}. \tag{10.83}$$

The particular form (10.82) of the operator $\hat{\mathrm{Z}}$ can also be found by applying the theorem 6, which guarantees the existence of the inverse operator $\hat{\mathrm{X}}_k^{-1}$ $(k = 1, 2)$. Multiplying both sides of (10.79) by $\hat{\Omega}_2^\dagger$ and using isometry (10.81) for the operator $\hat{\Omega}_2$, we find

$$\hat{\Omega}_2^\dagger \hat{\Omega}\, \hat{\mathrm{X}}_2 = \hat{1}. \tag{10.84}$$

If this latter expression is multiplied from the right by $\hat{\mathrm{X}}_2^{-1}$ and the relationship $\hat{\mathrm{X}}_k^{-1}\hat{\mathrm{X}}_k = \hat{1}$ is employed, we shall have

$$\hat{\mathrm{X}}_2^{-1} = \hat{\Omega}_2^\dagger \hat{\Omega}. \tag{10.85}$$

Further, multiplying (10.85) from the right by $\hat{\mathrm{X}}_1$ and employing (10.79) and (10.80) will yield the sought result (10.82) for the operator $\hat{\mathrm{Z}}$

$$\hat{\mathrm{X}}_2^{-1}\hat{\mathrm{X}}_1 = \hat{\mathrm{Z}} = \hat{\Omega}_2^\dagger \hat{\Omega} X_1 = \hat{\Omega}_2^\dagger \hat{\Omega}_1 \qquad \text{(QED)}. \tag{10.86}$$

When $\hat{\Omega}_1$ and $\hat{\Omega}_2$ are selected to be the Møller wave operators $\hat{\Omega}^+$ and $\hat{\Omega}^-$, respectively, then (10.78) acquires the form

$$\hat{\Omega}^+ = \hat{\Omega}^- \hat{\mathrm{Z}}. \tag{10.87}$$

After multiplication of (10.87) from the left by $\hat{\Omega}^{-\dagger}$, it follows

$$\hat{Z} = \hat{\Omega}^{-\dagger}\hat{\Omega}^{+}. \tag{10.88}$$

Recall that the standard definition of the scattering $\hat{S}-$operator is given by $\hat{S} = \hat{\Omega}^{-\dagger}\hat{\Omega}^{+}$ [7]. Therefore, for the mentioned choice of $\hat{\Omega}_k\,(k = 1, 2)$, the operator $\hat{Z}$ coincides with the scattering $\hat{S}-$operator

$$\hat{Z} = \hat{S} = \hat{\Omega}^{-\dagger}\hat{\Omega}^{+}. \tag{10.89}$$

In this special case, the commutator from (10.78) becomes

$$[\hat{Z}, \hat{U}_0(t)] = [\hat{S}, \hat{U}_0(t)] = \hat{0} \tag{10.90}$$

which is the mathematical statement for conservation of the total energy of the collisional system under study.

The operators $\hat{\Omega}_1$ and $\hat{\Omega}_2$ from the corollary 1 satisfy the following expression

$$\hat{\Omega}_1^{\dagger}\hat{U}(t)\hat{\Omega}_1 = \hat{\Omega}_2^{\dagger}\hat{U}(t)\hat{\Omega}_2. \tag{10.91}$$

This is obtained through multiplication of the intertwining relation $\hat{\Omega}_k\hat{U}_0(t) = \hat{U}(t)\hat{\Omega}_k$ from the left by by $\hat{\Omega}_k^{\dagger}$

$$\hat{U}_0(t) = \hat{\Omega}_1^{\dagger}\hat{U}(t)\hat{\Omega}_1 = \hat{\Omega}_1^{\dagger}\hat{U}(t)\hat{\Omega}_1. \tag{10.92}$$

Further, the similarity relation (10.92) can be used to prove the commutation relation from (10.78) i.e.

$$\begin{aligned}
[\hat{U}_0(t), \hat{Z}] &= \hat{\Omega}_1^{\dagger}\hat{U}(t)\hat{\Omega}_1\hat{Z} - \hat{Z}\hat{\Omega}_1^{\dagger}\hat{U}(t)\hat{\Omega}_1 \\
&= \hat{\Omega}_1^{\dagger}\hat{U}(t)\hat{\Omega}_1\hat{Z} - \hat{\Omega}_2^{\dagger}\hat{\Omega}_1\hat{\Omega}_1^{\dagger}\hat{U}(t)\hat{\Omega}_1 \\
&= \hat{\Omega}_1^{\dagger}\hat{U}(t)\hat{\Omega}_1\hat{Z} - \hat{\Omega}_2^{\dagger}\hat{\Omega}_1\hat{U}_0(t)\hat{\Omega}_1^{\dagger}\hat{\Omega}_1 \\
&= \hat{\Omega}_1^{\dagger}\hat{U}(t)\hat{\Omega}_1\hat{Z} - \hat{\Omega}_2^{\dagger}\hat{\Omega}_1\hat{U}_0(t) \\
&= \hat{\Omega}_1^{\dagger}\hat{U}(t)\hat{\Omega}_1\hat{Z} - \hat{\Omega}_2^{\dagger}\hat{U}(t)\hat{\Omega}_1 \\
&= \hat{\Omega}_1^{\dagger}\hat{U}(t)\hat{\Omega}_1\hat{Z} - \hat{U}_0(t)\hat{\Omega}_2^{\dagger}\hat{\Omega}_1 \\
&= \hat{\Omega}_1^{\dagger}\hat{U}(t)\hat{\Omega}_1\hat{Z} - \hat{\Omega}_1^{\dagger}\hat{U}(t)\hat{\Omega}_1\hat{Z} = \hat{0}
\end{aligned}$$

$$\therefore \qquad [\hat{U}_0(t), \hat{Z}] = \hat{0}. \tag{10.93}$$

Theorem 7: If $\hat{U}_0(t)$ and $\hat{U}(t)$ are the unitary representations of the additive group of real numbers $t \in \mathbb{R}$ associated with a given scattering system, and if $\hat{\Omega}_1$ is an intertwining operator, there exists, according to lemma 14, a limiting value when $t \to \mp\infty$ of an operator family of the type

$$\hat{\Omega}_1(t) = \hat{U}_0^{\dagger}(t)\hat{\Omega}_1\hat{U}_0(t). \tag{10.94}$$

In such a case, the following relation is valid

$$\hat{\Omega}_1(+\infty) = \hat{S}\hat{\Omega}_1(-\infty) \tag{10.95}$$

where $\hat{S}$ is the scattering operator.

Proof: According to lemma 15, there exists a unique, bounded and linear operator $\hat{X}$ such that

$$\hat{\Omega}_1 = \hat{\Omega}^+\hat{X}. \tag{10.96}$$

If we multiply this relation from the left by $\hat{U}_0^\dagger(t)$, as well as from the right by $\hat{U}_0(t)$ and use the fact that the operators $\hat{\Omega}_1$ and $\hat{U}_0(t)$ commute as per (10.53), it follows

$$\hat{\Omega}_1(t) = \hat{U}_0^\dagger(t)\hat{\Omega}^+\hat{X}\hat{U}_0(t) = \hat{U}_0^\dagger(t)\hat{\Omega}^+\hat{U}_0(t)\hat{X} = \hat{\Omega}^+(t)\hat{X} \tag{10.97}$$

$$\hat{\Omega}^+(t) = \hat{U}_0^\dagger(t)\hat{\Omega}^+\hat{U}_0(t). \tag{10.98}$$

Further, using the well-known relations $\hat{\Omega}^+(-\infty) = \hat{1}$ and $\hat{\Omega}^+(+\infty) = \hat{S}$ [7], we obtain

$$\hat{\Omega}_1(-\infty) = \hat{\Omega}^+(-\infty)\hat{X} = \hat{X} \qquad \hat{\Omega}_1(+\infty) = \hat{\Omega}^+(+\infty)\hat{X} = \hat{S}\,\hat{X}. \tag{10.99}$$

From the derived expressions $\hat{X} = \hat{\Omega}_1(-\infty)$ and $\hat{S}\hat{X} = \hat{\Omega}_1(+\infty)$, we have $\hat{\Omega}_1(+\infty) = \hat{S}\hat{\Omega}_1(-\infty)$ (QED). Thus, an immediate consequence of the theorem 7 is one of the alternative and equivalent definitions of the scattering $\hat{S}-$operator: if for the studied collisional system we are given the free and total (free+bound) dynamics via their evolution operators $\hat{U}_0(t)$ and $\hat{U}(t)$, respectively, then there exists the intertwining operator $\hat{\Omega}_1$ such that an alternative definition of the scattering $\hat{S}-$operator automatically follows from the relationship (10.95).

11

Four-body theories for fast ion-atom collisions

In chapters 11–19, we shall review the progress in solving problems encountered in non-relativistic ion (atom)-atom collisions with the participation of two active electrons. These processes involve e.g. (i) scattering between a bare nucleus (projectile) P of charge Z_P and a helium-like atomic system comprized of two electrons e_1 and e_2 that are initially bound to the target nucleus T of charge Z_T i.e. the $Z_P - (Z_T; e_1, e_2)_i$ collisions, (ii) scattering between two hydrogen-like atoms $(Z_P, e_1)_{i_1}$ and $(Z_T, e_2)_{i_2}$, etc. An adequate theoretical description of these collisional processes requires solutions to difficult four-body problems with four actively participating particles, including two electrons and two nuclei.

Considering different one- as well and two-electron transitions that can take place in these collisions, particular attention will be paid to double-electron capture, simultaneous transfer and ionization, simultaneous transfer and excitation, single electron detachment and single electron capture. For all these collisional phenomena, it is most appropriate to work in the four-body formalism of scattering theory, as adopted in chapters 11–19. As it is by now a well-established standard, we impose the correct Coulomb boundary conditions in both the entrance and exit channels to set the stage for the analysis of a number of the leading quantum mechanical methods. The correct boundary conditions do not consist solely of the scattering wave functions with the associated proper asymptotic behaviors. Additionally, such correctly behaving scattering states must be in full accordance with the corresponding perturbation potentials. Such an indispensable feature of the underlying asymptotic convergence problem will be strongly emphasized in the upcoming analysis.

The most prominent physical feature of four-body ion-atom collisions is the role of the electronic correlations that stem from the electrostatic Coulomb interaction between electrons. We shall examine both types of this key phenomenon i.e. static and dynamic dielectronic correlations, illustrating how they can be incorporated into four-body distorted wave methods. The former and latter correlations refer to spectroscopic and collisional few-body dynamics. Specifically, static correlations play a role in determining the eigenspectrum (binding energies of two electrons and the corresponding discrete-state wave functions) of a given unperturbed helium-like system, which is considered in isolation from the projectile or any other external influences

[58]–[83]. On the other hand, dynamic correlations originate from collisions of helium-like target with impinging projectiles, or in scattering phenomena involving two hydrogen-like atoms or ions. Of course, collisions are are not limited exclusively to dynamic correlations. Collisions can also encompass static correlations whenever a helium-like atom/ion is: (i) present from the outset as a target, (ii) created via single electron capture from a hydrogen-like target by a hydrogen-like projectile or (iii) formed through double charge exchange between a bare nucleus as a projectile and a two-electron target.

Performance of most of the reviewed methods will be illustrated in exhaustive comparisons with practically all the existing experimental data on the subject. In so doing, we shall be guided by both theoretical and practical rationale and considerations in selecting the illustrations. Theoretical rationale will point to those collisional particles that could offer a good possibility to assess the overall role of the chief ingredient of four-body collisions i.e. inter-electron correlations. Practical considerations shall illuminate those scattering aggregates that are important for wider inter-disciplinary applications of energetic four-body ion-atom collisions, ranging from thermonuclear fusion to medical accelerators for hadron radiotherapy.

11.1 Main features of interactive four-body dynamics

Determination of the interactive dynamics of atomic systems is still among the most fundamental challenges in physics. Since the interaction potentials in atomic systems are exactly known, any discrepancy between experimental measurements and theories can be attributed to inappropriate theoretical models for describing many particle systems and/or to unreliable experimental techniques. One of the central questions which arises in scattering problems involving many-electron systems concerns the influence of the electron-electron interaction on the overall dynamics in these collisional phenomena. Since the helium atom or a helium-like ion is the simplest many-electron target where one can assess the importance of electronic correlations, its investigation has attracted the most attention from both the theoretical and experimental sides. Collisional processes in which two nuclei and two electrons take part represent pure four-body problems [84]–[97].

One of the basic motivations for developing four-body theories to treat ion ion-atom collisions is to more thoroughly understand the role of the electron-electron correlations and phase coherences in such important processes. In atomic physics, electronic correlation effects originate from pure Coulombic interactions between active electrons. Phase coherences are interference patterns for competing mechanisms in four-body collisional transitions.

In ion-atom collisions, there are two kinds of electronic correlations: static and dynamic. Static correlations are built into multi-electron bound-state wave functions without any reference to collisions. Quantum mechanical bound states are "prepared" without the presence of an incident beam. Several methods for obtaining bound-state wave functions and the corresponding eigen-energies for two-electron atomic systems have recently been reviewed [83]. The dynamic correlations describe interactions between two electrons in the exit channel, if we deal with the $Z_{\rm P} - (Z_{\rm T}; e_1, e_2)_i$ collisions, or in the entrance channel, if the $(Z_{\rm P}, e_1)_{i_1} - (Z_{\rm T}, e_2)_{i_2}$ process is considered. The electronic interactions alone are capable of causing a transition of the entire collisional system from an initial to a final state. Such a dynamical effect automatically possesses both radial and angular correlations through the inclusion of the inter-electron Coulomb potential $1/r_{12}$ in the final interaction potential V_f appearing in the post form of the transition amplitude T^+_{if}, if the $Z_{\rm P} - (Z_{\rm T}; e_1, e_2)_i$ collisions are studied. The same potential $1/r_{12}$ appears in the initial perturbation potential V_i of the prior form of the transition amplitude T^-_{if}, if the $(Z_{\rm P}, e_1)_{i_1} - (Z_{\rm T}, e_2)_{i_2}$ collisions are investigated.

The majority of the theoretical studies that have considered the $Z_{\rm P} - (Z_{\rm T}; e_1, e_2)_i$ collisions employed the independent-particle model (IPM)[1] [98]–[122]. The basic feature of all these previous investigations within the IPM and its variants is the preservation of a pure three-body formalism, despite the fact that the studied four-body problems include two active electrons. Within the IPM itself, there are many ways of approximating the wave function of a helium-like atomic system. An approach in which an active electron of a two-electron atom or ion moves in an effective potential generated by the other nucleus and the passive electron has frequently been used. The term passive electron is used here in the sense that its interaction with the active electron does not contribute to the collisional process. Thus, in the IPM, the initial four-body problem is effectively reduced to a three-body problem. The main drawback of the IPM is that the dynamic correlation effects during the collisional phenomenon are completely ignored from the outset.

Hence, if we are to adequately assess the role of electron-electron correlations, we must deal with a four-body problem from the beginning. Guided by this argument, various quantum mechanical four-body methods have been proposed to study one- and two-electron transitions in scattering of completely stripped projectiles on helium-like atomic systems or in collisions between two hydrogen-like atoms or ions. In addition to four-body theories, which are the subject of chapters 11–19, the role of electronic correlations in energetic ion-atom collisions has also been investigated in Refs. [123]–[126].

The first formulation and implementation of the four-body continuum distorted wave (CDW-4B) method for double-electron capture was carried out by Belkić and Mančev [84, 85]. The CDW-4B method obeys the asymptotic convergence criteria of Dollard [41] for Coulomb potentials. These initial com-

---

[1] Recall that all acronyms are listed at the end of the book (see pp. 368–370).

putations [84, 85] on the formation of H^- in the $H^+ - He$ collisions yielded total cross sections that were in excellent agreement with available experimental data. Subsequently, the CDW-4B method was applied to other collisional systems [89]–[94], including two-electron capture into singly and doubly excited final states by multiply charged projectile ions. Further, an adequate description of simultaneous transfer and ionization has been devised using the CDW-4B method [127]–[130]. Studies of transfer ionization (TI) by means of the CDW-4B method indicate that dynamic electronic correlations in perturbation potentials are more important than the static ones. The substantial improvement of the CDW-4B method over e.g. the IPM has been attributed solely to the role of dynamic electron correlation effects.

Throughout chapters 11–19, emphasis is placed on the adequate solutions of the asymptotic convergence problem [41, 44] by requiring not only the correct asymptotic behaviors of all the scattering wave functions, but also their proper connections with the corresponding perturbation interactions. This strategy proves to be simultaneously fundamental (consistency of theory by reference to the first principles of physics) and practical (stringent scrutiny of theory through its systematic verification against experiment). A striking example which illustrates this issue is a four-body problem with single electron detachment from H^- by H^+. For this collision, the four-body eikonal Coulomb-Born (ECB-4B) method has been proposed by Gayet, Janev, and Salin [131] with the correct asymptotic behaviors of the initial and final scattering states. Yet, the ensuing total cross sections of this method overestimate the corresponding experimental data by some 2-3 orders of magnitude at all impact energies. As shown by Belkić [132, 133], the reason for this discrepancy was the lack of the proper link between the initial scattering state and the perturbation potential in the entrance channel. When this link has been properly established for the same collisional problem, the four-body modified Coulomb-Born (MCB-4B) method emerged [132, 133], exhibiting excellent agreement with the experimental data at all impact energies.

The MCB-4B method is a simplified version of the CDW-4B method for ionization in collisions of protons with the atomic negative hydrogen ion. The simplification is in replacing the full electronic Coulomb wave function from the entrance channel by its asymptotic form in the prior transition amplitude. Otherwise, the CDW-4B method for ionization is the extension of the corresponding three-body continuum distorted wave (CDW-3B) method of Belkić [134] for ionization in the pure three-body break-up collisions of nuclei and hydrogen-like atomic systems. In this latter ionization problem, Belkić [134] originally derived the scattering wave for the final state of three free particles as the product of three full Coulomb functions (later called the C3 function) which satisfies the correct boundary condition for three free charged particles in the exit channel. Afterwards in numerous studies, the C3 scattering wave function of Belkić [134] has repeatedly been rediscovered by e.g. Garibotti and Miraglia [135]–[137], also by Brauner, Briggs, and Klar [138], as well as by others [139]–[144]. Throughout the years [145]–[165] it was firmly established that the most adequate and successful theories for heavy-particle ion-atom ionization at high energies and multiply charge projectiles are the CDW-3B method [134] and its further approximation or a 'derivative' known as the three-body continuum distorted wave eikonal initial state (CDW-EIS-3B) method [151]. Of late, the CDW methodologies have been exported to neighboring research fields, such as medical physics for a more adequate description of the stopping power of multiply charged ions passing through matter, as encountered in applications to hadron radiotherapy [166]–[170].

A number of studies in the past literature on multiple scattering theories were based upon the impulse hypothesis and they dealt mainly with charge exchange and ionization in energetic ion-atom collisions [171]–[178]. For example, the three-body reformulated impulse approximation (RIA-3B) of Belkić [177]–[179], after resolving a long-lasting problem of the inadequacy of the corresponding impulse approximation (IA) for the total cross sections in the $H^+ - H$ charge exchange, has successfully been extended to four-body collisions. The cross sections of the four-body reformulated impulse approximation (RIA-4B) of Belkić for transfer ionization in the $H^+ - He$ collisions have been reported in a joint theoretical-experimental study [180]. The total cross sections of the RIA-4B for the TI process have clearly indicated the trend of the v^{-11} behavior at sufficiently large values of the impact velocity v. This asymptotic behavior, as the quantum mechanical counterpart of the corresponding Thomas classical double scattering [181], has been confirmed on the same collision by two subsequent measurements [182, 183].

As a further exploration of the CDW-4B method, simultaneous transfer and excitation (TE) have also constituted a subject of studies [184]–[188]. This process takes places when a target electron is captured by a non-bare projectile, while the initial electronic structure of the latter is excited at the same time. For the process of TE, where a double excited (auto-ionizing) state is formed on the projectile, two modes have been identified and termed as the resonant transfer excitation (RTE) and the non-resonant transfer excitation (NTE). In the RTE, excitation of the projectile is due to the dielectronic interaction between the projectile electron and the target electron which is captured. In the NTE, a target electron is transferred and excitation of the projectile comes from the interaction with the rest of the target. In addition to these two-electron transitions, the CDW-4B method has also been successfully applied to single electron capture [130, 189, 190] in a number of processes, such as the $H^+ - He$, $H^+ - Li^+$, $He^{2+} - He$ and the $Li^{3+} - He$ collisions. In the CDW-4B method, the electronic continuum intermediate states are included in both channels through the full Coulomb waves.

In the boundary-corrected four-body first Born (CB1-4B) approximation, pure electronic continuum intermediate states are not taken into account. Here, the scattering state vectors are given by the product of the unperturbed channel states and the logarithmic distortion phase factors due to the Coulomb long-range remainders of the perturbation potential. The CB1-4B method has initially been formulated and applied to double electron capture by Belkić

[87, 88]. This method has subsequently been used for single charge exchange in energetic collisions between two hydrogen-like atoms or ions [191, 192].

The four-body boundary-corrected continuum intermediate state (BCIS-4B) and the four-body Born distorted wave (BDW-4B) methods of Belkić [86, 89] have been introduced and used for investigation of both single and double electron capture. These two methods with asymmetric treatment of the initial and final channels can be applied and extended to any number of colliding particles so that the more generic acronyms BCIS and BDW can be used. Both methods employ the scattering wave functions from the CDW-4B method in one of the two channels, either in the entrance or the exit channel, for the initial or the final state, depending on whether the prior or post form of the transition amplitudes is used. For the other channel, the BCIS and BDW methods use the corresponding wave functions of the CB1 method. As a result, the distorting potentials that cause the transitions from the initial to the final states of the systems are different in the BCIS and BDW methods. These latter potentials are the usual electrostatic Coulomb interactions in the BCIS method (shared by the CB1 method), whereas they are the operator-valued potentials $\boldsymbol{\nabla} \cdot \boldsymbol{\nabla}$ in the BDW method (shared by the CDW method). Thus, if one wishes to make these remarks more transparent, the original acronym, such as BDW used by Belkić [89], could be relabelled as CDW-CB1. In particular, the notations for the post and prior BDW, or equivalently, CDW-CB1 can further be differentiated by highlighting the use of the boundary corrected first-order Born initial and final states (BIS, BFS). This has led to yet another equivalent set of acronyms CDW-BIS and CDW-BFS [193]–[196] for the post and the prior versions of the BDW method of Belkić [89]. Using the BCIS and BDW methods, the studies of Belkić [86, 89] have shown that double charge exchange is remarkably sensitive to the inclusion of the long-range Coulomb effects through the electronic continuum states. These latter states play an important role even at those incident energies at which the Thomas double scattering is not apparent. By means of the mentioned hybrid four-body approximations, one could study various mechanisms that can produce the Thomas peaks in the differential cross sections. Even for single charge exchange with helium-like targets, these methods deal explicitly with two active electrons from the outset and, therefore, they preserve the four-body nature of the original problem. The post and prior BDW methods (or equivalently, the CDW-BIS and CDW-BFS methods, respectively), have successfully been applied to compute both differential and total cross sections for single electron capture in collisions between bare projectiles and helium-like atoms or ions [193]–[196].

Additionally, there are other hybrid type approximations with the correct boundary conditions such as the CDW-EIS method [151] and the continuum distorted wave eikonal final state (CDW-EFS) methods [197, 198]. The CDW-EIS method has been introduced by Crothers and McCann [151] for ionization of hydrogen-like atomic systems by nuclei within a pure three-body formalism. In the framework of three- and four- or many-body formalisms, the CDW-

EIS method is always a further approximation to the corresponding CDW method for ionization. Both methods share the common C3 wave function for the exit channel. They differ in the entrance channel, where the CDW-EIS method introduces an additional approximation through the replacement of the electronic full Coulomb wave from the CDW method by the corresponding logarithmic phase as an eikonal factor. The same type of relationship also holds between the CDW and MCB methods.

In Ref. [198], the CDW-EIS and CDW-EFS methods for single electron capture from a two-electron target are reduced to a one-electron process. Here, the active (captured) electron has been described by an orbital of the frozen-core type (the so-called $5z$ function) from the Roothaan-Hartree-Fock (RHF) model [65]–[68]. The other non-captured electron is passive, since it is considered as being frozen in its initial state during the collision. Therefore, such versions of the CDW-EIS and CDW-EFS methods [197, 198] belong to the category of three-body approximations. As to pure four-body collisions with two active electrons, the four-body continuum distorted wave eikonal initial state (CDW-EIS-4B) method with the correct boundary conditions has also been introduced and applied to double capture from helium by alpha particles, but without any success [94]. The CDW-EIS and the CDW-EFS methods differ from the CDW-BIS and the CDW-BFS methods, since EIS and EFS are different from BIS and BFS, respectively. Specifically, the difference is in the dependent variables in the eikonal phase factors for the two sets of the invoked asymptotic states $\{\text{EIS}, \text{EFS}\}$ and $\{\text{BIS}, \text{BFS}\}$.

The dominant feature of most of the mentioned quantum mechanical four-body methods is that they show systematic agreement with the corresponding experimental data at intermediate and high impact energies. This is striking in view of the fact that the impact parameter versions of the investigated approximations often fail (and do so dramatically in some cases) in their attempts to reproduce experimental data. The clear implication of such an occurrence is that the dynamic correlation effects are of critical importance for two-electron transitions. One of the tasks of our analysis will be to highlight this latter feature and to assess its overall significance for energetic ion-atom collisions with two actively participating electrons. The major goal of this book is to critically evaluate the efficiency and the overall utility of the leading methods within the realm of four-body quantum mechanical scattering theory. Intermediate and high non-relativistic energies permit consistent extension of rigorous pure three-body distorted wave methods to their pure four-body counterparts without any significant additional approximation. This represents an excellent opportunity to estimate the relevance of the well-known asymptotic convergence problem from formal scattering theory for Coulomb potentials when more than three particles are actively involved. Such an opportunity will presently be taken by building on the past successful experience with similar challenges encountered in simpler three-body ion-atom rearrangement collisions for which Belkić *et al.* [44] have conclusively established the critical importance of the correct Coulomb boundary conditions in the most general case with the exact eikonal transition amplitude. Subsequent detailed numerical computations (with dramatic improvements relative to experimental data) in the CB1-4B method [199]–[207] and the boundary-corrected three-body second Born (CB2-3B) method [208]–[213] confirmed the validity of this theoretical concept which was then widely accepted and reviewed in several articles and books on the subject [45]–[50].

In chapters 11–19, atomic units will be used unless otherwise stated.

11.2 Notation and basic formulae

In chapters 11–19, we are interested in ion-atom collisions in which two electrons take part. Such processes involve (i) scattering between a bare nucleus projectile P of charge $Z_{\rm P}$ and a helium-like atomic system consisting of two electrons e_1 and e_2 initially bound to the target nucleus T of charge $Z_{\rm T}$ i.e. $Z_{\rm P} - (Z_{\rm T}; e_1, e_2)_i$ collisions, where the parentheses indicate the bound states, (ii) scattering between two hydrogen-like atoms $(Z_{\rm P}, e_1)_{i_1}$ and $(Z_{\rm T}, e_2)_{i_2}$, etc. We adopt the quantum mechanical non-relativistic spin-independent formalism, which permits consideration of the two electrons as being distinguishable from each other. Among various processes that can occur in such collisions, special attention will be paid to the following rearrangement collisions:

I. Collisions of nuclei with helium-like atoms or ions

double electron transfer

$$Z_{\rm P} + (Z_{\rm T}; e_1, e_2)_i \longrightarrow (Z_{\rm P}; e_1, e_2)_f + Z_{\rm T} \tag{11.1}$$

simultaneous electron transfer and ionization

$$Z_{\rm P} + (Z_{\rm T}; e_1, e_2)_i \longrightarrow (Z_{\rm P}, e_1)_f + Z_{\rm T} + e_2 \tag{11.2}$$

single electron transfer

$$Z_{\rm P} + (Z_{\rm T}; e_1, e_2)_i \longrightarrow (Z_{\rm P}, e_1)_{f_1} + (Z_{\rm T}, e_2)_{f_2} \tag{11.3}$$

single electron ionization or detachment

$$Z_{\rm P} + (Z_{\rm T}; e_1, e_2)_i \longrightarrow Z_{\rm P} + (Z_{\rm T}, e_2)_f + e_1 \tag{11.4}$$

II. Collisions of two hydrogen-like atoms or ions

single electron transfer

$$(Z_{\rm P}, e_1)_{i_1} + (Z_{\rm T}, e_2)_{i_2} \longrightarrow (Z_{\rm P}; e_1, e_2)_f + Z_{\rm T} \tag{11.5}$$

simultaneous electron transfer and excitation

$$(Z_{\rm P}, e_1)_{i_1} + (Z_{\rm T}, e_2)_{i_2} \longrightarrow (Z_{\rm P}; e_1, e_2)_f^{**} + Z_{\rm T} \tag{11.6}$$

where indices i, f, i_1, i_2, f_1 and f_2 represent the collective labels for the set of quantum numbers needed to describe the initial and final bound states, while the double asterisk denotes the doubly excited state.

Hereafter, the position vectors of the projectile nucleus, the target nucleus, and electrons $e_{1,2}$ relative to an arbitrary coordinate frame are denoted by $\boldsymbol{r}_1$, $\boldsymbol{r}_2$, $\boldsymbol{r}_3$ and $\boldsymbol{r}_4$, respectively. The kinetic energy operator is

$$H_0 = -\frac{1}{2M_\text{P}}\nabla^2_{r_1} - \frac{1}{2M_\text{T}}\nabla^2_{r_2} - \frac{1}{2}\nabla^2_{r_3} - \frac{1}{2}\nabla^2_{r_4} \tag{11.7}$$

where M_P and M_T are the masses of the projectile and the target, respectively. The respective position vectors of the electrons $e_{1,2}$ relative to Z_P and Z_T are labelled by $\boldsymbol{s}_{1,2}$ and $\boldsymbol{x}_{1,2}$. We denote by $\boldsymbol{R}$ the position vector of the projectile Z_P relative to Z_T, and by r_{12} the vector of the inter-electron distance so that $\boldsymbol{R} = \boldsymbol{r}_1 - \boldsymbol{r}_2$ and $\boldsymbol{r}_{12} = \boldsymbol{r}_3 - \boldsymbol{r}_4$

$$\boldsymbol{R} = \boldsymbol{x}_1 - \boldsymbol{s}_1 = \boldsymbol{x}_2 - \boldsymbol{s}_2 \qquad \boldsymbol{r}_{12} = \boldsymbol{x}_1 - \boldsymbol{x}_2 = \boldsymbol{s}_1 - \boldsymbol{s}_2. \tag{11.8}$$

11.3 The entrance channel

11.3.1 The $Z_\text{P} - (Z_\text{T}; e_1, e_2)_i$ collisional system

We first concentrate on the collisions of completely stripped projectiles with helium-like targets i.e. the $Z_\text{P} - (Z_\text{T}; e_1, e_2)_i$ collisions. Introducing $\boldsymbol{r}_i$ as a relative vector of Z_P with respect to the center of mass of the target $(Z_\text{T}; e_1, e_2)_i$, we have

$$\boldsymbol{r}_i = \boldsymbol{r}_1 - \frac{\boldsymbol{r}_3 + \boldsymbol{r}_4 + M_\text{T}\boldsymbol{r}_2}{M_\text{T} + 2}. \tag{11.9}$$

In addition to (11.7), it is convenient to express the Hamiltonian H_0 in an alternative form via another set of the independent variables such as $\{\boldsymbol{x}_1, \boldsymbol{x}_2, \boldsymbol{r}_i\}$

$$H_0 = -\frac{1}{2\mu_i}\nabla^2_{r_i} - \frac{1}{2b}\nabla^2_{x_1} - \frac{1}{2b}\nabla^2_{x_2} - \frac{1}{M_\text{T}}\boldsymbol{\nabla}_{x_1} \cdot \boldsymbol{\nabla}_{x_2} \tag{11.10}$$

where $\mu_i = M_\text{P}(M_\text{T} + 2)/(M_\text{P} + M_\text{T} + 2)$ and $b = M_\text{T}/(M_\text{T} + 1)$. The last term in (11.10) is the so-called mass polarization term, which can be neglected for heavy particles because $M_\text{T} \gg 1$.

The full Hamiltonian of the system under study, in the center of mass frame for the whole system, is given by

$$H = H_0 + V \tag{11.11}$$

where V represents the total interaction potential operator

$$V = \frac{Z_\text{P}Z_\text{T}}{R} - \frac{Z_\text{P}}{s_1} - \frac{Z_\text{P}}{s_2} - \frac{Z_\text{T}}{x_1} - \frac{Z_\text{T}}{x_2} + \frac{1}{r_{12}}. \tag{11.12}$$

As usual for rearranging collisions, the complete Hamiltonian from (11.11) can further be split as

$$H = H_i + V_i. \tag{11.13}$$

Here, H_i and V_i are the channel Hamiltonian and the perturbation potential in the entrance channel

$$H_i = H_0 - \frac{Z_{\mathrm{T}}}{x_1} - \frac{Z_{\mathrm{T}}}{x_2} + \frac{1}{r_{12}} \tag{11.14}$$

$$V_i = \frac{Z_{\mathrm{P}}Z_{\mathrm{T}}}{R} - \frac{Z_{\mathrm{P}}}{s_1} - \frac{Z_{\mathrm{P}}}{s_2}. \tag{11.15}$$

The unperturbed channel state Φ_i is defined by

$$(H_i - E)\Phi_i = 0 \qquad \Phi_i = \varphi_i(\boldsymbol{x}_1, \boldsymbol{x}_2)\mathrm{e}^{i\boldsymbol{k}_i\cdot\boldsymbol{r}_i}. \tag{11.16}$$

The function $\varphi_i(\boldsymbol{x}_1, \boldsymbol{x}_2)$ represents the two-electron bound state wave function of the atomic system $(Z_{\mathrm{T}}; e_1, e_2)_i$ whereas $\boldsymbol{k}_i$ is the initial wave vector. The wave function φ_i satisfies the following eigen-problem

$$(h_i - E_i)\varphi_i(\boldsymbol{x}_1, \boldsymbol{x}_2) = 0 \tag{11.17}$$

$$h_i = -\frac{1}{2b}\nabla^2_{x_1} - \frac{1}{2b}\nabla^2_{x_2} - \frac{Z_{\mathrm{T}}}{x_1} - \frac{Z_{\mathrm{T}}}{x_2} + \frac{1}{r_{12}} \tag{11.18}$$

where h_i is the electronic Hamiltonian and E_i is the electronic binding energy. The total energy E of the whole four-body system in the entrance channel is

$$E = \frac{k_i^2}{2\mu_i} + E_i. \tag{11.19}$$

The wave functions of two-electron atomic systems have been a subject of extensive studies over the years [58]–[83]. In the case of helium, the remarkable variational estimate $E_i = -2.903724377034105$ [82] via a fully correlated Hylleraas wave function, with an explicit allowance for the inter-electron coordinate r_{12} (through some 600 expansion terms), could be treated as practically the exact value.

The initial state Φ_i is distorted even at infinity, due to the presence of the asymptotic Coulomb repulsive potential

$$V_i^{\infty} = \frac{Z_{\mathrm{P}}(Z_{\mathrm{T}} - 2)}{R} \tag{11.20}$$

which represents the interaction between the projectile and the screened target nucleus. Notice that V_i^{∞} is the asymptotic value of the perturbation V_i

$$V_i = \frac{Z_{\mathrm{P}}Z_{\mathrm{T}}}{R} - \frac{Z_{\mathrm{P}}}{s_1} - \frac{Z_{\mathrm{P}}}{s_2} \longrightarrow \frac{Z_{\mathrm{P}}(Z_{\mathrm{T}} - 2)}{R} = V_i^{\infty} \qquad (r_i \to \infty). \tag{11.21}$$

Bearing in mind the long-range nature of the Coulomb interaction, the Hamiltonian H can be decomposed according to

$$H = H_i^d + V_i^d \tag{11.22}$$

$$H_i^d = -\frac{1}{2\mu_i}\nabla_{r_i}^2 + \frac{Z_{\rm P}(Z_{\rm T}-2)}{r_i} - \frac{1}{2b}\nabla_{x_1}^2 - \frac{1}{2b}\nabla_{x_2}^2 - \frac{Z_{\rm T}}{x_1} - \frac{Z_{\rm T}}{x_2} + \frac{1}{r_{12}} \tag{11.23}$$

$$V_i^d = \frac{Z_{\rm P}Z_{\rm T}}{R} - \frac{Z_{\rm P}(Z_{\rm T}-2)}{r_i} - \frac{Z_{\rm P}}{s_1} - \frac{Z_{\rm P}}{s_2}. \tag{11.24}$$

The potential V_i^d exhibits a short-range behavior when $R \to \infty$. The difference $1/R - 1/r_i$ is of the order of δ smaller than $[\boldsymbol{R} \cdot (\boldsymbol{x}_1 + \boldsymbol{x}_2)]/R^3$, where $\delta = 1/(M_{\rm T}+2)$, as can easily be checked by using the Taylor series expansion. Thus, neglecting the terms of the order of and smaller than $1/M_{\rm T}$, we have $\boldsymbol{r}_i \simeq \boldsymbol{R}$, so that V_i^d can be approximated as

$$V_i^d = \frac{2Z_{\rm P}}{R} - \frac{Z_{\rm P}}{s_1} - \frac{Z_{\rm P}}{s_2}. \tag{11.25}$$

Obviously, V_i^d tends to $\mathcal{O}(1/R^2)$ as $R \to \infty$. It should be emphasized that the perturbation V_i^d includes only the interactions between the projectile and the two electrons. The term $2Z_{\rm P}/R$ in (11.25), despite its R–dependent form, is not related at all to the inter-nuclear potential, but originates solely from the electron-projectile interaction. The asymptotic tail of the potential $-Z_{\rm P}/s_1$ is $-Z_{\rm P}/R$, since $s_1 \to R$ as $R \to \infty$. This can be seen by utilizing the Taylor expansion for $Z_{\rm P}/s_1$ around R. The small value of the x_1 coordinate in the entrance channel justifies this development. The same statement also holds true for the potential $-Z_{\rm P}/s_2$. It is important to note that, unlike the channel perturbation V_i^d, the corresponding perturbation V_i from (11.15) contains the inter-nuclear interaction $Z_{\rm P}Z_{\rm T}/R$. With the Hamiltonian H_i^d from (11.23), the eigen-problem in the entrance channel reads as

$$(H_i^d - E)\Phi_i^d = 0. \tag{11.26}$$

This is the counterpart of (11.16) when there is a remaining Coulomb potential in the asymptotic region. The solution of the eigen-problem for Φ_i^d is given by

$$\Phi_i^d = \varphi_i(\boldsymbol{x}_1, \boldsymbol{x}_2)\mathrm{e}^{i\boldsymbol{k}_i\cdot\boldsymbol{r}_i}\mathcal{N}^+(\nu_i)\,{}_1F_1(-i\nu_i, 1, ik_ir_i - i\boldsymbol{k}_i\cdot\boldsymbol{r}_i) \tag{11.27}$$

where $\mathcal{N}^+(\nu_i) = \mathrm{e}^{-\pi\nu_i/2}\Gamma(1+i\nu_i)$ and $\nu_i = Z_{\rm P}(Z_{\rm T}-2)/v$. The symbol ${}_1F_1(a,b,z)$ stands for the Kummer confluent hyper-geometric function. The Coulomb wave function Φ_i^d has the asymptotic form

$$\Phi_i^d(r_i \to \infty) \equiv \Phi_i^+ = \varphi_i(\boldsymbol{x}_1, \boldsymbol{x}_2)\mathrm{e}^{i\boldsymbol{k}_i\cdot\boldsymbol{r}_i + i\nu_i\ln(k_ir_i - \boldsymbol{k}_i\cdot\boldsymbol{r}_i)}. \tag{11.28}$$

Thus the function (11.27) satisfies the correct Coulomb boundary condition in the entrance channel.

We recall that the wave functions in the attractive Coulomb field $U = -\beta/r$ have the following forms [214]

$$\Psi_{\boldsymbol{k}}^{-} = N^{-}(\nu)\mathrm{e}^{i\boldsymbol{k}\cdot\boldsymbol{r}}\, {}_1F_1(-i\nu, 1, -ikr - i\boldsymbol{k}\cdot\boldsymbol{r}) \tag{11.29}$$

$$\Psi_{\boldsymbol{k}}^{+} = N^{+}(\nu)\mathrm{e}^{i\boldsymbol{k}\cdot\boldsymbol{r}}\, {}_1F_1(i\nu, 1, ikr - i\boldsymbol{k}\cdot\boldsymbol{r}) \tag{11.30}$$

where $N^{\pm}(\nu) = \mathrm{e}^{\pi\nu/2}\Gamma(1 \mp i\nu)$, $\nu = \mu\beta/k$ and μ is the reduced mass. At $r \to \infty$, this function behaves like

$$\Psi_{\boldsymbol{k}}^{-} \simeq \mathrm{e}^{i\boldsymbol{k}\cdot\boldsymbol{r} + i\nu\ln(kr + \boldsymbol{k}\cdot\boldsymbol{r})} + \mathcal{O}^{-}(1/r). \tag{11.31}$$

$$\Psi_{\boldsymbol{k}}^{+} \simeq \mathrm{e}^{i\boldsymbol{k}\cdot\boldsymbol{r} - i\nu\ln(kr - \boldsymbol{k}\cdot\boldsymbol{r})} + \mathcal{O}^{+}(1/r) \tag{11.32}$$

In the case of a repulsive field $U = \beta/r$, the incoming and outgoing waves are given by

$$\Psi_{\boldsymbol{k}}^{-} = \mathcal{N}^{-}(\nu)\mathrm{e}^{i\boldsymbol{k}\cdot\boldsymbol{r}}\, {}_1F_1(i\nu, 1, -ikr - i\boldsymbol{k}\cdot\boldsymbol{r}) \tag{11.33}$$

$$\Psi_{\boldsymbol{k}}^{+} = \mathcal{N}^{+}(\nu)\mathrm{e}^{i\boldsymbol{k}\cdot\boldsymbol{r}}\, {}_1F_1(-i\nu, 1, ikr - i\boldsymbol{k}\cdot\boldsymbol{r}) \tag{11.34}$$

where $\mathcal{N}^{\pm}(\nu) = \mathrm{e}^{-\pi\nu/2}\Gamma(1 \pm i\nu)$, so that the appropriate asymptotic forms read as

$$\Psi_{\boldsymbol{k}}^{-} \simeq \mathrm{e}^{i\boldsymbol{k}\cdot\boldsymbol{r} - i\nu\ln(kr + \boldsymbol{k}\cdot\boldsymbol{r})} + \mathcal{O}_{-}(1/r). \tag{11.35}$$

$$\Psi_{\boldsymbol{k}}^{+} \simeq \mathrm{e}^{i\boldsymbol{k}\cdot\boldsymbol{r} + i\nu\ln(kr - \boldsymbol{k}\cdot\boldsymbol{r})} + \mathcal{O}_{+}(1/r) \tag{11.36}$$

Here, the symbols $\mathcal{O}^{\pm}(1/r)$ and $\mathcal{O}_{\pm}(1/r)$ represent the remainders of the asymptotes for $\Psi_{\boldsymbol{k}}^{\pm}$ for the potentials $\mp\beta/r$, respectively. The two Coulomb wave functions $\Psi_{\boldsymbol{k}}^{\pm}$ for each of the two potentials are normalized to the usual δ-function, as in the corresponding plane waves

$$\langle\Psi_{\boldsymbol{k}'}^{\pm}|\Psi_{\boldsymbol{k}}^{\pm}\rangle = (2\pi)^3\delta(\boldsymbol{k}' - \boldsymbol{k}).$$

The normalization constant is chosen so that the corresponding wave function has unit amplitude.

11.3.2 The $(Z_{\mathrm{P}}, e_1)_{i_1} - (Z_{\mathrm{T}}, e_2)_{i_2}$ collisional system

The additive form of the Hamiltonian for this collision in the entrance channel is given by (11.22) with the redefinitions

$$\begin{aligned} V_i^d &= \frac{Z_{\mathrm{P}}Z_{\mathrm{T}}}{R} - \frac{Z_{\mathrm{P}}}{s_2} - \frac{Z_{\mathrm{T}}}{x_1} + \frac{1}{r_{12}} - \frac{(Z_{\mathrm{P}}-1)(Z_{\mathrm{T}}-1)}{r_i} \\ &\simeq \frac{Z_{\mathrm{T}} + Z_{\mathrm{P}} - 1}{R} - \frac{Z_{\mathrm{P}}}{s_2} - \frac{Z_{\mathrm{T}}}{x_1} + \frac{1}{r_{12}} \end{aligned} \tag{11.37}$$

$$H_i^d = H_0 - \frac{Z_{\rm P}}{s_1} - \frac{Z_{\rm T}}{x_2} + \frac{(Z_{\rm P}-1)(Z_{\rm T}-1)}{r_i} \tag{11.38}$$

$$H_0 = -\frac{1}{2\mu_i}\nabla_{r_i}^2 - \frac{1}{2a}\nabla_{s_1}^2 - \frac{1}{2b}\nabla_{x_2}^2 \tag{11.39}$$

where $\mu_i = (M_{\rm P}+1)(M_{\rm T}+1)/(M_{\rm P}+M_{\rm T}+2)$, $a = M_{\rm P}/(M_{\rm P}+1)$ and $b = M_{\rm T}/(M_{\rm T}+1)$. Here, $\boldsymbol{r}_i$ is the position vector between the centers of mass of the $(Z_{\rm T}, e_2)$ and $(Z_{\rm P}, e_1)$ systems. The set $\{\boldsymbol{s}_1, \boldsymbol{x}_2, \boldsymbol{r}_i\}$ represents the independent coordinates. The asymptotic channel state Φ_i^+ is defined by

$$(H_i^d - E)\Phi_i^+ = 0 \qquad \Phi_i^+ = \varphi_{i_1}(\boldsymbol{s}_1)\varphi_{i_2}(\boldsymbol{x}_2){\rm e}^{i\boldsymbol{k}_i\cdot\boldsymbol{r}_i + i\nu_i \ln(k_i r_i - \boldsymbol{k}_i\cdot\boldsymbol{r}_i)} \tag{11.40}$$

where $\nu_i = (Z_{\rm P}-1)(Z_{\rm T}-1)/v$. The functions $\varphi_{i_1}(\boldsymbol{s}_1)$ and $\varphi_{i_2}(\boldsymbol{x}_2)$ are the hydrogen-like initial states in the entrance channel for the $(Z_{\rm P}, e_1)_{i_1}$ and $(Z_{\rm T}, e_2)_{i_2}$ systems, respectively.

11.4 The exit channels

11.4.1 Double electron capture

In analogy to the vector $\boldsymbol{r}_i$ from (11.9), we can also consider $\boldsymbol{r}_f$ as the position vector of T with respect to the center of mass of the system $(Z_{\rm P}; e_1, e_2)_f$ via

$$\boldsymbol{r}_f = \boldsymbol{r}_2 - \frac{\boldsymbol{r}_3 + \boldsymbol{r}_4 + M_{\rm P}\boldsymbol{r}_1}{M_{\rm P}+2}. \tag{11.41}$$

With this, Hamiltonian H_0 can be written in terms of the independent variables $\{\boldsymbol{s}_1, \boldsymbol{s}_2, \boldsymbol{r}_f\}$ as

$$H_0 = -\frac{1}{2\mu_f}\nabla_{r_f}^2 - \frac{1}{2a}\nabla_{s_1}^2 - \frac{1}{2a}\nabla_{s_2}^2 - \frac{1}{M_{\rm P}}\boldsymbol{\nabla}_{s_1}\cdot\boldsymbol{\nabla}_{s_2} \tag{11.42}$$

where $\mu_f = M_{\rm T}(M_{\rm P}+2)/(M_{\rm P}+M_{\rm T}+2)$. The mass polarization term $(1/M_{\rm P})\boldsymbol{\nabla}_{s_1}\cdot\boldsymbol{\nabla}_{s_2}$ can be omitted in accordance with the mass limit $M_{\rm P} \gg 1$ for heavy particles. It is convenient to express the total Hamiltonian in a separable form $(H = H_f + V_f)$ where the channel Hamiltonian H_f and corresponding perturbation V_f are

$$H_f = H_0 - \frac{Z_{\rm P}}{s_1} - \frac{Z_{\rm P}}{s_2} + \frac{1}{r_{12}} \tag{11.43}$$

$$V_f = \frac{Z_{\rm P}Z_{\rm T}}{R} - \frac{Z_{\rm T}}{x_1} - \frac{Z_{\rm T}}{x_2}. \tag{11.44}$$

We introduce the unperturbed state Φ_f in the exit channel for double charge exchange as the solution of the eigen-problem

$$(H_f - E)\Phi_f = 0 \tag{11.45}$$

which yields

$$\Phi_f = \varphi_f(\boldsymbol{s}_1, \boldsymbol{s}_2)\mathrm{e}^{-i\boldsymbol{k}_f\cdot\boldsymbol{r}_f} \tag{11.46}$$

where $\varphi_f(\boldsymbol{s}_1, \boldsymbol{s}_2)$ is the bound state of the helium-like atom or ion $(Z_{\mathrm{P}}; e_1, e_2)_f$. This function satisfies the following eigen-problem

$$(h_f - E_f)\varphi_f(\boldsymbol{s}_1, \boldsymbol{s}_2) = 0 \tag{11.47}$$

$$\left(-\frac{1}{2a}\nabla^2_{s_1} - \frac{1}{2a}\nabla^2_{s_2} - \frac{Z_{\mathrm{P}}}{s_1} - \frac{Z_{\mathrm{P}}}{s_2} + \frac{1}{r_{12}} - E_f\right)\varphi_f(\boldsymbol{s}_1, \boldsymbol{s}_2) = 0 \tag{11.48}$$

where E_f is the binding energy of the final state. The total energy E in the exit channel is

$$E = \frac{k_f^2}{2\mu_f} + E_f \tag{11.49}$$

where $\boldsymbol{k}_f$ is the final wave vector, so that by reference to (11.19) it follows

$$E = \frac{k_i^2}{2\mu_i} + E_i = \frac{k_f^2}{2\mu_f} + E_f. \tag{11.50}$$

Of course, the total energy E is conserved in the asymptotic region of the entrance and exit channels when the collision is completed, and where all scattering experiments are performed (on-shell phenomena for real physical events). However, during the collisional event itself, in certain intermediate channels or stages in the interaction region, which is at any rate inaccessible to direct measurements, the total energy might not be conserved (off-shell mechanisms for undetectable, virtual events). In other words, certain intermediate off-shell matrix elements, as non-physical transition amplitudes, can be used during a theoretical analysis, but the physical transition amplitudes between the scattering aggregates' initial and final states, that are related to experimentally measurable quantities i.e. observables must ultimately conserve the total energy of the whole system. As such, the total energy $k_i^2/(2\mu_i) + E_i$ in the entrance channel ought to be equal to the total energy in the exit channel $k_f^2/(2\mu_f) + E_f$, as per (11.50). Intermediate-stage off-shell mechanisms can indirectly be revealed through some observables that are the results of a real physical phenomenon. For example, in a combined nuclear-atomic process called internal conversion, a nucleus is deexcited by releasing a virtual photon, which is absorbed by one of the $K-$shell electrons in the given multi-electron atom. The energized $K-$shell electron subsequently jumps to an excited state of the atom, which finally becomes deexcited through the photon emission. The latter photon is real, as it can be detected experimentally. Here, this very act of measurement, in fact, lends support to the assumption of the theory that the emergence of the real photon was made possible by the appearance of a virtual photon in the intermediate stage of the collision. In fast atomic collisions, such as the well-known billiard type Thomas-like double scattering processes, experimentally observed velocity distributions of recoil particles are broadened relative to the corresponding predictions of the classical physics. However, this detected broadening is a quantum mechanical phenomenon. It can be explained using quantum mechanical second-order theories by evoking the concept of energy non-conservation in the intermediate collisional states as illustrated in Ref. [215].

The final state is distorted even at asymptotically large separations due to the presence of an overall repulsive long-range Coulomb interaction between the target nucleus and the screened projectile

$$V_f^{\infty} = \frac{Z_{\rm T}(Z_{\rm P}-2)}{R} \tag{11.51}$$

similarly to (11.20). This suggests that the total Hamiltonian should be written in the following additive form

$$H = H_f^d + V_f^d \tag{11.52}$$

$$H_f^d = -\frac{1}{2\mu_f}\nabla_{r_f}^2 + \frac{Z_{\rm T}(Z_{\rm P}-2)}{r_f} - \frac{1}{2a}\nabla_{s_1}^2 - \frac{1}{2a}\nabla_{s_2}^2 - \frac{Z_{\rm P}}{s_1} - \frac{Z_{\rm P}}{s_2} + \frac{1}{r_{12}} \tag{11.53}$$

$$V_f^d = \frac{Z_{\rm P}Z_{\rm T}}{R} - \frac{Z_{\rm T}(Z_{\rm P}-2)}{r_f} - \frac{Z_{\rm T}}{x_1} - \frac{Z_{\rm T}}{x_2}. \tag{11.54}$$

Neglecting the terms of the order of and higher than $1/M_{\rm P}$, we have $\boldsymbol{r}_f \simeq -\boldsymbol{R}$, so that V_f^d is reduced to

$$V_f^d = \frac{2Z_{\rm T}}{R} - \frac{Z_{\rm T}}{x_1} - \frac{Z_{\rm T}}{x_2}. \tag{11.55}$$

The potential V_f^d is of a short-range, since it tends to $\mathcal{O}(1/R^2)$ when $R \to \infty$. We recall that

$$V_f = \frac{Z_{\rm P}Z_{\rm T}}{R} - \frac{Z_{\rm T}}{x_1} - \frac{Z_{\rm T}}{x_2} \longrightarrow \frac{Z_{\rm T}(Z_{\rm P}-2)}{R} = V_f^{\infty} \quad (r_f \to \infty). \tag{11.56}$$

The eigen-problem in the exit channel distorted by the residual Coulomb potential (11.51) reads as

$$(H_f^d - E)\Phi_f^d = 0 \tag{11.57}$$

with the solution

$$\Phi_f^d = \varphi_f(\boldsymbol{s}_1, \boldsymbol{s}_2)\mathrm{e}^{-i\boldsymbol{k}_f\cdot\boldsymbol{r}_f}\mathcal{N}^-(\nu_f)\,{}_1F_1(i\nu_f, 1, -ik_f r_f + i\boldsymbol{k}_f\cdot\boldsymbol{r}_f) \quad (11.58)$$

where $\mathcal{N}^-(\nu_f) = \mathrm{e}^{-\pi\nu_f/2}\Gamma(1 - i\nu_f)$ and $\nu_f = Z_\mathrm{T}(Z_\mathrm{P} - 2)/v$. The asymptotic form of Φ_f^d as $r_f \to \infty$ is

$$\Phi_f^d(r_f \to \infty) \equiv \Phi_f^- = \varphi_f(\boldsymbol{s}_1, \boldsymbol{s}_2)\mathrm{e}^{-i\boldsymbol{k}_f\cdot\boldsymbol{r}_f - i\nu_f\ln(k_f r_f - \boldsymbol{k}_f\cdot\boldsymbol{r}_f)} \quad (11.59)$$

which implies that the function (11.58) obeys the correct boundary condition in the exit channel.

For heavy particle collisions, we have $k_i^2/2\mu_i \gg |E_f - E_i|$. In this case, scattering takes place mainly in the forward direction, so that we can write $\hat{\boldsymbol{v}}_i \simeq \hat{\boldsymbol{v}}_f \equiv \hat{\boldsymbol{v}}$, where $\boldsymbol{v}_i = \boldsymbol{k}_i/\mu_i$, $\boldsymbol{v}_f = \boldsymbol{k}_f/\mu_f$ and $\hat{\boldsymbol{v}}_{i,f} = \boldsymbol{v}_{i,f}/v_{i,f}$. For double electron capture, it is readily verified that the following expressions are valid

$$\begin{aligned}\boldsymbol{k}_i\cdot\boldsymbol{r}_i + \boldsymbol{k}_f\cdot\boldsymbol{r}_f &= \boldsymbol{q}_\mathrm{P}\cdot(\boldsymbol{s}_1 + \boldsymbol{s}_2) + \boldsymbol{q}_\mathrm{T}\cdot(\boldsymbol{x}_1 + \boldsymbol{x}_2)\\ &= -2\boldsymbol{q}_\mathrm{P}\cdot\boldsymbol{R} - \boldsymbol{v}\cdot(\boldsymbol{x}_1 + \boldsymbol{x}_2) = 2\boldsymbol{q}_\mathrm{T}\cdot\boldsymbol{R} - \boldsymbol{v}\cdot(\boldsymbol{s}_1 + \boldsymbol{s}_2)\end{aligned} \quad (11.60)$$

$$2\boldsymbol{q}_\mathrm{P} = +\boldsymbol{\eta} - v^+\hat{\boldsymbol{v}} \qquad 2\boldsymbol{q}_\mathrm{T} = -\boldsymbol{\eta} - v^-\hat{\boldsymbol{v}} \qquad \boldsymbol{q}_\mathrm{P} + \boldsymbol{q}_\mathrm{T} = -\boldsymbol{v} \quad (11.61)$$

$$v^+ = v + \frac{E_f - E_i}{v} \qquad v^- = v - \frac{E_f - E_i}{v}. \quad (11.62)$$

The vector of the incident velocity $\boldsymbol{v}$ is chosen along the $Z-$axis i.e. $\hat{\boldsymbol{v}} = (0, 0, 1)$, whereas the vector $\boldsymbol{\eta}$ is the transverse momentum transfer in the XOY plane

$$\boldsymbol{\eta} = (\eta\cos\phi_\eta, \eta\sin\phi_\eta, 0) \qquad \boldsymbol{\eta}\cdot\boldsymbol{v} = 0. \quad (11.63)$$

11.4.2 Single electron capture

Keeping the same notation as in the case of double electron capture, we list the quantities that need to be redefined for single electron capture. Now, in the exit channel we have two hydrogen-like atomic systems, so it is convenient to introduce the vector $\boldsymbol{r}_f$ as the position vector between the centers of mass of the $(Z_\mathrm{P}, e_1)_{f_1}$ and $(Z_\mathrm{T}, e_2)_{f_2}$ systems. In this case, the set of the independent coordinates $\{\boldsymbol{s}_1, \boldsymbol{x}_2, \boldsymbol{r}_f\}$ can be used. Thus the kinetic energy operator in terms of these coordinates is given by

$$H_0 = -\frac{1}{2\mu_f}\nabla_{r_f}^2 - \frac{1}{2a}\nabla_{s_1}^2 - \frac{1}{2b}\nabla_{x_2}^2 \quad (11.64)$$

where $\mu_f = (M_\mathrm{P} + 1)(M_\mathrm{T} + 1)/(M_\mathrm{P} + M_\mathrm{T} + 2)$. It should be noted that the mass polarization term does not appear when coordinates $\{\boldsymbol{s}_1, \boldsymbol{x}_2, \boldsymbol{r}_f\}$ are employed in Hamiltonian H_0.

The channel Hamiltonian H_f and the associated perturbation V_f are

$$H_f = H_0 - \frac{Z_{\rm P}}{s_1} - \frac{Z_{\rm T}}{x_2} \tag{11.65}$$

$$V_f = \frac{Z_{\rm P}Z_{\rm T}}{R} - \frac{Z_{\rm T}}{x_1} - \frac{Z_{\rm P}}{s_2} + \frac{1}{r_{12}}. \tag{11.66}$$

Solving the pertinent eigen-value equation, $(H_f - E)\Phi_f = 0$, we obtain the unperturbed state Φ_f in the exit channel as

$$\Phi_f = \varphi_{f_1}(\boldsymbol{s}_1)\varphi_{f_2}(\boldsymbol{x}_2)\mathrm{e}^{-i\boldsymbol{k}_f\cdot\boldsymbol{r}_f} \tag{11.67}$$

where $\varphi_{f_1}(\boldsymbol{s}_1)$ and $\varphi_{f_2}(\boldsymbol{x}_2)$ are the hydrogen-like final states in the exit channel for the $(Z_{\rm P}, e_1)_{f_1}$ and $(Z_{\rm T}, e_2)_{f_2}$ systems, respectively. In the exit channel, the total energy E is

$$E = \frac{k_f^2}{2\mu_f} + E_f \qquad E_f = E_{f_1} + E_{f_2}$$

$$E_{f_1} = -\frac{Z_{\rm P}^2}{2n_{f_1}^2} \qquad E_{f_2} = -\frac{Z_{\rm T}^2}{2n_{f_2}^2}. \tag{11.68}$$

The distortion of the unperturbed state Φ_f in the case of single charge exchange is caused by the potential

$$V_f^\infty = \frac{(Z_{\rm P} - 1)(Z_{\rm T} - 1)}{R} \tag{11.69}$$

which represents the asymptotic form of the perturbation V_f. Here, the constituent two terms of the separable Hamiltonian $H = H_f^d + V_f^d$ are

$$H_f^d = -\frac{1}{2\mu_f}\nabla_{r_f}^2 + \frac{(Z_{\rm P} - 1)(Z_{\rm T} - 1)}{r_f} - \frac{1}{2a}\nabla_{s_1}^2 - \frac{1}{2b}\nabla_{x_2}^2 - \frac{Z_{\rm P}}{s_1} - \frac{Z_{\rm T}}{x_2} \tag{11.70}$$

$$V_f^d = \frac{Z_{\rm P}Z_{\rm T}}{R} - \frac{(Z_{\rm P} - 1)(Z_{\rm T} - 1)}{r_f} - \frac{Z_{\rm P}}{s_2} - \frac{Z_{\rm T}}{x_1} + \frac{1}{r_{12}}. \tag{11.71}$$

Using the eikonal approximation $r_f \simeq R$, we obtain the following approximate expression

$$V_f^d \simeq Z_{\rm P}\left(\frac{1}{R} - \frac{1}{s_2}\right) + (Z_{\rm T} - 1)\left(\frac{1}{R} - \frac{1}{x_1}\right) + \left(\frac{1}{r_{12}} - \frac{1}{x_1}\right). \tag{11.72}$$

With this, the solution of the eigen-value equation $H_f^d\Phi_f^d = E\Phi_f^d$ in the exit channel distorted by the Coulomb residual potential (11.69) becomes

$$\Phi_f^d = \varphi_{f_1}(\boldsymbol{s}_1)\varphi_{f_2}(\boldsymbol{x}_2)\mathrm{e}^{-i\boldsymbol{k}_f\cdot\boldsymbol{r}_f}\mathcal{N}^-(\nu_f)\,{}_1F_1(i\nu_f, 1, -ik_f r_f + i\boldsymbol{k}_f\cdot\boldsymbol{r}_f) \tag{11.73}$$

where $\mathcal{N}^-(\nu_f) = \mathrm{e}^{-\pi\nu_f/2}\Gamma(1-i\nu_f)$ and $\nu_f = (Z_\mathrm{P}-1)(Z_\mathrm{T}-1)/v$. The asymptotic form of Φ_f^d as $r_f \to \infty$ is correct, as evidenced via

$$\Phi_f^d(r_f \to \infty) \equiv \Phi_f^- = \varphi_{f_1}(\boldsymbol{s}_1)\varphi_{f_2}(\boldsymbol{x}_2)\mathrm{e}^{-i\boldsymbol{k}_f\cdot\boldsymbol{r}_f - i\nu_f\ln(k_f r_f - \boldsymbol{k}_f\cdot\boldsymbol{r}_f)}. \tag{11.74}$$

Employing a similar procedure as the one in the case of double electron capture, it can be shown that for heavy scattering aggregates the following relations are valid

$$\boldsymbol{k}_i\cdot\boldsymbol{r}_i + \boldsymbol{k}_f\cdot\boldsymbol{r}_f = \boldsymbol{\alpha}\cdot\boldsymbol{s}_1 + \boldsymbol{\beta}\cdot\boldsymbol{x}_1 = -\boldsymbol{\alpha}\cdot\boldsymbol{R} - \boldsymbol{v}\cdot\boldsymbol{x}_1 \tag{11.75}$$

$$\boldsymbol{\alpha} = \boldsymbol{\eta} - v^-\hat{\boldsymbol{v}} \qquad \boldsymbol{\beta} = -\boldsymbol{\eta} - v^+\hat{\boldsymbol{v}} \qquad \boldsymbol{\alpha}+\boldsymbol{\beta} = -\boldsymbol{v} \tag{11.76}$$

$$v^+ = \frac{v}{2} + \frac{E_i - E_f}{v} \qquad v^- = \frac{v}{2} - \frac{E_i - E_f}{v}. \tag{11.77}$$

11.4.3 Transfer ionization

During transfer ionization, one electron (e_1) is captured, while the other (e_2) is simultaneously ionized. Therefore, the unperturbed wave function for this process reads as

$$\Phi_f = \varphi_f(\boldsymbol{s}_1)\phi_f \tag{11.78}$$

where

$$\phi_f = (2\pi)^{-3/2}\mathrm{e}^{-i\boldsymbol{k}_f\cdot\boldsymbol{r}_f + i\boldsymbol{\kappa}\cdot\boldsymbol{x}_2}. \tag{11.79}$$

Here, $\boldsymbol{\kappa}$ represents the momentum vector of the ejected electron e_2 with respect to its parent nucleus T.

In the exit channel, the wave function which obeys the correct boundary conditions has the form

$$\Phi_f^d(r_f \to \infty, x_2 \to \infty) \equiv \Phi_f^- \tag{11.80}$$

with

$$\Phi_f^- = \mathrm{e}^{-i\nu_f\ln(k_f r_f - \boldsymbol{k}_f\cdot\boldsymbol{r}_f) + i(Z_\mathrm{T}/\kappa)\ln(kx_2 + \boldsymbol{k}\cdot\boldsymbol{x}_2)} \tag{11.81}$$

where $\nu_f = (Z_\mathrm{P}-1)(Z_\mathrm{T}-1)/v$.

The analysis in this chapter aims to emphasize the need to establish a proper connection between the long-range Coulomb distortion effects and the accompanying perturbation potentials. Otherwise unphysical results could easily be incurred, as has been shown by Belkić [132, 133] for ionization (detachment) of H^- by H^+.

12

Perturbation series with the correct boundary conditions

For a theory to be adequate, it must be reemphasized that the proper Coulomb boundary conditions in the entrance and exit channels are of crucial importance for ion-atom collisions. Experience has shown that if this requirement is disregarded, serious problems inevitably arise and the related models are unsatisfactory for a theoretically founded description of experimental data.

The dynamics of the entire four-body system are described by means of the Schrödinger equation

$$(H - E)\Psi^{\pm} = 0 \tag{12.1}$$

where $\Psi^{\pm}$ are the full scattering states with the outgoing and incoming boundary conditions, respectively

$$\Psi^{+} \longrightarrow \Phi_i^{+} \quad (r_i \to \infty) \qquad \Psi^{-} \longrightarrow \Phi_f^{-} \qquad (r_f \to \infty)\,. \tag{12.2}$$

The exact transition amplitude with the correct boundary conditions can be written in the post (+) and prior (−) forms as

$$T_{if}^{+} = \langle \Phi_f^{-} | V_f^d | \Psi_i^{+} \rangle \qquad T_{if}^{-} = \langle \Psi_f^{-} | V_i^d | \Phi_i^{+} \rangle. \tag{12.3}$$

Both forms are equivalent to each other on the energy shell i.e. the exact on-shell expressions are equal, $T_{if}^{+} = T_{if}^{-}$, for the transitions for which the total energy is conserved, as in (11.50).

Solving a scattering problem in which four bodies take part (two nuclei and two electrons) is extremely difficult. As usual, at intermediate and high impact energies, the powerful and versatile procedure of perturbation series expansions is frequently employed. To this end, it is convenient to convert the Schrödinger equation for a four-body problem into the corresponding integral equation such as the Lippmann-Schwinger equations or the associated distorted wave integral equations. Irrespective of whether one starts from the Born, Lippmann-Schwinger or Faddeev equations or their corresponding perturbation expansion series, the correct boundary conditions must always be imposed to the entrance and exit channels [41, 44]. Despite the widely accepted importance of such initial conditions [45]–[50], confusion and debates persisted in the literature for a long time on this very point [215]–[243]. Surprisingly, even in the most recent times some researchers continue to use methods that ignore the correct boundary conditions [244]–[246].

12.1 The Lippmann-Schwinger equations

We begin by introducing the total scattering function via

$$\Psi_i^+ \equiv i\varepsilon G^+ \Phi_i^+ \tag{12.4}$$

where G^+ is the full Green's operator and Φ_i^+ is the wave function defined by (11.28) and (11.40). Here, ε is an infinitesimally small positive number. In addition to the total Green operators $G^\pm$, we also define the initial $G_i^\pm$, the final $G_f^\pm$ and the free Green resolvent propagators $G_0^\pm$ as

$$G^\pm = (E - H \pm i\varepsilon)^{-1} \tag{12.5}$$

$$G_i^\pm = (E - H_i^d \pm i\varepsilon)^{-1} \tag{12.6}$$

$$G_f^\pm = (E - H_f^d \pm i\varepsilon)^{-1} \tag{12.7}$$

$$G_0^\pm = (E - H_0 \pm i\varepsilon)^{-1}. \tag{12.8}$$

These propagators are inter-related by the following Lippmann-Schwinger integral equations for the total Green functions

$$\begin{aligned} G^\pm &= G_i^\pm + G_i^\pm V_i^d G^\pm \\ G^\pm &= G_f^\pm + G_f^\pm V_f^d G^\pm \\ G^\pm &= G_0^\pm + G_0^\pm V G^\pm \end{aligned} \tag{12.9}$$

as can be checked. For example, if the ansatz

$$G^\pm = G_i^\pm + G_i^\pm V_i^d G^\pm$$

is multiplied from the left by $E - H_i^d \pm i\varepsilon$ and simultaneously from the right by $E - H \pm i\varepsilon$, it follows

$$E - H_i^d \pm i\varepsilon = E - H \pm i\varepsilon + V_i^d$$

in agreement with (11.22).

Applying the iteration procedure to (12.9), we obtain the following expansions for the total Green resolvent G^+ in terms of G_0^+, G_i^+ and G_f^+

$$G^+ = G_0^+ + G_0^+ V G_0^+ + G_0^+ V G_0^+ V G_0^+ + G_0^+ V G_0^+ V G_0^+ V G_0^+ + \cdots \tag{12.10}$$

$$G^+ = G_i^+ + G_i^+ V_i^d G_i^+ + G_i^+ V_i^d G_i^+ V_i^d G_i^+ + G_i^+ V_i^d G_i^+ V_i^d G_i^+ V_i^d G_i^+ + \cdots \tag{12.11}$$

$$G^+ = G_f^+ + G_f^+ V_f^d G_f^+ + G_f^+ V_f^d G_f^+ V_f^d G_f^+ + G_f^+ V_f^d G_f^+ V_f^d G_f^+ V_f^d G_f^+ + \cdots \,. \quad (12.12)$$

Inserting G^+ from (12.9) into (12.4), we have

$$\begin{aligned} \Psi_i^+ &= i\varepsilon G^+ \Phi_i^+ = i\varepsilon G_i^+ \Phi_i^+ + G_i^+ V_i^d i\varepsilon G^+ \Phi_i^+ \\ &= i\varepsilon G_i^+ \Phi_i^+ + G_i^+ V_i^d \Psi_i^+ \end{aligned} \quad (12.13)$$

where the first term can be written as $i\varepsilon G_i^+ \Phi_i^+ = \Phi_i^+$. This can be directly verified if $[i\varepsilon/(E - H_i^d + i\varepsilon)]\Phi_i^+ = \Phi_i^+$ is multiplied from the left by $E - H_i^d + i\varepsilon$. Thus we have

$$i\varepsilon \Phi_i^+ = (E - H_i^d + i\varepsilon)\Phi_i^+$$

in agreement with (11.40). In this way, we obtain the Lippmann-Schwinger equation for the total scattering wave function in the case of a four-body problem under study

$$\Psi_i^+ = \Phi_i^+ + G_i^+ V_i^d \Psi_i^+. \quad (12.14)$$

This is an inhomogeneous integral equation, since it explicitly contains the incident wave Φ_i^+. The integral equation (12.14) can formally be solved as

$$\begin{aligned} \Psi_i^+ &= \Phi_i^+ + G_i^+ V_i^d \Psi_i^+ \\ &= \Phi_i^+ + G_i^+ V_i^d (\Phi_i^+ + G_i^+ V_i^d \Psi_i^+) = \Phi_i^+ + G_i^+ V_i^d \Phi_i^+ + G_i^+ V_i^d G_i^+ V_i^d \Psi_i^+ \\ &= \Phi_i^+ + G_i^+ V_i^d \Phi_i^+ + G_i^+ V_i^d G_i^+ V_i^d \Phi_i^+ + G_i^+ V_i^d G_i^+ V_i^d G_i^+ V_i^d \Psi_i^+ \\ &= (1 + G_i^+ V_i^d + G_i^+ V_i^d G_i^+ V_i^d + G_i^+ V_i^d G_i^+ V_i^d G_i^+ V_i^d + \cdots)\Phi_i^+ \\ &= \left[1 + \sum_{n=1}^{\infty} (G_i^+ V_i^d)^n \right] \Phi_i^+ = (1 + G^+ V_i^d)\Phi_i^+. \end{aligned}$$

Here, the last term $G^+ V_i^d$ also follows from multiplication of (12.11) by V_i^d

$$\begin{aligned} G^+ V_i^d &= G_i^+ V_i^d + G_i^+ V_i^d G_i^+ V_i^d + G_i^+ V_i^d G_i^+ V_i^d G_i^+ V_i^d + \\ &\quad + G_i^+ V_i^d G_i^+ V_i^d G_i^+ V_i^d G_i^+ V_i^d + \cdots = \sum_{n=1}^{\infty} (G_i^+ V_i^d)^n . \end{aligned}$$

Hence, the formal solution of the four-body Lippmann-Schwinger equation in terms of the total Green operator G^+ is

$$\Psi_i^+ = \Phi_i^+ + G^+ V_i^d \Phi_i^+ = (1 + G^+ V_i^d)\Phi_i^+. \quad (12.15)$$

12.2 The Born expansions with the correct boundary conditions for four-body collisions

Inserting the formal solution (12.15) into (12.3) for the post form of the transition amplitude, it follows that

$$T_{if}^{+} = \langle \Phi_f^- | V_f^d | \Psi_i^+ \rangle = \langle \Phi_f^- | V_f^d (1 + G^+ V_i^d) | \Phi_i^+ \rangle. \tag{12.16}$$

This implies that by substituting G^+ from (12.10)–(12.12) into (12.16), we can write several different versions of the Born expansions with the correct boundary conditions

$$T_{if}^{+} = T_{if}^{(\mathrm{CB1})+} + \langle \Phi_f^- | V_f^d G_0^+ V_i^d | \Phi_i^+ \rangle + \langle \Phi_f^- | V_f^d G_0^+ V G_0^+ V_i^d | \Phi_i^+ \rangle + \cdots \tag{12.17}$$

$$T_{if}^{+} = T_{if}^{(\mathrm{CB1})+} + \langle \Phi_f^- | V_f^d G_i^+ V_i^d | \Phi_i^+ \rangle + \langle \Phi_f^- | V_f^d G_i^+ V_i^d G_i^+ V_i^d | \Phi_i^+ \rangle + \cdots \tag{12.18}$$

$$T_{if}^{+} = T_{if}^{(\mathrm{CB1})+} + \langle \Phi_f^- | V_f^d G_f^+ V_i^d | \Phi_i^+ \rangle + \langle \Phi_f^- | V_f^d G_f^+ V_f^d G_f^+ V_i^d | \Phi_i^+ \rangle + \cdots \tag{12.19}$$

$$T_{if}^{(\mathrm{CB1})+} = \langle \Phi_f^- | V_f^d | \Phi_i^+ \rangle. \tag{12.20}$$

Here, $T_{if}^{(\mathrm{CB1})+}$ is the post form of the first Born method with the correct boundary conditions for four-body collisions i.e. the CB1-4B method [87, 88]. As can be seen, the term $T_{if}^{(\mathrm{CB1})+}$ is identical in all the presented versions. In other words, the CB1-4B method can be obtained by replacing the total wave function Ψ_i^+ by the corresponding asymptotic channel state Φ_i^+. Two methods for an explicit calculation of the matrix elements in the CB1-4B method for double charge exchange have been devised and implemented by Belkić [87, 88] with the same numerical results.

Likewise, the n th Born method with the correct boundary conditions (CBn-4B) may be obtained by keeping the first n terms in the above perturbation expansion. For example, the four-body second Born method with the correct boundary conditions (CB2-4B) can be obtained in this way in the forms

$$T_{if;0}^{(\mathrm{CB2})+} = T_{if}^{(\mathrm{CB1})+} + \langle \Phi_f^- | V_f^d G_0^+ V_i^d | \Phi_i^+ \rangle \tag{12.21}$$

$$T_{if;i}^{(\mathrm{CB2})+} = T_{if}^{(\mathrm{CB1})+} + \langle \Phi_f^- | V_f^d G_i^+ V_i^d | \Phi_i^+ \rangle \tag{12.22}$$

$$T_{if;f}^{(\mathrm{CB2})+} = T_{if}^{(\mathrm{CB1})+} + \langle \Phi_f^- | V_f^d G_f^+ V_i^d | \Phi_i^+ \rangle. \tag{12.23}$$

Here, (12.21) in terms of G_0^+ is recognized as a direct extension of the corresponding CB2-3B method of Belkić [7, 208, 209]. Of course, many other versions of the Born expansion can be formulated by utilizing various possible iterative solutions for G^+. In other words, a unique Born series of the transition amplitude T_{if}^+ does not exist.

A similar procedure can be employed for the prior form of the transition amplitude. In this case, the time-independent wave function of the whole system in the exit channel is given by the following integral form

$$\Psi_f^- = \Phi_f^- + G_f^- V_f^d \Psi_f^- = (1 + G^- V_f^d)\Phi_f^-. \tag{12.24}$$

The corresponding prior form of the transition amplitude is

$$T_{if}^- = \langle \Phi_f^- | (1 + G^- V_f^d)^\dagger V_i^d | \Phi_i^+ \rangle. \tag{12.25}$$

Thus far, explicit computations of the CB1-4B method have been carried out for double electron capture by Belkić [87, 88] as well as for single charge exchange by Mančev [191, 192]. The CB2-4B method has not yet been used within the four-body formalism. However, it should be emphasized that much experience has been gained using the CB2-3B method [7], with its distinct improvement in the description of single charge exchange, when passing from the first to the second order perturbation theory, as demonstrated by Belkić [208]–[210]. Guided by this fact, here we recall the key aspects of the basic three-body problem with the two arbitrary nuclear charges Z_{P} and Z_{T}

$$Z_{\text{P}} + (Z_{\text{T}}, e) \longrightarrow (Z_{\text{P}}, e) + Z_{\text{T}}. \tag{12.26}$$

If one employs a pure Coulomb potential $V_{\text{T}} = -Z_{\text{T}}/x$ as the perturbation in the exit channel, together with the unperturbed wave functions Φ_i and Φ_f, where the plane waves describe the relative motion of heavy aggregates, one would obtain the three-body first-order Brinkman-Kramers (BK1-3B) approximation via its transition amplitude in the post form

$$T_{if}^{(\text{BK1}-3\text{B})+} = \langle \Phi_f | V_{\text{T}} | \Phi_i \rangle. \tag{12.27}$$

By including the second-order term $\langle \Phi_f | V_{\text{T}} G_0^+ V_{\text{P}} | \Phi_i \rangle$ from the Born expansion with the incorrect boundary conditions, where $V_{\text{P}} = -Z_{\text{P}}/s$ and G_0^+ is the free-particle Green function for process (12.26), one obtains the three-body second-order Brinkman-Kramers (BK2-3B) approximation

$$T_{if}^{(\text{BK2}-3\text{B})+} = \langle \Phi_f | V_{\text{T}} | \Phi_i \rangle + \langle \Phi_f | V_{\text{T}} G_0^+ V_{\text{P}} | \Phi_i \rangle. \tag{12.28}$$

If the initial and final perturbations V_{P} and V_{T} appearing in (12.27) and (12.28) are replaced by the corresponding full perturbations of the undistorted system in the entrance and exit channels

$$V_i = V_{\text{P}} + V_{\text{PT}} = -\frac{Z_{\text{P}}}{s} + \frac{Z_{\text{P}} Z_{\text{T}}}{R} \tag{12.29}$$

$$V_f = V_{\rm T} + V_{\rm PT} = -\frac{Z_{\rm T}}{x} + \frac{Z_{\rm P} Z_{\rm T}}{R} \tag{12.30}$$

then the three-body first- and second-order Jackson-Schiff (JS1-3B and JS2-3B) approximations are deduced

$$T_{if}^{(\rm JS1-3B)+} = \langle \Phi_f | V_{\rm T} + V_{\rm PT} | \Phi_i \rangle \tag{12.31}$$

$$\begin{aligned} T_{if}^{(\rm JS2-3B)+} &= \langle \Phi_f | V_{\rm T} + V_{\rm PT} | \Phi_i \rangle \\ &+ \langle \Phi_f | (V_{\rm T} + V_{\rm PT}) G_0^+ (V_{\rm P} + V_{\rm PT}) | \Phi_i \rangle. \end{aligned} \tag{12.32}$$

The BK1-3B [247, 248], BK2-3B [249]–[253], JS1-3B [254]–[257] and JS2-3B [258, 259] methods do not satisfy the correct boundary conditions. This is the case for the BK1-3B and BK2-3B methods for any value of nuclear charges $Z_{\rm T}$ and $Z_{\rm P}$ in process (12.26). The same applies to the JS1-3B and JS2-3B methods with only one exception for $Z_{\rm P} = 1 = Z_{\rm T}$ in process (12.26) when the correct boundary conditions are fortuitously satisfied. By contrast, the corresponding three-body first- and second-order Born theories i.e. the CB1-3B [199] and CB2-3B [208] methods always satisfy the correct boundary conditions for process (12.26) for arbitrary values of $Z_{\rm P}$ and $Z_{\rm T}$.

From the physical viewpoint, the BK2-3B model should be more adequate than the BK1-3B model, due to the addition of the second-order propagator $V_{\rm T} G_0^+ V_{\rm P}$. However, as has been shown by Belkić [209], the BK2-3B model gives a much worse description of the corresponding experimental data than the BK1-3B model, which itself is inadequate for both differential and total cross sections of charge exchange in the $\rm H^+ - H$ collisions, as well as in the more general process (12.26). This clearly shows that the inclusion of higher order terms from a perturbation series can easily deteriorate the overall description, if certain basic principles are disregarded from the outset, as in the BK1-3B and BK2-3B models. Similar inadequacies are also typical for the JS1-3B and JS2-3B models. For total cross sections these two latter models additionally suffer from an unphysical effect due to a non-zero contribution of the internuclear potential $V_{\rm PT} = Z_{\rm P} Z_{\rm T}/R$ in (12.29) and (12.30) for any $Z_{\rm P}$ and $Z_{\rm T}$ except $Z_{\rm P} = 1 = Z_{\rm T}$, as mentioned. This is at variance with the proof of Belkić *et al.* [44] who showed that the exact eikonal total cross sections for process (12.26) in the case of arbitrary $Z_{\rm P}$ and $Z_{\rm T}$ are the same, irrespective of whether $V_{\rm PT}$ is retained or not.

In sharp contrast to the BK2-3B model, it has been shown [7, 208] that the CB2-3B method yields very reliable results relative to the experimental data. Crucially, the CB2-3B method shows a significant improvement over the CB1-3B method [199]. This success of the CB2-3B method in describing charge exchange processes (12.26) has been attributed solely to the rigorous treatment of the correct boundary conditions, that represent the essential and distinct features of any scattering event [208, 209]. Subsequently, the CB2-3B method for single electron capture (12.26) has been used in Refs. [211]–[213].

13

The Dodd-Greider series for four-body collisions

In order to solve (12.1), we shall adopt the distorted wave formalism. By way of introduction, we will first recall the salient features of this theory [260, 261]. In the distorted wave formalism, instead of directly solving the Schrödinger equation (12.1) with rigidly determined interactions, one customarily considers a related model problem in which the real channel interactions V_i and V_f are replaced by certain distorting potential operators W_i and W_f. The following Green operators are associated with these latter two interaction potentials

$$g_i^+ = (E - H_i - W_i + i\varepsilon)^{-1} \tag{13.1}$$

$$g_f^- = (E - H_f - W_f - i\varepsilon)^{-1} \tag{13.2}$$

or equivalently

$$g_i^+ = (1 + g_i^+ W_i)\mathcal{G}_i^+ \equiv \omega^+ \mathcal{G}_i^+ \tag{13.3}$$

$$g_f^- = (1 + g_f^- W_f)\mathcal{G}_f^- \equiv \omega^- \mathcal{G}_f^- . \tag{13.4}$$

Here, $\mathcal{G}_i^+$ and $\mathcal{G}_f^-$ are the Green functions defined by

$$\mathcal{G}_i^+ = (E - H_i + i\varepsilon)^{-1} \tag{13.5}$$

$$\mathcal{G}_f^- = (E - H_f - i\varepsilon)^{-1} \tag{13.6}$$

where $\omega^\pm$ are the Møller wave operators. Next, in lieu of the full wave functions $\Psi_{i,f}^\pm = (1 + G^\pm V_{i,f}^d)\Phi_{i,f}^\pm$ we introduce the related distorted waves

$$\chi_{i,f}^\pm = (1 + g_{i,f}^\pm W_{i,f})\Phi_{i,f} = \omega^\pm \Phi_{i,f}. \tag{13.7}$$

The distorted waves χ_i^+ and χ_f^- satisfy the following equations in the limits $\varepsilon \to 0^\pm$

$$(E - H_i - W_i)\chi_i^+ = 0 \qquad (E - H_f - W_f)\chi_f^- = 0. \tag{13.8}$$

The connection of the model problem (13.8) with the original Schrödinger equation (12.1) is provided through the requirement that $\chi^{\pm}_{i,f}$ and $\Psi^{\pm}_{i,f}$ exhibit the same asymptotic behaviors as $r_{i,f} \to \infty$

$$\chi_i^+ \longrightarrow \Psi_i^+ \longrightarrow \Phi_i^+ \qquad r_i \to \infty \tag{13.9}$$

$$\chi_f^- \longrightarrow \Psi_f^- \longrightarrow \Phi_f^- \qquad r_f \to \infty\,. \tag{13.10}$$

The prior form of the transition amplitude which is defined by

$$T_{if}^- = \langle \Psi_f^- | V_i | \Phi_i \rangle = \langle \Phi_f | (1 + G^- V_f)^\dagger V_i | \Phi_i \rangle \equiv \langle \Phi_f | \Omega^{-\dagger} V_i | \Phi_i \rangle \tag{13.11}$$

can be expressed in terms of the model quantities in the entrance channel via

$$T_{if}^- = \langle \Phi_f | \Omega^{-\dagger} (V_i - W_i) \omega^+ + \Omega^{-\dagger} [1 - (V_i - W_i) g_i^+] W_i | \Phi_i \rangle. \tag{13.12}$$

This relation can be readily verified by employing the definitions for g_i^+ and ω^+ or by an algebraic derivation

$$\begin{aligned} V_i &= V_i(1 + g_i^+ W_i) - W_i(1 + g_i^+ W_i) + [1 - (V_i - W_i) g_i^+] W_i \\ &= (V_i - W_i)\omega^+ + [1 - (V_i - W_i) g_i^+] W_i. \end{aligned} \tag{13.13}$$

Using the well-known Chew-Goldberger operator identity $1/A - 1/B = (1/A)(B-A)(1/B)$ with $1/A = G^+$ and $1/B = g_i^+$ the second term in (13.12) becomes $(E + i\varepsilon - H_f)(\omega^+ - 1) = i\varepsilon(\omega^+ - 1)$. We write the transition amplitude T_{if}^- as follows

$$T_{if}^- = \langle \Phi_f | \Omega^{-\dagger} (V_i - W_i) \omega^+ | \Phi_i \rangle + T_{if}^d \tag{13.14}$$

$$T_{if}^d = \lim_{\varepsilon \to 0} i\varepsilon \langle \Phi_f | \omega^+ | \Phi_i \rangle = \lim_{\varepsilon \to 0} i\varepsilon \langle \Phi_f | \chi_i^+ \rangle. \tag{13.15}$$

Here, the contribution from the additive term T_{if}^d will vanish in the limit $\varepsilon \to 0$. The corresponding condition

$$\lim_{\varepsilon \to 0} i\varepsilon \langle \Phi_f | \chi_i^+ \rangle = 0 \tag{13.16}$$

is satisfied by choosing a distorting potential which leads only to elastic scattering in the considered channel and, as such, does not cause rearrangement. This can be achieved by choosing the distorting potential to depend only on the relative coordinate between the projectile and the target.

By a simple transformation, the wave operator Ω^- from (13.11), can be rewritten via the matrix element

$$\Omega^- = [1 + G^-(V_f - W_f)]\omega^-. \tag{13.17}$$

Then, by employing (13.16), it follows from (13.15)

$$\begin{aligned} T_{if}^- &= \langle \Phi_f | \Omega^{-\dagger} (V_i - W_i) \omega^+ | \Phi_i \rangle \\ &= \langle \Phi_f | \omega^{-\dagger} [1 + (V_f - W_f^\dagger) G^+] (V_i - W_i) \omega^+ | \Phi_i \rangle \\ &\equiv \langle \Phi_f | U_{if}^- | \Phi_i \rangle. \end{aligned} \tag{13.18}$$

We recall that the Hermitean conjugated relation to (13.17) is given by $\Omega^{-\dagger} = \omega^{-\dagger}[1 + (V_f - W_f^\dagger)G^+]$. Hence, the exact transition amplitude T_{if}^- in the distorted wave theory reads as

$$T_{if}^- = \langle \chi_f^- | (V_i - W_i) + (V_f - W_f^\dagger)G^+(V_i - W_i)|\chi_i^+\rangle. \tag{13.19}$$

Similarly, we can obtain the exact post form of the transition amplitude in the distorted wave formalism via

$$T_{if}^+ = \langle \chi_f^- | (V_f - W_f^\dagger) + (V_f - W_f^\dagger)G^+(V_i - W_i)|\chi_i^+\rangle \tag{13.20}$$

provided that

$$\lim_{\varepsilon\to 0} i\varepsilon\langle \Phi_f|\omega^{-\dagger}|\Phi_i\rangle = 0. \tag{13.21}$$

Using the expression

$$\Omega^- = [1 + \Omega^- \mathcal{G}_f^-(V_f - W_f)]\omega^- \tag{13.22}$$

the transition operator U_{if}^- introduced in (13.18) can be written as the following integral equation

$$U_{if}^- = \omega^{-\dagger}(V_i - W_i)\omega^+ + \omega^{-\dagger}(V_f - W_f^\dagger)\mathcal{G}_f^+ U_{if}^-. \tag{13.23}$$

This can alternatively be cast into the form

$$U_{if}^-(1 - \mathcal{K}) = \omega^{-\dagger}(V_i - W_i)\omega^+ \tag{13.24}$$

$$\mathcal{K} = \omega^{-\dagger}(V_f - W_f^\dagger)\mathcal{G}_f^+ \tag{13.25}$$

where $\mathcal{K}$ represents the so-called kernel i.e. the homogeneous term of the integral equation. Since ω^- is given by $\omega^- = 1 + g_f^- W_f$, it follows that the form of $\mathcal{K}$ is independent of the choice of the distortion in the entrance channel. Expanding U_{if}^- in powers of $\mathcal{K}$ i.e. in an infinite perturbation series, we obtain

$$U_{if}^- = I(1 + \sum_{n=1}^{\infty} \mathcal{K}^n) \tag{13.26}$$

$$I = \omega^{-\dagger}(V_i - W_i)\omega^+ \tag{13.27}$$

where I is an inhomogeneous term of the integral equation (13.23). However, this latter expansion diverges in the case of rearrangement collisions due to the existence of the so-called disconnected diagrams. These Feynman diagrams correspond to collisional paths describing three constituents interacting pairwise with each other in the presence of a fourth body as a freely propagating particle. In the impulse space, this physical situation is described by means of the Dirac δ-function in the kernel of the integral equation. The presence of this δ-function indicates the conservation of momentum. Such a kernel is said to contain the mentioned disconnected diagrams. The free motion is mediated via the free-particle Green resolvent G_0^+ which leads to the factor $1/(E - E_0 + i\varepsilon)$, where $E_0 = k^2/2$. Since in the T-matrix we have an integration over k in the whole space, it is clear that we may have a situation where $E = E_0$, and this causes divergence of the energy-dependent term $1/(E - E_0 + i\varepsilon)$ in the limit $\varepsilon \to 0$. The typical kernel $(V_f - W_f^\dagger)G_0^+(V_i - W_i)$ from the iterated transition T operator would not contain any disconnected diagrams if none of the two-body interactions in the perturbation $V_f - W_f^\dagger$ is repeated in $V_i - W_i$. This can be achieved with the introduction of a virtual intermediate channel "x" as originally suggested by Dodd and Greider [260, 261]. The Green operator associated with this virtual channel is

$$g_x^\pm = (E - H + V_x \pm i\varepsilon)^{-1} \tag{13.28}$$

where V_x is an appropriate channel potential. A conveniently chosen V_x can eliminate all disconnected diagrams. Using the relation

$$G^+ = g_x^+(1 + V_x G^+) \tag{13.29}$$

we obtain the integral equation

$$U_{if}^-(1 - \mathcal{K}_1) = \omega^{-\dagger}(V_i - W_i)\omega^+ + \omega^{-\dagger}(V_f - W_f^\dagger)g_x^+(V_i - W_i)]\omega^+ \tag{13.30}$$

where $\mathcal{K}_1$ is a new kernel

$$\mathcal{K}_1 = \omega^{-\dagger}(V_f - W_f^\dagger)g_x^+\mathcal{G}_f^+ . \tag{13.31}$$

By employing certain suitably chosen potentials V_x and W_f, the kernel $\mathcal{K}_1$ can become free from any disconnected diagram. An example of such a situation occurs when the potential V_x does not appear in $V_f - W_f^\dagger$, as in the CDW-4B method [84]. Inserting (13.29) into (13.19), we arrive at the expression

$$T_{if}^- = \langle \chi_f^- | (V_i - W_i) + (V_f - W_f^\dagger) g_x^+ (1 + V_x G^+)(V_i - W_i) | \chi_i^+ \rangle . \tag{13.32}$$

If we neglect the term with the Green operator G^+, we obtain a first-order approximation (for simplicity also denoted by T_{if}^-) for the prior form of the transition amplitude

$$T_{if}^- = \langle \chi_f^- | \{1 + g_x^-(V_f - W_f)\}^\dagger (V_i - W_i) | \chi_i^+ \rangle . \tag{13.33}$$

Introducing an auxiliary distorted wave ξ_f^- as follows

$$|\xi_f^-\rangle = \{1 + g_x^-(V_f - W_f)\}|\chi_f^-\rangle \tag{13.34}$$

we have

$$T_{if}^- = \langle \xi_f^- | V_i - W_i | \chi_i^+ \rangle . \tag{13.35}$$

13.1 Derivation of the distorted waves for the initial states

13.1.1 The $Z_{\rm P} - (Z_{\rm T}; e_1, e_2)_i$ collisional system

According to the requirement (13.9) and the correct asymptotic behavior of Ψ^+ which is

$$\Psi^+ \longrightarrow \varphi_i(\boldsymbol{x}_1, \boldsymbol{x}_2) {\rm e}^{i\boldsymbol{k}_i \cdot \boldsymbol{r}_i + i\nu_i \ln(k_i r_i - \boldsymbol{k}_i \cdot \boldsymbol{r}_i)} \equiv \Phi_i^+ \qquad (r_i \to \infty) \quad (13.36)$$

the following factorized form for the function χ_i^+ appears as optimal

$$\chi_i^+ = \varphi_i(\boldsymbol{x}_1, \boldsymbol{x}_2) \psi_i^+. \quad (13.37)$$

Here, ψ_i^+ is an unknown function to be determined according to a particular choice of the distorting potential. Inserting (13.37) into (13.8), we obtain

$$\varphi_i(E - E_i - H_0 - V_i)\psi_i^+ + \frac{1}{b}\sum_{k=1}^{2} \boldsymbol{\nabla}_{x_k}\varphi_i \cdot \boldsymbol{\nabla}_{x_k}\psi_i^+ + U_i\varphi_i\psi_i^+ + \psi_i^+(E_i - h_i)\varphi_i = 0 \quad (13.38)$$

$$U_i = V_i - W_i \,. \quad (13.39)$$

The term

$$(E_i - h_i)\varphi_i \equiv O_i(\boldsymbol{x}_1, \boldsymbol{x}_2) \equiv O_i \quad (13.40)$$

appearing in (13.38) is equal to zero only for the exact eigen-solutions $\varphi_i(\boldsymbol{x}_1, \boldsymbol{x}_2) \equiv \varphi_i$ and E_i of the target Hamiltonian h_i. However, since these latter solutions are unavailable, the term O_i should, in principle, be kept throughout, as suggested by Belkić [87] within the CB1-4B method. The explicit computations for double charge exchange [87] and transfer ionization [127] have shown that the contribution from this term is $\sim (10 - 15)\%$. This correction will presently be neglected, in which case (13.38) becomes

$$\varphi_i\left(\frac{k_i^2}{2\mu_i} - H_0 - \frac{Z_{\rm P}Z_{\rm T}}{R} + \frac{Z_{\rm P}}{s_1} + \frac{Z_{\rm P}}{s_2}\right)\psi_i^+ + \frac{1}{b}\sum_{k=1}^{2} \boldsymbol{\nabla}_{x_k}\varphi_i \cdot \boldsymbol{\nabla}_{x_k}\psi_i^+ + U_i\varphi_i\psi_i^+ = 0. \quad (13.41)$$

In general, the presence of the coupling term $\boldsymbol{\nabla}_{x_k}\varphi_i \cdot \boldsymbol{\nabla}_{x_k}\psi_i^+$ precludes a separation of the independent variables in (13.41). However, at the same time, there is a flexibility provided by the perturbation potential operator

$U_i = V_i - W_i$, which permits cancellation of the coupling term. An adequate choice has been made by Belkić and Mančev [84, 85] as

$$U_i\chi_i^+ = -\frac{1}{b}\sum_{k=1}^{2}\boldsymbol{\nabla}_{x_k}\varphi_i \cdot \boldsymbol{\nabla}_{x_k}\psi_i^+ . \tag{13.42}$$

Alternatively, the following choices for U_i can be implemented [189, 190]

$$U_i\chi_i^+ = Z_{\rm P}\left(\frac{1}{R} - \frac{1}{s_2}\right)\chi_i^+ - \frac{1}{b}\sum_{k=1}^{2}\boldsymbol{\nabla}_{x_k}\varphi_i \cdot \boldsymbol{\nabla}_{x_k}\psi_i^+ \tag{13.43}$$

$$U_i\chi_i^+ = \left[Z_{\rm P}\left(\frac{1}{R} - \frac{1}{s_2}\right) + \left(\frac{1}{R} - \frac{1}{s_1}\right)\right]\chi_i^+ - \frac{1}{b}\sum_{k=1}^{2}\boldsymbol{\nabla}_{x_k}\varphi_i \cdot \boldsymbol{\nabla}_{x_k}\psi_i^+ . \tag{13.44}$$

Although other choices are possible, in every single case the requirement that the function χ_i^+ has the correct asymptotic behavior must be fulfilled. It is seen that the distorting potentials (13.42)–(13.44), contain the term $-(1/b)\sum_{k=1}^{2}\boldsymbol{\nabla}_{x_k}\varphi_i \cdot \boldsymbol{\nabla}_{x_k}\psi_i^+$, which together with the eikonal approximation $\boldsymbol{R} \simeq -\boldsymbol{r}_f$ ($M_{\rm P,T} \gg 1$), provide the exact solution ψ_i^+ of the differential equation (13.41). In this case, a separation of the independent variables $\boldsymbol{s}_1$, $\boldsymbol{s}_2$ and $\boldsymbol{r}_f$ is possible i.e. we can write

$$\psi_i^+ = C_i^+ \mathcal{F}_1^+(\boldsymbol{s}_1)\mathcal{F}_2^+(\boldsymbol{s}_2)\mathcal{F}^+(\boldsymbol{r}_f) \tag{13.45}$$

where C_i^+ is a constant to be determined. We shall first find the distorted wave χ_i^+ for the distorting potential (13.42) following Belkić and Mančev [84, 85]. Inserting (13.45) into (13.41), we obtain

$$\left(\frac{1}{2a}\nabla_{s_k}^2 + \frac{Z_{\rm P}}{s_k} + \frac{p_k^2}{2a}\right)\mathcal{F}_k^+(\boldsymbol{s}_k) = 0 \qquad (k = 1, 2) \tag{13.46}$$

$$\left(\frac{1}{2\mu_f}\nabla_{r_f}^2 - \frac{Z_{\rm P}Z_{\rm T}}{r_f} + \frac{p_f^2}{2\mu_f}\right)\mathcal{F}^+(\boldsymbol{r}_f) = 0. \tag{13.47}$$

The exact solutions of these equations are

$$\mathcal{F}_k^+(\boldsymbol{s}_k) = N^+(\nu_{{\rm P}_k}){\rm e}^{i\boldsymbol{p}_k\cdot\boldsymbol{s}_k}\,{}_1F_1(i\nu_{{\rm P}_k}, 1, ip_k s_k - i\boldsymbol{p}_k\cdot\boldsymbol{s}_k) \tag{13.48}$$

$$\mathcal{F}^+(\boldsymbol{r}_f) = \mathcal{N}^+(\nu_{\rm PT}){\rm e}^{i\boldsymbol{p}_f\cdot\boldsymbol{r}_f}\,{}_1F_1(-i\nu_{\rm PT}, 1, ip_f r_f - i\boldsymbol{p}_f\cdot\boldsymbol{r}_f) \tag{13.49}$$

where $\mathcal{N}^+(\nu_{\rm PT}) = {\rm e}^{-\pi\nu_{\rm PT}/2}\Gamma(1+i\nu_{\rm PT})$, $\nu_{{\rm P}_k} = aZ_{\rm P}/p_k$ and $\nu_{\rm PT} = Z_{\rm P}Z_{\rm T}\mu_f/p_f$. The auxiliary vectors $\boldsymbol{p}_1$, $\boldsymbol{p}_2$ and $\boldsymbol{p}_f$ are determined from the conditions

$$E - E_i = \frac{p_1^2}{2a} + \frac{p_2^2}{2a} + \frac{p_f^2}{2\mu_f} \tag{13.50}$$

$$\boldsymbol{p}_1 \cdot \boldsymbol{s}_1 + \boldsymbol{p}_2 \cdot \boldsymbol{s}_2 + \boldsymbol{p}_f \cdot \boldsymbol{r}_f = \boldsymbol{k}_i \cdot \boldsymbol{r}_i \tag{13.51}$$

$$C_i^+ \exp[i\nu_{\mathrm{PT}} \ln(p_f r_f - \boldsymbol{p}_f \cdot \boldsymbol{r}_f) - i\sum_{k=1}^{2} \nu_{\mathrm{P}_k} \ln(p_k s_k - \boldsymbol{p}_k \cdot \boldsymbol{s}_k)] \longrightarrow$$
$$\longrightarrow \exp[i\nu_i \ln(k_i r_i - \boldsymbol{k}_i \cdot \boldsymbol{r}_i)] \qquad r_i \to \infty\,. \tag{13.52}$$

Relation (13.50) is introduced in order to obtain three separable equations from (13.41) for the independent variables $\boldsymbol{s}_1$, $\boldsymbol{s}_2$ and $\boldsymbol{r}_f$. Expressions (13.51) and (13.52) originate from the requirement that χ_i^+ must satisfy the correct boundary conditions (13.9). It is easily shown that

$$\boldsymbol{p}_k \simeq -\boldsymbol{v} \qquad (M_{\mathrm{P}} \gg 1) \qquad\qquad \boldsymbol{p}_f \simeq -\boldsymbol{k}_i \qquad (M_{\mathrm{T}} \gg 1). \tag{13.53}$$

With these values of vectors $\boldsymbol{p}_1$, $\boldsymbol{p}_2$ and $\boldsymbol{p}_f$, the energy conservation law is satisfied within the eikonal mass limit, so that $E - E_i = k_i^2/(2\mu_i)$ and, moreover, the constant C_i^+ is identified as

$$C_i^+ = \mu_i^{-2i\nu_{\mathrm{P}}}. \tag{13.54}$$

This result for the constant C_i^+, which is needed in (13.52), follows from the usual asymptotic limit

$$\frac{1}{\mu_i}\frac{k_i r_i - \boldsymbol{k}_i \cdot \boldsymbol{r}_i}{v s_k + \boldsymbol{v} \cdot \boldsymbol{s}_k} \longrightarrow 1 \qquad r_i \to \infty \qquad (k = 1, 2). \tag{13.55}$$

Hence, the solution for the function ψ_i^+ becomes

$$\psi_i^+ = \mu_i^{-2i\nu_{\mathrm{P}}} N^+(\nu)[N^+(\nu_{\mathrm{P}})]^2 \mathrm{e}^{i\boldsymbol{k}_i \cdot \boldsymbol{r}_i}\, {}_1F_1(-i\nu, 1, ik_i r_f + i\boldsymbol{k}_i \cdot \boldsymbol{r}_f)$$
$$\times\; {}_1F_1(i\nu_{\mathrm{P}}, 1, ivs_1 + i\boldsymbol{v} \cdot \boldsymbol{s}_1)\, {}_1F_1(i\nu_{\mathrm{P}}, 1, ivs_2 + i\boldsymbol{v} \cdot \boldsymbol{s}_2) \tag{13.56}$$

where $\nu_{\mathrm{K}} = Z_{\mathrm{K}}/v$ (K = P, T) and $\nu = Z_{\mathrm{P}} Z_{\mathrm{T}}/v$. Now, it is readily checked that the distorted wave $\chi_i^+ = \varphi_i \psi_i^+$ has the correct asymptotic behavior (13.9). It should be noted that the proof of the correctness of the boundary conditions of the continuum distorted wave methodologies is consistent with the concept of the strong limit from the formal scattering theory [7, 41]. This has been demonstrated within the three-body distorted wave formalism for single electron capture [262]. Such a demonstration is not hampered by the presence of the kinetic energy perturbation from the three-body symmetric eikonal (SE-3B) approximation, or by the Coulombic behavior of the perturbative potentials from the corresponding CDW-3B method [262]. The same conclusions can be verified to hold also true for the scattering vector χ_i^+ from (13.56), as encountered in double electron capture treated in the four-body distorted wave formalism. The role and the physical meaning of the strong limit is studied in several chapters on the formal scattering theory in the first part of the present book as well as in Ref. [7].

Proceeding as before, the choice of the distorting potential (13.43) provides a solution for the distorted wave χ_i^+ in the entrance channel in the following form

$$\chi_i^+ = \mu_i^{i\nu_\mathrm{P}} \mathcal{N}^+(\nu_i')N^+(\nu_\mathrm{P})\,\mathrm{e}^{i\boldsymbol{k}_i\cdot\boldsymbol{r}_i}\,{}_1F_1(-i\nu_i', 1, ik_i r_f + i\boldsymbol{k}_i\cdot\boldsymbol{r}_f) \\ \times\ {}_1F_1(i\nu_\mathrm{P}, 1, ivs_1 + i\boldsymbol{v}\cdot\boldsymbol{s}_1)\varphi_i(\boldsymbol{x}_1,\boldsymbol{x}_2) \tag{13.57}$$

where $\nu_i' = Z_\mathrm{P}(Z_\mathrm{T}-1)/v$. For the distorting potential given by (13.44) we obtain the same distorted wave as in (13.57), but the following quantities should be redefined accordingly: $\nu_i' \to \nu_i'' = [Z_\mathrm{P}(Z_\mathrm{T}-1)-1]/v$ and $\nu_\mathrm{P} \to \nu_\mathrm{P}'' = (Z_\mathrm{P}-1)/v$. The determination of the distorted waves in the exit channel depends on the considered collisional process.

13.1.2 The $(Z_\mathrm{P}, e_1)_{i_1} - (Z_\mathrm{T}, e_2)_{i_2}$ collisional system

Imposing the correct boundary conditions to the $(Z_\mathrm{P}, e_1)_{i_1} - (Z_\mathrm{T}, e_2)_{i_2}$ collisions via

$$\chi_i^+ \to \Psi_i^+ \to \Phi_i^+ \tag{13.58}$$

where

$$\Phi_i^+ = \varphi_{i_1}(\boldsymbol{s}_1)\varphi_{i_2}(\boldsymbol{x}_2)\mathrm{e}^{i\boldsymbol{k}_i\cdot\boldsymbol{r}_i + i\nu_i ln(k_i r_i - \boldsymbol{k}_i\cdot\boldsymbol{r}_i)} \tag{13.59}$$

we look for χ_i^+ in a separable form

$$\chi_i^+ = \varphi_{i_1}(\boldsymbol{s}_1)\varphi_{i_2}(\boldsymbol{x}_2)\zeta_i^+. \tag{13.60}$$

Inserting (13.60) into the defining equation $(E - H_i - W_i)\chi_i^+ = 0$ for χ_i^+, as per (13.8), we obtain the following equation for ζ_i^+

$$\left[E - E_i + \frac{1}{2\mu_f}\nabla_{r_f}^2 + \frac{1}{2a}\nabla_{s_1}^2 + \frac{1}{2a}\nabla_{s_2}^2 - \frac{Z_\mathrm{T}(Z_\mathrm{P}-1)}{r_f} + \frac{Z_\mathrm{P}-1}{s_2}\right]\zeta_i^+ = 0 \tag{13.61}$$

provided that the distorting potential is chosen according to

$$U_i\chi_i^+ = \left[Z_\mathrm{T}\left(\frac{1}{R} - \frac{1}{x_1}\right) - \frac{1}{s_2} + \frac{1}{r_{12}}\right]\chi_i^+ \\ -\frac{1}{a}\varphi_{i_1}(\boldsymbol{s}_1)\boldsymbol{\nabla}_{x_2}\varphi_{i_2}(\boldsymbol{x}_2)\cdot\boldsymbol{\nabla}_{s_2}\zeta_i^+ - \frac{1}{b}\varphi_{i_2}(\boldsymbol{x}_2)\boldsymbol{\nabla}_{s_1}\varphi_{i_1}(\boldsymbol{s}_1)\cdot\boldsymbol{\nabla}_{x_1}\zeta_i^+. \tag{13.62}$$

We can solve (13.61) exactly due to separation of the independent variables. As a net result for the distorted wave χ_i^+, we have

$$\chi_i^+ = N^+(\nu_\mathrm{P})\mathcal{N}^+(\nu)\,\varphi_{i_1}(\boldsymbol{s}_1)\varphi_{i_2}(\boldsymbol{x}_2)\,\mathrm{e}^{i\boldsymbol{k}_i\cdot\boldsymbol{r}_i}\,{}_1F_1(-i\nu, 1, ik_i r_i - i\boldsymbol{k}_i\cdot\boldsymbol{r}_i) \\ \times\ {}_1F_1(i\nu_\mathrm{P}, 1, ivs_2 + i\boldsymbol{v}\cdot\boldsymbol{s}_2)\varphi_i(\boldsymbol{x}_1,\boldsymbol{x}_2) \tag{13.63}$$

where

$$N^+(\nu_\mathrm{P}) = \Gamma(1 - i\nu_\mathrm{P})\mathrm{e}^{\pi\nu_\mathrm{P}/2} \qquad \mathcal{N}^+(\nu) = \Gamma(1 + i\nu)\mathrm{e}^{-\pi\nu/2} \\ \nu = \frac{Z_\mathrm{T}(Z_\mathrm{P}-1)}{v} \qquad \nu_\mathrm{P} = \frac{Z_\mathrm{P}-1}{v}. \tag{13.64}$$

14

Double electron capture

14.1 The CDW-4B method

In order to complete the expression for the transition amplitude in the distorted wave theory for double electron capture, we look for the distorted waves in the exit channel. First, we determine the auxiliary distorted wave ξ_f^- defined by (13.34). Letting $\varepsilon \to 0^+$, it is seen that, according to (13.34) and (13.28), ξ_f^- is the solution of

$$(E - H + V_x)\xi_f^- = (E - H + V_f - W_f + V_x)\chi_f^- \,. \tag{14.1}$$

Since χ_f^- satisfies the relation (13.8), it follows that (14.1) can be reduced to

$$(E - H + V_x)\xi_f^- = V_x\chi_f^- \,. \tag{14.2}$$

Under the assumption

$$V_x^\dagger \chi_f^- = 0 \tag{14.3}$$

it follows that (14.2) becomes solvable analytically. In such a case, we write ξ_f^- in the factored form

$$\xi_f^- = \varphi_f(\boldsymbol{s}_1, \boldsymbol{s}_2)\psi_f^- \tag{14.4}$$

where the unknown function ψ_f^- is the solution of the equation

$$\varphi_f(E - E_f - H_0 - V_f)\psi_f^- + \frac{1}{a}\sum_{k=1}^{2} \boldsymbol{\nabla}_{s_k}\varphi_f \cdot \boldsymbol{\nabla}_{s_k}\psi_f^- + V_x\varphi_f\psi_f^- = 0\,. \tag{14.5}$$

Choosing the potential V_x in the form used by Belkić and Mančev [84, 85]

$$V_x = -\frac{1}{a}\sum_{k=1}^{2} \boldsymbol{\nabla}_{s_k} \ln\varphi_f \cdot \boldsymbol{\nabla}_{s_k} \tag{14.6}$$

we have

$$(E - E_f - H_0 - V_f)\psi_f^- = 0 \tag{14.7}$$

so that the function ψ_f^- finally reads as

$$\psi_f^- = \mu_f^{2i\nu_\mathrm{T}}\mathcal{N}^-(\nu)[N^-(\nu_\mathrm{T})]^2\mathrm{e}^{-i\boldsymbol{k}_f\cdot\boldsymbol{r}_f}\,{}_1F_1(i\nu,1,-ik_fr_i-i\boldsymbol{k}_f\cdot\boldsymbol{r}_i)$$
$$\times\,{}_1F_1(-i\nu_\mathrm{T},1,-ivx_1-i\boldsymbol{v}\cdot\boldsymbol{x}_1)\,{}_1F_1(-i\nu_\mathrm{T},1,-ivx_2-i\boldsymbol{v}\cdot\boldsymbol{x}_2)\,. \quad (14.8)$$

Utilizing (13.42) for the distorting potential U_i and the corresponding distorted wave for the initial state $\chi_i^+ = \varphi_i\psi_i^+$, where the function ψ_i^+ is determined by (13.56), the transition amplitude for double electron capture in the CDW-4B method becomes [84]

$$T_{if}^- = -N^2\iiint \mathrm{d}\boldsymbol{x}_1\mathrm{d}\boldsymbol{x}_2\mathrm{d}\boldsymbol{r}_i\,\mathrm{e}^{i\boldsymbol{k}_i\cdot\boldsymbol{r}_i+i\boldsymbol{k}_f\cdot\boldsymbol{r}_f}\mathcal{L}(\boldsymbol{r}_i,\boldsymbol{r}_f)\varphi_f^*(\boldsymbol{s}_1,\boldsymbol{s}_2)$$
$$\times\,{}_1F_1(i\nu_\mathrm{T},1,ivx_1+i\boldsymbol{v}\cdot\boldsymbol{x}_1)\,{}_1F_1(i\nu_\mathrm{T},1,ivx_2+i\boldsymbol{v}\cdot\boldsymbol{x}_2)$$
$$\times\{\,{}_1F_1(i\nu_\mathrm{P},1,ivs_2+i\boldsymbol{v}\cdot\boldsymbol{s}_2)\boldsymbol{\nabla}_{x_1}\varphi_i(\boldsymbol{x}_1,\boldsymbol{x}_2)\cdot\boldsymbol{\nabla}_{s_1}\,{}_1F_1(i\nu_\mathrm{P},1,ivs_1+i\boldsymbol{v}\cdot\boldsymbol{s}_1)$$
$$+\,{}_1F_1(i\nu_\mathrm{P},1,ivs_1+i\boldsymbol{v}\cdot\boldsymbol{s}_1)\boldsymbol{\nabla}_{x_2}\varphi_i(\boldsymbol{x}_1,\boldsymbol{x}_2)\cdot\boldsymbol{\nabla}_{s_2}\,{}_1F_1(i\nu_\mathrm{P},1,ivs_2+i\boldsymbol{v}\cdot\boldsymbol{s}_2)\} \quad (14.9)$$

where $N = N^+(\nu_\mathrm{P})N^+(\nu_\mathrm{T})$ and

$$\mathcal{L}(\boldsymbol{r}_i,\boldsymbol{r}_f) = \mu_i^{-2i\nu_\mathrm{P}}\mu_f^{-2i\nu_\mathrm{T}}[\mathcal{N}^-(\nu)]^2$$
$$\times\,{}_1F_1(-i\nu,1,ik_ir_f+i\boldsymbol{k}_i\cdot\boldsymbol{r}_f)\,{}_1F_1(-i\nu,1,ik_fr_i+i\boldsymbol{k}_f\cdot\boldsymbol{r}_i). \quad (14.10)$$

A simplification of (14.10) follows from the eikonal approximation

$$[\mathcal{N}^-(\nu)]^2\,{}_1F_1(-i\nu,1,ik_ir_f+i\boldsymbol{k}_i\cdot\boldsymbol{r}_f)\,{}_1F_1(-i\nu,1,ik_fr_i+i\boldsymbol{k}_f\cdot\boldsymbol{r}_i)$$
$$\simeq(k_ir_f+\boldsymbol{k}_i\cdot\boldsymbol{r}_f)^{i\nu}(k_fr_i+\boldsymbol{k}_f\cdot\boldsymbol{r}_i)^{i\nu}\simeq(\mu_i\mu_f)^{i\nu}[(vR-\boldsymbol{v}\cdot\boldsymbol{R})(vR+\boldsymbol{v}\cdot\boldsymbol{R})]^{i\nu}$$
$$=(\mu_i\mu_f)^{i\nu}[v^2(R^2-Z^2)]^{i\nu}=(\mu_i\mu_f)^{i\nu}(v\rho)^{2i\nu}\simeq(\mu v\rho)^{2i\nu}$$

$$\therefore\qquad \mathcal{L}(\boldsymbol{r}_i,\boldsymbol{r}_f)\simeq\mu^{-2i(\nu_\mathrm{P}+\nu_\mathrm{T})}(\mu v\rho)^{2i\nu} \quad (14.11)$$

where $\mu = M_\mathrm{P}M_\mathrm{T}/(M_\mathrm{P}+M_\mathrm{T})$. Here, $\boldsymbol{\rho}$ is the projection of vector $\boldsymbol{R}$ to the XOY plane perpendicular to the Z-axis i.e. $\boldsymbol{\rho} = \boldsymbol{R} - \boldsymbol{Z}$ with $\boldsymbol{\rho}\cdot\boldsymbol{Z} = 0$, where vector $\boldsymbol{Z}$ represents the projection of vector $\boldsymbol{R}$ to the Z-axis. The phase factor $(\mu v\rho)^{2i\nu}$, which stems directly from the inter-nuclear potential $V_\mathrm{PT} = Z_\mathrm{P}Z_\mathrm{T}/R$, does not influence the total cross section, since

$$Q_{if}^-(a_0^2) = \frac{1}{(2\pi v)^2}\int \mathrm{d}\boldsymbol{\eta}\left|T_{if}^-(\boldsymbol{\eta})\right|^2 = \int \mathrm{d}\boldsymbol{\eta}\left|\frac{R_{if}^-(\boldsymbol{\eta})}{2\pi v}\right|^2 \quad (14.12)$$

where

$$R_{if}^-(\boldsymbol{\eta}) = -N^2\iiint \mathrm{d}\boldsymbol{x}_1\mathrm{d}\boldsymbol{x}_2\mathrm{d}\boldsymbol{r}_i\,\mathrm{e}^{i\boldsymbol{q}_\mathrm{P}\cdot(\boldsymbol{s}_1+\boldsymbol{s}_2)+i\boldsymbol{q}_\mathrm{T}\cdot(\boldsymbol{x}_1+\boldsymbol{x}_2)}\varphi_f^*(\boldsymbol{s}_1,\boldsymbol{s}_2)$$
$$\times\,{}_1F_1(i\nu_\mathrm{T},1,ivx_1+i\boldsymbol{v}\cdot\boldsymbol{x}_1)\,{}_1F_1(i\nu_\mathrm{T},1,ivx_2+i\boldsymbol{v}\cdot\boldsymbol{x}_2)$$
$$\times\{\,{}_1F_1(i\nu_\mathrm{P},1,ivs_2+i\boldsymbol{v}\cdot\boldsymbol{s}_2)\boldsymbol{\nabla}_{x_1}\varphi_i(\boldsymbol{x}_1,\boldsymbol{x}_2)\cdot\boldsymbol{\nabla}_{s_1}\,{}_1F_1(i\nu_\mathrm{P},1,ivs_1+i\boldsymbol{v}\cdot\boldsymbol{s}_1)$$
$$+\,{}_1F_1(i\nu_\mathrm{P},1,ivs_1+i\boldsymbol{v}\cdot\boldsymbol{s}_1)\boldsymbol{\nabla}_{x_2}\varphi_i(\boldsymbol{x}_1,\boldsymbol{x}_2)\cdot\boldsymbol{\nabla}_{s_2}\,{}_1F_1(i\nu_\mathrm{P},1,ivs_2+i\boldsymbol{v}\cdot\boldsymbol{s}_2)\}\,. \quad (14.13)$$

It is now obvious from (14.12) that the total cross section Q^-_{if} is independent of the inter-nuclear potential $Z_\mathrm{P}Z_\mathrm{T}/R$, as it should be [44]. The expression (14.13) for the basic matrix element R^-_{if} represents the main working expression for calculating the total cross sections. Such a CDW-4B method represents a strict generalization of the Cheshire's CDW-3B method [42] for purely three-body single charge exchange (12.26) (see also [43]). As per the derivation which followed the original work of Belkić and Mančev [84], the result (14.13) for R^-_{if} in the CDW-4B method represents the rigorous first-order term (13.33) in the four-body Dodd-Greider perturbation series. This is very important, in view of the absence of any disconnected diagrams in the Dodd-Greider expansion, a feature which precludes divergence of the series. Only non-divergent perturbation series have a chance to provide the mathematically meaningful first-order terms that, in turn, could capture the major physical effects. Such is the CDW-4B method which can be called the four-body first-order continuum distorted wave (CDW-4B1) method. A similar remark also applies to the CDW-EIS-4B method which, therefore, can alternatively be termed as the four-body first-order continuum distorted wave eikonal initial state (CDW-EIS-4B1) method.

The post form of the transition amplitude can also be established. Similarly to (13.33), we employ the first-order of the Dodd-Greider integral equation

$$T^+_{if} = \langle \Phi_f | \Omega^{-\dagger}_f (V_f - W^\dagger_f)[1 + g^+_x (V_i - W_i)]\Omega^+_i | \Phi_i \rangle. \tag{14.14}$$

The derivation of the transition amplitude T^+_{if} in the post form is carried out in a fashion similar to its prior counterpart, so that

$$\begin{aligned}
T^+_{if} &= -N^2 \iiint \mathrm{d}\boldsymbol{s}_1 \mathrm{d}\boldsymbol{s}_2 \mathrm{d}\boldsymbol{r}_f \, \mathrm{e}^{i\boldsymbol{k}_i\cdot\boldsymbol{r}_i + i\boldsymbol{k}_f\cdot\boldsymbol{r}_f} \mathcal{L}(\boldsymbol{r}_i, \boldsymbol{r}_f)\varphi_i(\boldsymbol{x}_1, \boldsymbol{x}_2) \\
&\times {}_1F_1(i\nu_\mathrm{P}, 1, ivs_1 + i\boldsymbol{v}\cdot\boldsymbol{s}_1)\, {}_1F_1(i\nu_\mathrm{P}, 1, ivs_2 + i\boldsymbol{v}\cdot\boldsymbol{s}_2) \\
&\times \left\{ {}_1F_1(i\nu_\mathrm{T}, 1, ivx_2 + i\boldsymbol{v}\cdot\boldsymbol{x}_2) \boldsymbol{\nabla}_{s_1}\varphi^*_f(\boldsymbol{s}_1, \boldsymbol{s}_2) \cdot \boldsymbol{\nabla}_{x_1} {}_1F_1(i\nu_\mathrm{T}, 1, ivx_1 + i\boldsymbol{v}\cdot\boldsymbol{x}_1) \right. \\
&\left. + {}_1F_1(i\nu_\mathrm{T}, 1, ivx_1 + i\boldsymbol{v}\cdot\boldsymbol{x}_1) \boldsymbol{\nabla}_{s_2}\varphi^*_f(\boldsymbol{s}_1, \boldsymbol{s}_2) \cdot \boldsymbol{\nabla}_{x_2} {}_1F_1(i\nu_\mathrm{T}, 1, ivx_2 + i\boldsymbol{v}\cdot\boldsymbol{x}_2) \right\}.
\end{aligned} \tag{14.15}$$

The post total cross sections is

$$Q^+_{if}(a_0^2) = \frac{1}{(2\pi v)^2} \int \mathrm{d}\boldsymbol{\eta} \left| T^+_{if}(\boldsymbol{\eta}) \right|^2 = \int \mathrm{d}\boldsymbol{\eta} \left| \frac{R^+_{if}(\boldsymbol{\eta})}{2\pi v} \right|^2 \tag{14.16}$$

where

$$\begin{aligned}
R^+_{if}(\boldsymbol{\eta}) &= -N^2 \iiint \mathrm{d}\boldsymbol{s}_1 \mathrm{d}\boldsymbol{s}_2 \mathrm{d}\boldsymbol{r}_f \, \mathrm{e}^{i\boldsymbol{q}_\mathrm{P}\cdot(\boldsymbol{s}_1+\boldsymbol{s}_2) + i\boldsymbol{q}_\mathrm{T}\cdot(\boldsymbol{x}_1+\boldsymbol{x}_2)} \varphi_i(\boldsymbol{x}_1, \boldsymbol{x}_2) \\
&\times {}_1F_1(i\nu_\mathrm{P}, 1, ivs_1 + i\boldsymbol{v}\cdot\boldsymbol{s}_1)\, {}_1F_1(i\nu_\mathrm{P}, 1, ivs_2 + i\boldsymbol{v}\cdot\boldsymbol{s}_2) \\
&\times \left\{ {}_1F_1(i\nu_\mathrm{T}, 1, ivx_2 + i\boldsymbol{v}\cdot\boldsymbol{x}_2) \boldsymbol{\nabla}_{s_1}\varphi^*_f(\boldsymbol{s}_1, \boldsymbol{s}_2) \cdot \boldsymbol{\nabla}_{x_1} {}_1F_1(i\nu_\mathrm{T}, 1, ivx_1 + i\boldsymbol{v}\cdot\boldsymbol{x}_1) \right. \\
&\left. + {}_1F_1(i\nu_\mathrm{T}, 1, ivx_1 + i\boldsymbol{v}\cdot\boldsymbol{x}_1) \boldsymbol{\nabla}_{s_2}\varphi^*_f(\boldsymbol{s}_1, \boldsymbol{s}_2) \cdot \boldsymbol{\nabla}_{x_2} {}_1F_1(i\nu_\mathrm{T}, 1, ivx_2 + i\boldsymbol{v}\cdot\boldsymbol{x}_2) \right\}.
\end{aligned} \tag{14.17}$$

The transition amplitude as a function of vector $\boldsymbol{\rho}$ can be obtained via

$$a_{if}^{\pm}(\boldsymbol{\rho}) = \frac{1}{2\pi v}\rho^{2i\nu}\int \mathrm{d}\boldsymbol{\eta}\,\mathrm{e}^{i\boldsymbol{\eta}\cdot\boldsymbol{\rho}}R_{if}^{\pm}(\boldsymbol{\eta}) . \tag{14.18}$$

Using the Parseval relation i.e. the convolution theorem for the Fourier integral in the total cross sections, we have

$$Q_{if}^{\pm}(a_0^2) = \int \mathrm{d}\boldsymbol{\rho}\,|a_{if}^{\pm}(\boldsymbol{\rho})|^2 . \tag{14.19}$$

The differential cross section in the CDW-3B and CDW-4B methods can be calculated directly from the expressions for $T_{if}^{\pm}$ [175, 263]. Alternatively, we can first carry out the Fourier integral according to (14.18), and then use the following expression for the differential cross section [264]–[266]

$$\frac{\mathrm{d}Q_{if}^{\pm}}{\mathrm{d}\Omega} = \left| i\mu v\int_0^{\infty}\mathrm{d}\rho\rho^{1+2i\nu}J_{m_i-m_f}\left(2\mu v\rho\sin\frac{\theta}{2}\right)a_{if}^{\pm}(\rho)\right|^2\left(\frac{a_0^2}{sr}\right) \tag{14.20}$$

where θ is the scattering angle in the center of mass frame of reference. Here, $J_\nu(z)$ is the Bessel function of the first order and νth kind, whereas m_i and m_f are the magnetic quantum numbers of the initial and final bound state, respectively.

Calculation of the matrix elements for double electron capture into the ground state $1s^2$ from any helium-like atom/ion has been carried out by Belkić and Mančev [84]. Their method of calculation provides the total cross section $Q_{if}^{\pm}$ through four-dimensional integrals that are subsequently computed by utilizing the scaled Gauss-Legendre and Gauss-Mehler quadratures [203, 267, 268]. It can be verified that in the symmetric resonant case ($i=f,\ Z_{\mathrm{P}}=Z_{\mathrm{T}}$), there is no post-prior discrepancy i.e. $R_{if}^{-}=R_{if}^{+}$, so that $Q_{if}^{-}=Q_{if}^{+}$.

14.2 The SE-4B method

The CDW-4B method for double charge exchange treats two electrons and two nuclei in an entirely symmetric manner in the entrance and exit channel. In particular, the two-electron full Coulomb wave functions are used to describe double continuum intermediate states that distort the initial and final unperturbed states Φ_i and Φ_f. This double electronic distortion function is given by the product of two Coulomb wave functions (the so-called C2 wave function) centered either on the projectile or target nucleus according to (13.56) and (14.8) in the entrance and exit channels, respectively. The behavior of each of the two Coulomb wave functions at large distances is given by the asymptotes from (11.30). In particular, for $|vs_k+\boldsymbol{v}\cdot\boldsymbol{s}_k|\gg 1$ $(k=1,2)$, the leading asymptotic term in the product of the two confluent hyper-geometric functions from the C2 wave function is the twofold Coulomb logarithmic phase factor in the entrance channel

$$[N^+(\nu_{\rm P})]^2\,{}_1F_1(i\nu_{\rm P},1,ivs_1+i\boldsymbol{v}\cdot\boldsymbol{s}_1)\,{}_1F_1(i\nu_{\rm P},1,ivs_2+i\boldsymbol{v}\cdot\boldsymbol{s}_2)$$
$$\approx \ln{(vs_1+\boldsymbol{v}\cdot\boldsymbol{s}_1)}^{-i\nu_{\rm P}}\ln{(vs_2+\boldsymbol{v}\cdot\boldsymbol{s}_2)}^{-i\nu_{\rm P}}. \qquad (14.21)$$

Likewise, in the exit channel we have for $|vx_k+\boldsymbol{v}\cdot\boldsymbol{x}_k|\gg 1\,(k=1,2)$

$$[N^-(\nu_{\rm T})]^2\,{}_1F_1(-i\nu_{\rm T},1,-ivx_1-i\boldsymbol{v}\cdot\boldsymbol{x}_1)\,{}_1F_1(-i\nu_{\rm T},1,-ivx_2-i\boldsymbol{v}\cdot\boldsymbol{x}_2)$$
$$\approx \ln{(vx_1+\boldsymbol{v}\cdot\boldsymbol{x}_1)}^{i\nu_{\rm T}}\ln{(vx_2+\boldsymbol{v}\cdot\boldsymbol{x}_2)}^{i\nu_{\rm T}}. \qquad (14.22)$$

If one makes the additional approximations (14.21) and (14.22) to the exact CDW-4B method simultaneously in the entrance and exit channels, one obtains the four-body symmetric eikonal (SE-4B) method for double capture in process (11.1). The initial scattering state wave function in the SE-4B methods is

$$\psi_i^+ = \mu_i^{-2i\nu_{\rm P}}\mathcal{N}^+(\nu)\mathrm{e}^{i\boldsymbol{k}_i\cdot\boldsymbol{r}_i}\,{}_1F_1(-i\nu,1,ik_ir_f+i\boldsymbol{k}_i\cdot\boldsymbol{r}_f)$$
$$\times\ln{(vs_1+\boldsymbol{v}\cdot\boldsymbol{s}_1)}^{-i\nu_{\rm P}}\ln{(vs_2+\boldsymbol{v}\cdot\boldsymbol{s}_2)}^{-i\nu_{\rm P}} \qquad (14.23)$$

whereas the final state vector reads

$$\psi_f^- = \mu_f^{2i\nu_{\rm T}}\mathcal{N}^-(\nu)\mathrm{e}^{-i\boldsymbol{k}_f\cdot\boldsymbol{r}_f}\,{}_1F_1(i\nu,1,-ik_fr_i-i\boldsymbol{k}_f\cdot\boldsymbol{r}_i)$$
$$\times\ln{(vx_1+\boldsymbol{v}\cdot\boldsymbol{x}_1)}^{i\nu_{\rm T}}\ln{(vx_2+\boldsymbol{v}\cdot\boldsymbol{x}_2)}^{i\nu_{\rm T}}. \qquad (14.24)$$

Thus, the SE-4B method also treats both electrons on the same footing in the entrance and exit channels, as does the CDW-4B method, except that the former method uses the logarithmic phases instead of the original full Coulomb waves from the latter method. This modification of the electronic C2 wave function from the CDW-4B method must simultaneously be accompanied by the associated change in the perturbation interactions through the appearance of the two-electron kinetic energy operators alongside the usual gradient-gradient potential operators. These kinetic energy operators represent the additional perturbations introduced by eikonalization of the full Coulomb wave functions. With these modifications, the prior form of the transition amplitude in the SE-4B method becomes

$$T_{if}^- = -\iiint \mathrm{d}\boldsymbol{x}_1\mathrm{d}\boldsymbol{x}_2\mathrm{d}\boldsymbol{r}_i\,\mathrm{e}^{i\boldsymbol{q}_{\rm P}\cdot(\boldsymbol{s}_1+\boldsymbol{s}_2)+i\boldsymbol{q}_{\rm T}\cdot(\boldsymbol{x}_1+\boldsymbol{x}_2)}\mathcal{L}(\boldsymbol{r}_i,\boldsymbol{r}_f)$$
$$\times\varphi_f^*(\boldsymbol{s}_1,\boldsymbol{s}_2)\ln{(vx_1+\boldsymbol{v}\cdot\boldsymbol{x}_1)}^{-i\nu_{\rm T}}\ln{(vx_2+\boldsymbol{v}\cdot\boldsymbol{x}_2)}^{-i\nu_{\rm T}}$$
$$\times\left[\frac{1}{2}\nabla_{s_1}^2+\frac{1}{2}\nabla_{s_2}^2+\boldsymbol{\nabla}_{x_1}\ln\varphi_i(\boldsymbol{x}_1,\boldsymbol{x}_2)\cdot\boldsymbol{\nabla}_{s_1}+\boldsymbol{\nabla}_{x_2}\ln\varphi_i(\boldsymbol{x}_1,\boldsymbol{x}_2)\cdot\boldsymbol{\nabla}_{s_2}\right]$$
$$\times\varphi_i(\boldsymbol{x}_1,\boldsymbol{x}_2)\ln{(vs_1+\boldsymbol{v}\cdot\boldsymbol{s}_1)}^{-i\nu_{\rm P}}\ln{(vs_2+\boldsymbol{v}\cdot\boldsymbol{s}_2)}^{-i\nu_{\rm P}}. \qquad (14.25)$$

Similarly, the post version of the $T-$matrix elements in the same method is

$$T_{if}^{+} = -\iiint \mathrm{d}\boldsymbol{s}_1 \mathrm{d}\boldsymbol{s}_2 \mathrm{d}\boldsymbol{r}_f \, \mathrm{e}^{i\boldsymbol{q}_{\mathrm{P}}\cdot(\boldsymbol{s}_1+\boldsymbol{s}_2)+i\boldsymbol{q}_{\mathrm{T}}\cdot(\boldsymbol{x}_1+\boldsymbol{x}_2)} \mathcal{L}(\boldsymbol{r}_i,\boldsymbol{r}_f)$$
$$\times \varphi_i(\boldsymbol{x}_1,\boldsymbol{x}_2) \ln(vs_1 + \boldsymbol{v}\cdot\boldsymbol{s}_1)^{-i\nu_{\mathrm{P}}} \ln(vs_2 + \boldsymbol{v}\cdot\boldsymbol{s}_2)^{-i\nu_{\mathrm{P}}}$$
$$\times \left[\frac{1}{2}\nabla_{x_1}^2 + \frac{1}{2}\nabla_{x_2}^2 + \boldsymbol{\nabla}_{s_1} \ln\varphi_f^*(\boldsymbol{s}_1,\boldsymbol{s}_2)\cdot\boldsymbol{\nabla}_{x_1} + \boldsymbol{\nabla}_{s_2}\ln\varphi_f^*(\boldsymbol{s}_1,\boldsymbol{s}_2)\cdot\boldsymbol{\nabla}_{x_2}\right]$$
$$\times \varphi_f^*(\boldsymbol{s}_1,\boldsymbol{s}_2) \ln(vx_1 + \boldsymbol{v}\cdot\boldsymbol{x}_1)^{-i\nu_{\mathrm{T}}} \ln(vx_2 + \boldsymbol{v}\cdot\boldsymbol{x}_2)^{-i\nu_{\mathrm{T}}}. \qquad (14.26)$$

Here, the function $\mathcal{L}(\boldsymbol{r}_i,\boldsymbol{r}_f)$ is taken from (14.10). In the consistent mass limit $M_{\mathrm{P,T}} \gg 1$, the eikonal form (14.11) can be used for $\mathcal{L}(\boldsymbol{r}_i,\boldsymbol{r}_f)$ omitting the unimportant phases of unit moduli e.g. $(\mu v)^{2i\nu}\mu^{-2i(\nu_{\mathrm{P}}+\nu_{\mathrm{T}})}$. The remaining phase factor $\rho^{2i\nu}$, as the only contribution from the inter-nuclear potential $V_{\mathrm{PT}} = Z_{\mathrm{P}}Z_{\mathrm{T}}/R$, is important for differential cross sections $\mathrm{d}Q_{if}^{\mp}/\mathrm{d}\Omega$ computed by the Fourier-Bessel transform of the $\rho-$dependent transition amplitudes $a_{if}^{\mp}(\boldsymbol{\rho})$ via (14.18). The matrix elements $R_{if}^{\mp}(\boldsymbol{\eta})$ from (14.12) and (14.16) differ from $T_{if}^{\mp}$ in (14.25) and (14.26) only in the absence of the functions $\mathcal{L}(\boldsymbol{r}_i,\boldsymbol{r}_f)$, respectively. The inter-nuclear phase $\rho^{2i\nu}$ gives a significant contribution primarily at larger scattering angles. The phase factor $\rho^{2i\nu}$ disappears altogether from total cross sections $Q_{if}^{\mp}$ computed from (14.12), (14.16) or (14.19). This is expected, since the inter-nuclear potential cannot give any contribution to total cross sections in the mass limits $M_{\mathrm{P,T}} \gg 1$ [44].

14.3 The CDW-EIS-4B method

The CDW-EIS-4B method is a hybrid asymmetrical model which treats the entrance and exit channels by the SE-4B and CDW-4B methods, respectively. Here, the two-electron eikonal initial state is employed, so that scattering wave function ψ_i^+ in the entrance channel is given by (14.23). The scattering wave function ψ_f^- in the final state for the exit channel is borrowed from the CDW-4B method via (14.8). Therefore, the transition amplitude in the CDW-EIS-4B method reads as

$$T_{if}^{+} = -[N^+(\nu_{\mathrm{T}})]^2 \iiint \mathrm{d}\boldsymbol{s}_1 \mathrm{d}\boldsymbol{s}_2 \mathrm{d}\boldsymbol{r}_f \, \mathrm{e}^{i\boldsymbol{q}_{\mathrm{P}}\cdot(\boldsymbol{s}_1+\boldsymbol{s}_2)+i\boldsymbol{q}_{\mathrm{T}}\cdot(\boldsymbol{x}_1+\boldsymbol{x}_2)} \mathcal{L}(\boldsymbol{r}_i,\boldsymbol{r}_f)$$
$$\times \varphi_i(\boldsymbol{x}_1,\boldsymbol{x}_2) \ln(vs_1 + \boldsymbol{v}\cdot\boldsymbol{s}_1)^{-i\nu_{\mathrm{P}}} \ln(vs_2 + \boldsymbol{v}\cdot\boldsymbol{s}_2)^{-i\nu_{\mathrm{P}}}$$
$$\times \left\{ {}_1F_1(i\nu_{\mathrm{T}},1,ivx_2 + i\boldsymbol{v}\cdot\boldsymbol{x}_2)\boldsymbol{\nabla}_{s_1}\varphi_f^*(\boldsymbol{s}_1,\boldsymbol{s}_2)\cdot\boldsymbol{\nabla}_{x_1}\, {}_1F_1(i\nu_{\mathrm{T}},1,ivx_1 + i\boldsymbol{v}\cdot\boldsymbol{x}_1)\right.$$
$$\left. + {}_1F_1(i\nu_{\mathrm{T}},1,ivx_1 + i\boldsymbol{v}\cdot\boldsymbol{x}_1)\boldsymbol{\nabla}_{s_2}\varphi_f^*(\boldsymbol{s}_1,\boldsymbol{s}_2)\cdot\boldsymbol{\nabla}_{x_2}\, {}_1F_1(i\nu_{\mathrm{T}},1,ivx_2 + i\boldsymbol{v}\cdot\boldsymbol{x}_2)\right\}. \qquad (14.27)$$

14.4 The CDW-EFS-4B method

The CDW-EFS-4B method is also a hybrid asymmetrical model, but here the entrance and exit channels are described by the CDW-4B and SE-4B methods, respectively. This time, the two-electron eikonal final state is used with the scattering wave function ψ_f^- in the exit channel given by (14.24). The scattering wave function ψ_i^+ in the initial state for the entrance channel is taken from the CDW-4B method through (13.56). Thus, the CDW-EFS-4B method is the mirror image of the CDW-EIS-4B method. The transition amplitude in the CDW-EFS-4B method is

$$T_{if}^- = -[N^+(\nu_{\rm P})]^2 \iiint \mathrm{d}\boldsymbol{x}_1 \mathrm{d}\boldsymbol{x}_2 \mathrm{d}\boldsymbol{r}_i \, \mathrm{e}^{i\boldsymbol{q}_{\rm P}\cdot(\boldsymbol{s}_1+\boldsymbol{s}_2)+i\boldsymbol{q}_{\rm T}\cdot(\boldsymbol{x}_1+\boldsymbol{x}_2)} \mathcal{L}(\boldsymbol{r}_i, \boldsymbol{r}_f)$$
$$\times \varphi_f^*(\boldsymbol{s}_1, \boldsymbol{s}_2) \ln{(vx_1 + \boldsymbol{v}\cdot\boldsymbol{x}_1)}^{-i\nu_{\rm T}} \ln{(vx_2 + \boldsymbol{v}\cdot\boldsymbol{x}_2)}^{-i\nu_{\rm T}}$$
$$\times \{ {}_1F_1(i\nu_{\rm P}, 1, ivs_2 + i\boldsymbol{v}\cdot\boldsymbol{s}_2) \boldsymbol{\nabla}_{x_1}\varphi_i(\boldsymbol{x}_1, \boldsymbol{x}_2)\cdot\boldsymbol{\nabla}_{s_1}\, {}_1F_1(i\nu_{\rm P}, 1, ivs_1 + i\boldsymbol{v}\cdot\boldsymbol{s}_1)$$
$$+ \, {}_1F_1(i\nu_{\rm P}, 1, ivs_1 + i\boldsymbol{v}\cdot\boldsymbol{s}_1) \boldsymbol{\nabla}_{x_2}\varphi_i(\boldsymbol{x}_1, \boldsymbol{x}_2)\cdot\boldsymbol{\nabla}_{s_2}\, {}_1F_1(i\nu_{\rm P}, 1, ivs_2 + i\boldsymbol{v}\cdot\boldsymbol{s}_2) \} . \tag{14.28}$$

Regarding differential as well as total cross sections, the same procedure from the CDW-4B or SE-4B methods also applies to the CDW-EIS-4B and CDW-EFS-4B methods. This amounts to using the generic expressions (14.18), (14.12), (14.16) and (14.19). In these latter formulae, the matrix elements $R_{if}^+(\boldsymbol{\eta})$ and $R_{if}^-(\boldsymbol{\eta})$ coincide with the transition amplitudes T_{if}^+ and T_{if}^- from (14.27) and (14.28) provided that the function $\mathcal{L}(\boldsymbol{r}_i, \boldsymbol{r}_f)$ from (14.11) due to the inter-nuclear distortion is set to unity, as justified by $M_{\rm P,T} \gg 1$.

We reiterate that the CDW-EIS-4B and CDW-EFS-4B methods are two different approximate variants to the CDW-4B method. The supplementary approximations consist of replacing the electronic C2 wave functions from the CDW-4B method by their asymptotic forms (Coulomb logarithmic phase factors) in one of the two channels (the entrance or exit channel in the CDW-EIS-4B or CDW-EFS-4B methods, respectively). However, these further approximations introduced by the CDW-EIS-4B and CDW-EFS-4B methods destroy the original symmetric treatments of two electrons and two nuclei in the CDW-4B method.

The above setting of the SE-4B (prior, post), CDW-EIS-4B and CDW-EFS-4B methods stems simply from making further approximations (of varying severity) to the already available expressions from the exact CDW-4B method. Such a setting directly establishes the connections among different methods. The found relationships facilitate comparisons among these methods, so that potentially notable differences in the obtained results could be interpreted in terms of the corresponding degrees of physical mechanisms invoked in different approximations. Alternatively, one can derive the SE-4B,

CDW-EIS-4B and CDW-EFS-4B methods without recourse to the CDW-4B method by making separate choices of distorting potentials and subsequently solving the resulting equations for the distorted waves. Such an analysis within the SE-4B, CDW-EIS-4B and CDW-EFS-4B methods might give the wrong impression that these approximations are unrelated to the CDW-4B method. However, irrespective of the way in which the derivation proceeds, one inevitably obtains the same results as given in the above succinct outlines (with no derivation whatsoever) by appropriately approximating the CDW-4B method. Hence the needed relationship.

The important question to ask is: why should one make the eikonal approximations to the electronic distorting functions in the initial or final states if these latter functions could be treated exactly? Do the supplementary approximations eventually simplify the computations by a sizeable factor? And, most importantly, is there any significant physical effect which is lost by these eikonalizations?

For three-body problems, the SE-3B method gives closed analytical expressions for single charge exchange, relative to the corresponding one-dimensional numerical quadrature in the CDW-3B method. Nevertheless, the difference in the computational effort invested to generate the needed tables and data bases for cross sections is negligible, since a single numerical integration is a trivial task by any standard. However, the SE-3B method irretrievably loses the important Thomas double scattering mechanism, which on the other hand is described by the CDW-3B method. This is manifested by the BK1-3B type $\propto v^{-12-2l_i-2l_f}$ –behavior of $Q_{if}^{(\mathrm{SE-3B})}$ in the limit of high impact velocities v for fixed values of the angular momenta l_i and l_f in the initial and final bound-state hydrogen-like wave functions. By contrast, the CDW-3B method and experiments give a completely different CB2-3B type $\propto v^{-11}$–behavior for arbitrary values of l_i and l_f. In the case of the SE-4B method for double capture, no analytical results can be obtained for the transition amplitude, let alone cross sections. In other words, numerical computations efforts are again comparable in the SE-4B and CDW-4B methods. As to the billiard-type scattering mechanisms of Thomas [181], the situation becomes even more aggravated with eikonalization of the full twofold electronic continua, since double capture is expected to exhibit three Thomas peaks [90] and none of them can be predicted by the SE-4B method.

Since the CDW-EIS-3B or CDW-EFS-3B methods use the eikonal continuum intermediate states in the entrance or exit channel, it is anticipated that these two approximations will preserve the mentioned unphysical features of the SE-3B method. By the same token, the CDW-EIS-3B or CDW-EFS-3B methods will inherit the good features of the CDW-3B method in the relevant parts concerned with the complementary exit or entrance channels, respectively. This can be seen in the corresponding high-velocity asymptotic formulae of the total cross sections $Q_{if}^{(\mathrm{CDW-EIS-3B})} \propto v^{-11-2l_i}$ (any l_f) and $Q_{if}^{(\mathrm{CDW-EFS-3B})} \propto v^{-11-2l_f}$ (any l_i) instead of the the correct asymptote

$Q_{if}^{(\mathrm{CDW-3B})} \propto v^{-11}$ (any l_i and l_f). Analogous and possibly more severe failures could occur by passing to four-body problems such as double capture when studied by means of the CDW-EIS-4B and CDW-EFS-4B methods. This will be analyzed in the section with illustrations for two-electron transfer in process (11.1).

14.5 The BDW-4B method

The CDW-4B method takes full account of the double Coulomb wave functions due to the potentials $V_{\mathrm{Pk}} = -Z_{\mathrm{P}}/s_k$ and $V_{\mathrm{T_k}} = -Z_{\mathrm{T}}/x_k$ $(k = 1, 2)$ for describing the continuum intermediate states of the two electrons e_1 and e_2 at all distances s_k and x_k (finite, in the interaction region, and infinitely large, in the asymptotic region) in the entrance and exit channel, respectively.

On the other hand, in the CB1-4B method, the motions for the same two electrons are distorted in a much simpler way by including only the twofold Coulomb logarithmic phases due to the electrons-nuclei potentials in the initial and final asymptotic regions $V_{\mathrm{P}}^{\infty} = -2Z_{\mathrm{P}}/R$ and $V_{\mathrm{T}}^{\infty} = -2Z_{\mathrm{T}}/R$, respectively. The potentials $V_{\mathrm{P1}} + V_{\mathrm{P2}}$ and $V_{\mathrm{T1}} + V_{\mathrm{T2}}$ tend to V_{P}^{∞} and V_{T}^{∞}, since in the initial and final asymptotic regions, where $s_k \longrightarrow \infty$ and $x_k \longrightarrow \infty$, we have $s_k \approx R$ and $x_k \approx R$ (for both $k = 1$ and $k = 2$) in the entrance and exit channel, respectively. As a result, the corresponding continua are included through the electron asymptotic distortion phase factors $\exp[-2\nu_{\mathrm{P}} \ln(vR - \boldsymbol{v} \cdot \boldsymbol{R})]$ and $\exp[2\nu_{\mathrm{T}} \ln(vR + \boldsymbol{v} \cdot \boldsymbol{R})]$ due to V_{P}^{∞} and V_{T}^{∞} where $\nu_{\mathrm{K}} = Z_{\mathrm{K}}/v$ $(\mathrm{K} = \mathrm{P}, \mathrm{T})$. Such phases remain in the computation for both differential and total cross sections in the CB1-4B method. Despite the explicit appearance of the vector $\boldsymbol{R}$, which happens to be the vector of the inter-nuclear distance, these electron asymptotic phases have nothing to do with the inter-nuclear repulsive potential itself, which is $V_{\mathrm{PT}} = Z_{\mathrm{P}} Z_{\mathrm{T}}/R$. The potential V_{PT} distorts the relative motion of the two nuclei P and T in both scattering channels, thus leading to the product of the associated asymptotic initial and final phases, $\exp[i\nu \ln(vR - \boldsymbol{v} \cdot \boldsymbol{R})]$ and $\{\exp[-\nu \ln(vR + \boldsymbol{v} \cdot \boldsymbol{R})]\}^*$ in the transition amplitude, where $\nu = Z_{\mathrm{P}} Z_{\mathrm{T}}/v$. The said product, which is equal to $(v\rho)^{2i\nu}$, is the only effect caused by the presence of V_{PT} in the exact eikonal four-body transition amplitude. With such an occurrence, it is easily shown that the ensuing exact eikonal total cross section is independent of $(v\rho)^{2i\nu}$ and, hence, of the inter-nuclear potential V_{PT}, as it ought to be on physical grounds [44]. Of course, this remains true for every particular approximation, provided that the correct boundary conditions are satisfied.

When in the entrance channel, the electronic and nuclear asymptotic phase factors $\exp[-2\nu_P \ln(vR - \boldsymbol{v} \cdot \boldsymbol{R})]$ and $\exp[i\nu \ln(vR - \boldsymbol{v} \cdot \boldsymbol{R})]$ are added together, the total phase follows via $\exp[i\nu_i \ln(vR - \boldsymbol{v} \cdot \boldsymbol{R})]$ where $\nu_i = Z_{\mathrm{P}}(Z_{\mathrm{T}} -$

$2)/v$. This latter composite phase indicates that, on the level of determining the distortion of the unperturbed state Φ_i in the entrance channel, the presence of the two electrons is felt, in effect, merely through a screening of $Z_{\rm T}$ to yield the effective nuclear charge $Z_{\rm T}-2$. The deduced $\boldsymbol{R}$−dependent total distortion phase factor is now recognized as being due to the asymptotic value $V_i^\infty = Z_{\rm P}(Z_{\rm T}-2)/R$ of the perturbation V_i in the entrance channel. Similarly in the exit channel, the electron and nuclear asymptotic phase factors $\exp[2\nu_{\rm P}\ln(vR+\boldsymbol{v}\cdot\boldsymbol{R})]$ and $\exp[-i\nu\ln(vR+\boldsymbol{v}\cdot\boldsymbol{R})]$ yield the overall distortion $\exp[-i\nu_f\ln(vR+\boldsymbol{v}\cdot\boldsymbol{R})]$ where $\nu_f = Z_{\rm T}(Z_{\rm P}-2)/v$. Here, the combined phase shows that the sole role for the electron distortion is to screen $Z_{\rm P}$ to $Z_{\rm P}-2$. Hence, such a reasoning on the level of the total $\boldsymbol{R}$−dependent distortion phase factor recovers the form of the asymptotic value $V_i^\infty = Z_{\rm P}(Z_{\rm T}-2)/R$ of the perturbation V_i in the entrance channel.

Double ionization dominates over double charge exchange at high energies. Therefore, to properly describe electron transfer to a final bound state, in the limit of high energies, the electronic continuum intermediate states should be included at all distances, and this is fully accomplished in the CDW-4B method. Conversely, at lower energies, two-electron transfer dominates over ionization. This time, the electronic continuum states represent a drawback, since they overweight the intermediate ionization paths of the studied reaction. Consequently, the CDW-4B method for double capture overestimates the corresponding experimental data at lower energies, as is also the case with single capture within the CDW-3B method [269]–[290].

The models that partially mitigate the over-account of continuum intermediate states at lower energies are certain hybrid approximations that combine the CDW-4B method in one channel with the CB1-4B method in the other channel. An example from this hybrid category is the BDW-4B method of Belkić [89]. Specifically, the BDW-4B method exactly coincides with the CDW-4B method in one channel and with the CB1-4B method in the other channel. As such, the BDW-4B method preserves the correct boundary conditions in both scattering channels, since both the CDW-4B and CB1-4B methods do so. Here, the wave function given by (13.56) from the CDW-4B method is employed for the entrance channel. We now determine the distorted wave χ_f^- in the exit channel. The boundary condition given by

$$\chi_f^- \underset{r_f\to\infty}{\longrightarrow} \Psi_f^- \underset{r_f\to\infty}{\longrightarrow} \Phi_f^- \tag{14.29}$$

suggests that χ_f^- should be sought in the following factorized form

$$\chi_f^- = \varphi_f(\boldsymbol{s}_1,\boldsymbol{s}_2)\zeta_f^- \,. \tag{14.30}$$

The equation from which we determine ζ_f^- reads as

$$\varphi_f^-(E-E_f-H_0)\zeta_f^- - W_f\varphi_f\zeta_f^- + \frac{1}{a}\sum_{k=1}^{2}\boldsymbol{\nabla}_{s_k}\varphi_f\cdot\boldsymbol{\nabla}_{s_k}\zeta_f^- = 0\,. \tag{14.31}$$

Here, the term $\zeta_f^-(h_f - E_f)\varphi_f$ is neglected as in the CDW-4B method. Potential W_f is chosen as $W_f = Z_{\rm T}(Z_{\rm P} - 2)/r_f$ which is the asymptotic form of the perturbation V_f in the exit channel [89]. Obviously, this choice of W_f implies that the function ζ_f^- will not explicitly contain the independent variables for the purely electronic coordinates, in which case the coupling term $\boldsymbol{\nabla}_{s_k}\varphi_f \cdot \boldsymbol{\nabla}_{s_k}\zeta_f^-$ from (14.31) will become identical to zero. Consequently, the remaining equation

$$(E - E_f - H_0 - W_f)\zeta_f^- = 0 \tag{14.32}$$

can be solved exactly with the result

$$\chi_f^- = \mathcal{N}^-(\nu_f)\,\mathrm{e}^{-i\boldsymbol{k}_f\cdot\boldsymbol{r}_f}\,{}_1F_1(i\nu_f, 1, -ik_f r_i - i\boldsymbol{k}_f\cdot\boldsymbol{r}_i)\varphi_f(\boldsymbol{s}_1, \boldsymbol{s}_2) \tag{14.33}$$

where $\mathcal{N}^-(\nu_f) = \mathrm{e}^{-\pi\nu_f/2}\Gamma(1 - i\nu_f)$ and $\nu_f = Z_{\rm T}(Z_{\rm P} - 2)/v$. Function (14.33) coincides with the previously derived expression (11.58).

Next, we neglect the term with the total Green operator G^+ in the exact expression for the transition amplitude (13.19). This yields the following transition amplitude in the prior form of the BDW-4B method

$$T_{if}^{(\rm BDW)-} = \langle\chi_f^-|V_i - W_i|\chi_i^+\rangle. \tag{14.34}$$

Inserting (13.42), (13.56) and (14.33) into (14.34), it follows that

$$T_{if}^{(\rm BDW)-} = -N_{\rm P}\iiint \mathrm{d}\boldsymbol{x}_1\mathrm{d}\boldsymbol{x}_2\mathrm{d}\boldsymbol{r}_i\,\mathrm{e}^{i\boldsymbol{k}_i\cdot\boldsymbol{r}_i + i\boldsymbol{k}_f\cdot\boldsymbol{r}_f}\mathcal{L}_1(\boldsymbol{r}_i, \boldsymbol{r}_f)\varphi_f^*(\boldsymbol{s}_1, \boldsymbol{s}_2)$$
$$\times\{{}_1F_1(i\nu_{\rm P}, 1, ivs_2 + i\boldsymbol{v}\cdot\boldsymbol{s}_2)\boldsymbol{\nabla}_{x_1}\varphi_i(\boldsymbol{x}_1, \boldsymbol{x}_2)\cdot\boldsymbol{\nabla}_{s_1}\,{}_1F_1(i\nu_{\rm P}, 1, ivs_1 + i\boldsymbol{v}\cdot\boldsymbol{s}_1)$$
$$+\,{}_1F_1(i\nu_{\rm P}, 1, ivs_1 + i\boldsymbol{v}\cdot\boldsymbol{s}_1)\boldsymbol{\nabla}_{x_2}\varphi_i(\boldsymbol{x}_1, \boldsymbol{x}_2)\cdot\boldsymbol{\nabla}_{s_2}\,{}_1F_1(i\nu_{\rm P}, 1, ivs_2 + i\boldsymbol{v}\cdot\boldsymbol{s}_2)\} \tag{14.35}$$

where $N_{\rm P} = [N^+(\nu_{\rm P})]^2$ and $\nu_{\rm P} = Z_{\rm P}/v$. Here, we have

$$\mathcal{L}_1(\boldsymbol{r}_i, \boldsymbol{r}_f) = \frac{\mathcal{N}_1}{\mu_i^{2i\nu_{\rm P}}}\,{}_1F_1(-i\nu, 1, ik_i r_f + i\boldsymbol{k}_i\cdot\boldsymbol{r}_f)$$
$$\times\,{}_1F_1(-i\nu_f, 1, ik_f r_i + i\boldsymbol{k}_f\cdot\boldsymbol{r}_i) \tag{14.36}$$

where $\mathcal{N}_1 = \mathcal{N}^+(\nu)\mathcal{N}^{-*}(\nu_f)$. Within the eikonal approximation, the following simplification is possible

$$\mathcal{L}_1(\boldsymbol{r}_i, \boldsymbol{r}_f) \simeq \mu_i^{i\nu_i}\mu_f^{i\nu_f}\,\mathrm{e}^{i\nu_f\,\ln(vR + \boldsymbol{v}\cdot\boldsymbol{R})}\,\mathrm{e}^{i\nu\,\ln(vR - \boldsymbol{v}\cdot\boldsymbol{R})}$$
$$\simeq \mu^{-i(\xi_{\rm P} + \xi_{\rm T})}(\mu v\rho)^{2i\nu}\,\mathrm{e}^{-i\xi_{\rm T}\,\ln(vR + \boldsymbol{v}\cdot\boldsymbol{R})}. \tag{14.37}$$

where

$$\xi_{\rm K} = 2\nu_{\rm K} = 2\frac{Z_{\rm K}}{v} \qquad (\mathrm{K} = \mathrm{P}, \mathrm{T}). \tag{14.38}$$

Then, the total cross section can be found from (14.12), with R_{if}^{-} replaced by $R_{if}^{(\mathrm{BDW})-}$ where

$$R_{if}^{(\mathrm{BDW})-}(\boldsymbol{\eta})=-N_{\mathrm{P}}\iiint \mathrm{d}\boldsymbol{x}_1\mathrm{d}\boldsymbol{x}_2\mathrm{d}\boldsymbol{R}\,\mathrm{e}^{i\boldsymbol{q}_{\mathrm{P}}\cdot(\boldsymbol{s}_1+\boldsymbol{s}_2)+i\boldsymbol{q}_{\mathrm{T}}\cdot(\boldsymbol{x}_1+\boldsymbol{x}_2)}$$
$$\times\varphi_f^*(\boldsymbol{s}_1,\boldsymbol{s}_2)(vR+\boldsymbol{v}\cdot\boldsymbol{R})^{-i\xi_{\mathrm{T}}}$$
$$\times\{\,{}_1F_1(i\nu_{\mathrm{P}},1,ivs_2+i\boldsymbol{v}\cdot\boldsymbol{s}_2)\boldsymbol{\nabla}_{x_1}\varphi_i(\boldsymbol{x}_1,\boldsymbol{x}_2)\cdot\boldsymbol{\nabla}_{s_1}\,{}_1F_1(i\nu_{\mathrm{P}},1,ivs_1+i\boldsymbol{v}\cdot\boldsymbol{s}_1)$$
$$+\,{}_1F_1(i\nu_{\mathrm{P}},1,ivs_1+i\boldsymbol{v}\cdot\boldsymbol{s}_1)\boldsymbol{\nabla}_{x_2}\varphi_i(\boldsymbol{x}_1,\boldsymbol{x}_2)\cdot\boldsymbol{\nabla}_{s_2}\,{}_1F_1(i\nu_{\mathrm{P}},1,ivs_2+i\boldsymbol{v}\cdot\boldsymbol{s}_2)\}\,. \tag{14.39}$$

An extension of the analysis to the post version of the formalism can also be accomplished. This time we choose the distorting potential in the form $W_i = Z_{\mathrm{P}}(Z_{\mathrm{T}}-2)/r_i$. Such a choice in the eikonal approximation gives the distorted waves χ_i^+ as

$$\chi_i^+ = \mathcal{N}^+(\nu_i)\,\mathrm{e}^{i\boldsymbol{k}_i\cdot\boldsymbol{r}_i}\,{}_1F_1(-i\nu_i,1,ik_ir_f+i\boldsymbol{k}_i\cdot\boldsymbol{r}_f)\varphi_i(\boldsymbol{x}_1,\boldsymbol{x}_2)\,. \tag{14.40}$$

In the exit channel, the potential U_f is chosen as in the CDW-4B method i.e.

$$U_f\chi_f^- = -\frac{1}{a}\sum_{k=1}^{2}\boldsymbol{\nabla}_{s_k}\varphi_f\cdot\boldsymbol{\nabla}_{x_k}\psi_f^-\,. \tag{14.41}$$

The distorted wave is given by $\chi_f^- = \varphi_f(\boldsymbol{s}_1,\boldsymbol{s}_2)\psi_f^-$ where the function ψ_f^- is determined by (14.8). The corresponding post form of the transition amplitude is obtained by neglecting the second term in (13.20), so that

$$T_{if}^{(\mathrm{BDW})+} = \langle\chi_f^-|V_f-W_f^\dagger|\chi_i^+\rangle = \langle\chi_f^-|U_f^\dagger|\chi_i^+\rangle. \tag{14.42}$$

Substituting (14.40), (14.41) and (14.8) into (14.42), we have

$$T_{if}^{(\mathrm{BDW})+}=-N_{\mathrm{T}}\iiint \mathrm{d}\boldsymbol{s}_1\mathrm{d}\boldsymbol{s}_2\mathrm{d}\boldsymbol{r}_f\mathrm{e}^{i\boldsymbol{k}_i\cdot\boldsymbol{r}_i+i\boldsymbol{k}_f\cdot\boldsymbol{r}_f}\mathcal{L}_2(\boldsymbol{r}_i,\boldsymbol{r}_f)\varphi_i(\boldsymbol{x}_1,\boldsymbol{x}_2)$$
$$\times\{\,{}_1F_1(i\nu_{\mathrm{T}},1,ivx_2+i\boldsymbol{v}\cdot\boldsymbol{x}_2)\boldsymbol{\nabla}_{s_1}\varphi_f^*(\boldsymbol{s}_1,\boldsymbol{s}_2)\cdot\boldsymbol{\nabla}_{x_1}\,{}_1F_1(i\nu_{\mathrm{T}},1,ivx_1+i\boldsymbol{v}\cdot\boldsymbol{x}_1)$$
$$+\,{}_1F_1(i\nu_{\mathrm{T}},1,ivx_1+i\boldsymbol{v}\cdot\boldsymbol{x}_1)\boldsymbol{\nabla}_{s_2}\varphi_f^*(\boldsymbol{s}_1,\boldsymbol{s}_2)\cdot\boldsymbol{\nabla}_{x_2}\,{}_1F_1(i\nu_{\mathrm{T}},1,ivx_2+i\boldsymbol{v}\cdot\boldsymbol{x}_2)\} \tag{14.43}$$

where

$$\mathcal{L}_2(\boldsymbol{r}_i,\boldsymbol{r}_f) = \frac{\mathcal{N}_2}{\mu_f^{2i\nu_{\mathrm{T}}}}\,{}_1F_1(-i\nu_i,1,ik_ir_f+i\boldsymbol{k}_i\cdot\boldsymbol{r}_f)$$
$$\times\,{}_1F_1(-i\nu,1,ik_fr_i+i\boldsymbol{k}_f\cdot\boldsymbol{r}_i) \tag{14.44}$$

with $N_{\mathrm{T}} = [N^{-*}(\nu_{\mathrm{T}})]^2$ and $\mathcal{N}_2 = \mathcal{N}^+(\nu_i)\mathcal{N}^{-*}(\nu)$ and $\nu = Z_{\mathrm{P}}Z_{\mathrm{T}}/v$. The function $\mathcal{L}_2(\boldsymbol{r}_i,\boldsymbol{r}_f)$ can also be expressed in the eikonal approximation via

$$\mathcal{L}_2(\boldsymbol{r}_i,\boldsymbol{r}_f) \simeq \mu_i^{i\nu_i}\mu_f^{i\nu_f}\mathrm{e}^{i\nu_i\,\ln(vR-\boldsymbol{v}\cdot\boldsymbol{R})}\,\mathrm{e}^{i\nu\,\ln(vR+\boldsymbol{v}\cdot\boldsymbol{R})}$$
$$\simeq \mu^{-i(\xi_{\mathrm{P}}+\xi_{\mathrm{T}})}(\mu v\rho)^{2i\nu}\,\mathrm{e}^{-i\xi_{\mathrm{P}}\,\ln(vR-\boldsymbol{v}\cdot\boldsymbol{R})}\,. \tag{14.45}$$

Therefore, the total cross section is given by

$$Q_{if}^{(\mathrm{BDW})\pm}(a_0^2) = \int \mathrm{d}\boldsymbol{\eta} \left| \frac{R_{if}^{(\mathrm{BDW})\pm}(\boldsymbol{\eta})}{2\pi v} \right|^2 \tag{14.46}$$

where

$$\begin{aligned} R_{if}^{(\mathrm{BDW})+}(\boldsymbol{\eta}) = & -N_{\mathrm{T}} \iiint \mathrm{d}\boldsymbol{s}_1 \mathrm{d}\boldsymbol{s}_2 \mathrm{d}\boldsymbol{R}\, \mathrm{e}^{i\boldsymbol{q}_{\mathrm{P}}\cdot(\boldsymbol{s}_1+\boldsymbol{s}_2)+i\boldsymbol{q}_{\mathrm{T}}\cdot(\boldsymbol{x}_1+\boldsymbol{x}_2)} \\ & \times \varphi_i(\boldsymbol{x}_1,\boldsymbol{x}_2)(vR - \boldsymbol{v}\cdot\boldsymbol{R})^{-i\xi_{\mathrm{P}}} \\ & \times \{ {}_1F_1(i\nu_{\mathrm{T}},1,ivx_2 + i\boldsymbol{v}\cdot\boldsymbol{x}_2)\boldsymbol{\nabla}_{s_1}\varphi_f^*(\boldsymbol{s}_1,\boldsymbol{s}_2)\cdot\boldsymbol{\nabla}_{x_1}\, {}_1F_1(i\nu_{\mathrm{T}},1,ivx_1 + i\boldsymbol{v}\cdot\boldsymbol{x}_1) \\ & + {}_1F_1(i\nu_{\mathrm{T}},1,ivx_1+i\boldsymbol{v}\cdot\boldsymbol{x}_1)\boldsymbol{\nabla}_{s_2}\varphi_f^*(\boldsymbol{s}_1,\boldsymbol{s}_2)\cdot\boldsymbol{\nabla}_{x_2}\, {}_1F_1(i\nu_{\mathrm{T}},1,ivx_2+i\boldsymbol{v}\cdot\boldsymbol{x}_2)\}\,. \end{aligned} \tag{14.47}$$

Notice that $R_{if}^{(\mathrm{BDW})-}$ can be obtained directly from $R_{fi}^{(\mathrm{BDW})+}$ by making the transformations $\boldsymbol{s}_1 \longleftrightarrow \boldsymbol{s}_2$ and $\boldsymbol{x}_1 \longleftrightarrow \boldsymbol{x}_2$ in both (14.39) and (14.47). This is possible because the vector $\boldsymbol{R}$ is invariant under this latter transformation. In such a case, these transformations will map the first of the two terms in $R_{if}^{(\mathrm{BDW})\pm}$ into the second term and vice versa. In other words, the contributions to $R_{if}^{(\mathrm{BDW})\pm}$ coming from $\boldsymbol{\nabla}_1\cdot\boldsymbol{\nabla}_1$ and $\boldsymbol{\nabla}_2\cdot\boldsymbol{\nabla}_2$ are identical to each other. Hence, these expressions can be rewritten as follows

$$\begin{aligned} R_{if}^{(\mathrm{BDW})-}(\boldsymbol{\eta}) &= -2N_{\mathrm{P}} \iiint \mathrm{d}\boldsymbol{x}_1 \mathrm{d}\boldsymbol{x}_2 \mathrm{d}\boldsymbol{R}\, \mathrm{e}^{i\boldsymbol{q}_{\mathrm{P}}\cdot(\boldsymbol{s}_1+\boldsymbol{s}_2)+i\boldsymbol{q}_{\mathrm{T}}\cdot(\boldsymbol{x}_1+\boldsymbol{x}_2)} \\ & \quad \times (vR + \boldsymbol{v}\cdot\boldsymbol{R})^{-i\xi_{\mathrm{T}}}\varphi_f^*(\boldsymbol{s}_1,\boldsymbol{s}_2)\, F_{\mathrm{P}}(\boldsymbol{x}_1,\boldsymbol{x}_2;\boldsymbol{s}_1,\boldsymbol{s}_2) \end{aligned} \tag{14.48}$$

and

$$\begin{aligned} R_{if}^{(\mathrm{BDW})+}(\boldsymbol{\eta}) &= -2N_{\mathrm{T}} \iiint \mathrm{d}\boldsymbol{s}_1 \mathrm{d}\boldsymbol{s}_2 \mathrm{d}\boldsymbol{R}\, \mathrm{e}^{i\boldsymbol{q}_{\mathrm{P}}\cdot(\boldsymbol{s}_1+\boldsymbol{s}_2)+i\boldsymbol{q}_{\mathrm{T}}\cdot(\boldsymbol{x}_1+\boldsymbol{x}_2)} \\ & \quad \times (vR - \boldsymbol{v}\cdot\boldsymbol{R})^{-i\xi_{\mathrm{P}}}\varphi_i(\boldsymbol{x}_1,\boldsymbol{x}_2)\, F_{\mathrm{T}}(\boldsymbol{x}_1,\boldsymbol{x}_2;\boldsymbol{s}_1,\boldsymbol{s}_2) \end{aligned} \tag{14.49}$$

where

$$\begin{aligned} F_{\mathrm{P}}(\boldsymbol{x}_1,\boldsymbol{x}_2;\boldsymbol{s}_1,\boldsymbol{s}_2) &= {}_1F_1(i\nu_{\mathrm{P}},1,ivs_2 + i\boldsymbol{v}\cdot\boldsymbol{s}_2) \\ & \quad \times \boldsymbol{\nabla}_{x_1}\varphi_i(\boldsymbol{x}_1,\boldsymbol{x}_2)\cdot\boldsymbol{\nabla}_{s_1}\, {}_1F_1(i\nu_{\mathrm{P}},1,ivs_1 + i\boldsymbol{v}\cdot\boldsymbol{s}_1) \end{aligned} \tag{14.50}$$

$$\begin{aligned} F_{\mathrm{T}}(\boldsymbol{x}_1,\boldsymbol{x}_2;\boldsymbol{s}_1,\boldsymbol{s}_2) &= {}_1F_1(i\nu_{\mathrm{T}},1,ivx_2 + i\boldsymbol{v}\cdot\boldsymbol{x}_2) \\ & \quad \times \boldsymbol{\nabla}_{s_1}\varphi_f^*(\boldsymbol{s}_1,\boldsymbol{s}_2)\cdot\boldsymbol{\nabla}_{x_1}\, {}_1F_1(i\nu_{\mathrm{T}},1,ivx_1 + i\boldsymbol{v}\cdot\boldsymbol{x}_1). \end{aligned} \tag{14.51}$$

The physical interpretation of the prior form of the T-matrix element in the BDW-4B method can be done in the following plausible manner. The incident particle scatters on each of the three constituents of the target $(Z_{\mathrm{T}}; e_1, e_2)$.

In the entrance channel, collision between the projectile $Z_{\rm P}$ and target nucleus $Z_{\rm T}$ results in accumulation of the Coulombic phase factor $\exp[(i/v)Z_{\rm P}Z_{\rm T} \times \ln(vR-\boldsymbol{v}\cdot\boldsymbol{R})]$. On the other hand, in the exit channel, the target nucleus $Z_{\rm T}$ interacts with the newly formed atom or ion $(Z_{\rm P},2e)_f$ considered as the point charge $(Z_{\rm P}-2)$, thus accumulating the phase factor $\exp[-(i/v)Z_{\rm T}(Z_{\rm P}-2) \times \ln(vR+\boldsymbol{v}\cdot\boldsymbol{R})]$ due to the asymptotic residual Coulombic interaction $W_f = Z_{\rm T}(Z_{\rm P}-2)/R = V_f^{\infty}$. Thus, the nucleus T sees the two electrons as playing a role of screening the nuclear charge $Z_{\rm P}$ to its effective value $Z_{\rm P}-2$ in the helium-like atomic system $(Z_{\rm P},2e)_f$. In contrast, in the entrance channel, the BDW-4B method allows the projectile to separately distort the nuclear and electronic motions through the additive three Coulombic interactions. Thus, the interaction of $Z_{\rm P}$ with the electrons e_1 and e_2 leads to double ionization of the target $(Z_{\rm T};e_1,e_2)_i$. The ionized electrons propagate in the Coulomb field of $Z_{\rm P}$ in a particular eikonal direction with the momenta $\boldsymbol{\kappa}_1 \approx \boldsymbol{\kappa}_2 \approx \boldsymbol{v}$. Finally, capture of the two electrons occurs from these intermediate ionizing states (capture from continuum), because these electrons are travelling along each other, as well as together with the projectile in the same direction, such that the attractive potential between $Z_{\rm P}$ and e_k $(k=1,2)$ is sufficient to bind them together into the new helium-like atomic system $(Z_{\rm P};e_1,e_2)_f$. This is a quantum version of the well-known Thomas classical double scattering [181]. An analogous situation can also be pictured in the case of the post form $R_{if}^{(\rm BDW)+}$ of the transition amplitude. Explicit calculations of the matrix elements $R_{if}^{(\rm BDW)\pm}$ have been carried out by Belkić [89]. He has shown that the matrix elements $R_{if}^{(\rm BDW)\pm}$ can be reduced to four-dimensional real numerical integrations from 0 to 1. The ensuing total cross sections in the BDW-4B method are obtained by a five-dimensional quadratures. It should be noted that the integrands in the prior and post forms have the functions $[\tau_k/(1-\tau_k)]^{i\nu_{\rm P,T}}$ $(k=1,2)$ which originate from the standard integral representation of the two confluent hyper-geometric functions. These functions possess integrable branch-points singularities at $\tau_{1,2}=0$ and $\tau_{1,2}=1$, as well as simple poles at points $\tau_{1,2}=0$. Therefore, the Cauchy regularization of the whole integrand should be performed before applying the usual Gauss-Legendre quadratures [89].

When computing differential cross sections, a very favorable computational circumstance occurs using the BDW-4B approximation in the prior form for $Z_{\rm P}=2$ or in the post version for $Z_{\rm T}=2$. In such a special case with the H^{2+} projectile impinging upon any two-electron target (or an arbitrary projectile $Z_{\rm P}$ and helium target), the Sommerfeld parameter $\nu_f = Z_{\rm T}(Z_{\rm P}-2)/v$ or $\nu_f = Z_{\rm P}(Z_{\rm T}-2)/v)$ is zero in (14.36) or in (14.44), so that the Coulomb logarithmic phase factors $(vR-\boldsymbol{v}\cdot\boldsymbol{R})^{i\xi_{\rm T}}$ or $(vR+\boldsymbol{v}\cdot\boldsymbol{R})^{i\xi_{\rm P}}$ is the only $\boldsymbol{R}-$dependent function (without any phase in terms of ρ) in $T_{if}^{(\rm BDW)-}$ or $T_{if}^{(\rm BDW)+}$. In other words, the distorting function $\mathcal{L}^-(\boldsymbol{r}_i,\boldsymbol{r}_f)$ is reduced to the following form

$$T_{if}^{(\rm BDW)-}: \qquad \mathcal{L}_1(\boldsymbol{r}_i,\boldsymbol{r}_f) = \mu_i^{-i\xi_{\rm T}}\mathcal{N}^+(\xi_{\rm T})\,{}_1F_1(-i\xi_{\rm T},1,ik_ir_f+i\boldsymbol{k}_i\cdot\boldsymbol{r}_f)$$

$$\simeq (vR - \boldsymbol{v}\cdot\boldsymbol{R})^{i\xi_{\rm T}}$$
$$\xi_{\rm T} = 2\nu_{\rm T} = 2Z_{\rm T}/v \qquad (Z_{\rm P} = 2\ ,\ \text{any}\ Z_{\rm T}) \tag{14.52}$$

and a similar simplification is obtained for $\mathcal{L}_2(\boldsymbol{r}_i, \boldsymbol{r}_f)$ via

$$T_{if}^{({\rm BDW})+}: \qquad \mathcal{L}_2(\boldsymbol{r}_i,\boldsymbol{r}_f) = \mu_f^{-i\xi_{\rm P}}\mathcal{N}^+(\xi_{\rm P})\,{}_1F_1(-i\xi_{\rm P},1,ik_f r_i + i\boldsymbol{k}_f\cdot\boldsymbol{r}_i)$$
$$\simeq (vR + \boldsymbol{v}\cdot\boldsymbol{R})^{i\xi_{\rm P}}$$
$$\xi_{\rm P} = 2\nu_{\rm P} = 2Z_{\rm P}/v \qquad (Z_{\rm T} = 2\ ,\ \text{any}\ Z_{\rm P}). \tag{14.53}$$

Under these particular circumstances, the computationally difficult and time-consuming Fourier-Bessel transform (as an integral over $\rho-$dependent transition amplitude), is not needed in the BDW-4B method, since the angular distribution $({\rm d}/{\rm d}\Omega)Q^{({\rm BDW})-}$ or $({\rm d}/{\rm d}\Omega)Q^{({\rm BDW})+}$ can be obtained directly via

$$\frac{{\rm d}Q^{({\rm BDW})-}}{{\rm d}\Omega} = \left|\frac{\mu}{2\pi}T_{if}^{({\rm BDW})-}\right|^2 \qquad (Z_{\rm P} = 2,\ \text{any}\ Z_{\rm T}) \tag{14.54}$$

and

$$\frac{{\rm d}Q^{({\rm BDW})+}}{{\rm d}\Omega} = \left|\frac{\mu}{2\pi}T_{if}^{({\rm BDW})+}\right|^2 \qquad (Z_{\rm T} = 2,\ \text{any}\ Z_{\rm P}) \tag{14.55}$$

where μ is the reduced mass of P and T given by $\mu = M_{\rm P}M_{\rm T}/(M_{\rm P}+M_{\rm T})$. The BDW-4B method has first been formulated by Belkić [89] who illustrated this method for double electron capture in the He^{2+} – He collisions. Subsequently, the BDW-4B method has been applied by Mančev [193]–[196] to single electron capture by fast nuclei from helium-like targets.

14.6 The BCIS-4B method

Recall that the three-body continuum intermediate state (CIS-3B) approximation has been introduced by Belkić [275] for process (12.26). The CIS-3B method was aimed to treat asymmetric collisions, such that its prior and post versions are adapted for $Z_T \gg Z_P$ and $Z_P \gg Z_T$, respectively. This method satisfies the correct boundary condition only in the channel with the stronger potential [42, 275]. In the channel with the weaker potential, the CIS-3B method uses the unperturbed state which does not have the proper asymptotic behavior for process (12.26). Such a drawback can be rectified via distortion of the unperturbed state from the entrance channel (which has a weaker potential) by an additional $\boldsymbol{R}-$dependent phase due to the electron-nucleus interaction at infinitely large distances. As a consequence, the correct boundary condition also becomes satisfied in the channel with the weaker potential. This leads to the three-body boundary-corrected intermediate state

(BCIS-3B) method for single electron capture (12.26) in a general case with arbitrary nuclear charges $Z_{\rm P}$ and $Z_{\rm T}$.

For a more complicated process, such as double electron capture (11.1), a proper extension of the BCIS-3B method is needed to treat four-body collisions. This generalization is known as the BCIS-4B method which has been formulated and implemented by Belkić [86]. The BCIS-4B method takes full account of the twofold electronic continuum intermediate states in one channel (entrance or exit, depending upon whether the prior or post form of the transition amplitudes is considered). The matrix element from the transition amplitudes in the prior and the post versions of the BCIS-4B method are [86]

$$
\begin{aligned}
T_{if}^{\rm (BCIS)-}(\boldsymbol{\eta}) = & Z_{\rm P}N_{\rm T}\iiint {\rm d}\boldsymbol{s}_1{\rm d}\boldsymbol{s}_2{\rm d}\boldsymbol{R}\,{\rm e}^{i\boldsymbol{k}_i\cdot\boldsymbol{r}_i+i\boldsymbol{k}_f\cdot\boldsymbol{r}_f}\mathcal{L}_1(\boldsymbol{r}_i,\boldsymbol{r}_f)\\
&\times\varphi_f^*(\boldsymbol{s}_1,\boldsymbol{s}_2)\left(\frac{2}{R}-\frac{1}{s_1}-\frac{1}{s_2}\right)\varphi_i(\boldsymbol{x}_1,\boldsymbol{x}_2)\\
&\times{}_1F_1(i\nu_{\rm T},1,ivx_1+i\boldsymbol{v}\cdot\boldsymbol{x}_1)\,{}_1F_1(i\nu_{\rm T},1,ivx_2+i\boldsymbol{v}\cdot\boldsymbol{x}_2)
\end{aligned}
\tag{14.56}
$$

and

$$
\begin{aligned}
T_{if}^{\rm (BCIS)+}(\boldsymbol{\eta}) = & Z_{\rm T}N_{\rm P}\iiint {\rm d}\boldsymbol{x}_1{\rm d}\boldsymbol{x}_2{\rm d}\boldsymbol{R}\,{\rm e}^{i\boldsymbol{k}_i\cdot\boldsymbol{r}_i+i\boldsymbol{k}_f\cdot\boldsymbol{r}_f}\mathcal{L}_2(\boldsymbol{r}_i,\boldsymbol{r}_f)\\
&\times\varphi_f^*(\boldsymbol{s}_1,\boldsymbol{s}_2)\left(\frac{2}{R}-\frac{1}{x_1}-\frac{1}{x_2}\right)\varphi_i(\boldsymbol{x}_1,\boldsymbol{x}_2)\\
&\times{}_1F_1(i\nu_{\rm P},1,ivs_1+i\boldsymbol{v}\cdot\boldsymbol{s}_1)\,{}_1F_1(i\nu_{\rm P},1,ivs_2+i\boldsymbol{v}\cdot\boldsymbol{s}_2)
\end{aligned}
\tag{14.57}
$$

where the functions $\mathcal{L}_1(\boldsymbol{r}_i,\boldsymbol{r}_f)$ and $\mathcal{L}_2(\boldsymbol{r}_i,\boldsymbol{r}_f)$ are the same as those in (14.36) and (14.44), respectively. The two full Coulomb waves from (14.36) and (14.44) for the relative motion of heavy particles could, in principle, be kept in the calculations throughout. This would include the contributions of the order of or smaller than $1/\mu_i$ and $1/\mu_f$. Numerically, these latter contributions are negligibly small, since they amount to keeping all the terms that are of the order of or less than 10^{-4} relative to 1. Of course, this is totally unnecessary within a consistent application of the eikonal approximation, in which the full Coulomb wave functions for the relative motions of heavy particles should be systematically replaced by their logarithmic Coulomb phase factors as in (14.37) and (14.45). Under such circumstances, the post and prior total cross section in the BCIS-4B methods are given by

$$
Q_{if}^{\rm (BCIS)\pm}(a_0^2) = \int {\rm d}\boldsymbol{\eta}\left|\frac{R_{if}^{\rm (BCIS)\pm}(\boldsymbol{\eta})}{2\pi v}\right|^2 \tag{14.58}
$$

where $R_{if}^{\rm (BCIS)+}$ and $R_{if}^{\rm (BCIS)-}$ are independent of the inter-nuclear potential

$$R_{if}^{(\text{BCIS})-}(\boldsymbol{\eta})=Z_{\text{P}}N_{\text{T}}\iiint \text{d}\boldsymbol{s}_1\text{d}\boldsymbol{s}_2\text{d}\boldsymbol{R}\,\text{e}^{i\boldsymbol{q}_{\text{P}}\cdot(\boldsymbol{s}_1+\boldsymbol{s}_2)+i\boldsymbol{q}_{\text{T}}\cdot(\boldsymbol{x}_1+\boldsymbol{x}_2)}$$
$$\times(vR-\boldsymbol{v}\cdot\boldsymbol{R})^{-i\xi_{\text{P}}}\varphi_f^*(\boldsymbol{s}_1,\boldsymbol{s}_2)\left(\frac{2}{R}-\frac{1}{s_1}-\frac{1}{s_2}\right)\varphi_i(\boldsymbol{x}_1,\boldsymbol{x}_2)$$
$$\times{}_1F_1(i\nu_{\text{T}},1,ivx_1+i\boldsymbol{v}\cdot\boldsymbol{x}_1)\,{}_1F_1(i\nu_{\text{T}},1,ivx_2+i\boldsymbol{v}\cdot\boldsymbol{x}_2) \qquad (14.59)$$

$$R_{if}^{(\text{BCIS})+}(\boldsymbol{\eta})=Z_{\text{T}}N_{\text{P}}\iiint \text{d}\boldsymbol{x}_1\text{d}\boldsymbol{x}_2\text{d}\boldsymbol{R}\,\text{e}^{i\boldsymbol{q}_{\text{P}}\cdot(\boldsymbol{s}_1+\boldsymbol{s}_2)+i\boldsymbol{q}_{\text{T}}\cdot(\boldsymbol{x}_1+\boldsymbol{x}_2)}$$
$$\times(vR+\boldsymbol{v}\cdot\boldsymbol{R})^{-i\xi_{\text{T}}}\varphi_f^*(\boldsymbol{s}_1,\boldsymbol{s}_2)\left(\frac{2}{R}-\frac{1}{x_1}-\frac{1}{x_2}\right)\varphi_i(\boldsymbol{x}_1,\boldsymbol{x}_2)$$
$$\times{}_1F_1(i\nu_{\text{P}},1,ivs_1+i\boldsymbol{v}\cdot\boldsymbol{s}_1)\,{}_1F_1(i\nu_{\text{P}},1,ivs_2+i\boldsymbol{v}\cdot\boldsymbol{s}_2)\,. \qquad (14.60)$$

It should be noted that in $R_{if}^{(\text{BCIS})\pm}$ the electronic continuum intermediate states are included in the same way as in $R_{if}^{(\text{BDW})\mp}$. The essential difference between $R_{if}^{(\text{BDW})\mp}$ and $R_{if}^{(\text{BCIS})\pm}$ lies in the perturbation potentials. In $R_{if}^{(\text{BCIS})\mp}(\boldsymbol{\eta})$, these potentials are given by scalar operators $[Z_{\text{P}}(2/R-1/s_1-1/s_2)]$ and $[Z_{\text{T}}(2/R-1/x_1-1/x_2)]$, whereas in the case $R_{if}^{(\text{BDW})\pm}(\boldsymbol{\eta})$ we have the sum of the two typical vectorial differential (gradient) operator potentials $[(1/a)\sum_{k=1}^{2}\boldsymbol{\nabla}_{s_k}\ln\varphi_f^*(\boldsymbol{s}_1,\boldsymbol{s}_2)\cdot\boldsymbol{\nabla}_{x_k}]$ and $[(1/b)\sum_{k=1}^{2}\boldsymbol{\nabla}_{x_k}\ln\varphi_i(\boldsymbol{x}_1,\boldsymbol{x}_2)\cdot\boldsymbol{\nabla}_{s_k}]$, which are familiar from the CDW-4B method.

Two alternative methods have been developed by Belkić [86] for calculation of the matrix elements in the BCIS-4B method. One of these gives the matrix elements $R_{if}^{\pm}(\boldsymbol{\eta})$ in terms of a three-dimensional numerical integration over real variables from 0 to 1. This method provides the basic quantities $R_{if}^{\pm}(\boldsymbol{\eta})$ in the form of the four-dimensional numerical quadratures over real variables. Both methods have been found to give the same numerical results [86].

Similarly to the BDW-4B method, it is also possible in the BCIS-4B method to alleviate altogether the cumbersome Fourier-Bessel transform, and thus perform direct computations of differential cross sections $(\text{d}/\text{d}\Omega)Q^{(\text{BCIS})-}$ or $(\text{d}/\text{d}\Omega)Q^{(\text{BCIS})+}$ for a special case $Z_{\text{T}}=2$ or $Z_{\text{P}}=2$, respectively. The only difference is that the particular case $Z_{\text{T}}=2$ or $Z_{\text{P}}=2$ relates to the post $(\text{d}/\text{d}\Omega)Q^{(\text{BCIS})+}$ or prior $(\text{d}/\text{d}\Omega)Q^{(\text{BCIS})-}$ angular distributions, respectively (the reverse assignment takes place in the BDW-4B method)

$$\frac{\text{d}Q^{(\text{BCIS})-}}{\text{d}\Omega}=\left|\frac{\mu}{2\pi}T_{if}^{(\text{BCIS})-}\right|^2 \qquad (Z_{\text{T}}=2,\ \text{any } Z_{\text{P}}) \qquad (14.61)$$

$$\frac{\text{d}Q^{(\text{BCIS})+}}{\text{d}\Omega}=\left|\frac{\mu}{2\pi}T_{if}^{(\text{BCIS})+}\right|^2 \qquad (Z_{\text{P}}=2,\ \text{any } Z_{\text{T}}). \qquad (14.62)$$

Double electron capture (11.1) has also been investigated by Purkait [291] and Purkait *et al.* [292]. They used the unperturbed wave function Φ_i for the initial scattering state which, therefore, disregards the correct boundary condition in the entrance channel, in the case of the general values of $Z_{\rm T}$ and $Z_{\rm P}$. Specifically, only for $Z_{\rm T}=2$ and any $Z_{\rm P}$, with no residual Coulomb potential $V_i^\infty = Z_{\rm P}(Z_{\rm T}-2)/R = 0$, the unperturbed state Φ_i from Refs. [291, 292] possesses the correct asymptotic behavior, because exclusively in this particular case we have $\Phi_i^+ = \Phi_i$. In the exit channel the scattering wave function from Refs. [291, 292] is adequate, and it coincides with the corresponding state from the BCIS-4B method [86]. Had the correct boundary conditions for the entrance channel also been included by Purkait [291] and Purkait *et al.* [292] from the outset in the general case of arbitrary values of $Z_{\rm P}$ and $Z_{\rm T}$ for process (11.1), these authors would have rediscovered the BCIS-4B method of Belkić [86] according to (14.56) and (14.57). Explicit computations from Refs. [291, 292] were concerned with double charge exchange in the He^{2+} – He, Li^{3+} – He and B^{5+} – He collisions. In this particular case with a helium target ($Z_{\rm T}=2$), the Sommerfeld parameter $\nu_f = Z_{\rm P}(Z_{\rm T}-2)$ becomes zero which leads to $\mathcal{N}^{-*}(\nu_f)_1F_1(-i\nu_f,1,ik_f r_i + i\boldsymbol{k}_f\cdot\boldsymbol{r}_i) = 1$ in $\mathcal{L}_1(\boldsymbol{r}_i,\boldsymbol{r}_f)$ from (14.36). Fortuitously, this implies that $\Phi_i^+ = \Phi_i$ in which case the matrix elements from Refs. [291, 292] coincides with (14.56) from the BCIS-4B method. In the computations from Refs. [291, 292] the remaining full Coulomb wave function $\chi_{\rm Coul} \equiv \mathcal{N}^+(\nu)_1F_1(-i\nu,1,ik_i r_f + i\boldsymbol{k}_i\cdot\boldsymbol{r}_f)$ with $\nu = Z_{\rm P}Z_{\rm P}/v$ for the relative motion of heavy particles was used instead of its consistent eikonal expressions $\chi_{\rm eik} \equiv \mu_i^{i\nu}\exp\left[i\nu\ln(vR-\boldsymbol{v}\cdot\boldsymbol{R})\right]$. Nevertheless, as already stated, the difference between the corresponding results for the total cross sections obtained using the said full Coulomb wave and its eikonal limit i.e. $\chi_{\rm Coul}$ and $\chi_{\rm eik}$ within the BCIS-4B method for double electron capture in e.g. the $Z_{\rm P}$ – He collisions should be in the 3rd or 4th decimal places. The two such explicit sets of the total cross sections in the BCIS-4B method for the He^{2+} – He collisions based upon $\chi_{\rm eik}$ and $\chi_{\rm Coul}$ in the entrance channel have been published by Belkić [86] and Purkait [291], respectively. The results from these two studies should be indistinguishable from each other within the accuracy $10^{-3} - 10^{-4}$. However, this is not the case, pointing to the existence of some computational errors in the work of Purkait [291], given the veracity of the results of Belkić [86] from the BCIS-4B method, which has been thoroughly checked to yield the identical cross sections using two completely different methods for all the matrix elements.

14.7 The CB1-4B method

Numerous investigations and comparisons with experiments have confirmed that the CB1-3B method is an accurate theory for rearrangement collisions at intermediate and high impact energies [199]–[207]. Therefore, it is reasonable to extend this approximation to four-body collisions with one or two active electrons. Such an extension for double capture has been done by [87, 88] through the introduction of the CB1-4B method. The transition amplitudes in the CB1-4B method for double charge exchange within the prior (T_{if}^{-}) and the post (T_{if}^{+}) forms are

$$T_{if}^{-} = \langle \Phi_f^{-} | V_i^d | \Phi_i^{+} \rangle \qquad T_{if}^{+} = \langle \Phi_f^{-} | V_f^d | \Phi_i^{+} \rangle. \tag{14.63}$$

Here, Φ_i^{+} and Φ_f^{-} are defined by (11.28) and (11.59), whereas V_i^d and V_f^d are the same as in (11.25) and (11.55), respectively. Explicitly, the transition amplitudes taken from Ref. [88] without the term $(v\rho)^{2iZ_{\rm P}Z_{\rm T}/v}$ are given by

$$T_{if}^{\pm}(\boldsymbol{\eta}) = \int \mathrm{d}\boldsymbol{R} \mathrm{e}^{\mp 2i\boldsymbol{q}_{\rm P,T}\cdot\boldsymbol{R}} (vR + \boldsymbol{v}\cdot\boldsymbol{R})^{-i\xi} \mathcal{F}^{\pm}(\boldsymbol{R}) \tag{14.64}$$

$$\begin{aligned} \mathcal{F}^{-}(\boldsymbol{R}) = Z_{\rm P} \iint & \mathrm{d}\boldsymbol{s}_1 \mathrm{d}\boldsymbol{s}_2 \varphi_f^{*}(\boldsymbol{s}_1, \boldsymbol{s}_2) \mathrm{e}^{-i\boldsymbol{v}\cdot(\boldsymbol{s}_1+\boldsymbol{s}_2)} \\ & \times \left(\frac{2}{R} - \frac{1}{s_1} - \frac{1}{s_2} \right) \varphi_i(\boldsymbol{x}_1, \boldsymbol{x}_2) \end{aligned} \tag{14.65}$$

$$\begin{aligned} \mathcal{F}^{+}(\boldsymbol{R}) = Z_{\rm T} \iint & \mathrm{d}\boldsymbol{x}_1 \mathrm{d}\boldsymbol{x}_2 \varphi_f^{*}(\boldsymbol{s}_1, \boldsymbol{s}_2) \mathrm{e}^{-i\boldsymbol{v}\cdot(\boldsymbol{x}_1+\boldsymbol{x}_2)} \\ & \times \left(\frac{2}{R} - \frac{1}{x_1} - \frac{1}{x_2} \right) \varphi_i(\boldsymbol{x}_1, \boldsymbol{x}_2) \end{aligned} \tag{14.66}$$

where $\xi = 2(Z_{\rm T} - Z_{\rm P})/v$. In order to derive these results, we used (11.60) and (14.38) together with vectors $\boldsymbol{q}_{\rm T}$ and $\boldsymbol{q}_{\rm P}$ from (11.61). A complete analytical calculation of the matrix elements $T_{if}^{\pm}(\boldsymbol{\eta})$, as a two-dimensional integral, has been carried out by Belkić [88]. The method used is general in the sense that it can be applied to hetero-nuclear (asymmetric) [88], as well as homo-nuclear (symmetric) [87] collisions in which double charge exchange occurs. This has been substantiated for the symmetric He^{2+} – He collision, for which the algorithm of Belkić [88] reproduced exactly the results from the corresponding previous study [87]. This cross-validation is important, since Belkić [87] presented a completely different way of calculating the matrix elements. It should be mentioned that the partial wave analysis of the transition amplitude in the CB1-4B method has also been performed for double electron capture in collisions of alpha particles and helium [293]. If the necessary convergence over the partial waves has been achieved, the numerical results of Gulyás and Szabo [293] would be the same as those obtained without the partial wave analysis. However, this is not the case (as will be shown later in Fig. 14.4), possibly due to the inclusion of merely three partial waves and/or because of some computational errors made by Gulyás and Szabo [293].

14.8 Comparison between theories and experiments

14.8.1 Double electron capture into the ground state

We first analyze the total cross sections in the CDW-4B method for double electron capture from He by fast H^+ and He^{2+} ions

$$H^+ + He(1s^2) \longrightarrow H^-(1s^2) + He^{2+} \tag{14.67}$$

$$He^{2+} + He(1s^2) \longrightarrow He(1s^2) + He^{2+}. \tag{14.68}$$

In order to investigate the sensitivity of the prior and post forms of the total cross sections to the choice of the ground-state wave functions for He and H^-, we employ four two-electron functions: a one-parameter uncorrelated functions of Hylleraas [58], a radially correlated two-parameter orbital of Silverman *et al.* [71], a three-parameter function of Green *et al.* [70], and a four-parameter function of Löwdin [69]. As shown by Belkić and Mančev [84], the post-prior discrepancy for reaction (14.67) for all four wave functions is within at most 40% at impact energies where the CDW-4B method is expected to be most adequate ($E \geq 100$ keV). In the case of the wave function of Löwdin [69], the difference between the prior and post cross section does not exceed 20% at $E \geq 100$ keV.

Total cross sections in their post forms for process (14.67) are given in Figs. 14.1 and 14.2 using the wave functions of Hylleraas [58] for the initial and final helium-like states. It has been verified in the CDW-4B method [85] that the high-energy cross sections computed using the orbitals of Silverman *et al.* [71], Green *et al.* [70] and Löwdin [69] are very close to those associated with the wave function of Hylleraas [58]. The cross sections of the CDW-4B method are seen in Fig. 14.1 to be in excellent agreement with the available experimental data at impact energies $E \geq 100$ keV. However, the results of the CB1-4B method [294], also given in Fig. 14.1, markedly overestimate the experimental data. The cross sections from the three-state two-center close coupling atomic orbital (AO) method used by Lin [295] considerably underestimate the measured data at lower energies ($E \leq 45$ keV), with precisely the reversed pattern above 120 keV.

In Fig. 14.2 the cross sections obtained by Belkić [294] using the CB1-4B and the four-body first-order Jackson-Schiff (JS1-4B) methods are compared with each other for process (14.67). As can be seen from this figure, noticeable differences exist between these two methods below 200 keV pointing to the importance of the correct boundary conditions that are preserved in the CB1-4B method and ignored in the JS1-4B method. Also shown in Fig. 14.2 are the earlier results of Gerasimenko [296]. He aimed at using the JS1-4B method, but his results are wrong and, as such, denoted by $JS1^{\#}$-4B to avoid confusion with the corresponding exact results from the JS1-4B method [294].

We can conclude that, in the case of reaction (14.67) at impact energies $E \geq 100\,\text{keV}$, the CDW-4B method is not strongly dependent upon the choice of the bound-state wave functions from Refs. [58]–[71]. Hence, the simplest one-parameter orbital of Hylleraas [58] can confidently be used in subsequent applications regarding process (14.67). At energies $E \geq 100\,\text{keV}$, where the CDW-4B method is assessed to be adequate, the prior and post cross sections are in satisfactory mutual agreement and, furthermore, they provide an excellent interpretation of the existing experimental data on the $H^+ - He$ double charge exchange.

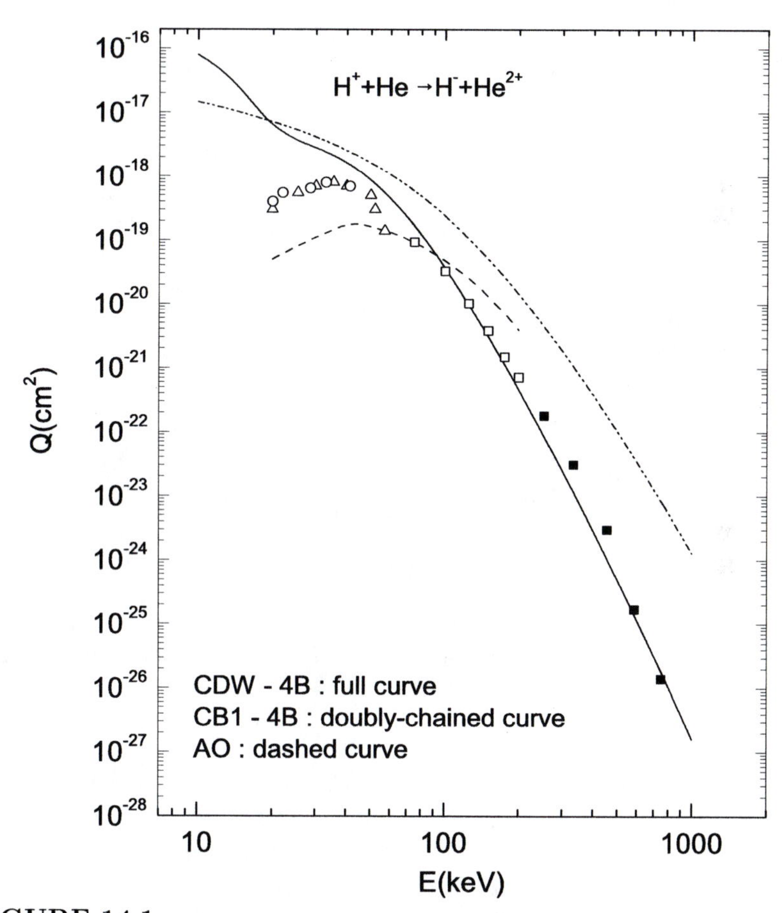

FIGURE 14.1

Total cross sections Q as a function of the incident energy E for process (14.67). Theory: full curve (CDW-4B method [84]), doubly-chained curve (CB1-4B method [294]) and dashed curve (AO method [295]). Experiment: ■ [297], □ [298], △ [299] and ∘ [300].

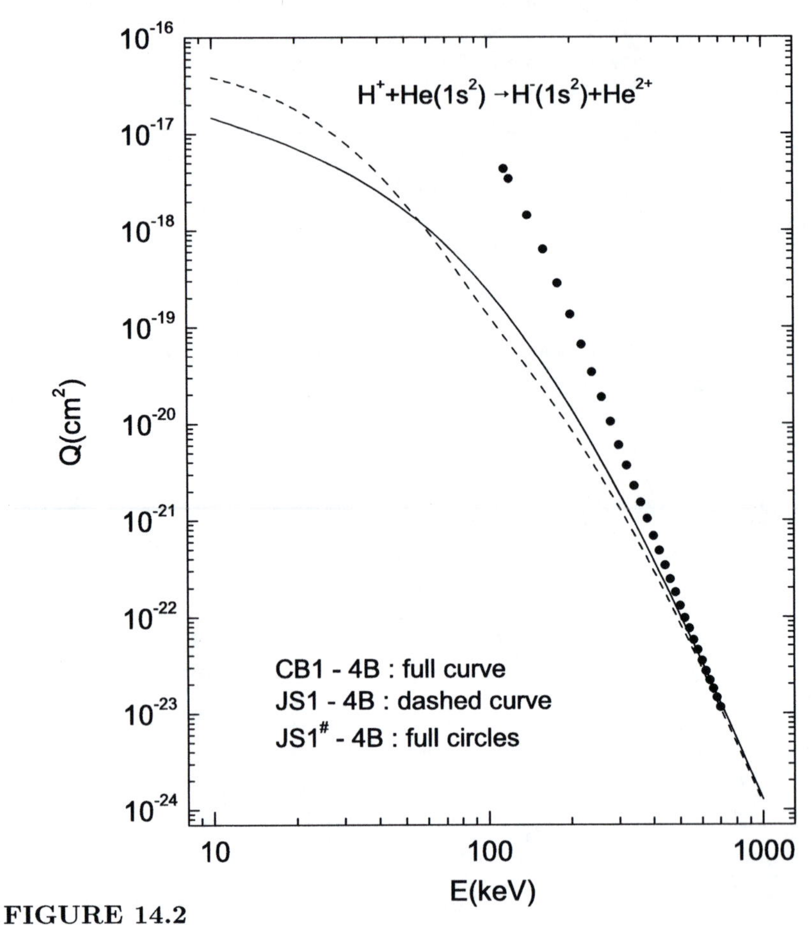

FIGURE 14.2
Total cross sections Q as a function of the incident energy E for process (14.67). Theory: full curve (CB1-4B method [294]), dashed curve (JS1-4B method [294]) and filled circles • (JS1$^{\#}$-4B method [296]).

The cross sections for process (14.68) in the CDW-4B method have first been reported in Refs. [89, 90]. There is no post-prior discrepancy for this reaction, so that $Q_{if}^{-} = Q_{if}^{+} \equiv Q_{if}$. It has been found by Belkić *et al.* [90] that the dependence of the total cross sections for the $He^{2+} - He$ collisions upon the bound-state wave functions is not strong and, therefore, similar to that in the $H^{+} - He$ collisions. The first-order theories for process (14.68) are illustrated in Figs. 14.3–14.9 using the CB1-4B, BDW-4B, BCIS-4B, CDW-4B1 and CDW-EIS-4B1 methods. Figure 14.10 presents certain additional results with approximate second-order contributions from a distorted wave perturbation expansion [94] which is not the series of Dodd and Greider [261]. Surprisingly, as seen in Fig. 14.3, the results from the CDW-4B1 method [89, 90] for process (14.68) do not reproduce most of the experimental data, except those at the two largest energies 4 and 6 MeV.

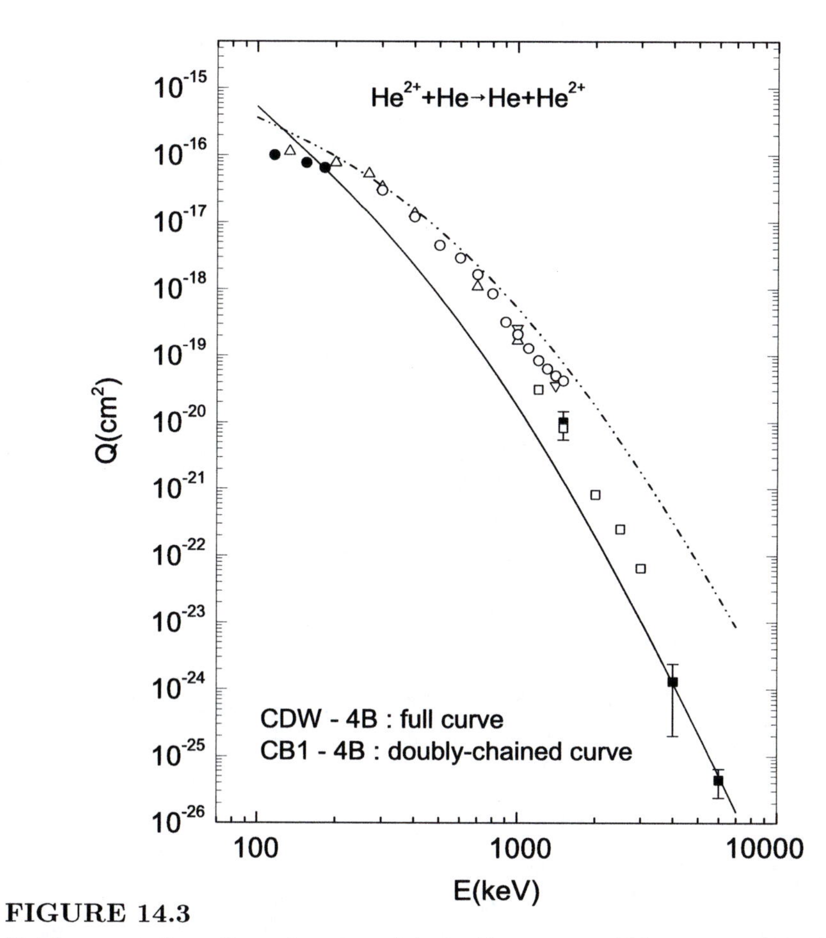

FIGURE 14.3

Total cross sections Q as a function of the incident energy E for process (14.68). Theory: full curve (CDW-4B method [89]) and doubly-chained curve (CB1-4B method [87, 88]). Experiment: ■ [301], ○ [302], ▽ [303], △ [304], □ [305] and • [306].

In the same Fig. 14.3 for process (14.68) we also give the cross sections from the CB1-4B method [87, 88]. Below 1.5 MeV, the CDW-4B method agrees well with the experimental data, but gives too large cross sections above 1.5 MeV. Fortuitously, and only for the particular process (14.68) where $Z_{\rm P} = 2 = Z_{\rm T}$, the CB1-4B and JS1-4B method coincide with each other. However, since in

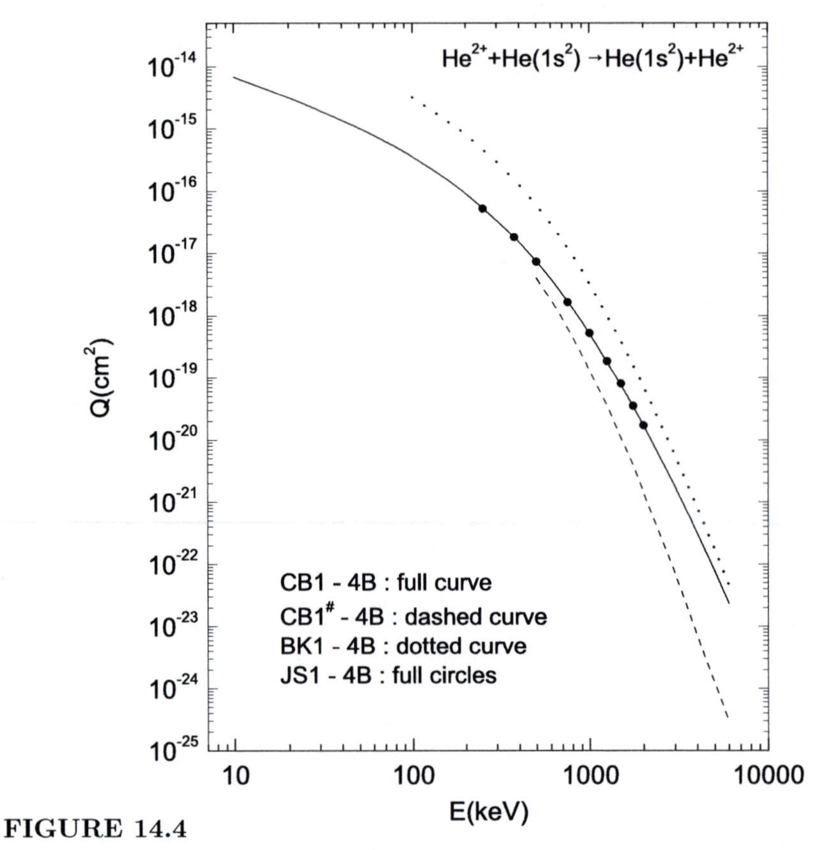

FIGURE 14.4

Total cross sections Q as a function of the incident energy E for process (14.68). Theory: full curve (CB1-4B method [294]), dashed curve (CB1$^{\#}$-4B method [293]), dotted curve (BK1-4B [294]) and full circles (JS1-4B method [307]).

this special case, both the initial and final Coulomb phases for the relative motions of the nuclei disappear altogether, the channel wave functions in the BK1-4B method also possess the proper asymptotic behaviors. Nevertheless, the BK1-4B method still disobeys the correct boundary conditions, even in the case of process (14.68), due to the appearance of the inadequate perturbation potentials. For example, in the exit channel the perturbation potential within the BK1-4B method has only the long-range interactions $Z_{\rm T}(-1/x_1 - 1/x_2)$ instead of the physical short-range potential $Z_{\rm T}(2/R - 1/x_1 - 1/x_2)$, which is required by the Dollard's asymptotic convergence problem [41, 44]. Therefore, the importance of the correct boundary conditions can also be inferred by comparing the CB1-4B and BK1-4B methods. Moreover, such a comparison would reveal the relative importance of the contributions from the terms $Z_{\rm T}(-1/x_1 - 1/x_2)$ and $2Z_{\rm T}/R$. This is illustrated in Fig. 14.4 where the cross

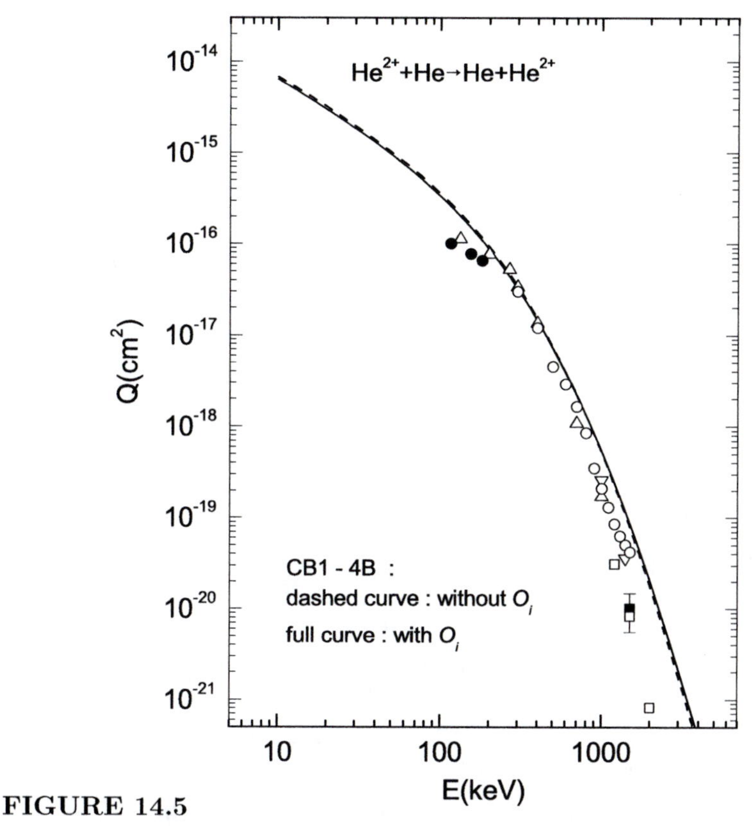

FIGURE 14.5
Total cross sections Q as a function of the incident energy E for process (14.68). Theory: dashed curve (CB1-4B method without perturbation O_i [87, 88]) and full curve (CB1-4B method with perturbation O_i [88]). Experiment: ∘ [302], ▽ [303], △ [304], □ [305] and • [306].

sections of the BK1-4B method are seen to markedly overestimate the corresponding results from the CB1-4B method. Also shown in Fig. 14.4 are the results of Gulyás and Szabo [293] who used the partial wave analysis within the CB1-4B method. Their results markedly underestimate the true cross sections of the CB1-4B method [87, 88], and this discrepancy increases when the energy is augmented, indicating that the three partial waves used in Ref. [293] are totally insufficient for achieving convergence. It is well-known that the number of partial waves needed for convergence increases significantly with the increased impact energy, and this is at variance with keeping only 3 partial waves throughout, as done by Gulyás and Szabo [293]. The erroneous data given by Gulyás and Szabo [293] are denoted by CB1$^{\#}$-4B in Fig. 14.4 in order to avoid confusion with the corresponding exact cross sections from

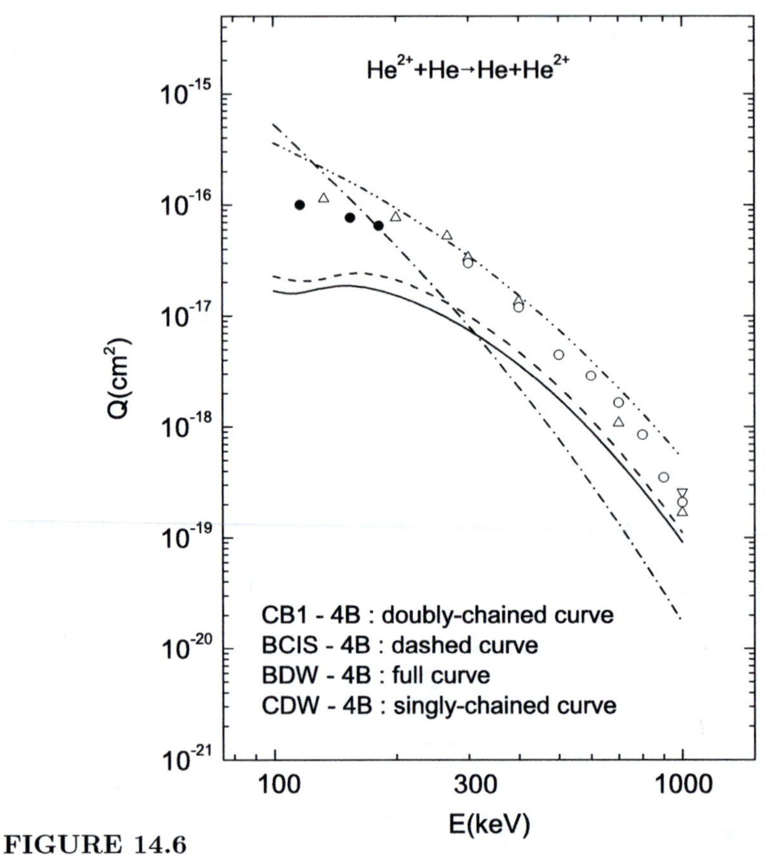

FIGURE 14.6

Total cross sections Q as a function of the incident energy E for process (14.68). Theory: doubly-chained curve (CB1-4B method [87, 88]), dashed curve (BCIS-4B method [86]), full curve (BDW-4B method [89]) and singly-chained curve (CDW-4B method [89]). Experiment: $\circ$ [302], ∇ [303], $\triangle$ [304] and $\bullet$ [306].

the CB1-4B method obtained by Belkić [87, 88]. Note that the earlier findings from the CB1-4B method reported by Gerasimenko and Rosentsveig [307] are seen in Fig. 14.4 to be in perfect agreement with the corresponding results of Belkić [87, 88].

In Fig. 14.5 comparison is made between the results from the CB1-4B method [87] with and without the perturbation O_i for process (14.68). This additional term in the perturbation potential which is given by (13.40) stems from the non-existence of the exact helium-like wave function $\varphi_i(\boldsymbol{x}_1, \boldsymbol{x}_2)$ for the initial state. It is seen in Fig. 14.5 that the contribution of the perturbation O_i is very small and, as such, can be neglected throughout.

One of the inadequacies of the CDW-4B1 method is the use of unnormalized total scattering wave functions in both the entrance and exit channels. Of course, the same drawback of the CDW-3B method is also encountered for single charge exchange (12.26) [308], but without a significant consequence at at impact energies satisfying the usual validity condition [44]

$$\text{Incident energy } E(\text{keV/amu}) \geq 80 \max\{|E_i|, |E_f|\} \tag{14.69}$$

where E_i and E_f are the initial and the final orbital energies of the captured electron, respectively[1]. The discrepancy between the CDW-4B1 method and experiments for reaction (14.68) may indicate that the same type of inadequacies invoked in theories of rearrangement collisions could be more serious for double than for single charge exchange. Total cross sections for high-energy two-electron transfer are smaller than the corresponding results for one electron capture by at least two orders of magnitude. Therefore, it is not surprising that double charge exchange, as a much weaker effect than single electron transfer, appears to be very sensitive to any (even apparently small) inadequacies of the theory. Nevertheless, this normalization problem is not expected to be the main cause for the lack of agreement between the CDW-4B1 method and experiment below 4 MeV in Fig. 14.6. This could be inferred from the work of Martínez *et al.* [94]. Their results from the CDW-EIS-4B1 method (which uses the normalized eikonal scattering wave function in the entrance channel) underestimate both the experiments and the CDW-4B1 method for the same process (14.68), as seen in Figs. 14.8–14.10.

As was initially conjectured by Belkić *et al.* [90], an alternative reason for the fact that the CDW-4B1 method is satisfactory only at the highest energies (4 and 6 MeV) in Fig. 14.3 could be neglect of the second-order contribution from a perturbation series. Subsequently, Martínez *et al.* [94] used the sum of the $T-$matrix element from the CDW-4B1 method and an approximate on-shell second-order term from a perturbation expansion. They called the sum of these two latter $T-$matrix elements the four-body second-order continuum distorted wave (CDW-4B2) approximation. The usual eikonalization of the two full electronic continua in the entrance channel introduces a further approximation to the CDW-4B2 method known as the four-body second-order continuum distorted wave eikonal initial state (CDW-EIS-4B2) method [94]. Recall that the CDW-4B1 and CDW-EIS-4B1 methods are the first-orders to the perturbation series expansion of Dodd-Greider [261] without disconnected diagrams. Likewise, the proper CDW-4B2 and CDW-EIS-4B2 methods, as the second-orders to the same perturbation development of Dodd-Greider [261], would be obtained by including the second terms in this series. This is not what has been done by Martínez *et al.* [94]. Instead, they added a second-order propagator from an ordinary distorted wave expansion (with disconnected diagrams) to the transition $T-$operator of the CDW-4B1 method.

[1]According to (14.69), the expected limit of the validity of the CDW-4B1 method for a He^{2+} projectile is 0.45 MeV, whereas for a Li^{3+} ion impact it is above 2 MeV, and for a B^{5+} projectile it is above 9.7 MeV.

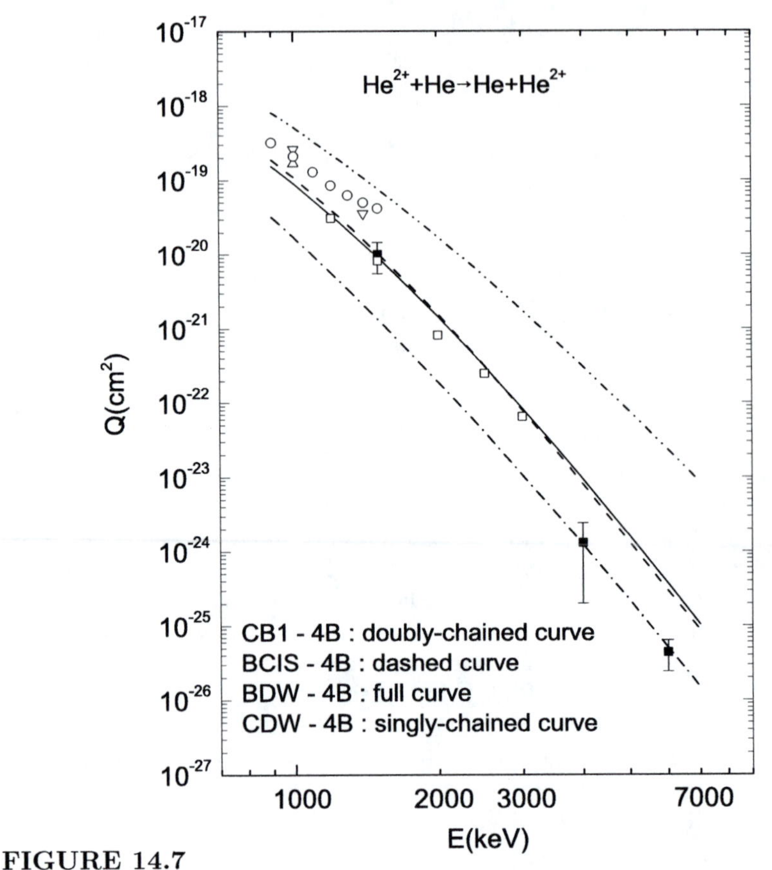

FIGURE 14.7

Total cross sections Q as a function of the incident energy E for process (14.68). Theory: doubly-chained curve (CB1-4B method [87, 88]), dashed curve (BCIS-4B method [86]), full curve (BDW-4B method [89]) and singly-chained curve (CDW-4B method [89]). Experiment: ■ [301], ∘ [302], ▽ [303], △ [304] and □ [305].

Such a mixing of two terms from two different series with connected and disconnected diagrams is obviously inconsistent. As a consequence, acronyms CDW-4B2 and CDW-EIS-4B2 used by Martínez *et al.* [94] are misleading, since they do not refer to the second-orders of the Dodd-Greider series [261], as opposed to what one would be inclined to think by extrapolating the terminology with the CDW-4B1 and CDW-EIS-4B1 methods to the next (second) order in the expansion. For this reason in the conclusion to a recent state-of-the-art review of Belkić *et al.* [97], the second-order model of Martínez *et al.* [94] was called the 'augmented continuum distorted wave method' and the 'augmented continuum distorted wave eikonal initial state method'. Likewise, hereafter we shall interchangeably use the abbreviations CDW-4B2 and 'augmented CDW-4B' (as well as CDW-EIS-4B2 and 'augmented CDW-EIS-4B').

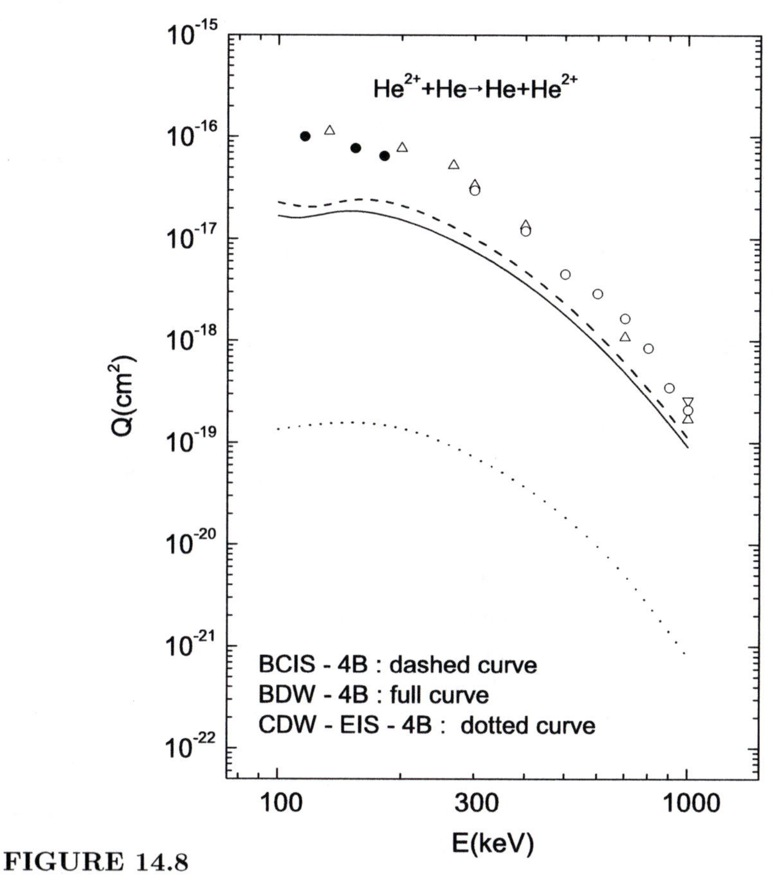

FIGURE 14.8
Total cross sections Q as a function of the incident energy E for process (14.68). Theory: dashed curve (BCIS-4B method [86]), full curve (BDW-4B method [89]) and dotted curve (CDW-EIS-4B method [94]). Experiment: $\circ$ [302], ∇ [303], Δ [304] and $\bullet$ [306].

However, the reader should always bear in mind the mentioned remark about the genuine second-order Dodd-Greider methods [261], especially in interpreting the results of comparisons between the CDW-4B1 and CDW-4B2 methods on the one hand, as well as between the CDW-EIS-4B1 and CDW-EIS-4B2 methods on the other hand.

The CDW-4B2 method yields the cross sections that agree favorably with most of the available experimental data in Fig. 14.10, except at 4 and 6 MeV (precisely the reverse pattern relative to the CDW-4B1 method). However, as seen in Fig. 14.10, the CDW-EIS-4B2 method gives the results that overestimate both the cross sections from the CDW-4B2 method and the experimental data for process (14.68). Regarding the CDW-EIS-4B1 method, Figs. 14.8

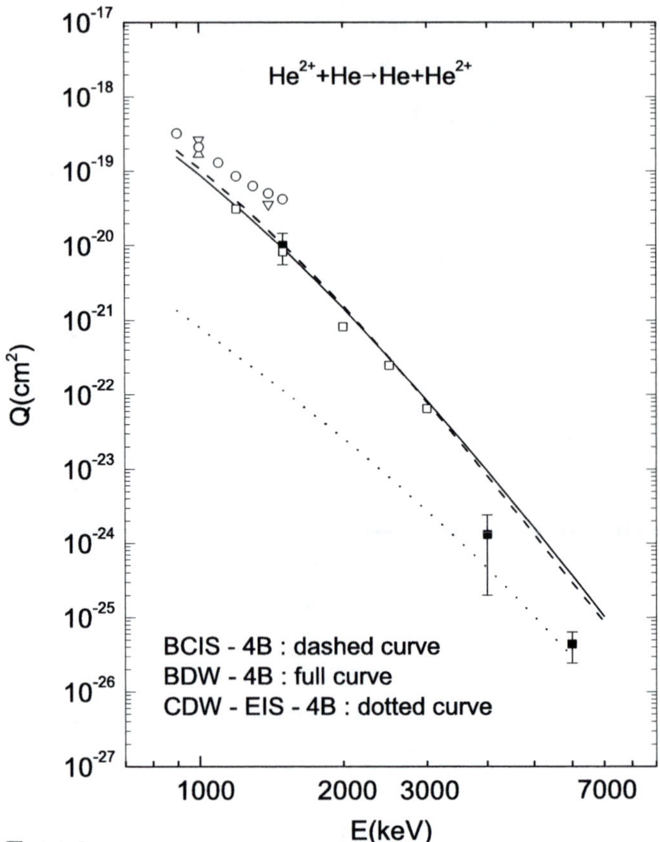

FIGURE 14.9
Total cross sections Q as a function of the incident energy E for process (14.68). Theory: dashed curve (BCIS-4B method [86]), full curve (BDW-4B method [89]) and dotted curve (CDW-EIS-4B method [94]). Experiment: ■ [301], ○ [302], ▽ [303], △ [304] and □ [305].

and 14.9 show that throughout the interval 100–3000 keV, $Q_{if}^{(\mathrm{CDW-EIS-4B1})}$ markedly underestimates the experimental data. For example, this underestimation is by 2–3 orders of magnitude in the range 100–2000 keV. This latter energy range is well within the expected validity domain of the CDW-EIS-4B1 method. On the other hand, in Fig. 14.10, the cross sections $Q_{if}^{(\mathrm{CDW-EIS-4B2})}$ overestimate considerably all the available experimental data in the whole range under consideration. In particular, at impact energies 100–1000 keV, the values of the ratio $Q_{if}^{(\mathrm{CDW-EIS-4B2})}/Q_{if}^{(\mathrm{CDW-EIS-4B1})}$ obtained by Martínez *et al.* [94] are enormous, ranging from 10^3 to 10^4.

The same on-shell Green propagator for a second-order contribution has also been used within the IA by Gravielle and Miraglia [95] for process (14.68).

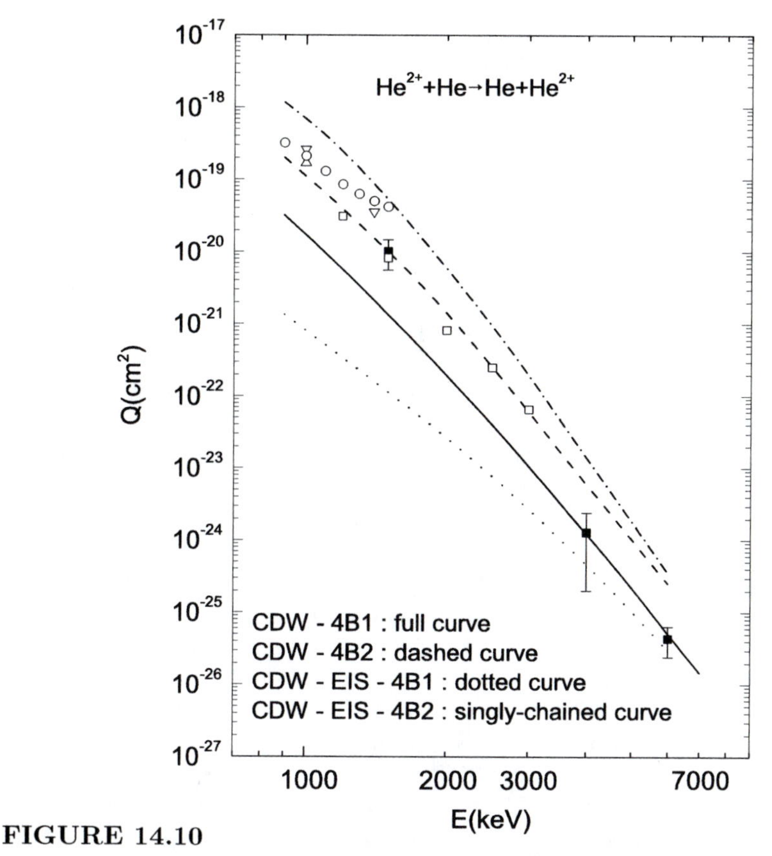

FIGURE 14.10

Total cross sections Q as a function of the incident energy E for process (14.68). Theory: full curve (CDW-4B1 method [89, 94]) dashed curve (CDW-4B2 method [94]) dotted curve (CDW-EIS-4B1 method [94]) and singly-chained curve (CDW-EIS-4B2 method [94]). Experiment: ■ [301], ∘ [302], ▽ [303], △ [304] and □ [305].

They omitted the first-order IA, but this is unjustified as they have not shown that this ignored term is indeed negligible. Moreover, such an omission is not supported by the CDW-4B2 method, which contains a significant contribution from the corresponding first-order term provided by the CDW-4B1 method, as seen via the full and dashed curves in Fig. 14.10. It is pertinent to recall that the IA for three- and four-body collisions does not obey the correct boundary conditions. This most serious drawback has been rectified by Belkić with the emergence of the the RIA-3B and the RIA-4B [177]–[180].

Overall, as opposed to single electron capture, it is physically plausible that the second-order term in a perturbation expansion could play an important role for double electron capture, since in this latter process two electrons participate actively in the collision. This is evidenced by large differences between the CDW-4B1 and CDW-4B2 methods, on the one hand, and between the CDW-EIS-4B1 and CDW-EIS-4B2 methods, on the other hand.

These initial assessments of the second-order terms in a perturbation series are encouraging. Nevertheless, it would be very important to extend such computations by including both the on- and off-shell second-order contributions in the CDW-4B2 and CDW-EIS-4B2 methods for process (14.68). Furthermore, it would be indispensable to assess the convergence rate in the spectral representation of the Green function from a second-order propagator. This latter spectral representation from Martínez *et al.* [94] is inconclusive, as it takes into account only the two hydrogen-like ground states centered on the projectile and target nucleus, without the necessary assessment of the contribution from any of the other, ignored intermediate states.

Further, as can be seen from Figs. 14.6–14.9, the BCIS-4B and BDW-4B methods yield very similar values for the displayed total cross sections. These two approximations use normalized scattering wave functions in one channel. The total cross sections from the BCIS-4B and BDW-4B methods are much smaller that those from the CB1-4B method throughout the energy range under consideration. For example, the difference between the findings of the BDW-4B and CB1-4B methods increases as the impact energy is augmented, reaching two orders of magnitude at 6 MeV. Importantly, the BDW-4B and BCIS-4B methods are in good agreement with most of the available experimental data even without any second-order term from a distorted wave series[2]. However, at the two largest energies (4 and 6 MeV), the BDW-4B and BCIS-4B methods are seen in Fig. 14.7 to overestimate the measurements.

Next, we analyze the results of Belkić [88] obtained by means of the CB1-4B method for the following asymmetric reaction

$$^{7}\mathrm{Li}^{3+} + {}^{4}\mathrm{He}(1s^2) \longrightarrow {}^{7}\mathrm{Li}^{+}(1s^2) + {}^{4}\mathrm{He}^{2+}. \tag{14.70}$$

The results for the total cross sections in both the post and prior versions for process (14.70) are depicted in Fig. 14.11. As can be seen from this figure, the post cross sections are slightly larger than the prior ones. The post-prior discrepancy appears to be somewhat more pronounced at lower than at higher energies. A comparison between the CB1-4B method and the experimental data of Shah and Gilbody [309] is also shown in Fig. 14.11. The results from the CB1-4B method are seen to be in satisfactory agreement with the experimental data. Thus, considering only the $1s^2 \longrightarrow 1s^2$ transition in process (14.70), it follows that the CB1-4B method compares more favorably with the measurement than the corresponding CDW-4B method, as evidenced by Figs. 14.6 and 14.11.

The total cross sections reported by Purkait *et al.* [292] for process (14.70) are close to the results from the CB1-4B method and the experimental data

[2]Incidently, this casts doubt on the CDW-EIS-4B1 method which has been claimed [94] to be inadequate for double charge exchange because of neglect of the second-order propagator in the expansion of the total Green operator for the whole system.

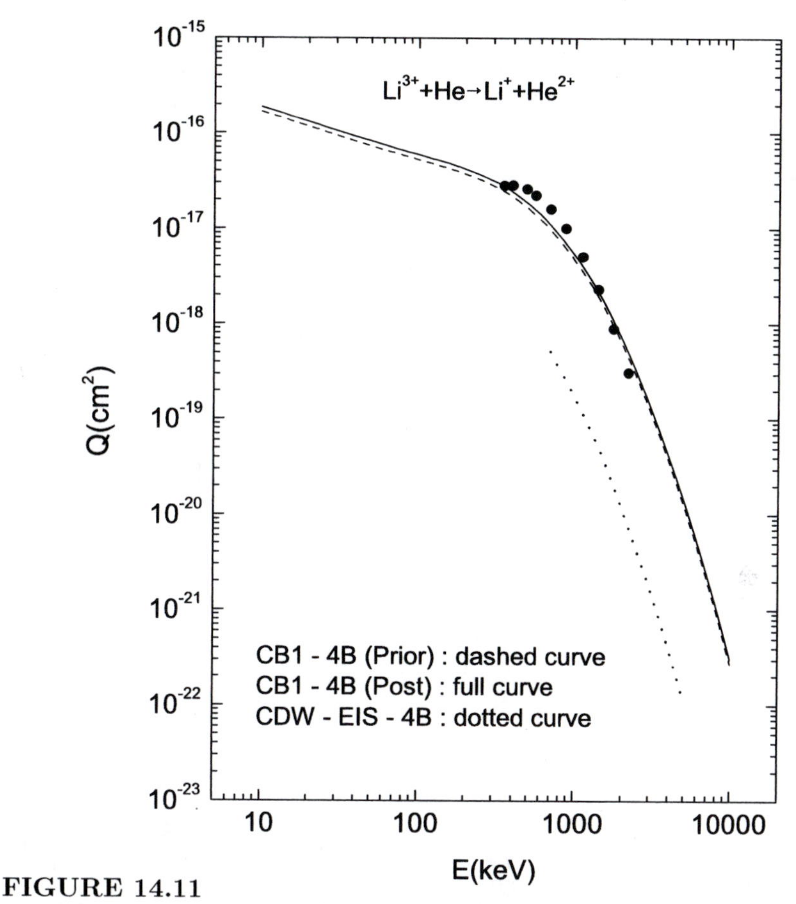

FIGURE 14.11

Total cross sections Q as a function of the incident energy E for process (14.70). Theory: dashed curve (prior CB1-4B method [88]), full curve (post CB1-4B method [88]) and dotted curve (CDW-EIS-4B method [96]). Experiment: • [309].

from Fig. 14.11. Nevertheless, this needs to be reassessed by taking into account the final excited states of the Li^+ ion in process (14.70), since they have been neglected by both Belkić [86] and Purkait *et al.* [292].

Gayet *et al.* [96] have also studied process (14.70). They used the CDW-EIS-4B method at impact energies 700-5000 keV (recall that CDW-EIS-4B ≡CDW-EIS-4B1). Their total cross sections grossly underestimate the experimental data from Fig. 14.11 by two orders of magnitude. This fact and a similar observation, which we have already made for process (14.68), indicate that the CDW-EIS-4B method is inadequate for double electron capture in heavy particle collisions.

It is also instructive to consider the IPM for reaction (14.68). According to the IPM, the transition amplitude for double electron capture is given as a product of the amplitudes for single electron capture [104]–[107]. The differential cross section for double electron capture in the IPM version of the CDW method (denoted by CDW-IPM) is the Fourier-Bessel transform

$$\frac{\mathrm{d}Q_{if}^{(\mathrm{CDW-IPM})}}{\mathrm{d}\Omega} = \left| i v\mu \int_0^\infty \mathrm{d}\rho\, \rho^{1+2i\nu} \left[a_{if}^{(\mathrm{CDW-3B})}(\rho) \right]^2 J_0(\eta\rho) \right|^2 \left(\frac{a_0^2}{sr} \right). \quad (14.71)$$

Here, $a_{if}^{(\mathrm{CDW-3B})}(\rho)$ is the transition amplitude as a function of ρ in the CDW-3B method [42, 44] for single electron capture (12.26). The expression for $a_{if}^{(\mathrm{CDW-3B})}(\rho)$ in the prior form is given by Belkić and Salin [265] as

$$a_{if}^{(\mathrm{CDW-3B})}(\rho) = \frac{32i}{v} (Z_\mathrm{P} Z_\mathrm{T})^{5/2} N^+(\nu_\mathrm{P}) N^+(\nu_\mathrm{T}) [(1 - i\nu_\mathrm{T}) I_0 + i\nu_\mathrm{T} I_1]$$

where $N^+(\nu_\mathrm{K}) = \Gamma(1 - i\nu_\mathrm{K}) \exp(\pi\nu_\mathrm{K}/2)$, $\nu_\mathrm{K} = Z_\mathrm{K}/v\, (\mathrm{K} = \mathrm{P}, \mathrm{T})$ and

$$I_0 = \int_0^\infty \mathrm{d}\kappa\, \kappa J_0(\kappa\rho) \frac{A}{C^2} \left(1 - \frac{\omega_1}{\kappa^2 + \gamma^2} \right)^{-1-i\nu_\mathrm{P}} \left(1 - \frac{\omega_2}{\kappa^2 + \delta^2} \right)^{-i\nu_\mathrm{T}}$$

$$I_1 = \int_0^\infty \mathrm{d}\kappa\, \kappa J_0(\kappa\rho) \frac{A + iB}{C^2} \left(1 - \frac{\omega_1}{\kappa^2 + \gamma^2} \right)^{-1-i\nu_\mathrm{P}} \left(1 - \frac{\omega_2}{\kappa^2 + \delta^2} \right)^{-1-i\nu_\mathrm{T}}$$

$$A = \kappa^2 + \delta^2 - (Z_\mathrm{P} + iv)(Z_\mathrm{P} - i\alpha) \qquad B = v(Z_\mathrm{P} - i\alpha)$$

$$\gamma = \alpha^2 + Z_\mathrm{P}^2 \qquad \alpha = \frac{v}{2} - \frac{E_i - E_f}{v} \qquad \omega_1 = 2v(\alpha + iZ_\mathrm{P})$$

$$\delta = \beta^2 + Z_\mathrm{T}^2 \qquad \beta = \frac{v}{2} + \frac{E_i - E_f}{v} \qquad \omega_2 = 2v(\beta + iZ_\mathrm{T})$$

$$C = (\kappa^2 + \gamma^2)(\kappa^2 + \delta^2) \qquad \gamma = \delta.$$

The total cross sections $Q_{if}^{(\mathrm{CDW-IPM})}$ are computed from (14.71) by standard integration over ρ in the interval $[0, \infty]$. The ensuing results have been reported by Belkić *et al.* [90]. These results are in satisfactory agreement with measurements in the energy range 0.9–7 MeV (not shown in Fig. 14.10 to avoid clutter, but similar results from CDW-IPM will be displayed in Fig. 14.12). Additional computations have been performed in the CDW-IPM using the Hylleraas wave function [58] for the final bound state φ_f, and the RHF orbital [65]–[67] for φ_i, which is expressed in an analytical form as given by Clementi and Roetti [68]. These results, denoted by $Q_{if}^{(\mathrm{CDW-IPM})(\mathrm{ii})}$ have been compared by Belkić *et al.* [90] with the cross sections $Q_{if}^{(\mathrm{CDW-IPM})(\mathrm{i})}$ computed by means of the orbital of Hylleraas [58] for both the initial and final bound states. In Ref. [90] it was found that at lower and intermediate energies 100–1000 keV the results for $Q_{if}^{(\mathrm{CDW-IPM})(\mathrm{ii})}$ are smaller than those for $Q_{if}^{(\mathrm{CDW-IPM})(\mathrm{i})}$ by a factor $\gamma' \equiv Q_{if}^{(\mathrm{CDW-IPM})(\mathrm{ii})}/Q_{if}^{(\mathrm{CDW-IPM})(\mathrm{i})}$ with the numerical values confined to the interval $\gamma' \in [0.90, 0.52]$. Such a pattern is precisely reversed at higher energies from 1 to 7 MeV at which $\gamma' \in [0.90, 1.65]$. The difference between the results $Q_{if}^{(\rm CDW-IPM)(i)}$ and $Q_{if}^{(\rm CDW-IPM)(ii)}$ is a well-known consequence of electronic correlations. Radial correlations are abundantly present in the RHF orbital [68], whereas they are ignored in the Hylleraas wave function[3]. The IPM and the related independent event model (IEM) [310] completely ignore the dynamic correlation effects that make double charge exchange fundamentally different from single electron transfer. Nevertheless, both the IPM and IEM can be amended by incorporating static correlations. This has been shown by Crothers and McCarroll [310] who used the IEM within the CDW method (as denoted by CDW-IEM) to study double electron capture in the $\rm He^{2+} - He(1s^2)$ collisions. They included the static electron correlation effects in the target through the wave function of Pluvinage [60] with the explicit appearance of the inter-electronic coordinate r_{12}. Deco and Grün [107] used the CDW-IPM with target static correlation effects included by means of the configuration interaction (CI) wave functions (also called linear superposition of configurations).

14.8.2 Double electron capture into excited states

The prediction of the contributions from excited states requires a convenient description of singly and doubly excited states of a helium-like atomic system. One possibility is to describe the final state $\varphi_f(\boldsymbol{s}_1, \boldsymbol{s}_2)$ by means of the CI wave function. For these CI functions, the procedure to calculate the bound-free form factors as the matrix elements in the CDW-4B method has previously been devised by Belkić and Mančev [84] in a general manner, which is applicable to both the ground and excited states of helium-like atoms or ions. This can be done by employing a basis set of mono-electronic functions such as Slater type orbitals (STO) or hydrogen-like orbitals with a nuclear charge $Z_{\rm P}$ [81, 93]. Such functions are particularly convenient for describing singly or doubly excited states. When the final state is auto-ionizing, only the bound components of $\varphi_f(\boldsymbol{s}_1, \boldsymbol{s}_2)$ are kept throughout, since its decay occurs much after the collision has been completed. The use of these CI wave functions [81] within the said procedure of Belkić and Mančev [84] regarding bound-free form factors facilitates the calculation of the matrix elements in the transition amplitude for double electron capture into excited states [93]. These latter calculations were restricted to the singly excited states ($1snl$) with $n \leq 3$ and $l \leq n-1$ and doubly excited states ($2l2l'$) with $l, l' \leq 1$.

Although other singly excited states should be included, this procedure could provide an initial indication about the contribution from excited states

[3]Of course, a kind of radial correlation in the Hylleraas wave function of helium-like atoms might be considered as being roughly taken into account through the Slater screening by which the nuclear charge $Z_{\rm T}$ is replaced with $Z_{\rm T} - 5/16 = Z_{\rm T} - 0.3125$.

relative to the ground state. The scattering integrals that appear in the calculation with wave functions for excited state are of the type considered by Nordsieck [311]. The explicit methods that bypass the cumbersome and implicit differentiation for calculating the most general cases of these scalar and vectorial bound-free form factors, have been developed by Belkić [312]–[317] in both parabolic and spherical coordinates.

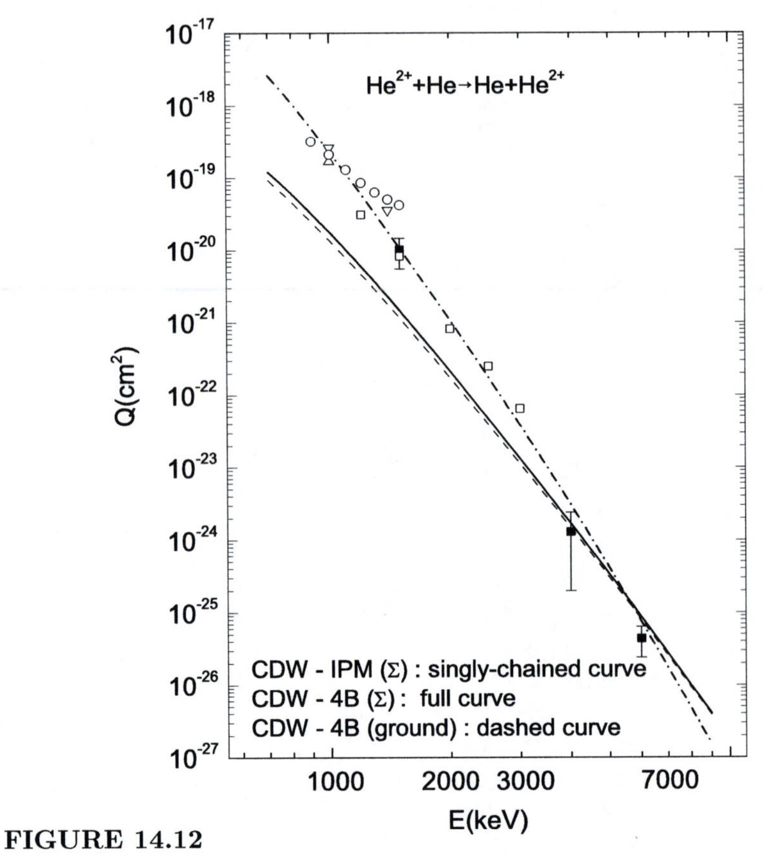

FIGURE 14.12

Total cross sections Q as a function of the incident energy E for process He^{2+} + $He(1s^2) \to He(\Sigma) + He^{2+}$. Theory: singly-chained curve (CDW-IPM method [92] including the sum of the ground and all excited final states), dashed curve (CDW-4B method [93] including the final ground state alone), and full curve (CDW-4B method [93] including the sum of the ground and all excited final states). Experiment: ■ [301], ∘ [302], ∇ [303], △ [304] and □ [305].

Total cross sections in the CDW-4B method for double electron capture from He by He^{2+} including several excited states have been reported in Ref.

[93]. These results are based upon the wave function of Löwdin [69] for helium in the entrance channel and the CI wave function [81] for the final state in the exit channel. From the outset, capture into excited states is not expected to play a significant role in process (14.68), because of the dominance of the ground-to-ground state transition which is resonant. The explicit computations from Ref. [93] confirm this expectation, such that the sum from all the considered final bound states of helium is very close to the full curve in Fig. 14.10 from the CDW-4B method for the ground-to-ground state transition alone considered by Belkić [89]. This is evidenced directly in Fig. 14.12. Here, e.g. at the incident energies 1, 2 and 5 MeV the total cross sections from the ground-to-ground state transition (and the sum of this latter contribution with the corresponding yield from the excited states, written in the parentheses) are: $1.48 \times 10^{-20}\text{cm}^2$ $(1.97 \times 10^{-20}\text{cm}^2)$, $1.89 \times 10^{-22}\text{cm}^2$ $(2.33 \times 10^{-22}\text{cm}^2)$ and $3.25 \times 10^{-25}\text{cm}^2$ $(3.60 \times 10^{-25}\text{cm}^2)$, respectively. Moreover, the cross sections due to doubly excited state are smaller than those for singly excited ones, especially at high impact energies [93].

The CDW-4B method used in Ref. [93] has also been applied to double electron capture in the $\text{Li}^{3+}-\text{He}$ and $\text{B}^{5+}-\text{He}$ collisions. Here, it is anticipated that the contributions from the excited states are important, as confirmed in Figs. 14.13 and 14.14. This occurs because the ground-to-ground transitions for the $\text{Li}^{3+}-\text{He}$ and $\text{B}^{5+}-\text{He}$ double charge exchange are non-resonant and, therefore, excited states could yield a sizeable contribution. At smaller impact energies, the main contribution to total cross section originates from singly excited states, while the ground state $(1s^2)\,^1S$ provides about 40% of the total cross section. In all the cases under consideration, the difference between the cross sections for the ground state and singly excited states diminishes with increased impact energy. It is clear that the ground state contribution dominates at very high impact energies.

The cross sections for formation of doubly excited states are one order of magnitude smaller than the cross sections for singly excited states in the investigated energy range. Furthermore, for the $\text{Li}^{3+}-\text{He}$ and $\text{B}^{5+}-\text{He}$ collisions, the contributions from the states $(1s3s)\,^1S$ and $(1s3p)\,^1P$ become of the order of or even significantly larger than the ones from the states $(1s2s)\,^1S$ and $(1s2p)\,^1P$ in the investigated energy region [93]. The results from the CDW-IPM [92] are also shown in Figs. 14.13 and 14.14. These results correspond to double electron capture into all final states of the $(Z_{\text{P}}; e_1, e_2)_f$ system, and they are seen to overestimate the experimental data.

The main goal of this subsection is to assess the contribution of excited states from the $(Z_{\text{P}}; e_1, e_2)_f$ system in the exit channel of process (11.1). It is found that these latter contributions can be important, provided that the studied transitions are non-resonant when the target is in the ground state, as usual. Moreover, the inclusion of excited states into the computation can noticeably improve agreement between the CDW-4B method and experimental data, as in the case of the $\text{Li}^{3+}-\text{He}$ and $\text{B}^{5+}-\text{He}$ collisions [92] (although in the former case the results from the CDW-4B method still lie considerably

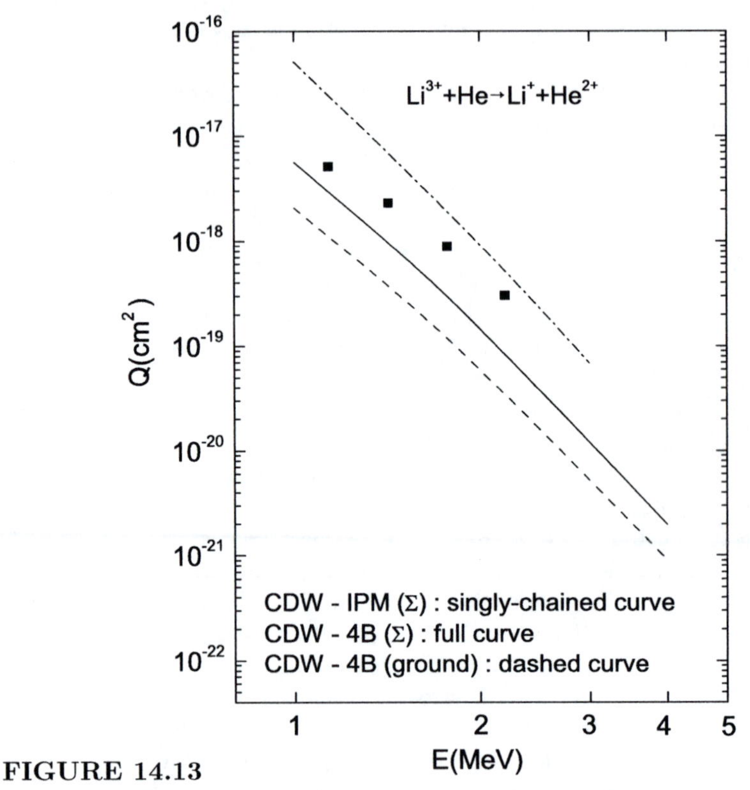

FIGURE 14.13
Total cross sections Q as a function of the incident energy E for process Li^{3+} + $\mathrm{He}(1s^2) \to \mathrm{Li}^+(\Sigma) + \mathrm{He}^{2+}$. Theory: singly-chained curve (CDW-IPM method [92] including the sum of the ground and all excited final states), dashed curve (CDW-4B method [93] including the final ground state alone), and full curve (CDW-4B method [93] including the sum of the ground and all excited final states). Experiment: ■ [309].

below the measured data, as seen in Fig. 14.13). However, this is definitely not the case for the $\mathrm{He}^{2+} - \mathrm{He}$ collision [92], since the ground-to-ground state transition in process (14.68) is dominant due to resonance. Note that for formation of H^- in the $\mathrm{H}^+ - \mathrm{He}$ double charge exchange [84, 85], there are no excited states in the exit channel. Hence, it can be concluded that the CDW-4B1 method provides relatively reliable predictions for double electron capture at intermediate and high impact energies for the $\mathrm{H}^+ - \mathrm{He}$, $\mathrm{Li}^{3+} - \mathrm{He}$ and the $\mathrm{B}^{5+} - \mathrm{He}$ collisions, but not for the the $\mathrm{He}^{2+} - \mathrm{He}$ collision, for which an approximate version of the CDW-4B2 method yields good agreement with experiments (see Fig. 14.10).

It should be noted that the CDW-4B method can also be used with multi-parameter highly correlated wave functions, such as those from Refs. [74]–[77]

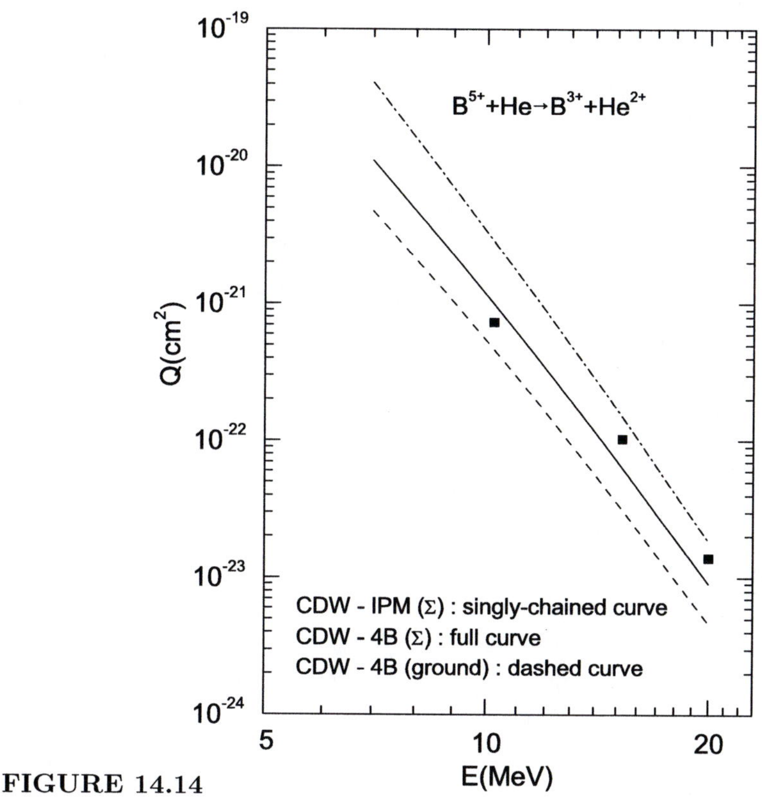

FIGURE 14.14

Total cross sections Q as a function of the incident energy E. for process $B^{5+} + He(1s^2) \to B^{3+}(\Sigma) + He^{2+}$. Theory: singly-chained curve (CDW-IPM method [92] including the sum of the ground and all excited final states), dashed curve (CDW-4B method [93] including the final ground state alone), and full curve (CDW-4B method [93] including the sum of the ground and all excited final states). Experiment: ■ [318].

that include a number of the CI terms ranging from 12 to 108. These latter orbitals are capable of including most of the radial and angular correlations, despite the fact that such functions do not explicitly contain the inter-electronic coordinate r_{12}. The wave functions from Refs. [76] and [77] have been extensively used by Belkić [7, 49, 132, 133] for single electron detachment from $H^-(1s^2)$ by impact of H^+ studied by using the MCB-4B method.

In order to illustrate the validity of the presented distorted wave approximations, angular distributions $dQ/d\Omega$ should also be reviewed. To this end, we analyze the differential cross sections for double electron capture in collisions between the He^{2+} ions and He atoms for reaction (14.68) at $E = 1.5$ MeV. Recall that for this latter symmetric and resonant process, there is no post-prior discrepancy. The results for $dQ/d\Omega$ from the CB1-4B method with and without the initial state perturbation O_i given by (13.40) are displayed on Fig. 14.15. These two sets of differential cross sections from the CB1-4B method computed using the initial and final helium wave functions of Hylleraas [58] are very close to each other. As expected, this is in accordance with

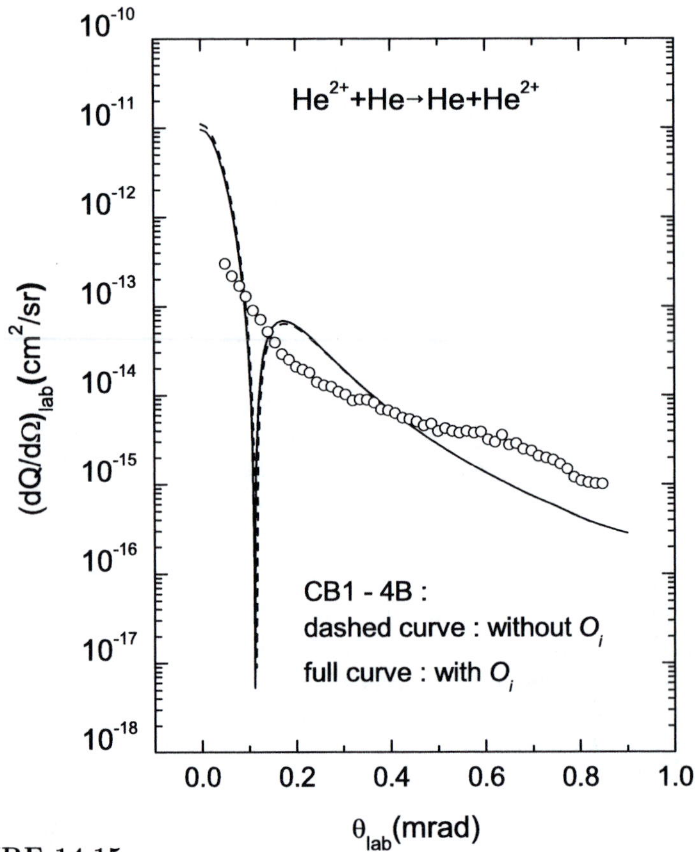

FIGURE 14.15
Differential cross sections $dQ/d\Omega$ as a function of the scattering angle θ at the incident energy E=1.5 MeV for process (14.68). Theory: dashed curve (CB1-4B method without perturbation O_i [87, 88, 90]) and full curve (CB1-4B method with perturbation O_i [88, 90]). Experiment: $\circ$ [301].

the similar situation already encountered in Fig. 14.5 regarding the total cross sections for process (14.68) treated in the CB1-4B method. Thus the corrective perturbation O_i can be ignored for both $dQ/d\Omega$ and Q. It is seen in Fig. 14.15, that the CB1-4B method exhibits an unphysical and experimentally unobserved dip at $\theta_{\rm lab} \simeq 0.112\,{\rm mrad}$. This extremely sharp dip is due to a strong cancellation of the opposite contributions coming from the repulsive $(2Z_{\rm T}/R)$ and the attractive $(-Z_{\rm T}/x_1 - Z_{\rm T}/x_2)$ potentials in (14.66). In a narrow cone near the forward direction $\theta_{\rm lab} \simeq 0\,\text{mrad}$, the differential cross sections in the CB1-4B method markedly overestimate the experimental findings. On the other hand, as seen in Fig. 14.15 the CB1-4B method underestimates the experimental data at larger scattering angles.

The differential cross sections obtained using the BDW-4B method are shown in Fig. 14.16, where a comparison is made with the BCIS-4B and CB1-

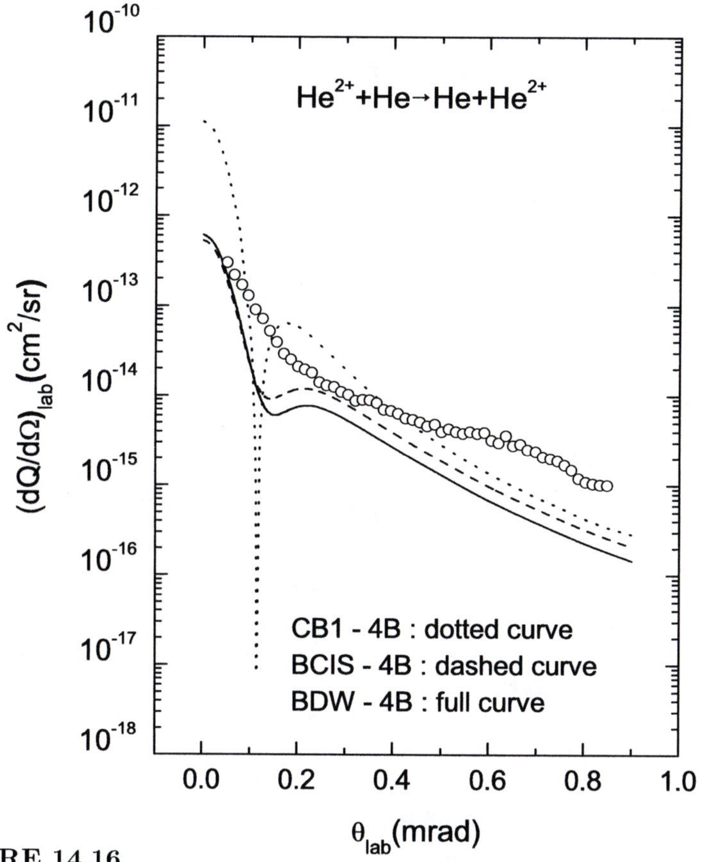

FIGURE 14.16

Differential cross sections $dQ/d\Omega$ as a function of the scattering angle θ at the incident energy E=1.5 MeV for process (14.68). Theory: dotted curve (CB1-4B method [87, 88]), dashed curve (BCIS-4B method [86]) and full curve (BDW-4B method [89, 90]). Experiment: $\circ$ [301].

4B methods, as well as with the experimental data of Schuch *et al.* [301]. The

CB1-4B, BDW-4B and BCIS-4B methods exhibit the proper asymptotic behaviors at large inter-aggregate separations in both the entrance and the exit channels. However, unlike the CB1-4B method, the BDW-4B and BCIS-4B methods take full account of the Coulomb continuum intermediate states of both electrons in one channel. Hence, by comparing these two theories with the CB1-4B method, one could learn about the relative importance of the intermediate electronic ionization continua. As seen in Fig. 14.16, the BCIS-4B method provides a substantial improvement over the CB1-4B method. First, in the BCIS-4B method, the dip in the angular distribution disappears, and near the dip region the angular distribution exhibits only a minimum at $\theta_{\rm lab} \simeq 0.121\,{\rm mrad}$, followed by a neighboring broader maximum (the Thomas peak), despite the fact that the same perturbation potential is used as in the CB1-4B method. The behavior of the angular distribution obtained in the BDW-4B method is altogether quite similar to that in the BCIS-4B method.

It should be recalled that the BDW-4B and BCIS-4B methods differ only in the perturbation potentials, such that the former contains the two gradient operators, whereas the latter uses the scalar (Coulomb) potentials. As can be seen from Fig. 14.16, the overall agreement of the BDW-4B and BCIS-4B methods with the experimental data can be considered fairly good. Nevertheless, at larger scattering angles, despite the proper inclusion of the Rutherford scattering, both the BDW-4B and BCIS-4B methods yield differential cross sections that are considerably lower than the corresponding experimental data [301]. Note that the measured findings relate to double electron capture into all the states of He, whereas the considered theoretical methods include only the ground-to-ground state transition ($1s^2 \to 1s^2$). The main purpose of Fig. 14.16 is to demonstrate the influence of electronic intermediate ionization continua to the differential cross sections by direct comparisons among the results of the analyzed four-body methods. None of the theoretical results displayed in Fig. 14.16 were folded with the experimental resolution function. Further, using (14.71), Belkić *et al.* [90] also computed differential cross sections by means of the CDW-IPM for reaction (14.68). Their results in the CDW-IPM are in good agreement with the experimental data at small and intermediate scattering angles (not shown in Fig. 14.16 to avoid clutter). At larger values of θ, the CDW-IPM underestimates the measured data. Unlike the CDW-4B method and the experiment, the CDW-IPM show some undulations in angular distribution at larger scattering angles [90].

At sufficiently high impact energies, it should be possible to predict three maximae in the differential double capture cross sections. These maximae result from different higher-order contributions as predicted by Belkić *et al.* [90]. Applying purely classical arguments, one expects to find the customary Thomas double scattering peak at the angle $\theta^{(1)}_{\rm lab} = (1/M_{\rm P})\sin 60^\circ = (1/M_{\rm P})\sqrt{3}/2 \approx 0.118$ mrad$= 0.0068^\circ$. This peak corresponds to two consecutive events: (i) one electron is captured through the direct first-order mechanism, and (ii) the other electron is captured through the Thomas double scattering. The next similar structure should occur at the angle $\theta_{\rm lab}^{(2)} = 2\theta_{\rm lab}^{(1)} = (2/M_{\rm P})\sin 60^\circ = (1/M_{\rm P})\sqrt{3} \approx 0.236$ mrad$= 0.0136^\circ$. In this case, when both electrons are treated classically, they are supposed to be in the same place at the same time to exhibit the cumulative Thomas double scattering. Each electron first scatters elastically on the projectile through 60° towards its parent nucleus. Subsequent scattering of each electron on the target nucleus is also elastic through the next 60°. The two electrons are then ejected from the target with the velocity of the projectile in the incident beam direction. Then the attractive potential between $Z_{\rm P}$ and $2e$ suffices to bind these three particles together into the $(Z_{\rm P}; 2e)_f$ system. These two Thomas peaks have also been analyzed in Ref. [104] within the IPM version of the CDW-EIS method (as acronymed by CDW-EIS-IPM). Their theoretical results for differential cross sections at 400 MeV clearly show the appearance of the structures at $\theta_{\rm lab}^{(1)}$ and $\theta_{\rm lab}^{(2)}$, associated with the mentioned intermediate double scattering processes.

The third peak can be expected at the angle $\theta_{\rm lab}^{(3)} = (1/M_{\rm P})\sqrt{2}\sin 45^\circ = (1/M_{\rm P}) \approx 0.136$ mrad$= 0.0078^\circ$, which is situated between $\theta_{\rm lab}^{(1)}$ and $\theta_{\rm lab}^{(2)}$. This time, one electron (say, e_1) is first scattered on the projectile through 45° towards the other electron e_2, thus acquiring velocity $v_1 = v\sqrt{2}$. Then, e_1 collides with e_2 elastically and finds itself deflected through another 45° in the incident beam direction with the velocity $v_1' = v$. The consequence of such an event on e_2 is manifested in the recoil of this second electron with speed $v_2 = v$ through 90° perpendicular to the incident direction. In the final step, e_2 scatters elastically on the target nucleus through another 90° with $v_2' = v$ in the projectile direction. Then both electrons travel in the incident beam direction and are, therefore, captured by the projectile. This event producing the peak at $\theta_{\rm lab}^{(3)}$ represents a genuine third-order effect. The peak at $\theta_{\rm lab}^{(3)}$ is a pure four-body effect due to dynamic correlations. Since these latter correlations are absent from treatments involving independent particles, the peak at $\theta_{\rm lab}^{(3)}$ has not been obtained by Martínez *et al.* [104] in the CDW-EIS-IPM. In order to adequately describe these higher-order phenomena, it would probably be the most appropriate to use the CB2-4B and CB3-4B methods. In the case of the CB3-4B method, one would encounter multi-dimensional numerical quadratures that could be optimally computed by the Monte Carlo algorithm VEGAS [319]–[321], as has been shown by Belkić [7, 49, 177, 178].

New experimental data are required at higher impact energies to provide a check of these theoretically predicted peaks in the angular distributions. In addition to the experimental results at 1.5 MeV that represent the first measurement of differential cross sections for double electron capture in the $\mathrm{He}^{2+}-\mathrm{He}$ collisions, there are also state-selective differential cross sections for the same process at energies 0.25–0.75 MeV [322, 323] obtained by using a powerful and versatile atomic microscope type technique known as cold target recoil ion momentum spectroscopy (COLTRIMS). The same COLTRIMS technique has also been used more recently to measure differential cross sections at impact energies ranging from 0.75 to 1.5 MeV/amu [324]. The COLTRIMS technique offers a unique, albeit indirect, but nevertheless extremely precise way to determine the final state of the projectile, including its scattering angles. Instead of energy losses and scattering angles of the projectile itself, COLTRIMS simultaneously determines the longitudinal and transverse momenta of the recoil ion (He^{2+} in the discussed case). Since there are only two particles in the final state, the momentum change of the projectile must be compensated exactly by the momentum change of the recoil ion, as per the total momentum conservation law. Thus analyzing the longitudinal momentum (in the beam direction) of the recoil ion is equivalent to the customary translational spectroscopy of the projectile. Moreover, the determination of the recoil ion transverse momentum is equivalent to measuring the scattering angle of the projectile. Although the achieved scattering angle resolution is better than $\pm 10^{-2}$ mrad [322, 323], no structure from the Thomas type mechanisms has been found in these experiments. This indicates that higher impact energies than those considered by Dörner *et al.* [322, 323] seem to be necessary to detect these Thomas peaks unambiguously in the measurements.

15

Simultaneous transfer and ionization

15.1 The CDW-4B method

In this chapter we consider a process called transfer ionization, as already abbreviated by TI, where according to (11.2) simultaneous electron capture and target ionization take place

$$Z_{\rm P} + (Z_{\rm T}; e_1, e_2)_i \longrightarrow (Z_{\rm P}, e_1)_f + Z_{\rm T} + e_2(\boldsymbol{\kappa}) \tag{15.1}$$

where $\boldsymbol{\kappa}$ is the momentum vector of the free electron in the target frame. Here, the analysis of the entrance channel is the same as for the corresponding double electron capture (11.1). Therefore, we only need to focus on the exit channel for reaction (15.1). To this end we first determine the distorted wave ξ_f^- which satisfies the following equation [127]

$$(E - H + V_x - i\varepsilon)|\xi_f^-\rangle = -(i\varepsilon - V_x)|\chi_f^-\rangle \tag{15.2}$$

which is obtained from (13.34). Choosing the intermediate channel potential V_x in such a way that the constraint

$$V_x|\chi_f^-\rangle = 0 \tag{15.3}$$

is automatically satisfied, we have in the limit $\varepsilon \longrightarrow 0^+$

$$(E - H + V_x)|\xi_f^-\rangle = 0. \tag{15.4}$$

Writing ξ_f^- in a factored form similar to the distorted wave χ_i^+

$$\xi_f^- = \varphi_f \mathcal{G}_f^- \tag{15.5}$$

we arrive at

$$\mathcal{G}_f^-(E_f - H_0 - V_{\rm P})\varphi_f + \varphi_f(E - E_f - H_0 - V_f)\mathcal{G}_f^- \\ + \frac{1}{a_1}\boldsymbol{\nabla}_{s_1}\varphi_f \cdot \boldsymbol{\nabla}_{x_1}\mathcal{G}_f^- + V_x\xi_f^- = 0 \tag{15.6}$$

where $a_1 = M_{\rm P}/(M_{\rm P}+1)$. It is important to realize that (15.6) can be solved without any further approximations [127] if the model potential V_x is

$$V_x = Z_{\rm P}\left(\frac{1}{R} - \frac{1}{s_2}\right) - \left(\frac{1}{x_1} - \frac{1}{r_{12}}\right) - \frac{1}{a_1}\boldsymbol{\nabla}_{s_1}\varphi_f \cdot \boldsymbol{\nabla}_{x_1} \circ \frac{1}{\varphi_f}. \tag{15.7}$$

Thus, using the mass limit $M_{\mathrm{P,T}} \gg 1$, it follows that (15.6) is reduced to

$$\left[E - E_f - H_0 + \frac{Z_{\mathrm{T}} - 1}{x_1} + \frac{Z_{\mathrm{T}}}{x_2} - \frac{Z_{\mathrm{P}}(Z_{\mathrm{T}} - 1)}{r_i}\right] \mathcal{G}_f^- = 0 \tag{15.8}$$

where the independent variables are separated, and this permits the exact solution. The possible nodes of φ_f would render V_x singular in (15.7). To bypass this difficulty, we introduce the symbol $\circ$ in (15.7) to indicate that V_x acts only on those functions that contain φ_f in the factored form, as in (15.5). In other words, the symbol $\circ$ determines the domain of the definition of the operator V_x, which is allowed to act only on a subspace of the complete Hilbert space containing wave functions with a factored hydrogen-like bound state φ_f. This will be the case if we seek $\mathcal{G}_f^-$ in a factored form, such as

$$\mathcal{G}_f^- = C_f^- \varphi_{\boldsymbol{q}_1}^-(\boldsymbol{x}_1)\varphi_{\boldsymbol{q}_2}^-(\boldsymbol{x}_2)\varphi_{\boldsymbol{q}_i}^-(\boldsymbol{r}_i) \tag{15.9}$$

which permits the exact solution in the form of the C3 function

$$\varphi_{\boldsymbol{q}_1}^-(\boldsymbol{x}_1) = \Gamma(1 + i\nu_{\mathrm{T}}')\mathrm{e}^{\pi\nu_{\mathrm{T}}'/2 + i\boldsymbol{q}_1\cdot\boldsymbol{x}_1}\, {}_1F_1(-i\nu_{\mathrm{T}}', 1, -iq_1x_1 - i\boldsymbol{q}_1\cdot\boldsymbol{x}_1) \tag{15.10}$$

$$\varphi_{\boldsymbol{q}_2}^-(\boldsymbol{x}_2) = \Gamma(1 + i\zeta')\mathrm{e}^{\pi\zeta'/2 + i\boldsymbol{q}_2\cdot\boldsymbol{x}_2}\, {}_1F_1(-i\zeta', 1, -iq_2x_2 - i\boldsymbol{q}_2\cdot\boldsymbol{x}_2) \tag{15.11}$$

$$\varphi_{\boldsymbol{q}_i}^-(\boldsymbol{r}_i) = \Gamma(1 - i\nu'')\mathrm{e}^{-\pi\nu''/2 + i\boldsymbol{q}_i\cdot\boldsymbol{r}_i}\, {}_1F_1(i\nu'', 1, -iq_ir_i - i\boldsymbol{q}_i\cdot\boldsymbol{r}_i) \tag{15.12}$$

where C_f^- is a constant, $\nu_{\mathrm{T}}' = (Z_{\mathrm{T}} - 1)a_1/q_1$, $\zeta' = Z_{\mathrm{T}}a_2/q_2$, $\nu'' = Z_{\mathrm{P}}(Z_{\mathrm{T}} - 1)\mu_i/q_i$ and $a_2 = a_1 = M_{\mathrm{P}}/(M_{\mathrm{P}} + 1)$. The unknown vectors $\boldsymbol{q}_1$, $\boldsymbol{q}_2$ and $\boldsymbol{q}_i$ can be determined by imposing the required simultaneous constraints

$$E - E_f = \frac{q_1^2}{2a_1} + \frac{q_2^2}{2a_2} + \frac{q_i^2}{2\mu_i} \tag{15.13}$$

$$\boldsymbol{q}_1\cdot\boldsymbol{x}_1 + \boldsymbol{q}_2\cdot\boldsymbol{x}_2 + \boldsymbol{q}_i\cdot\boldsymbol{r}_i = -\boldsymbol{k}_f\cdot\boldsymbol{r}_f + \boldsymbol{\kappa}\cdot\boldsymbol{x}_2. \tag{15.14}$$

Then, (15.13) and (15.14) together with the relation $\boldsymbol{r}_f = -a\boldsymbol{r}_i - b(\boldsymbol{x}_1 + \boldsymbol{x}_2)/\mu_f$ ($a = M_{\mathrm{P}}/[M_{\mathrm{P}} + 2]$ and $b = M_{\mathrm{T}}/[M_{\mathrm{T}} + 2]$), as well as with the mass limit $M_{\mathrm{P,T}} \gg 1$ lead to

$$\boldsymbol{q}_1 = \frac{a}{\mu_f}\boldsymbol{k}_f \simeq \frac{1}{\mu_f}\boldsymbol{k}_f \simeq \boldsymbol{v} \tag{15.15}$$

$$\boldsymbol{q}_2 = \frac{a}{\mu_f}\boldsymbol{k}_f + \boldsymbol{\kappa} \simeq \frac{1}{\mu_f}\boldsymbol{k}_f + \boldsymbol{\kappa} \simeq \boldsymbol{p} \qquad \boldsymbol{q}_i = a\boldsymbol{k}_f \simeq \boldsymbol{k}_f \tag{15.16}$$

$$\boldsymbol{p} = \boldsymbol{\kappa} + \boldsymbol{v}\,. \tag{15.17}$$

In this way, the distorted wave ξ_f^- from (15.5) is obtained as

$$\xi_f^- = N^-(\zeta)N^-(\nu_\mathrm{T})\mathcal{N}^-(\nu)\phi_f\varphi_f(\boldsymbol{s}_1)\,{}_1F_1(-i\zeta,1,-ipx_2-i\boldsymbol{p}\cdot\boldsymbol{x}_2)$$
$$\times\,{}_1F_1(-i\nu_\mathrm{T},1,-ivx_1-i\boldsymbol{v}\cdot\boldsymbol{x}_1)\,{}_1F_1(i\nu,1,-ik_fr_i-i\boldsymbol{k}_f\cdot\boldsymbol{r}_i) \quad (15.18)$$

where the function ϕ_f is defined by (11.78) and $N^-(\zeta) = \Gamma(1+i\zeta)\mathrm{e}^{\pi\zeta/2}$, $N^-(\nu_\mathrm{T}) = \Gamma(1+i\nu_\mathrm{T})\mathrm{e}^{\pi\nu_\mathrm{T}/2}$, $\zeta = Z_\mathrm{T}/p$, $\nu_\mathrm{T} = (Z_\mathrm{T}-1)/v$ and $\nu = Z_\mathrm{P}Z_\mathrm{T}/v$. Using (13.43), (13.57) and (15.18), the expression for the prior form of the transition amplitude, which is defined by

$$T_{if}^- = \langle\xi_f^-|U_i|\chi_i^+\rangle \quad (15.19)$$

becomes

$$T_{if}^- = \bar{N}\iiint \mathrm{d}\boldsymbol{s}_1\mathrm{d}\boldsymbol{s}_2\mathrm{d}\boldsymbol{R}\,\mathrm{e}^{i\boldsymbol{\alpha}\cdot\boldsymbol{s}_1+i\boldsymbol{\beta}\cdot\boldsymbol{x}_1-i\boldsymbol{\kappa}\cdot\boldsymbol{x}_2}R_\nu(\boldsymbol{r}_i,\boldsymbol{r}_f)\varphi_f^*(\boldsymbol{s}_1)$$
$$\times\,{}_1F_1(i\nu_\mathrm{T},1,ivx_1+i\boldsymbol{v}\cdot\boldsymbol{x}_1)\,{}_1F_1(i\zeta,1,ipx_2+i\boldsymbol{p}\cdot\boldsymbol{x}_2)$$
$$\times\,[Z_\mathrm{P}(1/R-1/s_2)\,{}_1F_1(i\nu_\mathrm{P},1,ivs_1+i\boldsymbol{v}\cdot\boldsymbol{s}_1)\varphi_i(\boldsymbol{x}_1,\boldsymbol{x}_2)$$
$$-\,\boldsymbol{\nabla}_{x_1}\varphi_i(\boldsymbol{x}_1,\boldsymbol{x}_2)\cdot\boldsymbol{\nabla}_{s_1}\,{}_1F_1(i\nu_\mathrm{P},1,ivs_1+i\boldsymbol{v}\cdot\boldsymbol{s}_1)$$
$$-\,{}_1F_1(i\nu_\mathrm{P},1,ivs_1+i\boldsymbol{v}\cdot\boldsymbol{s}_1)O_i(\boldsymbol{x}_1,\boldsymbol{x}_2)]$$
$$\equiv T_{if;\nu}^-(\boldsymbol{\eta}) \quad (15.20)$$

where $\bar{N} = (2\pi)^{-3/2}N^+(\nu_\mathrm{P})N^{-*}(\nu_\mathrm{T})N^{-*}(\zeta)$. Here O_i is from (13.40) and

$$R_\nu(\boldsymbol{r}_i,\boldsymbol{r}_f) = \mathcal{N}^{-*}(\nu)\mathcal{N}^+(\nu)$$
$$\times\,{}_1F_1(-i\nu,1,ik_fr_i+i\boldsymbol{k}_f\cdot\boldsymbol{r}_i)\,{}_1F_1(-i\nu,1,ik_ir_f+i\boldsymbol{k}_i\cdot\boldsymbol{r}_f) \quad (15.21)$$

with

$$\boldsymbol{k}_i\cdot\boldsymbol{r}_i+\boldsymbol{k}_f\cdot\boldsymbol{r}_f = \boldsymbol{\alpha}\cdot\boldsymbol{s}_1+\boldsymbol{\beta}\cdot\boldsymbol{x}_1$$
$$\boldsymbol{\alpha} = \boldsymbol{\eta}-\left(\frac{v}{2}-\frac{E_i-E_f-E_\kappa}{v}\right)\hat{\boldsymbol{v}}$$
$$\boldsymbol{\beta} = -\boldsymbol{\eta}-\left(\frac{v}{2}+\frac{E_i-E_f-E_\kappa}{v}\right)\hat{\boldsymbol{v}} \quad (15.22)$$

where $E_\kappa = \kappa^2/2$. The difference $E_i-(E_f+E_\kappa)\equiv\tilde{Q}$ from (15.22) between the initial and final electronic energies represents the so-called inelasticity factor, or equivalently, the $\tilde{Q}-$factor[1]. This observable is of key importance for translational spectroscopy, which through measurement of the inelasticity factor determines the energy gain or loss of the scattered projectiles.

[1]The $\tilde{Q}-$factor is a quantitative measure of the degree of departure of a given process from its elastic counterpart in which no energy loss or gain occurs. This is more commonly known as the $Q-$factor, but we use the label $\tilde{Q}$ instead, in order to avoid confusion with cross sections that are presently denoted by Q, as usual.

By analogy, the post form of the transition amplitude T_{if}^{+} can be derived with the final result [127]

$$\begin{aligned}T_{if}^{+} &= \bar{N}\iiint \mathrm{d}\boldsymbol{x}_1\mathrm{d}\boldsymbol{x}_2\mathrm{d}\boldsymbol{R}\,\mathrm{e}^{i\boldsymbol{\alpha}\cdot\boldsymbol{s}_1+i\boldsymbol{\beta}\cdot\boldsymbol{x}_1-i\boldsymbol{\kappa}\cdot\boldsymbol{x}_2}R_\nu(\boldsymbol{r}_i,\boldsymbol{r}_f)\varphi_i(\boldsymbol{x}_1,\boldsymbol{x}_2)\\ &\times\,{}_1F_1(i\nu_{\mathrm{P}},1,ivs_1+i\boldsymbol{v}\cdot\boldsymbol{s}_1)\,{}_1F_1(i\zeta,1,ipx_2+i\boldsymbol{p}\cdot\boldsymbol{x}_2)\\ &\times\{[Z_{\mathrm{P}}(1/R-1/s_2)+(1/r_{12}-1/x_1)]\,{}_1F_1(i\nu_{\mathrm{T}},1,ivx_1+i\boldsymbol{v}\cdot\boldsymbol{x}_1)\varphi_f^*(\boldsymbol{s}_1)\\ &-\boldsymbol{\nabla}_{s_1}\varphi_f^*(\boldsymbol{s}_1)\cdot\boldsymbol{\nabla}_{x_1}\,{}_1F_1(i\nu_{\mathrm{T}},1,ivx_1+i\boldsymbol{v}\cdot\boldsymbol{x}_1)\}\\ &\equiv T_{if;\nu}^{+}(\boldsymbol{\eta})\,. \end{aligned}\tag{15.23}$$

Here, a useful simplification can be made, similarly to (14.11) for double electron capture, by using the following eikonal approximation

$$R_\nu(\boldsymbol{r}_i,\boldsymbol{r}_f)\simeq(\mu v\rho)^{2i\nu}\,. \tag{15.24}$$

As before, it can be shown that the phase factor $(\mu v\rho)^{2i\nu}=(\mu v\rho)^{2iZ_{\mathrm{P}}(Z_{\mathrm{T}}-1)/v}$ does not contribute to the total cross sections $Q_{if}^{\pm}$, as expected [44]. The triple differential cross sections for simultaneous transfer and ionization can be obtained from

$$\begin{aligned}Q_{if}^{\pm}(\boldsymbol{\kappa}) \equiv \frac{d^3Q_{if}^{\pm}}{d\boldsymbol{\kappa}} &= \int \mathrm{d}\boldsymbol{\eta}\left|\frac{T_{if;\nu}^{\pm}(\boldsymbol{\eta})}{2\pi v}\right|^2\\ &= \int \mathrm{d}\boldsymbol{\eta}\left|\frac{T_{if;0}^{\pm}(\boldsymbol{\eta})}{2\pi v}\right|^2\\ &\equiv \int \mathrm{d}\boldsymbol{\eta}\left|\frac{R_{if}^{\pm}(\boldsymbol{\eta})}{2\pi v}\right|^2\end{aligned}\tag{15.25}$$

where

$$R_{if}^{\pm}(\boldsymbol{\eta}) = T_{if;0}^{\pm}(\boldsymbol{\eta}) \tag{15.26}$$

so that the corresponding total cross sections are given by

$$Q_{if}^{\pm} = \int \mathrm{d}\boldsymbol{\kappa}\,Q_{if}^{\pm}(\boldsymbol{\kappa}). \tag{15.27}$$

In the above derivation of the transition amplitudes in the CDW-4B method, the ionized electron e_2 is described by the Coulomb wave

$$\varphi_{\boldsymbol{\kappa}}^{-}(\boldsymbol{x}_2)\equiv N^{-}(\zeta)\phi_{\boldsymbol{\kappa}}(\boldsymbol{x}_2)\,{}_1F_1(-i\zeta,1,-ipx_2-i\boldsymbol{p}\cdot\boldsymbol{x}_2) \tag{15.28}$$

$$\phi_{\boldsymbol{\kappa}}(\boldsymbol{x}_2) = (2\pi)^{-3/2}\mathrm{e}^{i\boldsymbol{\kappa}\cdot\boldsymbol{x}_2} \tag{15.29}$$

where $\boldsymbol{p}=\boldsymbol{\kappa}+\boldsymbol{v}$, as in (15.17). Even though the appropriate starting ansatz in the undistorted scattering state Φ_f is given by the plane wave $\phi_{\boldsymbol{\kappa}}(\boldsymbol{x}_2)$ centered on T, the present four-body analysis establishes a distortion of $\phi_{\boldsymbol{\kappa}}(\boldsymbol{x}_2)$ by function $N^-(\zeta)\,{}_1F_1(-i\zeta, 1, -ipx_2 - i\boldsymbol{p}\cdot\boldsymbol{x}_2)$. This latter quantity is a function of the composite electron momentum $\boldsymbol{\kappa}+\boldsymbol{v}\equiv\boldsymbol{p}$, and not merely of $\boldsymbol{\kappa}$, which one would expect in the corresponding standard first Born approximation. The presence of the vector $\boldsymbol{v}$ in the momentum $\boldsymbol{p}$ of the ejected electron has physical meaning, which points to the possibility of describing the well-known electron capture to continuum (ECC) with a characteristic cusp effect in the emission spectra [325]–[334]. This was observed experimentally in the TI process for the $\mathrm{He}^{2+}-\mathrm{He}$ collision [335, 336].

Various choices have been made in the literature for the wave function of the final state of the emitted electron [337]–[349]. For example, in the case of ionization of helium by ion impact, this function has been defined as the product of the $e-$P continuum state with the electron-residual-target continuum state [332, 333]. As such, the fields of P and T act simultaneously on the electron ejected from helium (the two-center problem), as in the study of Belkić [134] in the first application of the CDW-3B method to ionization of atomic hydrogen by protons. Note that the final-state wave function of the electron ejected from helium can explicitly contain the inter-electron distance [263]. Overall, multiple continua represent the main difficulty in the TI process, since a proper description of the final state for the ejected electron requires the inclusion of the simultaneous influence of the Coulomb potentials due to the projectile, target and captured electron.

The presented prior and post forms T_{if}^- and T_{if}^+ have a common perturbation, $\Delta V_{\mathrm{P}2}\equiv\Delta V_{\mathrm{P}2}(R,s_2)=Z_{\mathrm{P}}(1/R-1/s_2)$. Of course, considered outside the $T-$matrix, the potential $V_{\mathrm{P}2}=-Z_{\mathrm{P}}/s_2$ represents the direct Coulomb interaction between e_2 and Z_{P}. Its asymptotic form $V_{\mathrm{P}2}^{\infty}(R)$ at large distances s_2 is given by $-Z_{\mathrm{P}}/R$, since $s_2\longrightarrow R$ as $R\longrightarrow\infty$. Hence, the term $\Delta V_{\mathrm{P}2}$ is precisely the difference between the finite and asymptotic values of the same potential $\Delta V_{\mathrm{P}2}=V_{\mathrm{P}2}(s_2)-V_{\mathrm{P}2}^{\infty}(R)$. As such, $\Delta V_{\mathrm{P}2}$ is a short-range interaction, in accordance with the correct boundary condition for perturbation potentials [41]. However, when placed in the $T-$matrices (15.20) and (15.23), the potential $V_{\mathrm{P}2}$ plays the role of a perturbation which causes capture of the electron e_1. This could only occur through some kind of underlying correlation between e_2 and e_1. For example, a part of the energy received by the electron e_2 in its collision with Z_{P} could be sufficient to accomplish the transfer of e_1 to the projectile. The post form T_{if}^+ contains an additional term

$$\Delta V_{12}\equiv\Delta V(r_{12},x_1)=\frac{1}{r_{12}}-\frac{1}{x_1} \tag{15.30}$$

which is completely absent from the corresponding prior form T_{if}^-. Through ΔV_{12} the dielectronic interaction $1/r_{12}$ appears explicitly in (15.30). In combination with the initial and final distorted functions on both centers Z_{P} and Z_{T}, this dielectronic interaction describes the Thomas $\mathrm{P}-e_1-e_2$ scattering. Due to the perturbation ΔV_{12}, the cross sections in the post form should be more adequate than its prior counterparts. The calculation of the matrix elements for TI has been carried out by Belkić *et al.* [127]. The total cross sections in the CDW-4B method for transfer ionization are given in terms of seven-dimensional integrals over real variables. The number of integration points per axis is gradually and systematically increased until convergence to two decimal places has been reached [127].

Transfer ionization has also been investigated in the CDW-IEM [350]. From the numerical point of view the CDW-IEM also encounters seven-dimensional integrals when dealing with the total cross sections for a TI process. The main difference between the CDW-IEM and CDW-4B method is in the treatment of electron correlations. The CDW-IEM from Ref. [350] includes the static electron correlations (SEC) in the target by using the bound state wave function of Pluvinage [60]

$$\varphi_i(\boldsymbol{x}_1, \boldsymbol{x}_2) = c(k)\frac{Z_{\mathrm{T}}^3}{\pi}\,\mathrm{e}^{-Z_{\mathrm{T}}(x_1+x_2)}\,\mathrm{e}^{-ikr_{12}}\,{}_1F_1(1-i\gamma\,,\,2\,,\,2ikr_{12}) \qquad (15.31)$$

where $\gamma = 1/(2k)$ and $c(k)$ is the normalization constant, with k being a non-linear variational parameter. The corresponding lowest binding energy $E_{i,\mathrm{Pluv}} = -2.878$ for the ground state ${}^1S$ of helium is obtained for $k = 0.41$, in which case $c(k) = 0.603366$. The wave function (15.31) contains two entirely uncorrelated hydrogen-like wave functions (with the unscreened charge Z_{T}) that are multiplied with a corrective $r_{12}-$dependent term of the form $\exp(-ikr_{12})\,{}_1F_1(1-i\gamma, 2, 2ikr_{12})$. Another wave function of helium [351] might also be useful, with the continuum wave function for the $e_1 - e_2$ interaction replaced by a simpler ansatz, which seems equally adequate as that of Pluvinage [60]. In the CDW-IEM [350], the dynamic electron correlations (DEC) are completely neglected, whereas the CDW-4B method explicitly includes the DEC through the dielectronic interaction $1/r_{12}$ in the transition $T-$operator. It should be noted that the SEC can also be fully included in the CDW-4B method by using the corresponding wave function [75]–[77], but it is ignored in the present illustrations with the purpose of providing an unambiguous assessment of the DEC alone.

15.2 Comparison between theories and experiments

In order to illustrate the validity of the CDW-4B method of Belkić *et al.* [127] for the TI process, we examine the total cross sections for the following symmetric reaction

$$\mathrm{He}^{2+} + \mathrm{He}(1s^2) \longrightarrow \mathrm{He}^{+}(1s) + \mathrm{He}^{2+} + e\,. \qquad (15.32)$$

We also analyze the asymmetric transfer ionization which has subsequently been studied by Mančev [130] for the reaction

$$\mathrm{Li}^{3+} + \mathrm{He}(1s^2) \longrightarrow \mathrm{Li}^{2+}(1s) + \mathrm{He}^{2+} + e\,. \tag{15.33}$$

The results for TI process (15.32) in the energy interval from 30 to 1000 keV/amu are displayed in Figs. 15.1–15.4. The influence of the initial state corrective perturbation O_i from (13.40) within the CDW-4B method in its

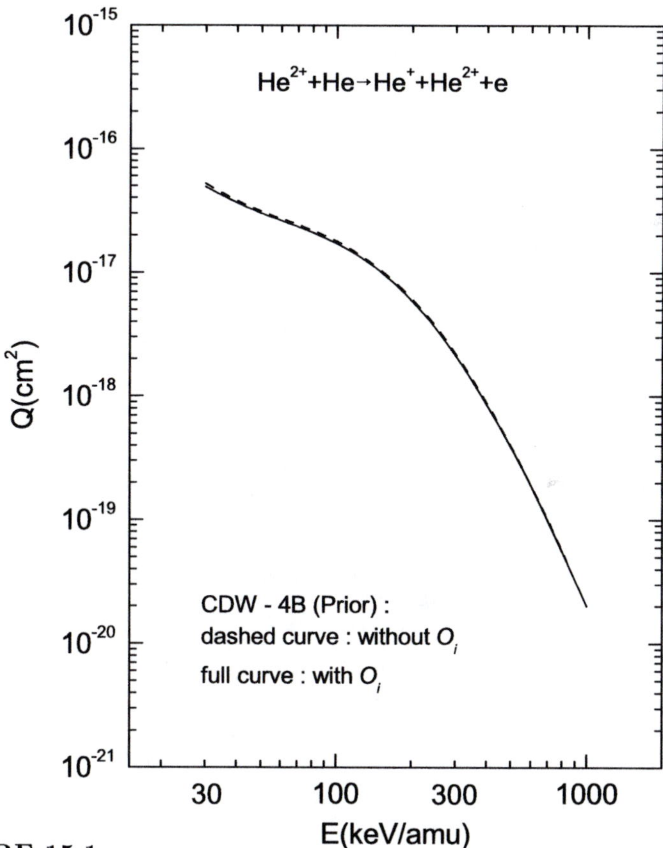

FIGURE 15.1
Total cross sections Q as a function of the incident energy E for process (15.32). Theory: dashed curve (prior CDW-4B method without perturbation O_i [127]) and full curve (prior CDW-4B method with perturbation O_i [127]).

prior version (15.20) for transfer ionization is seen in Fig. 15.1 to be negligible, similarly to double capture treated by means of the CB1-4B method (see Fig.

14.5). Therefore, the contribution from O_i can also be ignored for two-electron transitions involving transfer ionization.

The role of the perturbation $\Delta V_{P2} = Z_P(1/R - 1/s_2)$ in the prior CDW-4B method is illustrated in Fig. 15.2. The corresponding cross sections computed with and without the term ΔV_{P2} are seen in Fig. 15.2 to be in close mutual agreement. Thus, the contribution from ΔV_{P2} in the prior CDW-4B method can be ignored, especially at impact energies above 40 keV/amu.

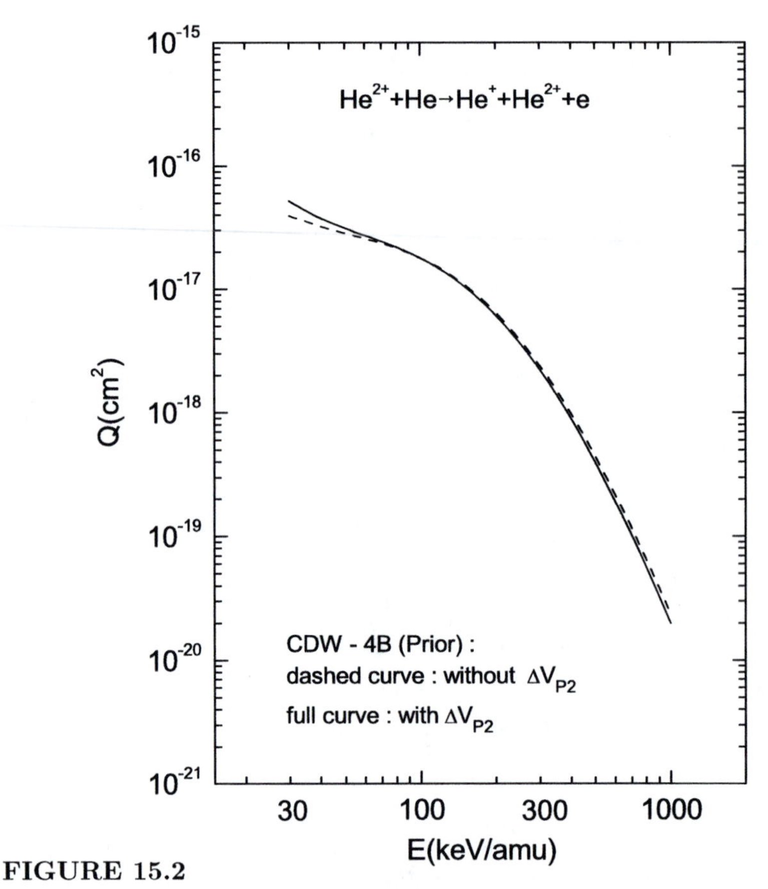

FIGURE 15.2
Total cross sections Q as a function of the incident energy E for process (15.32). Theory: dashed curve (prior CDW-4B method without perturbation ΔV_{P2} [127]) and full curve (prior CDW-4B method with perturbation ΔV_{P2} [127]).

Further, we evaluate the post-prior discrepancy which arises from the unequal perturbation potentials used in the $T-$matrices (15.20) and (15.23). In Fig. 15.3, we display the prior (Q_{if}^-) and post (Q_{if}^+) cross sections in the

CDW-4B method [127]. In this figure, the prior cross sections do not include the correction O_i, which is negligibly small by reference to Fig. 15.1. The comparison in Fig. 15.3 reveals that the post-prior discrepancy is significant throughout the energy range 50–1000 keV/amu. The post cross sections Q_{if}^+

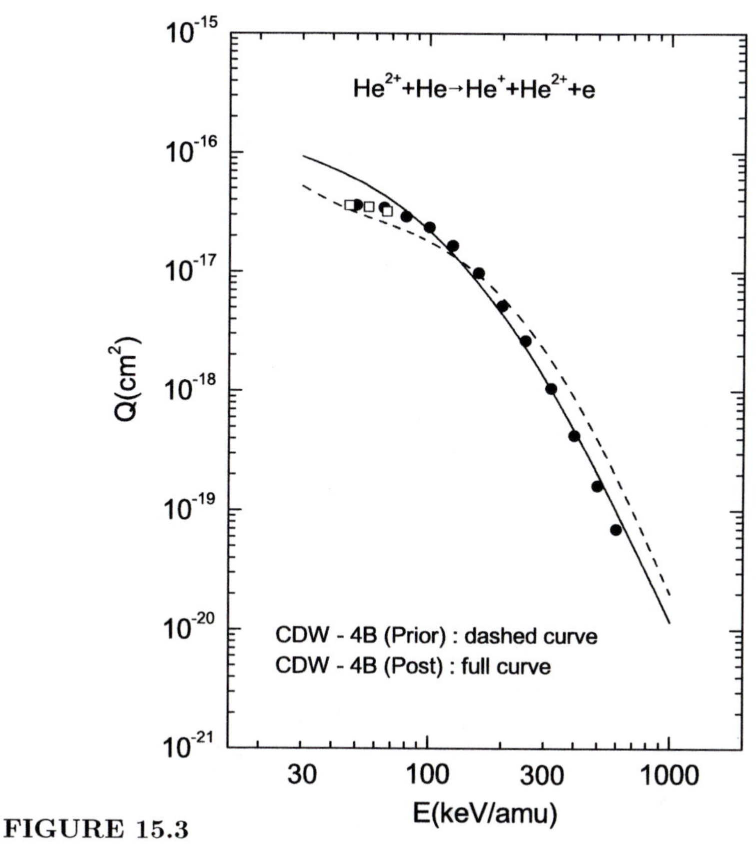

FIGURE 15.3

Total cross sections Q as a function of the incident energy E for process (15.32). Theory: dashed curve (prior CDW-4B method without the complete perturbation V_i [127]) and full curve (post CDW-4B method with the complete perturbation V_f [127]). Experiment: • [309] and □ [352].

are seen in Fig. 15.3 to be larger by nearly 50% than the corresponding prior results Q_{if}^- at lower energies, with precisely the opposite pattern at higher energies. This considerable difference is attributed to the role of the dielectronic repulsion $1/r_{12}$. We recall that the difference between Q_{if}^- and Q_{if}^+ is due solely to the potential $\Delta V_{12} = 1/r_{12} - 1/x_1$, which is present in the post and absent from the prior cross sections. In the same Fig. 15.3, the theoretical cross sections are compared with the experimental data for reaction (15.32). It can be seen from Fig. 15.3 that, in contrast to the prior variant Q_{if}^{-}, the post version Q_{if}^{+} of the CDW-4B method is in good agreement with the measurements at impact energies $E \geq 80$ keV/amu. At lower energies, the results

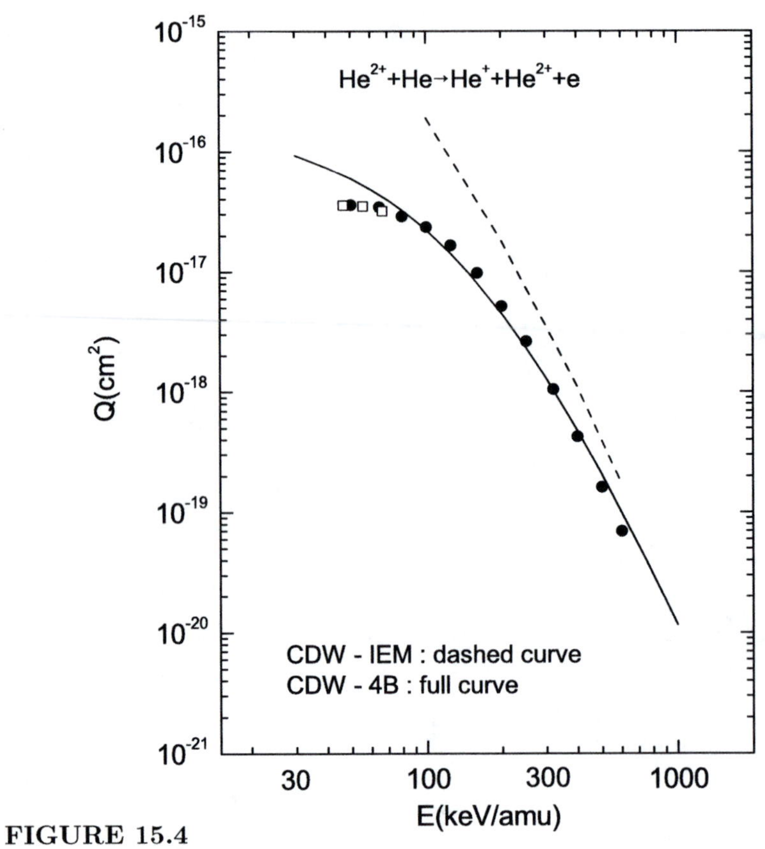

FIGURE 15.4
Total cross sections Q as a function of the incident energy E for process (15.32). Theory: dashed curve (post CDW-IEM [350]) and full curve (post CDW-4B method with the complete perturbation V_f [127]). Experiment: • [309] and □ [352].

for Q_{if}^{+} are larger than the experimental values, as expected from (14.69), since the CDW-4B method is a high-energy approximation. The superiority of the post over the prior version of this method is due to the presence of the electron-electron interaction $1/r_{12}$ in the perturbation potential U_f in the former variant.

Next, in Fig. 15.4, a comparison is made between the CDW-4B method [127] and CDW-IEM [350]. The relative role of the SEC and DEC is at once apparent from the two curves (full and dashed) associated with the CDW-IEM and the CDW-4B method, respectively. A comparison of these two theories with the experimental data in Fig. 15.4 clearly shows that the CDW-4B method represents a substantial improvement over the CDW-IEM. Therefore, it can be concluded that the DEC plays a more important role than the SEC for TI in the process (15.32) at all the impact energies from Fig. 15.4.

The post and the prior total cross sections for TI in reaction (15.33), derived with the full perturbations are plotted in Fig. 15.5, where two sets of experimental data are also displayed. The CDW-4B method used by Mančev

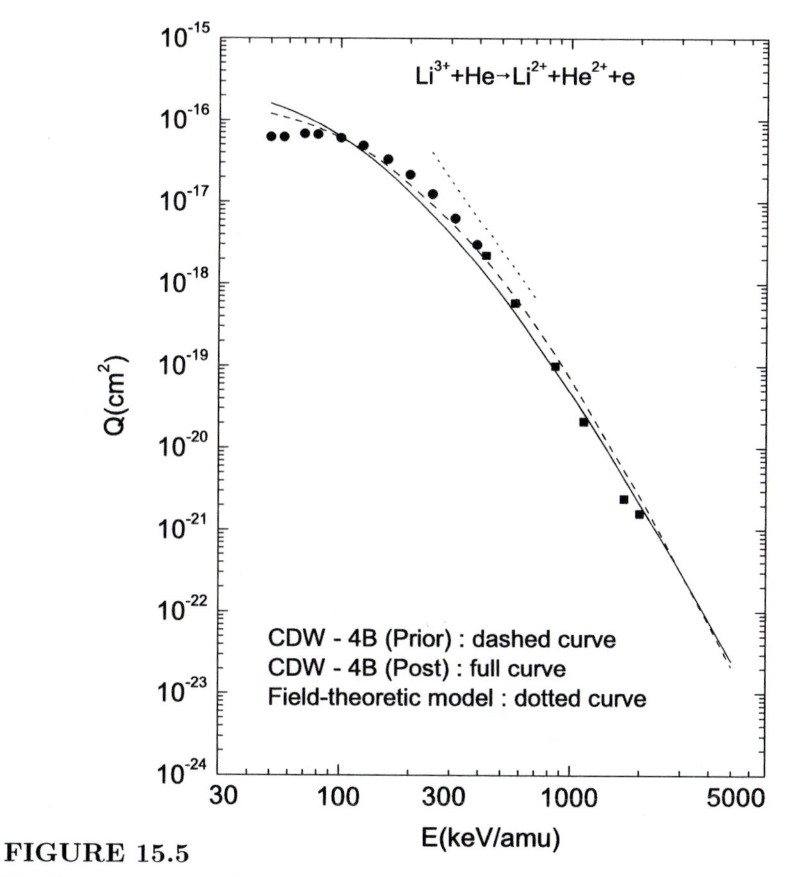

FIGURE 15.5
Total cross sections Q as a function of the incident energy E for process (15.33). Theory: dashed curve (prior CDW-4B method with the complete perturbation V_i [130] and full curve (post CDW-4B method with the complete perturbation V_f [130]) and dotted curve (a second-order theoretic field method [353]). Experiment: • [309] and ■ [354].

[130] is found to be in good agreement with the experimental data. The post cross sections lie below the prior ones at impact energies between 100 and 3000 keV/amu (with the reverse behavior outside this latter interval). These computations have been performed for electron transfer to the ground state alone. The agreement between the CDW-4B method and measurements especially at lower impact energies needs to be reassessed by including a contribution from all the excited states.

The theoretical results of Bhattacharyya *et al.* [353] are also depicted in Fig. 15.5. Their cross sections have been obtained using a relativistically covariant field-theoretic approach via a second-order term from Feynman's diagram series. As can be seen from Fig. 15.5, the results from Ref. [353] markedly overestimate the experimental data.

In the case of the TI process (15.33), it has been shown by Mančev [130], that the prior total cross sections computed with and without the term $\Delta V_{\rm P2}$ differ from each other by a considerable amount which can reach 67% at energies 50–5000 keV/amu. This is in sharp contrast to a negligible role of $\Delta V_{\rm P2}$ for the TI process (15.32), as seen in Fig. 15.2. Physically, the potential $-Z_{\rm P}/s_2$ from $\Delta V_{\rm P2}$ can cause capture of electron e_1 through the $e_1 - e_2$ correlation with its origin in the target. This points to a potential role of the SEC for process (15.33). The fact that the perturbation $\Delta V_{\rm P2}$ is less important in process (15.32) than in its counterpart (15.33) is partially due to the difference in the values of the projectile nuclear charge $Z_{\rm P} = 2$ and $Z_{\rm P} = 3$, respectively.

As shown above, the CDW-4B method is in good accordance with the experimental data on TI for the $\text{He}^{2+}-\text{He}$ and $\text{Li}^{3+}-\text{He}$ collisions. Regarding total cross sections for the TI process

$$\text{H}^+ + \text{He}(1s^2) \longrightarrow \text{H}(1s) + \text{He}^{2+} + e \tag{15.34}$$

the CDW-4B method overestimates some of the measured data from Refs. [180]–[183] and [355] at high energies, whereas at intermediate energies satisfactory agreement is found with the experimental findings of Shah and Gilbody [309] (not shown). Over the last five years, there has been increasing interest in studying process (15.34), as reported in Refs. [182, 183] and [355]–[371]. One of the motivations for these investigations has been the fact that multiply differential cross sections for fragmentation processes can provide valuable information on the nature of electronic correlations in atomic systems. When protons are used as projectiles, an atomic hydrogen is formed in the exit channel. In such a case, due to the absence of the post-collisional Coulomb interactions with the scattered projectile, target correlation effects are expected to be manifested in a more straightforward manner.

At high impact energies, two qualitatively different TI mechanisms contribute to the total cross sections by a comparable amount [183]. These are the so-called kinematic TI (KTI) and the Thomas TI (TTI) processes. As usual, the Thomas process can be understood as two consecutive binary collisions:

first by the projectile with one of the target electrons, and second between this latter electron and either the target nucleus (the Thomas P$-e-$T scattering) or another target electron (the Thomas P$-e_1-e_2$ scattering). The P$-e_1-e_2$ mechanism was first identified in the experiments of Pálinkás *et al.* [369, 370] by a coincident detection of the atomic hydrogen, and analysis of the energy as well as angular distributions of emitted electrons for scattering of 1 MeV protons on helium. Pálinkás *et al.* [369, 370] found a peak in the double differential cross section $\mathrm{d}^2Q_{if}/(\mathrm{d}E_e\mathrm{d}\theta_e)$ at $\theta_e = 90°$ and $E_e = 600\,\mathrm{eV}$. In such a P $- e_1 - e_2$ mechanism, the projectile-electron scattering is accompanied by the electron-electron encounter, thus leaving one of the electrons with a velocity which is nearly the same as that of the projectile. Since the projectile is much more massive than the electron, it continues to move at almost the same velocity as before the collision. The other electron is emitted into the continuum with a momentum magnitude equal to that of the projectile, but at an angle of 90° with respect to the direction of the projectile [371, 372].

Electron capture can proceed via different reaction channels called kinematic capture and P $- e_2 -$ T capture. Kinematic capture (also known as the Brinkman-Kramers mechanism) is mediated by the projectile-electron interaction in the field of the target ion. In order for the TI to take place, capture (either kinematic or P$-e_2-$T) must be accompanied by an additional process such as a shake-off or an independent ionization of the second electron by the projectile. In the shake-off process, one electron is removed from the atom by the projectile, whereas the other electron has a finite probability of ending up in the continuum. This mechanism becomes important for small values of Z_P/v. Hence, the sudden removal of one electron can lead to emission of the second electron into the continuum. For a transition of the second electron to an excited state, a shake-up process can occur. Calculations of the probability of these shake-type processes can be carried out by using the so-called sudden approximation [373].

In order to distinguish these different channels in an experiment, the projectile momentum transfer (both the transverse and longitudinal components) must be measured with an extremely high resolution to within 0.3 a.u. which is a very small fraction of about 10^{-5} of the actual projectile momentum. This is not feasible with the standard techniques such as translational spectroscopy or the like, but it is achievable with COLTRIMS. Any momentum change of the projectile must be compensated by summing the momenta of the recoiling ion and the emitted electrons. Measuring the impact projectile position on the detector (with a typical resolution less than 0.1 mm), as well as the time of flight of the fragments (from the moment of fragmentation up to the instant when the detector is hit), it becomes possible to determine the particle momenta after fragmentation.

Moreover, COLTRIMS can distinguish between the mentioned two mechanisms for capture, KTI and TTI, by measuring the longitudinal recoil ion momentum for each of the TI events [180, 182, 183, 355]. Namely, for the KTI mechanism, the expected longitudinal recoil ion momentum, which is determined by the energy and momentum conservation laws, is given by $p_{\|} = -v/2 - \tilde{Q}/v$. In the case of the TTI mechanism, the helium nucleus takes no part in the collision, so that the expected longitudinal recoil ion momentum for this mechanism is zero. In principle, the Thomas $\mathrm{P} - e_1 - e_2$ mechanism could occur with two initially unbound electrons. In other words, this process can proceed even when the two free electrons are at rest prior to scattering. With this latter assumption, the recoil momentum becomes zero, since the target nucleus does not participate in the process and, moreover, the same nucleus is at rest before the collision.

Recently, Schmidt *et al.* [183] have experimentally investigated the TI process in the $\mathrm{H}^+ - \mathrm{He}$ collisions in the earlier inaccessible high-energy range 1.4–5.8 MeV. They found that for the highest energy (5.8 MeV), the TTI mechanism yields the dominant contribution to the total TI cross section. On the other hand, at the lower end of the energy range considered in Ref. [183], it was found that the KTI mechanism represents the dominant pathway of the total TI.

Whenever the KTI and TTI processes can be separated [180, 182, 183, 355], it is possible to extract the quantity $P_{\mathrm{S}} = Q_{\mathrm{KTI}}/(Q_{\mathrm{SC}} + Q_{\mathrm{KTI}}) < 1$, where Q_{SC} denotes the cross section for single capture (SC). By definition, P_{S} can be interpreted as the probability for the shake-off of the second electron after the first electron has been captured by the projectile via the KTI. The Thomas $\mathrm{P} - e - e$ part is excluded here since the continuum electron is obviously not shaken off in that process. Schmidt *et al.* [183] have investigated the shake-off probability P_{S} as a function of the projectile velocity, and found a rather weak decrease over the energy range of their measurements (1.4–5.8 MeV). At higher impact energies these experimental data approach the expected shake-off limit of 1.63%, which was also found in the ratio of double- and single-ionization of He by photons or protons [373].

In the experiment of Mergel *et al.* [180], a velocity dependence $Q \sim v^{-7.4\pm1.0}$ of the total cross section for the Thomas $\mathrm{P} - e_1 - e_2$ double scattering was found in a rather limited interval of 0.3–1.4 MeV. Schmidt *et al.* [182] carried out a measurement on the TI process in the $\mathrm{H}^+ - \mathrm{He}$ collisions at higher energies 2.5–4.5 MeV and subsequently for an extended energy interval 1.4–5.8 MeV [183]. In the latter experiment, the total cross section for the TI process was estimated to exhibit the asymptotic behavior $Q \sim v^{-10.78\pm0.27}$. This velocity dependence is in agreement with the corresponding prediction of the classical model of Thomas [181] for the expected asymptotic v^{-11} behavior, which has also been obtained by the peaking approximation of the BK2 model [371]. However, these two theoretical models are valid at $v \gg v_e$ and, as such, they are not suitable for a quantitative comparison with the experimental data from Refs. [182, 183, 355]. On the other hand, detailed quantum mechanical computations of the cross sections for the TI by using the RIA-4B of Belkić [177, 178] show excellent agreement with the experimental data of Mergel *et al.* [180] as well as Schmidt *et al.* [182, 183, 355]. The RIA-4B also predicts the v^{-11} behavior at sufficiently high energies [7].

The differential cross sections for TI in process (15.34) have been measured by means of COLTRIMS [361]–[363] at impact energies ranging from 0.15 to 1.4 MeV. The azimuthal scattering angle θ of the formed hydrogen atom was as small as 0.1–0.5 mrad. These scattering angles are about 100 times smaller than those from earlier experiments of this type. The experiments from Refs. [361]–[363] found that: (i) the ejected electron is predominantly emitted into the backward direction and always opposite to the scattered hydrogen atom, (ii) the captured electron, the recoil He^{2+} ion and the ejected electron always share comparable momenta, and (iii) the direction of the maximum emission is insensitive to the impact energy, but contains some dependence upon the momentum transfer. Furthermore, from the final-state momentum pattern of H, He^{2+} and the electron it is possible to reveal a part of the initial momentum wave function, which is dominated by the non-s^2 contributions [362, 363]. On the basis of these experimental findings, it has been contemplated by Schmidt-Böcking *et al.* [362, 363] that the reaction (15.34) in the range of extremely small scattering angles θ can be used for obtaining information on the structure of the wave function in the momentum representation. However, this latter idea has become a subject of some controversy [364, 374]. In particular, the theoretical analysis by Popov *et al.* [364] has questioned the consistency of the hypothesis of Schmidt-Böcking *et al.* [362, 363] and, therefore, this theme needs further clarifications.

The most recent development and refinement within COLTRIMS provides a coincident multi-fragment imaging technique for the eV and sub-eV fragment detection [357]. In the experiments from Refs. [358] and [360], the TI process (15.34) has been examined at a single proton energy (630 keV), by detecting the ionized electron perpendicularly to the direction of incidence. These authors proposed a simple theoretical model which could explain qualitatively certain observed effects in the measured triple differential cross sections for TI in the $H^+ - He$ collision. They pointed out that the TI process offers a unique opportunity to study radial and angular correlations in the helium target. Specifically, angular electron correlations in the ground state of helium yield a broad peak in the electron emission spectra in the backward direction relative to the incoming projectile beam. As discussed, the TI event proceeds through different channels, such as the KTI and TTI. It is known that these mechanisms are sensitive to the collision energy. However, the experimentally observed features [360] were found to be virtually insensitive to the collision energy. This prompted Godunov *et al.* [360] to assume that the target correlation plays the leading role. They further argued that the fully differential cross sections for TI in the $H^+ - He$ collisions can be used as a sensitive probe for the target correlation whenever post-collisional effects are either non-existent or negligibly small. Such conjectures necessitate further theoretical studies to clarify more fully the role of these correlation effects in differential cross sections for the TI process.

In addition to simultaneous transfer and ionization in collisions by protons and other heavier nuclei with a helium-like target, the TI process can also occur when positrons are used as the projectiles. In such a case, formation of positronium takes place with one of the two electrons, while the other electron from the target is ionized [375]. It should be pointed out that in ion-atom collisions various processes of interest can occur e.g. simultaneous double electron capture and ionization in the He^{2+} – Ar collisions [376]. A proper description of such a process necessitates five-body models that are not the main subject of the present book.

16

Single electron detachment

The correct boundary conditions are also critical for a proper description of ionizing processes of type (11.4) in the case of the wave function for the final state of the emitted electron. In fact, this is best appreciated already in the related simpler pure three-body problem with ionization in e.g. the $Z_{\rm P}-(Z_{\rm T},e)$ collisional process

$$Z_{\rm P}+(Z_{\rm T},e)_i \longrightarrow Z_{\rm P}+Z_{\rm T}+e. \tag{16.1}$$

It is well-known that the correct Coulomb boundary conditions are much more difficult to fulfill for ionization (16.1) than for the corresponding electron capture (12.26). This difficulty is due to the presence of three charged particles in the exit channel of process (16.1). The main task is to use the total Schrödinger equation to consistently derive the final state scattering wave function with the exact asymptotic behavior when all the three free charged particles are far apart from each other. In the part of the asymptotic region where the three free charged particles are simultaneously present at infinitely large mutual separations, the exact form of the total wave function is known, and it is given by the product of three Coulomb-distorted plane waves (dressed or clothed plane waves). Each plane wave is modified by a multiplicative logarithmic Coulomb phase factor. Therefore, there will be altogether three such multiplicative distortions of the product of three plane waves in the exact asymptotic final wave function which must be matched by an adequate description of the exit channel in process (16.1). This exact Coulomb boundary condition for the final state in process (16.1) was fulfilled by Belkić [134] who was the first to derive the C3 function (via the product of three full separable Coulomb wave functions) from the complete Schrödinger equation in the distorted wave formalism[1]. The three Coulomb waves from this C3 wave function have the Sommerfeld parameters $Z_{\rm P}Z_{\rm T}/v$, $Z_{\rm P}/|\boldsymbol{\kappa}-\boldsymbol{v}|$ and $Z_{\rm T}/\kappa$ that are due to the pairwise separate interactions $Z_{\rm P}-Z_{\rm T}$, $Z_{\rm P}-e$ and $Z_{\rm T}-e$, respectively. Here, $\boldsymbol{\kappa}$ and $\boldsymbol{p}=\boldsymbol{\kappa}-\boldsymbol{v}$ are the

[1]The asymptotic form of the C3 wave function has been said to be given by P.J. Redmond in an unpublished work (cited by Rosenberg [337]), but without any derivation. Rosenberg [337] simply wrote down an *ad hoc* product of n Coulomb logarithmic phase factors for n free charged particles, admitting that this was not a result of any derivation, but rather a direct extension of the well-known two-particle case.

electron momentum vectors in the target and projectile frame, respectively, whereas $\boldsymbol{v}$ is the incident velocity, as usual.

In the same study, Belkić [134] also imposed the correct boundary condition to the entrance channel of process (16.1). This was again accomplished through the usage of the full distorted wave Schrödinger equation which gave Φ_i^+ as the product of the target bound state and the so-called C2 function. Here, the C2 function is the product of two full Coulomb waves for electron e and nucleus $Z_{\rm T}$, each centered on the projectile nuclear charge $Z_{\rm P}$. Such a distorted wave treatment of the entrance channel is the same for three-body electron capture and ionization processes (12.26) and (16.1), respectively.

The ensuing theory of Belkić [134] represents the CDW-3B method with the correct boundary conditions in the entrance and exit channels of ionization in the general three-body process (16.1). Precisely as in the case of electron capture (12.26) treated using the CDW-3B method, the transition amplitude for ionization (16.1) in the same theory, has the product of the initial and final Coulomb waves for the relative motion of the two nuclei as the entire contribution from the inter-nuclear potential $V_{\rm PT} = Z_{\rm P}Z_{\rm T}/R$. The standard eikonal result for this latter product is the well-known phase factor $(\mu v \rho)^{2iZ_{\rm P}Z_{\rm T}/v}$, which disappears from the corresponding total cross section $Q_{if}^{\rm (CDW-3B)}$ for process (16.1). Consequently, the inter-nuclear potential $V_{\rm PT}$ gives zero contribution to the total cross section $Q_{if}^{\rm (CDW-3B)}$ for ionization in the usual mass limit $\mathrm{M_P} \gg 1$ and $\mathrm{M_T} \gg 1$ which is amply justified for processes (12.26) and (16.1). This is required for every ion-atom collision involving heavy nuclei [44]. Such a physically indispensable result of the CDW-3B method is a direct consequence of the symmetric treatment of the relative motion of the projectile and target nucleus in both the entrance and exit channels.

As to the emitted electron e, the C3 wave function accounts fully for the simultaneous presence of the two Coulomb centers located at the nuclear charges $Z_{\rm P}$ and $Z_{\rm T}$. Consequently, the ejected electron moves in the field of $Z_{\rm P}$ as well as $Z_{\rm T}$, and this constitutes the so-called two-center effect. When $\kappa \ll v$, the electron is mainly in close vicinity to its parent nucleus $Z_{\rm T}$. These small values of electron momentum κ give the main contribution to total cross sections at high energies. However, the influence of the other Coulomb center, $Z_{\rm P}$, becomes dominant for the electrons emitted nearly in the direction of the scattered projectiles ($\boldsymbol{\kappa} \approx \boldsymbol{v}$). Such electrons are viewed as being 'captured' by $Z_{\rm P}$, albeit not in a bound state, but rather into a continuum state. This is the ECC effect, which is obviously a resonance phenomenon. In addition, the CDW-3B method predicts the forward and binary effects in the angular distributions of scattered projectiles $Z_{\rm P}$. The forward effect is manifested in the appearance of a peak near zero emission angle in the differential cross section for the ejected electrons. Similarly, the binary effect is seen as a peak when $\boldsymbol{\kappa} \approx 2\boldsymbol{v}$. The forward, ECC and binary peaks have all been observed experimentally and, moreover, they were found to be in quantitative agreement with the corresponding predictions from the CDW-3B method of

Belkić [134]. Finally, it should be emphasized that the CDW-3B method is also computationally appealing, since the transition amplitude $T_{if}^{\rm (CDW-3B)}$ for process (16.1) has been obtained in Ref. [134] in a purely analytical, closed form. Similarly to charge exchange, the CDW-3B method for ionization overestimates the experimentally measured total cross sections at lower incident energies below and close to the Massey peak [377]. This is primarily due to the presence of the normalization constant $N^+(\nu_{\rm P}) \equiv N^+(Z_{\rm P}/v)$ of the electronic full Coulomb wave function in the entrance channel. Namely, the factor $|N^+(Z_{\rm P}/v)|^2 = (2\pi Z_{\rm P}/v)/[1 - \exp{(-2\pi Z_{\rm P}/v)}]$ is enhanced with decreasing impact velocity v. In other words, the modified (scaled) cross sections $Q_{if}^{\rm (CDW-3B)}/|N^+(Z_{\rm P}/v)|^2$ will be significantly reduced below the Massey peak [377] relative to $Q_{if}^{\rm (CDW-3B)}$. The simplest way of eliminating the normalization $N^+(Z_{\rm P}/v)$ is to approximate the mentioned full Coulomb wave from the CDW-3B method by its asymptotic form. This simplification of the CDW-3B method [134] gives the CDW-EIS-3B method [151]. All told, the CDW-3B and CDW-EIS-3B methods share the same final scattering state and the $\boldsymbol{\nabla} \cdot \boldsymbol{\nabla}$ perturbation in the transition amplitude. They differ only in the description of the entrance channel, where the CDW-EIS-3B method simplifies the CDW-3B method through the replacement of the full Coulomb wave function by its asymptotic form given by the corresponding logarithmic phase factor. As expected from the discussed behavior of the auxiliary quantities $Q_{if}^{\rm (CDW-3B)}/|N^+(Z_{\rm P}/v)|^2$, explicit computations show that indeed below the Massey peak [377], the cross sections $Q_{if}^{\rm (CDW-EIS-3B)}$ are always substantially smaller than the corresponding values $Q_{if}^{\rm (CDW-3B)}$. This, in turn, considerably improves the agreement between theory and measurement at intermediate and lower energies.

In a subsequent study of the same process (16.1), Garibotti and Miraglia [135] rederived the C3 wave function of Belkić [134] in the exit channel. However, they failed to satisfy the correct boundary condition in the entrance channel for which they used the undistorted wave function Φ_i. Therefore, their so-named multiple scattering (MS) method[2] as a whole is inadequate, since it disregards the proper boundary conditions that must be simultaneously fulfilled in the entrance and exit channels [44, 134]. There is another severe drawback of the MS method [135], and that is a non-zero contribution to the total cross section Q from the inter-nuclear potential $V_{\rm PT} = Z_{\rm P}Z_{\rm T}/R$, which yields the mentioned Coulomb wave function for the relative motion of $Z_{\rm P}$ and $Z_{\rm T}$. Otherwise, this latter Coulomb wave is one of the three Coulomb waves from the C3 wave function of Belkić [134] in the exit channel of process (16.1). By contrast to the MS method [135], the inter-nuclear potential $V_{\rm PT}$

[2]The name 'multiple scattering' method presumably comes from using the Coulomb wave for the relative motion of heavy nuclei in the exit channel instead of a plane wave which is employed in the first Born approximation for the same motion.

disappears altogether from the total cross section in the CDW-3B method [134], as stated. This is due to multiplication of two Coulomb waves for the relative motions of nuclei $Z_{\rm P}$ and $Z_{\rm T}$ from the entrance and exit channel, such that merely the well-known phase factor $(\mu v\rho)^{2iZ_{\rm P}Z_{\rm T}/v}$ survives as the only remainder of the influence of $V_{\rm PT}$ on the transition amplitude. Finally, the phase $(\mu v\rho)^{2iZ_{\rm P}Z_{\rm T}/v}$ disappears from the total cross section in the CDW-3B method and this is the signature of zero contribution of the inter-nuclear potential $V_{\rm PT}$ to $Q_{if}^{\rm (CDW-3B)}$. Ironically, in addition to its basic faults, the MS method [135] is computationally demanding, since its transition amplitude cannot be calculated analytically, as opposed to the CDW-3B method [134]. To alleviate this difficulty, Garibotti and Miraglia [135] approximated their $T-$matrix element by an expression derived from the additional peaking approximation[3]. Needless to say, all these deficiencies of the MS method of Garibotti and Miraglia [135] should have been obvious by mere reference to the earlier study by Belkić [134] within the CDW-3B method.

More recently, the C3 wave function has been reinvented by Brauner, Briggs and Klar [138] and others [139]–[144]. Concretely, Brauner *et al.* [138] adapted the derivation of the C3 wave function of Belkić to ionization of atoms by electron impact without due citation of the original work [134]. However, such an adaptation from Ref. [138] is a matter of trivial specification of the required masses and this could have been done directly in the already known C3 wave function [134]. Otherwise, the C3 wave function [134] adapted to ionization by impact of electrons [138, 139, 164, 165] and photons [141, 142] has been shown to be accurate in comparison with experimental data [139]–[144], as is the case with the CDW-3B method [134] and its simplification, which is the CDW-EIS-3B method [151].

Extensive literature exists on many important applications of the CDW-3B method [134] to a variety of ionizing collisions [145]–[150]. Over the last three decades, regarding both differential and total cross sections for single electron emission from atoms by multiply charged ions, the CDW-3B method [134] has been firmly established as the most successful high-energy theory of ionization valid above 100 keV/amu in accordance with the associated limit (14.69), which has been found empirically for electron capture [44]. Naturally, the CDW-3B theory is not adequate below its lower limit of validity, but here the CDW-EIS-3B method [151] comes to rescue the situation by providing total cross sections that are in excellent agreement with experimental data even at energies smaller than 100 keV/amu.

The CDW-3B method can be extended to the CDW-4B method for single ionization of a helium-like atomic system by a bare nucleus in process (11.4), as done by Belkić [368]. The prototype of this latter process is single electron

[3]The peaking approximation from Ref. [135] is similar to that from the previously introduced Vainstein-Presnyakov-Sobelman (VPS) approximation [338]–[340].

detachment from H^- by H^+

$$H^+ + H^-(1s^2) \longrightarrow H^+ + H + e. \tag{16.2}$$

This is an important example of ionizing four-body collisions, since it offers the possibility to analyze the dependence of cross sections on inter-electron correlations that are known to be strong whenever H^- is involved.

16.1 The MCB-4B method

Process (16.2) has been studied intensively over the years both theoretically and experimentally [378]–[391]. Use was made of the four-body plane wave Born (PWB-4B) approximation as well as the ECB-4B method by Gayet, Janev and Salin [131], the four-body first Born (B1-4B) method by Bell *et al.* [381], the molecular orbital (MO) method by Sidis *et al.* [382], the AO method by Ermolaev [384], the MCB-4B method by Belkić [132, 133], etc[4]. The main focus of our analysis will be on the PWB, B1, ECB and MCB methods, but the results of the AO and MO methods will also be illustrated.

The $T-$matrices in the prior versions of the ECB and MCB methods for process (16.2) read as[5]

$$\begin{aligned} T_{if}^{(\mathrm{ECB})-} &= \langle\chi_f^-|V_i^{(\mathrm{ECB})}|\chi_i^+\rangle \\ &= \tilde{N}^{-*}(\zeta)\iiint \mathrm{d}\boldsymbol{s}_1\mathrm{d}\boldsymbol{x}_1\mathrm{d}\boldsymbol{x}_2\,\mathrm{e}^{i\boldsymbol{q}\cdot\boldsymbol{s}_1-i(\boldsymbol{\kappa}+\boldsymbol{q})\cdot\boldsymbol{x}_1}\varphi_f^*(\boldsymbol{x}_2) \\ &\times {}_1F_1(i\zeta,1,ips_1+i\boldsymbol{p}\cdot\boldsymbol{s}_1)(vs_1+\boldsymbol{v}\cdot\boldsymbol{s}_1)^{-i\nu_{\mathrm{P}}}V_i^{(\mathrm{ECB})}\varphi_i(\boldsymbol{x}_1,\boldsymbol{x}_2) \end{aligned} \tag{16.3}$$

$$\begin{aligned} T_{if}^{(\mathrm{MCB})-} &= \langle\chi_f^-|V_i^{(\mathrm{MCB})}|\chi_i^+\rangle \\ &= \tilde{N}^{-*}(\zeta)\iiint \mathrm{d}\boldsymbol{s}_1\mathrm{d}\boldsymbol{x}_1\mathrm{d}\boldsymbol{x}_2\,\mathrm{e}^{i\boldsymbol{q}\cdot\boldsymbol{s}_1-i(\boldsymbol{\kappa}+\boldsymbol{q})\cdot\boldsymbol{x}_1}\varphi_f^*(\boldsymbol{x}_2) \\ &\times {}_1F_1(i\zeta,1,ips_1+i\boldsymbol{p}\cdot\boldsymbol{s}_1)(vs_1+\boldsymbol{v}\cdot\boldsymbol{s}_1)^{-i\nu_{\mathrm{P}}}V_i^{(\mathrm{MCB})}\varphi_i(\boldsymbol{x}_1,\boldsymbol{x}_2) \end{aligned} \tag{16.4}$$

$$V_i^{(\mathrm{ECB})} = -\frac{1}{s_1} + \Delta V_{\mathrm{P}2} \tag{16.5}$$

$$V_i^{(\mathrm{MCB})} = [\Delta V_{\mathrm{P}2} + u_i] + \frac{\nu_{\mathrm{P}}}{s_1}\frac{1+i(\boldsymbol{v}s_1+v\boldsymbol{s}_1)\cdot\boldsymbol{\nabla}_{x_1}}{vs_1+\boldsymbol{v}\cdot\boldsymbol{s}_1} \tag{16.6}$$

[4] For brevity, in this section we will drop the part 4B from the acronyms of the listed method.
[5] In the case of process (11.4), the only remaining influence of the screened inter-nuclear potential $Z_{\mathrm{P}}(Z_{\mathrm{T}}-1)/R$ is through the eikonal phase factor $(\mu v\rho)^{2iZ_{\mathrm{P}}(Z_{\mathrm{T}}-1)/v}$. However, for electron detachment (16.2) this latter phase is equal to unity, since here $Z_{\mathrm{T}}=1$.

$$\tilde{N}^{-}(\zeta) = (2\pi)^{-3/2}\Gamma(1+i\zeta)\,\mathrm{e}^{\pi\zeta/2} \qquad \zeta = \frac{1}{p}$$

$$\Delta V_{\mathrm{P2}} = \frac{1}{R} - \frac{1}{s_2} \qquad u_i\varphi_i = (h_i - E_i)\varphi_i \equiv O_i$$

$$\boldsymbol{q} = \boldsymbol{k}_f - \boldsymbol{k}_i = \boldsymbol{\eta} + \frac{E_i - E_f - E_\kappa}{v}\hat{\boldsymbol{v}} \qquad \boldsymbol{\eta}\cdot\boldsymbol{v} = 0$$

$$E_\kappa = \frac{\kappa^2}{2} \qquad \nu_{\mathrm{P}} = \frac{1}{v} \qquad \boldsymbol{p} = \boldsymbol{\kappa} - \boldsymbol{v} \tag{16.7}$$

where h_i is the target Hamiltonian and $\boldsymbol{\kappa}$ is the momentum of the electron in the target frame. As seen from (16.5) and (16.6), the only difference between the ECB and MCB methods is in the perturbation potentials $V_i^{(\mathrm{ECB})}$ and $V_i^{(\mathrm{MCB})}$. Otherwise, both methods have the same initial χ_i^+ and final χ_f^- distorted wave functions

$$\chi_i^+ = \varphi_i(\boldsymbol{x}_1, \boldsymbol{x}_2)\mathrm{e}^{i\boldsymbol{k}_i\cdot\boldsymbol{r}_i}(vs_1 + \boldsymbol{v}\cdot\boldsymbol{s}_1)^{-i\nu_{\mathrm{P}}} \tag{16.8}$$

$$\chi_f^- = \tilde{N}^{-}(\zeta)\mathrm{e}^{i\boldsymbol{k}_f\cdot\boldsymbol{r}_i + i\boldsymbol{\kappa}\cdot\boldsymbol{x}_1}\varphi_f(\boldsymbol{x}_2)\,{}_1F_1(-i\zeta, 1, -ips_1 - i\boldsymbol{p}\cdot\boldsymbol{s}_1). \tag{16.9}$$

The functions (16.8) and (16.9) possess the correct asymptotic behaviors. However, the initial perturbation in the ECB method is incorrect, since $V_i^{(\mathrm{ECB})}$ from (16.5) is unrelated to χ_i^+. In sharp contrast, the expression (16.6) for $V_i^{(\mathrm{MCB})}$ is correct, since it follows directly and uniquely from the application of the full Schrödinger operator $H-E$ to χ_i^+, as per the definition of a generic perturbation. Here, H and E are the total Hamiltonian and the energy of the whole system, respectively. The interaction V_{P2} between the projectile P and electron e_2 indirectly leads to ionization of e_1 via the inter-electronic static correlation i.e. the SEC in the target wave function. The contribution of V_{P2} is found to be negligible and this term is ignored in computations by means of both the ECB and MCB methods. The additional potential operator O_i in $T_{if}^{(\mathrm{MCB})-}$ from (16.4) reflects the unavailability of the exact wave function φ_i of the two-electron bound state of H^-. It has been shown for double capture [87] and transfer ionization [127] for processes (14.68) and (15.32) that the term O_i does not play a significant role (see also Figs. 14.5, 14.15 and 15.1). We verified that the same conclusion also applies to single electron detachment in process (16.2). As such, the term O_i can be ignored in $T_{if}^{(\mathrm{MCB})-}$, as has been done in previous studies [132, 133].

The fact that the integrals over $\boldsymbol{x}_1$ and $\boldsymbol{x}_2$ in (16.3) and (16.4) encompass only bound-state wave functions allows one to employ the most elaborate $\varphi_i(\boldsymbol{x}_1, \boldsymbol{x}_2)$ with any degree of correlations. We optimally explore this circumstance by using the best wave functions of the ground state of $\mathrm{H}^-(^1S)$ e.g. the many-parameter correlated CI type wave functions of Tweed [76], as well as Joachain and Terao [77]. These benchmark computations with highly correlated wave functions in terms of some 21-61 variational parameters are compared with the results of the simple CI orbitals of Silverman *et al.*

[71] with 2 and 3 variational parameters for the $(1s1s')$ and $\{(1s1s'),(2p)^2\}$ descriptions of φ_i, respectively. In all these computations, we consider the electron e_2 as being passive in the exit channel in the sense of residing only in the ground state $f = 1s$ of the target rest H$(1s)$. In such a case, every term from the prior form $T_{if}^{(\mathrm{MCB})-}$ in (16.4), which contains the CI orbitals with non-zero values of the angular momentum quantum number $(l_i \neq 0)$ will be effectively equal to zero, due to the orthogonality of the spherical harmonics via $\int \mathrm{d}\hat{\boldsymbol{x}}_2 Y_{0,0}(\hat{\boldsymbol{x}}_2) Y^*_{l_i m_i}(\hat{\boldsymbol{x}}_2) = \delta_{l_i,0}\delta_{m_i,0}$. Nevertheless, the presence of the electronic angular correlations in $\varphi_i(\boldsymbol{x}_1,\boldsymbol{x}_2)$ is indirectly felt in (16.4), since the set of the variational parameters together with the binding energy E_i have different values when using the CI orbitals with and without the non-zero value of the angular momentum l_i. For example, the 2-parameter CI wave function of Silverman *et al.* [71] for the $(1s1s')$ description of the ground state ^1S of H$^-$ given by

$$\begin{aligned} \varphi_i(\boldsymbol{x}_1,\boldsymbol{x}_2) &= 0.03146105 \left[\mathrm{e}^{-1.039230\,x_1 - 0.2832215\,x_2} + (x_1 \longleftrightarrow x_2)\right] \\ E_i &= -0.51330289 \end{aligned} \tag{16.10}$$

incorporates only the radial correlations through a pure $s-$wave. However, the 3-parameter $\{(1s1s'),(2p)^2\}$ CI orbital for H$^-(^1S)$ [71], which includes both radial and angular correlations ($s-$ and $p-$waves), reads as

$$\begin{aligned} \varphi_i(\boldsymbol{x}_1,\boldsymbol{x}_2) &= 0.036902815 \left[\mathrm{e}^{-1.03556\,x_1 - 0.323563\,x_2} + (x_1 \longleftrightarrow x_2)\right] \\ &\quad - 0.074119614 \sum_{m_i'=-1}^{1} \varphi_{21,m_i'}(\boldsymbol{x}_1)\varphi_{21,-m_i'}(\boldsymbol{x}_2) \\ E_i &= -0.5245743 \end{aligned} \tag{16.11}$$

where $\varphi_{21,m_i'}(\boldsymbol{x}_k) = x_k\,\mathrm{e}^{-0.998504\,x_k} Y_{1,m_i'}(\hat{\boldsymbol{x}}_k)$ $(k = 1,2)$. Orthogonality of $Y_{1,m_i'}$ with $Y_{0,0}$ makes all the three terms $(m_i' = 0\,, \pm 1)$ vanish in the summation in (16.11) used for calculation of the $T-$matrix $T_{if}^{(\mathrm{MCB})-}$. Therefore, precisely the same algorithm built for the $(1s1s')$ configuration (16.10) can also be used for $\{(1s1s'),(2p)^2\}$ from (16.11) by simply changing the values of the set of the variational parameters as well as E_i. Similar arguments about $l_i \neq 0$ hold true for the CI wave function of Joachain and Terao [77] given by

$$\begin{aligned} \varphi_i(\boldsymbol{x}_1,\boldsymbol{x}_2) &= \frac{1}{4\pi}\sum_{l_i'=0}^{3}\left\{\sum_{m_i'}\sum_{m_i''} A^{(l_i')}_{m_i' m_i''} x_1^{l_i'} x_2^{l_i'} (x_1^{m_i'} x_2^{m_i''} + x_1^{m_i''} x_2^{m_i'})\,\mathrm{e}^{-\lambda(x_1+x_2)}\right\} \\ &\quad \times P_{l_i'}(\cos\theta_{12}). \end{aligned} \tag{16.12}$$

The wave function of Tweed [76] is of the same form, except for the redefinition $x_1^{l_i'} x_2^{l_i'} \equiv 1$ in (16.12). Here, $P_{l_i'}(\cos\theta_{12})$ is the Legendre polynomial, the coefficients $A^{(l_i')}_{m_i' m_i''}$ are the linear variational parameters and $\theta_{12} = \cos^{-1}(\hat{\boldsymbol{x}}_1 \cdot \hat{\boldsymbol{x}}_2)$.

Only the pure $s-$waves $P_0(\cos\theta_{12})$ from (16.12) need to be explicitly retained in the analytical calculation of $T_{if}^{(\mathrm{MCB})-}$, but the remaining terms with $P_{l'_i\neq 0}(\cos\theta_{12})$ are implicitly present in the final results, since they contribute significantly in determining the linear $\{A^{(l'_i)}_{m'_i m''_i}\}$ and non-linear (λ) coefficients as well as the binding energy E_i through the standard Rayleigh-Ritz variational principle. In the case of e.g. the 61-parameter CI wave function (16.12) from Ref. [77], we have $E_i = -0.5272225$. The corresponding 'exact' value $E_i = -0.527751016544203$ is due to the Hylleraas $r_{12}-$dependent wave function with some 616 parameters used by Drake [82].

The dynamic correlations are not contained in the prior transition amplitude $T_{if}^{(\mathrm{MCB})-}$ from (16.4), since the dielectronic interaction $V_{12} = 1/x_{12} \equiv 1/|\boldsymbol{x}_1 - \boldsymbol{x}_2|$ is not in the perturbation $V_i^{(\mathrm{MCB})}$, which directly causes the transition in process (16.2). Of course, V_{12} could also appear in $T_{if}^{(\mathrm{MCB})-}$ if we retain the small eigen-value correction O_i for any approximate wave function φ_i, since $O_i = (h_i - E_i)\varphi_i = (E_i + \nabla_1^2/2 + \nabla_2^2/2 + 1/x_1 + 1/x_2 - 1/x_{12})\varphi_i \neq 0$. Nevertheless, such a presence of V_{12} in $V_i^{(\mathrm{MCB})}$ would still represent a static correlation, since the term O_i giving rise to $1/x_{12}$ is due to a non-exact structural description of the target H^- and not to collisional dynamics.

The genuine dynamic correlations resulting from the dielectronic interaction during the collision in process (16.2) is contained in the post form of the transition amplitude $T_{if}^{(\mathrm{MCB})+}$ which reads as

$$\begin{aligned} T_{if}^{(\mathrm{MCB})+} &= \langle \chi_f^- | V_f^{(\mathrm{MCB})} | \chi_i^+ \rangle \\ &= \tilde{N}^{-*}(\zeta) \iiint \mathrm{d}\boldsymbol{s}_1 \mathrm{d}\boldsymbol{x}_1 \mathrm{d}\boldsymbol{x}_2 \, \mathrm{e}^{i\boldsymbol{q}\cdot\boldsymbol{s}_1 - i(\boldsymbol{\kappa}+\boldsymbol{q})\cdot\boldsymbol{x}_1} \varphi_f^*(\boldsymbol{x}_2) \\ &\times {}_1F_1(i\zeta, 1, ips_1 + i\boldsymbol{p}\cdot\boldsymbol{s}_1)(vs_1 + \boldsymbol{v}\cdot\boldsymbol{s}_1)^{-i\nu_{\mathrm{P}}} V_f^{(\mathrm{MCB})} \varphi_i(\boldsymbol{x}_1, \boldsymbol{x}_2) \end{aligned} \quad (16.13)$$

$$V_f^{(\mathrm{MCB})} = \Delta V_{12} + \Delta V_{\mathrm{P}2} \quad (16.14)$$

$$\Delta V_{12} = \frac{1}{x_{12}} - \frac{1}{x_1} \qquad \Delta V_{\mathrm{P}2} = \frac{1}{R} - \frac{1}{s_2}. \quad (16.15)$$

Additionally, the post matrix element $T_{if}^{(\mathrm{MCB})+}$ contains the static correlations through φ_i and $\Delta V_{\mathrm{P}2}$. As mentioned, $\Delta V_{\mathrm{P}2}$ can be ignored, since it plays a negligible role in process (16.2). In such a case, the integrals $\int \mathrm{d}\boldsymbol{s}_1(\cdots)$ and $\int\int \mathrm{d}\boldsymbol{x}_1 d\boldsymbol{x}_2(\cdots)$ in (16.3), (16.4) and (16.13) are independent of each other. These integrals over $\boldsymbol{x}_1$, $\boldsymbol{x}_2$ and $\boldsymbol{s}_1$ can be calculated analytically by employing the corresponding real integral representation from Ref. [392] for the functions $(vs_1 + \boldsymbol{v}\cdot\boldsymbol{s}_1)^{-i\nu}$ and $(vs_1 + \boldsymbol{v}\cdot\boldsymbol{s}_1)^{-i\nu-1}$.

Notice that if one introduces a $T-$matrix element between any preselected initial χ_i^+ and final χ_f^- on-shell distorted waves as $T_{if} = \langle \chi_f^- | H - E | \chi_i^+ \rangle$, then one would be free to apply the operator $H - E$ to either χ_i^+ or χ_f^-. Of course, if the wave function to which the operator $H - E$ acts is the exact on-shell scattering state, a zero state vector $\emptyset$ would be obtained. Otherwise, for approximate wave functions, the resulting $T-$matrix elements $T_{if}^- = \langle \chi_f^- | \xi_i^+ \rangle$ and $T_{if}^+ = \langle \xi_f^- | \chi_i^+ \rangle$ with $\xi_i^+ \equiv (H - E)\chi_i^+ \neq \emptyset$ and $\xi_f^- \equiv (H - E)\chi_f^- \neq \emptyset$ are unequal, $T_{if}^- \neq T_{if}^+$, whenever the initial or final bound state wave function is not exact. This is precisely the case in the entrance channel of process (16.2) and, therefore, the post-prior discrepancy is expected to occur in every particular approximation. In order to assess the extent of such a discrepancy within the MCB theory, we shall discuss the results for the total cross sections obtained by using the prior (16.4) and post (16.13) forms of the transition amplitude.

Alongside the ECB method, Gayet, Janev and Salin [131] used the PWB method. In the PWB method, the relative motions of the heavy nuclei in both the entrance channel and exit channel are described by the plane waves. The plane wave is employed also for the ejected electron in the final state. Therefore, the PWB method has incorrect boundary conditions for the initial and final scattering states in process (16.2). Formally, the transition amplitude $T_{if}^{(\mathrm{PWB})-}$ in the PWB method can be obtained by setting $\nu_{\mathrm{P}} = 0 = \zeta$ in (16.3) from the ECB method, so that

$$\begin{aligned} T_{if}^{(\mathrm{PWB})-} &= (2\pi)^{-3/2} \iiint \mathrm{d}\boldsymbol{x}_1 \mathrm{d}\boldsymbol{x}_2 \mathrm{d}\boldsymbol{R}\, \mathrm{e}^{i\boldsymbol{q}\cdot\boldsymbol{s}_1 - i(\boldsymbol{\kappa}+\boldsymbol{q})\cdot\boldsymbol{x}_1} \varphi_f^*(\boldsymbol{x}_2) \\ &\quad \times \left(\frac{1}{R} - \frac{1}{s_1} - \frac{1}{s_2}\right) \varphi_i(\boldsymbol{x}_1, \boldsymbol{x}_2). \end{aligned} \tag{16.16}$$

Since both ν_{P} and ζ tend to zero with $E \to \infty$, it follows that at asymptotically high impact energies, we must have $Q_{if}^{(\mathrm{ECB})-} \longrightarrow Q_{if}^{(\mathrm{PWB})-}$

$$Q_{if}^{(\mathrm{ECB})-} \approx Q_{if}^{(\mathrm{PWB})-} \qquad (v \gg 1). \tag{16.17}$$

Gayet, Janev and Salin [131] used the 2-parameter CI wave function from Ref. [71] for the initial state $\varphi_i(\boldsymbol{x}_1, \boldsymbol{x}_2)$. This function is not orthogonal to the plane wave final state

$$\begin{aligned} &\varphi_f(\boldsymbol{x}_1, \boldsymbol{x}_2) = \phi_{\boldsymbol{\kappa}}(\boldsymbol{x}_1)\varphi_f(\boldsymbol{x}_2) \qquad \phi_{\boldsymbol{\kappa}}(\boldsymbol{x}_1) = (2\pi)^{-3/2}\mathrm{e}^{-\boldsymbol{\kappa}\cdot\boldsymbol{x}_1} \\ &\langle \varphi_f | \varphi_i \rangle \neq 0. \end{aligned} \tag{16.18}$$

In such a case, integration over $\boldsymbol{R}$ reduces (16.16) to the expression

$$\begin{aligned} T_{if}^{(\mathrm{PWB})-} &= \frac{2}{q^2}(2\pi)^{-1/2} \iint \mathrm{d}\boldsymbol{x}_1 \mathrm{d}\boldsymbol{x}_2\, \varphi_f^*(\boldsymbol{x}_2) \\ &\quad \times \left(1 - \mathrm{e}^{-i\boldsymbol{q}\cdot\boldsymbol{x}_1} - \mathrm{e}^{-i\boldsymbol{q}\cdot\boldsymbol{x}_2}\right) \varphi_i(\boldsymbol{x}_1, \boldsymbol{x}_2). \end{aligned} \tag{16.19}$$

This can further be calculated analytically yielding a simple closed formula for $T_{if}^{(\mathrm{PWB})-}$. The significance of (16.19) is best appreciated by examining its high energy limit via the Maclaurin expansion of both exponentials $\exp(-\boldsymbol{q}\cdot\boldsymbol{x}_k)$ $(k=1,2)$ and the subsequent retension of only the first two terms. Then, the term $1-\sum_{k=1}^{2}\exp(-\boldsymbol{q}\cdot\boldsymbol{x}_k)$ in (16.19) simplifies as $1-\sum_{k=1}^{2}\exp(-\boldsymbol{q}\cdot\boldsymbol{x}_k)\approx -1+i\boldsymbol{q}\cdot(\boldsymbol{x}_1+\boldsymbol{x}_2)$. Thus, the asymptotic behavior of the transition amplitude in the PWB method is

$$\begin{aligned} T_{if}^{(\mathrm{PWB})-} \approx \tilde{T}_{if}^{(\mathrm{PWB})-} &\approx \frac{2}{q^2}(2\pi)^{-1/2}\int\int \mathrm{d}\boldsymbol{x}_1\mathrm{d}\boldsymbol{x}_2\,\varphi_f^*(\boldsymbol{x}_2) \\ &\times[-1+i\boldsymbol{q}\cdot(\boldsymbol{x}_1+\boldsymbol{x}_2)]\,\varphi_i(\boldsymbol{x}_1,\boldsymbol{x}_2) \qquad (v\gg 1). \end{aligned} \tag{16.20}$$

Therefore, at sufficiently high impact energies, the total cross sections $Q_{if}^{(\mathrm{PWB})-}$ and $\tilde{Q}_{if}^{(\mathrm{PWB})-}$ computed using (16.16) and (16.20), respectively, will coincide

$$Q_{if}^{(\mathrm{PWB})-} \approx \tilde{Q}_{if}^{(\mathrm{PWB})-} \qquad (v\gg 1). \tag{16.21}$$

The above procedure of searching the asymptotic limit of total cross sections for ionization is reminiscent of the dipole-type approximation. However, any valid variant of the dipole approximation for ionization must yield the high-energy limit $Q_{if}\propto(1/E)\ln(E)$, since this formula is provided solely by the dipole term $i\boldsymbol{q}\cdot(\boldsymbol{x}_1+\boldsymbol{x}_2)$. Instead, $\tilde{Q}_{if}^{(\mathrm{PWB})-}$ contains the function $i\boldsymbol{q}\cdot(\boldsymbol{x}_1+\boldsymbol{x}_2)-1$ through the usage of (16.20), where the second contribution (unity) dominates the dipole term at high energies. Due to this non-zero contribution from the said constant term, the cross sections $\tilde{Q}_{if}^{(\mathrm{PWB})-}$ and $Q_{if}^{(\mathrm{PWB})-}$ will not have the required high-energy asymptote $Q_{if}\propto(1/E)\ln(E)$. The same failure is shared by $Q_{if}^{(\mathrm{ECB})-}$ due to (16.17). The explicit computations by Gayet, Janev and Salin [131] confirms this anticipation.

If the initial $\varphi_i(\boldsymbol{x}_1,\boldsymbol{x}_2)$ and final $\varphi_f(\boldsymbol{x}_1,\boldsymbol{x}_2)$ states of the H^- ion were orthogonal, the unit term contribution would be equal to zero. Orthogonality can be easily achieved by the Gramm-Schmidt orthogonalization procedure via construction of the final state $\varphi_f(\boldsymbol{x}_1,\boldsymbol{x}_2)$ of the H^- ion as

$$\begin{aligned} \varphi_f(\boldsymbol{x}_1,\boldsymbol{x}_2) &= \Upsilon_f(\boldsymbol{x}_1,\boldsymbol{x}_2)-\langle\varphi_i|\Upsilon_f\rangle\varphi_i(\boldsymbol{x}_1,\boldsymbol{x}_2) \\ \Upsilon_f(\boldsymbol{x}_1,\boldsymbol{x}_2) &= \frac{1}{2}\left[\phi_{\boldsymbol{\kappa}}(\boldsymbol{x}_1)\varphi_f(\boldsymbol{x}_2)+\phi_{\boldsymbol{\kappa}}(\boldsymbol{x}_2)\varphi_f(\boldsymbol{x}_1)\right] \\ \langle\varphi_f|\varphi_i\rangle &= 0. \end{aligned} \tag{16.22}$$

As can be seen from (16.5) and (16.16), the ECB and PWB method have the same perturbation potential $V_i^{(\mathrm{PWB})}\equiv 1/R-1/s_1-1/s_2=V_i^{(\mathrm{ECB})}$. In the asymptotic region where the scattering aggregates are far apart from each other, we have $R\approx s_2$, so that here $\Delta V_{\mathrm{P}2}\approx 0$ and, therefore $V_i^{(\mathrm{PWB})}=V_i^{(\mathrm{ECB})}\approx -1/s_1$. In other words, in the asymptotic scattering region in which measurement is made, the common perturbation potential in the ECB and PMB methods is effectively reduced to the Coulomb potential between the projectile proton and the ejected electron $(-1/s_1)$. This prevents the asymptotic freedom from taking place and, therefore, the asymptotic convergence problem of Dollard [41] is disregarded. Such an occurrence is due to the fact that wave packets in a Coulomb field cannot be associated with free particles, since a Coulomb or Coulomb-like potential always distorts the unperturbed plane wave, even at asymptotically large interparticle distances. The notion of free particles in the asymptotic scattering region (the region of measurement of collisional observables such as cross sections) is necessary to ensure that one can differentiate with reasonable certainty the situation "before" and "after" collision. The situation "before" collision would correspond to the stage of experimental preparation of the initial state of the target in the entrance channel with the projectile beam switched off. Likewise, the situation "after" collision is associated with the measurement of configurations in the exit channel when the scattering is completed i.e. when all the interactions among the particles have ceased to exist. Both the ECB and PWB methods violate the concept of asymptotic freedom and it is important to assess the consequences of such a violation on the overall performance of these two models.

If the entrance channel in process (16.2) is treated in the same way as in the PWB method, but with the description of the single continuum state of the H^- ion in the exit channel improved by inclusion of the Coulomb wave function for the emitted electron, the standard B1 approximation would be obtained. This was done by Bell *et al.* [381] who used the prior version of the B1 method for process (16.2) with a wave function of the final state of the H^- ion given as a symmetrized function relative to the electronic coordinates $\boldsymbol{x}_1$ and $\boldsymbol{x}_2$. This latter final state wave function was constructed to be orthogonal to $\varphi_i(\boldsymbol{x}_1, \boldsymbol{x}_2)$ through the Gramm-Schmidt orthogonalization of the type (16.22). However, because of the usage of a plane wave for the relative inter-aggregate motion of two charged particles H^+ and H^- in the entrance channel, the B1 approximation disregards the correct boundary conditions and, hence, it is theoretically unsound.

Alternative studies of process (16.2) would also be important to perform by means of other theories, such as the CDW-4B method, which is an extension of the CDW-3B method of Belkić [134]. In so doing, one would obtain the following prior and post transition amplitudes for process (16.2)

$$\begin{aligned} T_{if}^{(\mathrm{CDW})-} &= \tilde{N}^{-*}(\zeta)N^{+}(\nu_{\mathrm{P}}) \iiint \mathrm{d}\boldsymbol{s}_1 \mathrm{d}\boldsymbol{x}_1 \mathrm{d}\boldsymbol{x}_2 \, \mathrm{e}^{i\boldsymbol{q}\cdot\boldsymbol{s}_1 - i(\boldsymbol{\kappa}+\boldsymbol{q})\cdot\boldsymbol{x}_1} \varphi_f^*(\boldsymbol{x}_2) \\ &\times {}_1F_1(i\zeta, 1, ips_1 + i\boldsymbol{p}\cdot\boldsymbol{s}_1) \left[(\Delta V_{\mathrm{P2}} + O_i)\, \varphi_i(\boldsymbol{x}_1, \boldsymbol{x}_2) \right. \\ &\times {}_1F_1(i\nu_{\mathrm{P}}, 1, ivs_1 + i\boldsymbol{v}\cdot\boldsymbol{s}_1) \\ &- \left. \boldsymbol{\nabla}_{s_1} {}_1F_1(i\nu_{\mathrm{P}}, 1, ivs_1 + i\boldsymbol{v}\cdot\boldsymbol{s}_1) \cdot \boldsymbol{\nabla}_{x_1} \varphi_i(\boldsymbol{x}_1, \boldsymbol{x}_2) \right] \end{aligned} \tag{16.23}$$

$$\begin{aligned} T_{if}^{(\mathrm{CDW})+} &= \tilde{N}^{-*}(\zeta)N^{+}(\nu_{\mathrm{P}}) \iiint \mathrm{d}\boldsymbol{s}_1 \mathrm{d}\boldsymbol{x}_1 \mathrm{d}\boldsymbol{x}_2 \, \mathrm{e}^{i\boldsymbol{q}\cdot\boldsymbol{s}_1 - i(\boldsymbol{\kappa}+\boldsymbol{q})\cdot\boldsymbol{x}_1} \varphi_f^*(\boldsymbol{x}_2) \\ &\times {}_1F_1(i\zeta, 1, ips_1 + i\boldsymbol{p}\cdot\boldsymbol{s}_1)\, (\Delta V_{\mathrm{P2}} + \Delta V_{12})\, \varphi_i(\boldsymbol{x}_1, \boldsymbol{x}_2) \\ &\times {}_1F_1(i\nu_{\mathrm{P}}, 1, ivs_1 + i\boldsymbol{v}\cdot\boldsymbol{s}_1) \end{aligned} \tag{16.24}$$

where $N^+(\nu_{\rm P}) = \Gamma(1 - i\nu_{\rm P})\,{\rm e}^{\pi\nu_{\rm P}/2}$ and $\nu_{\rm P} = 1/v$. A further approximation to $T_{if}^{\rm (CDW)+}$ can be made along with the prescription of Crothers and McCann [151] yielding the CDW-EIS model. These authors replace the wave function $N^+(\nu_{\rm P})_1F_1(i\nu_{\rm P}, 1, ivs_1 + \boldsymbol{v}\cdot\boldsymbol{s}_1)$ from the CDW method by the corresponding asymptotic form $(vs_1 + \boldsymbol{v}\cdot\boldsymbol{s}_1)^{-i\nu_{\rm P}}$, which is valid only when $|vs_1 + \boldsymbol{v}\cdot\boldsymbol{s}_1| \gg 1$. For process (16.2), the ensuing $T-$matrix element $T_{if}^{\rm (CDW-EIS)}$ formally coincides with the transition amplitude $T_{if}^{\rm (MCB)+}$ from (16.13). As an alternative to the derivation of Crothers and McCann [151], the MCB method in either its prior or post form from (16.4) or (16.13) is autonomously derived for process (16.2) without any needed reference to the CDW approximation.

The triple differential ${\rm d}^3Q_{if}^{\pm}/{\rm d}\boldsymbol{\kappa}$ and total $Q_{if}^{\pm}$ cross sections for process (16.2) are defined by

$$\frac{{\rm d}^3Q_{if}^{\pm}}{{\rm d}\boldsymbol{\kappa}} = \int {\rm d}\boldsymbol{\eta} \left|\frac{T_{if}^{\pm}(\boldsymbol{\eta})}{2\pi v}\right|^2 \tag{16.25}$$

$$Q_{if}^{\pm} = \int {\rm d}\boldsymbol{\kappa}\, \frac{{\rm d}^3Q_{if}^{\pm}}{{\rm d}\boldsymbol{\kappa}}. \tag{16.26}$$

In the present non-relativistic spin-independent formalism, the final results for ${\rm d}^3Q_{if}^{\pm}/{\rm d}\boldsymbol{\kappa}$ and $Q_{if}^{\pm}$ are multiplied by a factor of 2, since either of the electrons e_1 or e_2 could be detached from the target $\rm H^-$.

16.2 Comparison between theories and experiments

In all the present computations, we consider only the ground state of the atomic hydrogen H(1s) in the exit channel of process (16.2). On the other hand, the corresponding measurements from Refs. [18] and [387]–[389] relate to all the states (bound and continuous) of the target rest H. Therefore, for a more complete comparison of theory and experiment, inclusion of the discrete and continuous spectrum of atomic hydrogen is, in principle, necessary in theoretical models. Simultaneous detachment and excitation (or detachment and ionization) in process (16.2) would be both interesting and important to consider. Nevertheless, in the case of process (16.2), probabilities for these two-electron processes are expected to be small relative to single electron detachment accompanied by the emergence of the H(1s) atom as the target rest. As such, it is justified to compare the mentioned experimental data for process (16.2) with the theoretical cross sections for a simpler reaction

$$\rm H^+ + H^-(1s^2) \longrightarrow H^+ + H(1s) + \mathit{e}. \tag{16.27}$$

Total cross sections $Q_{if}^{(\text{ECB})-}$ for process (16.27), computed by Gayet, Janev and Salin [131] using the 2-parameter radially correlated CI wave function of Silverman *et al.* [71] for $\text{H}^-(1s^2)$ overestimate the experimental data by some astonishing 2-3 orders of magnitude, as shown in Fig. 16.1. Moreover, $Q_{if}^{(\text{ECB})-}$ saturates to a peculiar constant for large values of the impact energy E, at variance with the corresponding proper Bethe asymptotic limit, $Q_{if} \propto (1/E)\ln(E)$. Such flagrant deficiencies of the ECB method could have

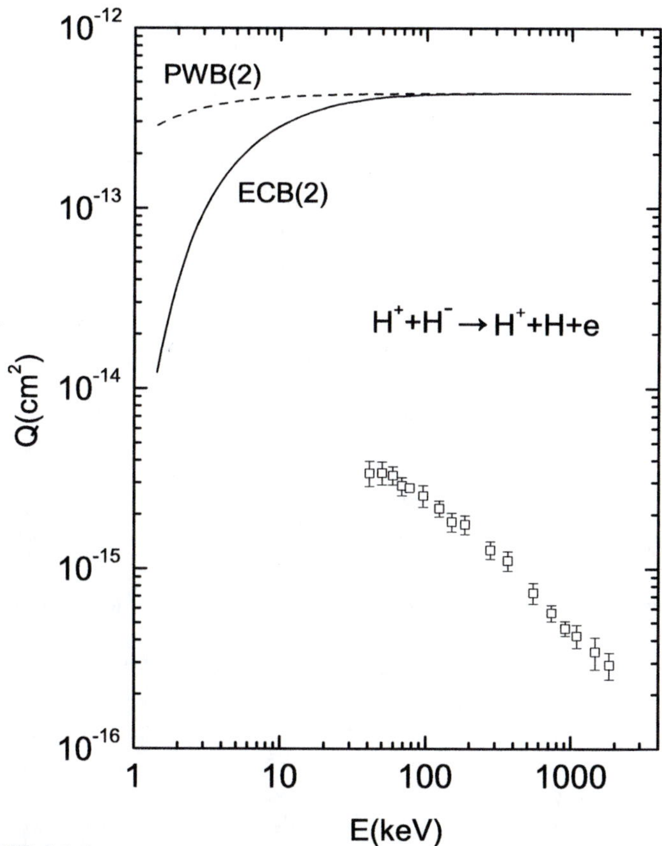

FIGURE 16.1
Total cross sections Q as a function of the incident energy E for process (16.2). Theory: full curve (ECB method [131]) and dashed curve (PWB [131]). Integer 2 with the acronyms is the number of variational parameters in the target CI wave function of Silverman *et al.* [71]. Experiment: □ [387]. The original data of Peart *et al.* [387] are for electron impact, and here they are scaled to the equivalent proton impact energy.

been detected by reference to the earlier study of Peart *et al.* [387] who reported the corresponding experimental data using electrons instead of protons as projectiles. It is well-known that close agreement exists between total cross sections for fast electron and proton impact ionization/detachment processes at the equivalent/scaled energies (or at the same incident velocities). Only the measurement from Ref. [387] is quoted in Fig. 16.1, since these experimental data were available at the time of publication of the work of Gayet *et al.* [131]. Alternatively, Gayet *et al.* [131] could have relied upon the scaled results of the Coulomb-Born (CB) method reported by Belly and Schwartz [380] for $e - \mathrm{H}^-$ detachment to realize that their ECB method is unphysical. Of course, the first hint that something was utterly inadequate in the computations from Ref. [131] was a plateau in the cross sections $Q_{if}^{(\mathrm{ECB})-}$ attained at high energies (see Fig. 16.1) instead of having the required and ubiquitous Bethe asymptotic behavior $Q_{if} \propto (1/E)\ln(E)$. Being unaware of the measurement from Ref. [387], the study of Gayet *et al.* [131] was misguided by the PWB approximation. However, the PWB method is seen in Fig. 16.1 to give the wrong cross sections $Q_{if}^{(\mathrm{PWB})-}$ that at high energies also tend to the same peculiar constant already encountered in the ECB method. This is expected from our asymptotic analysis [367, 368] where we demonstrated analytically that the PWB and ECB of Gayet, Janev and Salin [131] cannot reduce to the Bethe approximation at high energies.

Interestingly, an early study of Geltman [378] in the plane wave Born-Oppenheimer (PWBO) approximation on detachment in the $e-\mathrm{H}^-$ collisions gave the total cross sections that tend to a constant at high impact energies instead of the Bethe asymptotic limit $Q_{if} \propto (1/E)\ln(E)$, as reminiscent of the situation with the ECB method in Fig. 16.1. Subsequently, McDowell and Williamson [379] tried to attribute this unphysical result of Geltman [378] to the lack of orthogonality between the initial and final states of the H^- ion. They stated that, due to non-orthogonality of these latter states, the usual dipole approximation breaks down. Further, such a breakdown was said [379] to occur because the constant term (unity) in the series expansion of the exponential function $\exp(i\boldsymbol{q}\cdot\boldsymbol{r}) \approx 1 + i\boldsymbol{q}\cdot\boldsymbol{r}$ gives a dominant high-energy contribution relative to the dipole term $(i\boldsymbol{q}\cdot\boldsymbol{r})$. Recall that the high-energy asymptote $Q_{if} \propto (1/E)\ln(E)$ as the Bethe limit of total cross sections for ionization treated in the first Born approximation originates entirely from the dipole term. The argument of McDowell and Williamson [379] highlights a spurious contribution of the said constant term in a series expansion of the exponential. Nevertheless, these authors were wrong in concluding that the initial and final states of H^- must necessarily be orthogonal in order to eliminate this spurious contribution. Such a conclusion misled a number of subsequent studies that systematically tried to impose/force orthogonality of the initial and final states. Of course, it is good if one could construct the initial (bound) and (final) continuum states of the H^- ion in such a way that they are orthogonal to each other. However, scattering theory does not require orthogonality between the scattering wave functions of the whole entrance and exit channel (nor of the unperturbed initial and final states). Quite the contrary, formal scattering theory permits non-orthogonality of the eigenvectors of two different channel Hamiltonians $H_i \neq H_f$ for $i \neq f$ (see Section 2 in Ref. [44]). In other words, the initial and final states can be uniquely defined without any reference to their orthogonality.

Hence, while McDowell and Williamson [379] were correct in questioning the contribution of the non-dipole (constant) term, they were incorrect when drawing their inference on the necessity of imposing orthogonality of the initial and final states. One does not need to resort to the additional dipole approximation to any given method for ionization to secure the high-energy Bethe limit. This latter limit should naturally follow by imposing the correct boundary conditions to both the entrance and exit channels. The ECB method fails to fulfill this requirement in the entrance channel due to the wrong perturbation potential. Consequently, $Q_{if}^{(\mathrm{ECB})-}$ levels out at high energies instead of reaching the Bethe limit, but this has nothing to do with non-orthogonality of the initial and final states. As discussed, the dipole-type approximation to $T_{if}^{(\mathrm{ECB})-}$ at high energies shows that the unit term dominates over the dipole contribution in the mentioned exponential function, just as was the case in McDowell and Williamson's [379] analysis of Geltman's work [378]. But the reason for this is not exclusively in the lack of orthogonality between the initial and final states. Such a conjecture would be proven if one could obtain the proper Bethe limit for total cross sections by using the same non-orthogonal initial and final wave function as in the ECB method. This is indeed possible in the MCB method [132, 133] in which $Q_{if}^{(\mathrm{MCB})-}$ has the proper Bethe limit.

Recall that the MCB and ECB methods share the common initial and final states, but only the former method has the proper perturbation potential $V_i^{(\mathrm{MCB})}$ in the transition amplitude $T_{if}^{(\mathrm{MCB})-}$. Explicitly, $V_i^{(\mathrm{MCB})}$ is given by the difference $V_i^{(\mathrm{ECB})} - \Delta V_i$ where ΔV_i itself is $\Delta V_i = V_i^{(\mathrm{ECB})} - \Delta V_i'$ with $\Delta V_i'$ being the rhs of (16.5). Therefore, by imposing the correct boundary conditions to the initial state through its proper link to the corresponding perturbation potential which causes the transition in $T_{if}^{(\mathrm{MCB})-}$, the offending term $V_i^{(\mathrm{ECB})}$ is cancelled out in $V_i^{(\mathrm{MCB})}$ thus yielding the perturbation $V_i^{(\mathrm{MCB})} = V_i^{(\mathrm{ECB})} - [V_i^{(\mathrm{ECB})} - \Delta V_i'] = \Delta V_i'$ which, in turn, leads to the Bethe limit of the high-energy cross sections $Q_{if}^{(\mathrm{MCB})-}$. Moreover, the dipole approximation to $T_{if}^{(\mathrm{MCB})-}$ shows that the constant unit term from the expansion of the exponential function is cancelled, thus securing the dominance of the dipole term with the ensuing Bethe limit for $Q_{if}^{(\mathrm{MCB})-}$ as it should be (QED).

The primary reason for the fundamental inadequacy of the ECB model of Gayet, Janev and Salin [131] has been found by Belkić [132, 133]. Moreover, this was accomplished using the same 2-parameter wave function of Silverman *et al.* [71] used by Gayet *et al.* [131]. The reason was in disregard of the correct boundary conditions in the ECB method despite having both scattering wave functions with the proper asymptotic behaviors. As stated, the initial scattering wave function χ_i^+ in the ECB method is unrelated to the perturbation potential $V_i^{(\mathrm{ECB})}$. We emphasize again that the correct boundary conditions are satisfied only if the initial as well as final scattering wave functions possess the proper asymptotic behaviors, provided that they are both consistent with the corresponding perturbation potentials in the entrance and exit channel, respectively.

The explanation from Refs. [132, 133] for the main problem in the ECB model [131] simultaneously provides the key to a solution of this problem. To this end, it was sufficient to establish consistency between the initial scattering wave function and the associated entrance channel perturbation interaction. As mentioned earlier, such consistency follows automatically from the mere application of the Schrödinger operator $H - E$ to the otherwise correct ansatz χ_i^+ from the ECB model. The ensuing result uniquely defines the proper perturbation interaction $V_i^{(\mathrm{MCB})}$ which sharply differs from $V_i^{(\mathrm{ECB})}$. This sole modification constitutes the MCB method [132, 133]. The final scattering wave function χ_f^- and the corresponding exit channel perturbation potential in the ECB model are correct, and they are consistently inter-related. Therefore, the whole description of the exit channel in the ECB method is unaltered when passing to the MCB method. In other words, as stated before, the only difference between the prior transition amplitudes $T_{if}^{(\mathrm{ECB})-}$ and $T_{if}^{(\mathrm{MCB})-}$ from (16.3) and (16.4) in the ECB and MCB methods is in the perturbation interactions $V_i^{(\mathrm{ECB})}$ and $V_i^{(\mathrm{MCB})}$ from (16.5) and (16.6), respectively.

As shown by Belkić [7, 49], the discussed fundamental failure of the ECB method remains incurable by enhancing the angular correlations in the target CI wave function via the inclusion of further terms with a larger number ($N = 3 - 61$) of variational parameters. This is also evidenced in Figs. 16.2 and 16.3. It is obvious from these two figures that, irrespective of the degree of static correlations invoked, saturation of the high-energy cross sections $Q_{if}^{(\mathrm{ECB})-}$ to some constant values always occurs. Moreover, the plateaus of $Q_{if}^{(\mathrm{ECB})-}$ are different for unequal number of the variational parameters in the CI wave functions of the H^- target. In these computations three different sets of the CI ground-state wave functions $\varphi_i(\boldsymbol{x}_1, \boldsymbol{x}_2)$ for the $\mathrm{H}^-(^1S)$ target were employed. They differ only in the number of variationally determined parameters. These encompass the simpler 2- and 3-parameter orbitals of Silverman *et al.* [71]. Also included were more elaborated 21-41 parameter orbitals of Tweed [76], as well as the 61-parameter orbital of Joachain and Terao [77]. The conclusion from these computations illustrated in Figs. 16.2 and 16.3 is that no appreciable improvement is gained even with the highly-correlated 21-61 CI wave functions for which the cross sections $Q_{if}^{(\mathrm{ECB})-}$ still overestimate the experimental data by 2-3 orders of magnitude. This was the case in the work of Gayet *et al.* [131] who used the simple 2-parameter CI orbital from

Ref. [71] (see Fig. 16.1). A similar failure also persists in the PWB method as illustrated in Fig. 16.3 for $N = 21$ with the like pattern for $N = 31 - 61$.

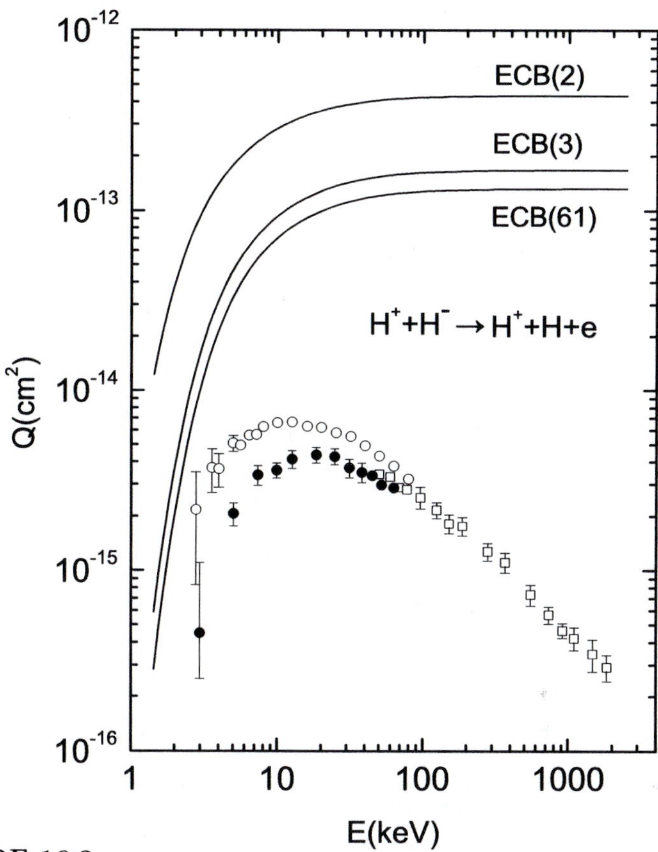

FIGURE 16.2
Total cross sections Q as a function of the incident energy E for process (16.2). Theory: full curves (ECB method [131]). Integer N with the acronyms is the number of variational parameters in the target CI wave function from Silverman *et al.* [71] ($N = 2, 3$), as well as from Joachain and Terao [77] ($N = 61$). Experiment: $\circ$ [18], $\square$ [387] and $\bullet$ [389].

As to the experiments, this time Fig. 16.2 shows other experimental data in addition to those of Peart *et al.* [387]. These include the experimental data of Melchert *et al.* [18] and Peart *et al.* [389] both measured directly for process (16.2) i.e. with proton impact. At higher energies, these two sets of proton impact data [18, 389] merge smoothly into the corresponding electron impact data [387] scaled to the equivalent proton energies, as expected. The error bars at intermediate and high energies on all the experimental data shown on Fig. 16.2 are relatively small, but noticeably larger uncertainties in the measurements from Refs. [18] and [389] are seen at lower energies.

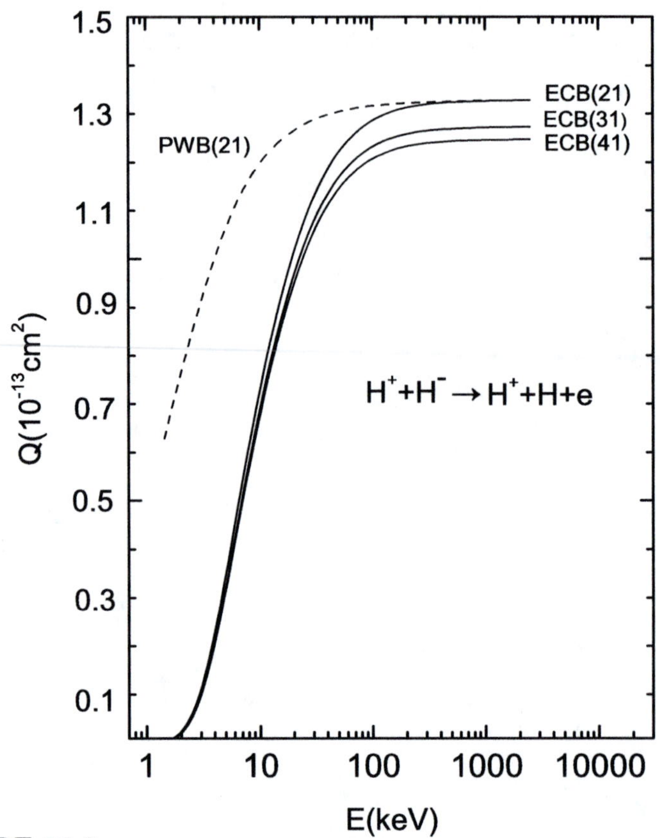

FIGURE 16.3
Total cross sections Q as a function of the incident energy E for process (16.2). Theory: full curves (ECB method [131]) and dashed curve (PWB [131]). Integer N with the acronyms is the number of variational parameters in the target CI wave function from Tweed [76] ($N = 21, 31, 41$).

Before we directly compare the results of the ECB and MCB methods, we shall first deal with the latter theory itself by examining post-prior discrepancy in Fig. 16.4, as well as convergence properties with an increasing degree of static inter-electron correlations, as illustrated in Figs. 16.5 and 16.6.

In Fig. 16.4, we compare the cross sections in the prior Q_{if}^{-} and post Q_{if}^{-} forms from the MCB method for process (16.27) by using the 2-parameter CI wave function of Silverman *et al.* [71]. First to be noticed here is a proper decline of the cross sections as the impact energy E increases, in striking contrast to the plateau in the ECB method from Fig. 16.1.

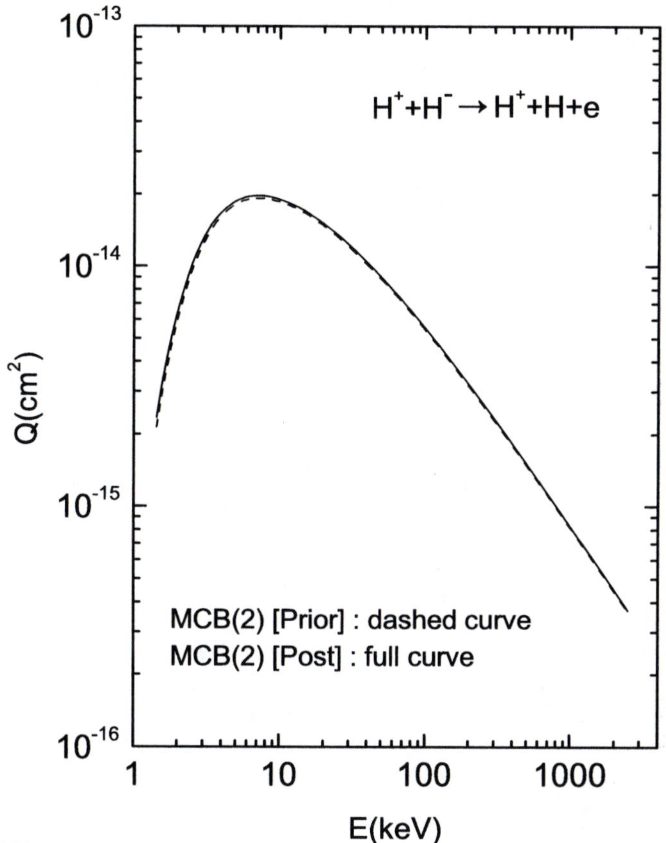

FIGURE 16.4
Total cross sections Q as a function of the incident energy E for process (16.2). Theory: dashed curves (prior MCB method [132, 133]) and dashed curves (post MCB method [132, 133]). Integer 2 with the acronyms is the number of variational parameters in the target CI wave function from Silverman *et al.* [71].

As will be seen later, this decline of $Q_{if}^{(\mathrm{MCB})\pm}$ closely follows the Bethe asymptotic limit $\propto (1/E)\ln(E)$ at large values of E, as it should be from the physical viewpoint. Further, Fig. 16.4 shows that there is perfect agreement between the prior $Q_{if}^{(\mathrm{MCB})-}$ and post $Q_{if}^{(\mathrm{MCB})+}$ total cross sections at all impact energies. This is an excellent feature of the MCB method despite the obvious post-prior asymmetry of the perturbation potentials $V_i^{(\mathrm{MCB})}$ and $V_f^{(\mathrm{MCB})}$ from (16.6) and (16.14) in the transition amplitudes $T_{if}^{(\mathrm{MCB})-}$ and $T_{if}^{(\mathrm{MCB})+}$ given by the matrix elements (16.4) and (16.13), respectively. The lack of any significant post-prior discrepancy in Fig. 16.4 is the guarantee that one can safely rely only upon one version of the MCB method in extensive computations with more elaborated wave functions of H^-. In the transition amplitude $T_{if}^{(\mathrm{MCB})-}$, we neglect the small perturbations $\Delta V_{\mathrm{P}2}$ and O_i that do not contain the inter-electron potential $1/x_{12}$ in the role of dynamic correlations. With this, $T_{if}^{(\mathrm{MCB})-}$ becomes a computationally easier to handle than $T_{if}^{(\mathrm{MCB})+}$. Therefore, in this section, the remaining illustrations based upon our computations will deal exclusively with the prior total cross sections.

The convergence rate of the cross sections $Q_{if}^{(\mathrm{MCB})-}$ is investigated in Figs. 16.5 and 16.6 where the experimental data are also shown. It is seen from Fig. 16.5 that the results for the 2-parameter $(1s1s')$ radially-correlated wave function of Silverman *et al.* [71] differ greatly from the other two curves which include both radial and angular correlations. The angular correlation terms $(l_i' \neq 0)$ of the wave functions with $N = 3 - 61$ parameters [71, 76, 77] do not enter explicitly into the $T_{if}^{(\mathrm{MCB})-}$ from (16.4). Nevertheless, they are implicitly incorporated through the variational parameters that are different for (i) the wave functions built from the pure s–functions $(l_i' = 0)$ and (ii) a mixture of orbitals with $l_i' = 0$ and $l_i' \neq 0$. There is a striking improvement in the cross sections $Q_{if}^{(\mathrm{MCB})-}$ when passing from the 2- to 3-parameter CI wave functions $(1s1s')$ (radial correlations only) and $\{(1s1s'), (2p)^2\}$ (both radial and angular correlations) of Silverman *et al.* [71] (see Fig. 16.5). This indicates a remarkable role of the angular correlations. Such an effect is further illuminated by employing the highly-correlated CI wave functions of Tweed [76] with 21-, 31- and 41-parameters, as well as of Joachain and Terao [77]. The results shown in Figs. 16.5 and 16.6 imply fast convergence of the cross sections $Q_{if}^{(\mathrm{MCB})-}$ as a function of the systematically increased degree of correlations. This is yet another attractive feature of the MCB method.

In Fig. 16.5, comparisons with measurements shows that the cross sections $Q_{if}^{(\mathrm{MCB})-}$ due to the 2-parameter radially correlated orbital [71] overestimate the experimental data by a relatively large factor ranging from 2.9 to 1.6 at $E \in [26.03, 918.06]$ keV. Nevertheless, the overall shape of the curve is still adequate at all energies, leading to qualitative agreement of theory and experiments. However, a substantial improvement in $Q_{if}^{(\mathrm{MCB})-}$ yielding highly satisfactory and quantitative agreement with the experimental data is obtained by using merely the 3-parameter radially and angularly correlated orbital [71]. It is indeed remarkable that this simple $\{(1s1s'), (2p)^2\}$ CI wave function [71] is capable of bringing the MCB method into good agreement with the measurements, as seen in Fig. 16.5. Otherwise, both the 2- and 3-parameter orbitals [71] produce the cross sections $Q_{if}^{(\mathrm{MCB})-}$ that are either in qualitative (2-parameter) or quantitative (3-parameter) agreement with the Bethe asymptote $\propto (1/E)\ln(E)$ at high impact energies E.

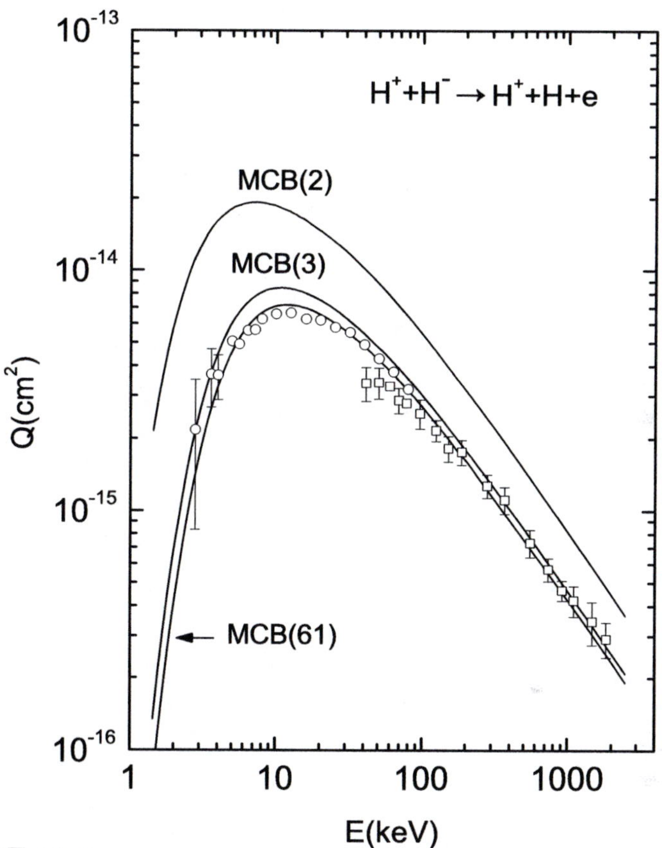

FIGURE 16.5
Total cross sections Q as a function of the incident energy E for process (16.2). Theory: full curves (MCB method [132, 133]). Integer N with the acronyms is the number of variational parameters in the target CI wave function from Silverman *et al.* [71] ($N = 2, 3$), as well as from Joachain and Terao [77] ($N = 61$). Experiment: $\circ$ [18] and $\square$ [387].

Nevertheless, there is still room for quantitative improvement of the MCB method, especially in a quite extended region (5-100 keV) around the Massey peak [377]. To this end, the experience with the 3-parameter orbital [71] motivates the extension of computations by means of the MCB method to also encompass the highly correlated many parameter (21-61) CI wave functions [76, 77]. The ensuing results shown in Figs. 16.5 and 16.6 efficiently achieve this goal. The fully converged cross sections $Q_{if}^{(\mathrm{MCB})-}$ obtained with the 61-parameter orbital [77] are in perfect agreement with all the experimental data from the threshold, through the whole region of the Massey maximum [377], and including the Bethe asymptote $\propto (1/E)\ln(E)$ at high impact energies.

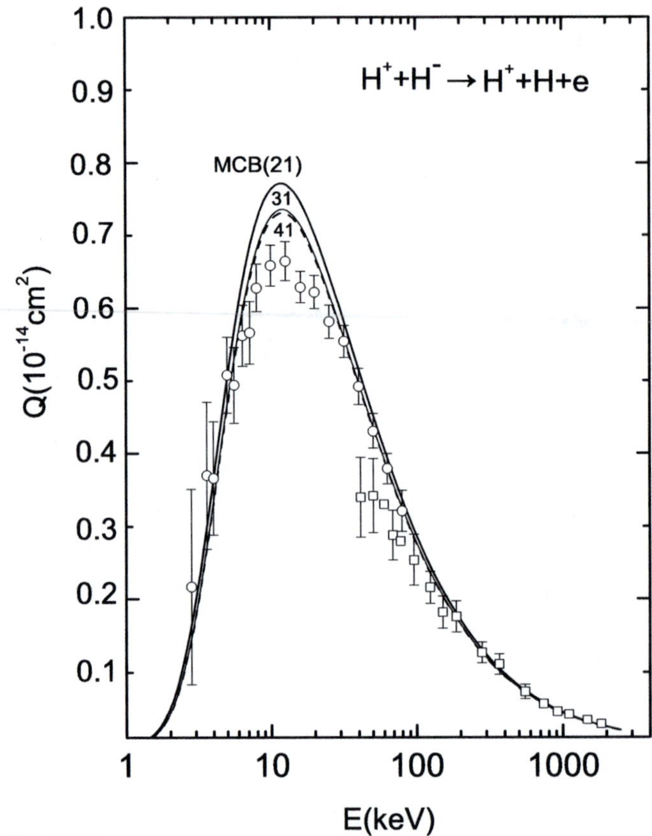

FIGURE 16.6
Total cross sections Q as a function of the incident energy E for process (16.2). Theory: full curves (MCB method [132, 133] with $N = 21, 31$) and dashed curve (MCB method [132, 133] with $N = 41$). Integer N with the acronyms is the number of variational parameters in the target CI wave function from Tweed [76]). Experiment: ○ [18] and □ [387].

After analyzing and discussing separately the overall performance of the ECB and MCB methods, we now carry out their direct comparisons. These comparisons are performed in Figs. 16.7 and 16.8 that also include the corresponding total cross sections obtained using the PWB, B1, AO and MO methods. Of course, the experimental data are plotted also in 16.7 and 16.8 mainly to see the overall performance of the remaining theoretical models i.e. the PWB, B1, AO and MO methods.

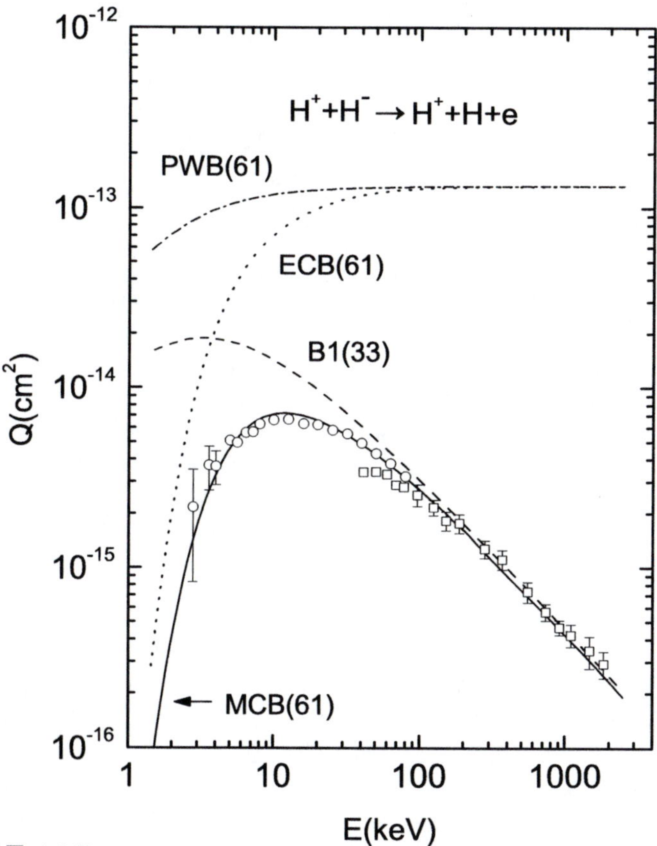

FIGURE 16.7
Total cross sections Q as a function of the incident energy E for process (16.2). Theory: singly-chained curve (PMB method [131] with $N = 61$), dotted curve (ECB method [131] with $N = 61$), dashed curve (B1 method [381] with $N = 33$) and full curves (MCB method [132, 133] with $N = 61$). Integer N with the acronyms is the number of variational parameters in the target wave function from Joachain and Terao [77] ($N = 61$) as well as from Rotenberg [78] ($N = 33$). Experiment: $\circ$ [18] and $\square$ [387].

The net quantitative result of including all the Coulombic effects of free charged particles in both channels of process (16.27) is that the total prior cross sections $Q_{if}^{(\mathrm{ECB})-}$ computed with the 61-parameter (1s1s′) radially correlated wave function $\varphi_i(\boldsymbol{x}_1, \boldsymbol{x}_2)$ [77] are considerably smaller than $Q_{if}^{(\mathrm{PWB})-}$ at low energies, as can be seen from Fig. 16.7.

Previously, Geltman [378] arrived at a similar conclusion in the PWBO approximation empirically modified to include the simplest form of Coulombic effects through the Coulomb normalization factor in the entry channel. Also seen in Fig. 16.7 is that at intermediate and high energies, the results $Q_{if}^{(\mathrm{ECB})-}$ and $Q_{if}^{(\mathrm{PWB})-}$ are very close to each other and they both have the same constant for their asymptotic limit, at variance with the correct Bethe asymptotic behavior $\propto (1/E)\ln(E)$ for large E. This is the first severe repercussion of the inconsistency between scattering states and perturbation potentials encountered in the ECB method. In 1976 the experimental data became available from the measurement carried out by Peart *et al.* [389] for single electron detachment from H^- by proton impact as in process (16.2). It was then revealed that the ECB model overestimates the measured total cross sections Q by two orders of magnitude at energies E between 2.98 and 70.40 keV (see Fig. 16.2). As mentioned, this could have also been inferred using the earlier experimental data of Peart *et al.* [387] from 1970 on the $e + \mathrm{H}^- \longrightarrow e + \mathrm{H} + e$ collisions through rescaling the incident energies as done in Figs. 16.1, 16.2, 16.5 and 16.7. The measurement from Ref. [389] extends all the way up to 918.06 keV of the equivalent proton energy, at which the ECB model exceeds the experimental findings by three orders of magnitude, as can be seen in Figs. 16.2, 16.5 and 16.7. The reason for such a huge discrepancy was not known until 1997 when the problem was reinvestigated by Belkić [132, 133] who found that the distorted wave χ_i^+ and the distorting potential $V_{\mathrm{ECB}} = -1/s_1$ (with or without ΔV_{P2}) in the entrance channel are not consistent with each other. As a result, the MCB method emerged [132, 133] as the most adequate theory to date for detachment process (16.2). The total cross sections $Q_{if}^{(\mathrm{MCB})-}$ obtained with the same 61-parameter wave function φ_i from Ref. [77] are seen in Fig. 16.7 to be 2 to 3 orders of magnitude smaller than $Q_{if}^{(\mathrm{ECB})-}$. Moreover, the two cross sections exhibit a completely different dependence on the impact energy E. The discrepancy between these models is most dramatic at larger values of E for which $Q_{if}^{(\mathrm{MCB})-}$ possesses the correct behavior $\propto (1/E)\ln(E)$, in sharp contrast to the constant limiting value of $Q_{if}^{(\mathrm{ECB})-}$ at high energies. Crucially, the MCB is in perfect agreement with the experimental data at all energies, including the experimental data of Melchert *et al.* [18] from 1999. Simultaneously, the ECB completely fails despite the usage of the most correlated 61-parameter wave function. This finding from Figs. 16.2 and 16.7 disproves the claim by Bell *et al.* [381] that the ECB is unsuccessful relative to measurements because Gayet *et al.* [131] employed the simple 2-parameter wave function [71]. Also shown in Fig. 16.7 are the cross sections $Q_{if}^{(\mathrm{B1})-}$ in the B1 method reported by Bell *et al.* [381]. The results of the MCB and B1 methods agree closely with each other at sufficiently high energies (above 150 keV), as expected, since they both contain the correct Bethe limit. The curve from the B1 method of Bell *et al.* [381]

was obtained using the highly correlated 33-parameter wave function φ_i of Rotenberg and Stein [78]. The cross sections $Q_{if}^{(\mathrm{B1})-}$ from Fig. 16.7 markedly overestimate the experimental data at E below 30 keV, where the electrostatic attractive Coulomb potential between the projectile H^+ and target H^- is strongest. This is because a slow incident proton spends more time in the vicinity of the target. Therefore, in order to overcome the weakness of the B1 method, especially at lower and intermediate energies, it is indispensable to include at least the Coulomb effects between H^+ and H^- in the entrance channel. The MCB method comes to rescue the situation in a remarkably complete manner (compare the B1 and MCB method in Fig. 16.7). In short, the MCB method is seen as being highly adequate due to the full reproduction of the two mutually coherent sets of experimental data. This is a direct consequence of the internal consistency of the MCB method, particularly in providing (i) the initial and final scattering states with the correct asymptotic behaviors, as well as (ii) the adequate distorting perturbation potentials in the entrance and exit channels with their proper connections to the corresponding wave functions. If only part (i) of the correct boundary conditions in the entrance channel is satisfied, the ECB method is obtained, but the net result of the missing part (ii) is the flagrant disagreement of this theoretical model with the experimental data. However, when the requirements (i) and (ii) are simultaneously fulfilled, the MCB method emerges with the ensuing striking success at all impact energies. The worst situation is in the B1 method in which neither the initial scattering wave function Φ_i nor the perturbation interaction $V_i^{(\mathrm{B1})}$ in the entrance channel is correct. This is because at $R \to \infty$, the interaction $V_i^{(\mathrm{B1})} \equiv 1/R - 1/s_1 - 1/s_2 = \Delta V_{\mathrm{P2}} - 1/s_1 \approx -1/s_1$ which causes the transition in the prior transition amplitude $T_{if}^{(\mathrm{B1})-}$ in the B1 method, represents an attractive long-range Coulomb potential $(-1/s_1)$ between the incident proton and the active electron e_1 to be ejected. This is inconsistent with the fact that the entrance channel wave function in the B1 method is merely the unperturbed state $\Phi_i = \varphi_i \exp(i\boldsymbol{k}_i \cdot \boldsymbol{r}_i)$, which could be adequate only if $V_i^{(\mathrm{B1})}$ were a short-range potential.

Finally, in Fig. 16.8, we compare the MCB method with the AO and MO methods. The latter two methods, as the close coupling approximations, employ reasonably large atomic or molecular basis sets of the expansion functions. Thus Ermolaev [384] employed 29 and 36 atomic orbitals, whereas Sidis *et al.* [382] used 12 molecular orbitals in solving the close-coupling system of ordinary differential equations. The AO method [384] with 29 orbitals approximately reproduces only the measured cross sections at very low energies near the threshold, whereas the experimental data at intermediate and large values of E are underestimated by the AO method with 36 orbitals (see Fig. 16.8). Additionally, the results of the AO method with 29 and 36 orbitals do not converge to each other in the overlapping energy region around the Massey maximum of the cross sections. The AO and MCB methods are in good mutual agreement only in a very limited energy range 1.5–4 keV. On the other hand, the MO method grossly underestimates the experimental data at all the energies from 3 to 20 keV that were considered in the computations from Ref. [382]. The curves of the MO and MCB methods are nearly parallel at most of the overlapping energies, but the cross sections of the former method are lower than those due to the latter method (and, as such, underestimate the measured data) by a factor ranging from approximately 7 at 3 keV to about 2 at 10 keV.

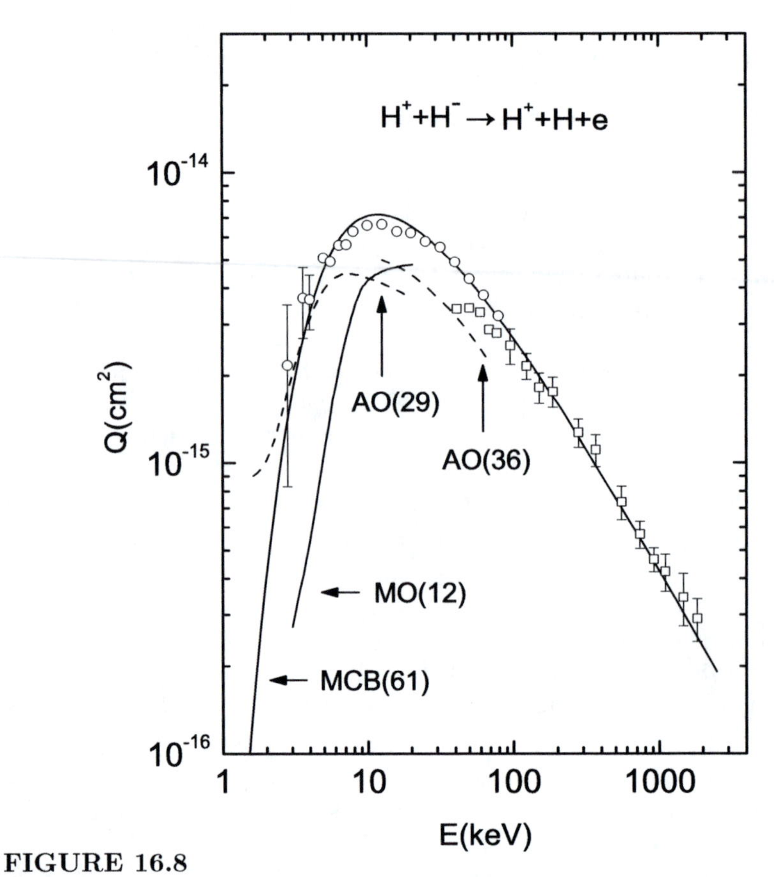

FIGURE 16.8
Total cross sections Q as a function of the incident energy E for process (16.2). Theory: dashed curves (AO method [384] with $N = 29, 36$), full curve $E \in [3, 20]$ keV (MO method [382] with $N = 12$) and full curve $E \in [1.4, 2498]$ keV (MCB method [132, 133] with $N = 61$). Integer N with the acronyms is the number of variational parameters in the target CI wave function from Joachain and Terao [77] ($N = 61$) as well as from atomic [384] ($N = 29, 36$) and molecular [382] ($N = 12$) orbitals. Experiment: $\circ$ [18] and $\Box$ [387].

Overall, this analysis of process (16.2) conclusively demonstrates the need for great care in establishing a proper connection between the long-range Coulomb distortion effects and the accompanying perturbation potential. Otherwise, unphysical results could easily be incurred, as in the work of Gayet, Janev and Salin [131]. In addition, the discussed results clearly show that perfect quantitative agreement between the MCB theory and the available experimental data at all impact energies is possible by using the bound-state wave functions of the $H^-(1s^2)$ target with a high degree of static inter-electron correlations (see Fig. 16.5).

17

Single electron capture

17.1 The CDW-4B method

The transition amplitudes for single electron capture from a helium-like system $(Z_{\rm T}; e_1, e_2)_i$ by $Z_{\rm P}$ i.e. for process (11.3), treated within the CDW-4B method [189] in the prior and post versions without the term $(v\rho)^{2iZ_{\rm P}(Z_{\rm T}-1)/v}$ are given by

$$\begin{aligned} T_{if}^{(\mathrm{CDW})-} &= N^+(\nu_{\rm P})N^{-*}(\nu_{\rm T}) \iiint \mathrm{d}\boldsymbol{x}_1 \mathrm{d}\boldsymbol{x}_2 \mathrm{d}\boldsymbol{R}\, \mathrm{e}^{i\boldsymbol{\alpha}\cdot\boldsymbol{s}_1 + i\boldsymbol{\beta}\cdot\boldsymbol{x}_1} \varphi_{f_1}^*(\boldsymbol{s}_1)\varphi_{f_2}^*(\boldsymbol{x}_2) \\ &\times {}_1F_1(i\nu_{\rm T}, 1, ivx_1 + i\boldsymbol{v}\cdot\boldsymbol{x}_1)[(\Delta V_{\rm P2} + O_i)\varphi_i(\boldsymbol{x}_1, \boldsymbol{x}_2) \\ &\times {}_1F_1(i\nu_{\rm P}, 1, ivs_1 + i\boldsymbol{v}\cdot\boldsymbol{s}_1) \\ &- \boldsymbol{\nabla}_{x_1}\varphi_i(\boldsymbol{x}_1, \boldsymbol{x}_2)\cdot\boldsymbol{\nabla}_{s_1}\, {}_1F_1(i\nu_{\rm P}, 1, ivs_1 + i\boldsymbol{v}\cdot\boldsymbol{s}_1)] \end{aligned} \tag{17.1}$$

and

$$\begin{aligned} T_{if}^{(\mathrm{CDW})+} &= N^+(\nu_{\rm P})N^{-*}(\nu_{\rm T}) \iiint \mathrm{d}\boldsymbol{x}_1 \mathrm{d}\boldsymbol{x}_2 \mathrm{d}\boldsymbol{R}\, \mathrm{e}^{i\boldsymbol{\alpha}\cdot\boldsymbol{s}_1 + i\boldsymbol{\beta}\cdot\boldsymbol{x}_1} \varphi_i(\boldsymbol{x}_1, \boldsymbol{x}_2)\varphi_{f_2}^*(\boldsymbol{x}_2) \\ &\times {}_1F_1(i\nu_{\rm P}, 1, ivs_1 + i\boldsymbol{v}\cdot\boldsymbol{s}_1)\left\{[\Delta V_{\rm P2} + \Delta V_{12}]\varphi_{f_1}^*(\boldsymbol{s}_1) \right. \\ &\times {}_1F_1(i\nu_{\rm T}, 1, ivx_1 + i\boldsymbol{v}\cdot\boldsymbol{x}_1) \\ &\left. - \boldsymbol{\nabla}_{s_1}\varphi_{f_1}^*(\boldsymbol{s}_1)\cdot\boldsymbol{\nabla}_{x_1}\, {}_1F_1(i\nu_{\rm T}, 1, ivx_1 + i\boldsymbol{v}\cdot\boldsymbol{x}_1)\right\} \end{aligned} \tag{17.2}$$

with

$$\Delta V_{\rm P2} = Z_{\rm P}\left(\frac{1}{R} - \frac{1}{s_2}\right) \qquad \Delta V_{12} = \left(\frac{1}{r_{12}} - \frac{1}{x_1}\right) \tag{17.3}$$

where $N^-(\nu_{\rm T}) = \Gamma(1 + i\nu_{\rm T})\mathrm{e}^{\pi\nu_{\rm T}/2}$, $N^+(\nu_{\rm P}) = \Gamma(1 - i\nu_{\rm P})\mathrm{e}^{\pi\nu_{\rm P}/2}$, $\nu_{\rm P} = Z_{\rm P}/v$ and $\nu_{\rm T} = (Z_{\rm T} - 1)/v$. The two momentum transfers $\boldsymbol{\alpha}$ and $\boldsymbol{\beta}$ in (17.1) and (17.2) read as

$$\begin{aligned} &\boldsymbol{\alpha} = \boldsymbol{\eta} - \left(\frac{v}{2} - \frac{\Delta E}{v}\right)\hat{\boldsymbol{v}} \qquad \boldsymbol{\beta} = -\boldsymbol{\eta} - \left(\frac{v}{2} + \frac{\Delta E}{v}\right)\hat{\boldsymbol{v}} \\ &\Delta E = E_i - E_f \qquad E_f = E_{f_1} + E_{f_2} \end{aligned} \tag{17.4}$$

where E_i and $E_{f_{1,2}}$ are the initial (helium-like) and final (hydrogen-like) binding energies, respectively. As can be identified from (17.1) and (17.2), the entrance and exit channel perturbations are

$$V_i = \Delta V_{\mathrm{P}2} + O_i - \boldsymbol{\nabla}_{x_1} \ln \varphi_i(\boldsymbol{x}_1, \boldsymbol{x}_2) \cdot \boldsymbol{\nabla}_{s_1} \tag{17.5}$$

$$V_f = \Delta V_{\mathrm{P}2} + \Delta V_{12} - \boldsymbol{\nabla}_{s_1} \ln \varphi^*_{f_1}(\boldsymbol{s}_1) \cdot \boldsymbol{\nabla}_{x_1}. \tag{17.6}$$

In a test computation using (17.1), we checked that the initial eigen-problem corrective perturbation O_i from (13.40) yields a small contribution relative to those from the other terms. Therefore, the term O_i will be ignored in all the illustrations that follow without incurring significant errors.

In the term $\Delta V_{\mathrm{P}2} = Z_{\mathrm{P}}(1/R - 1/s_2)$ from (17.5) and (17.6), the interaction $-Z_{\mathrm{P}}/s_2$ represents the Coulomb potential between the projectile Z_{P} and the passive electron e_2. The asymptotic tail of this potential is $-Z_{\mathrm{P}}/R$ so that the difference $\Delta V_{\mathrm{P}2}$ between these two latter interactions is of short range as $R \longrightarrow \infty$. The active electron e_1 can be captured by the projectile without its direct interaction with Z_{P}. This occurs when the passive electron e_2 interacts directly with Z_{P} and the subsequent transfer of e_1 to the projectile is made possible through the $e_1 - e_2$ correlations in the bound state $\varphi_i(\boldsymbol{x}_1, \boldsymbol{x}_2)$ of the target $(Z_{\mathrm{T}}; e_1, e_2)_i$. This latter effect is known as static correlation, because it exists in helium-like systems as one of their spectroscopic features without any reference to collision. The interaction $\Delta V_{\mathrm{P}2}$ yields a negligibly small contribution in the forward direction which predominantly determines the total cross sections. This potential is significant only at larger scattering angles in differential cross sections. In any case, in the CDW-4B method, the contribution from the potential $\Delta V_{\mathrm{P}2}$ to both differential and total cross sections rapidly diminishes with increasing values of the impact energy E.

The second term ΔV_{12} in (17.6) represents the dynamic inter-electronic correlation through the corresponding interaction $1/x_{12}$, which is screened at large distances $x_1 \gg x_2$ by the potential $V_1 = 1/x_1$. The $e_1 - e_2$ potential V_{12} must be a constituent part of the interaction potential V_f, since ΔV_{12} emerges in the definition of the exit channel perturbation through the difference between the total interaction $V = Z_{\mathrm{P}}Z_{\mathrm{T}}/R - Z_{\mathrm{P}}/s_1 - Z_{\mathrm{P}}/s_2 - Z_{\mathrm{T}}/x_1 - Z_{\mathrm{T}}/x_2 + 1/x_{12}$ and the binding potentials in the non-interacting hydrogen-like atomic systems $(Z_{\mathrm{P}}, e_1)_{f_1}$ and $(Z_{\mathrm{T}}, e_2)_{f_2}$. The mentioned residual potential $1/x_1$, as the limiting value of V_{12} at infinitely large x_1 and finite x_2, also enters the expression for V_f from (17.6). This is because at infinitely large x_1, the active electron e_1 from the $(Z_{\mathrm{P}}, e_1)_{f_1}$ system cannot discern the individual constituents in the $(Z_{\mathrm{T}}, e_2)_{f_2}$ system which is, therefore, conceived as the net point charge $Z_{\mathrm{T}} - 1$. In order to account for this correctly screened nuclear charge, the genuine potential $-Z_{\mathrm{T}}/x_1$ is written as $-Z_{\mathrm{T}}/x_1 \equiv -(Z_{\mathrm{T}} - 1)/x_1 - 1/x_1$. Here, the term $-(Z_{\mathrm{T}} - 1)/x_1$ is used to yield the distortion $N^-(\nu_{\mathrm{T}})_1F_1(i\nu_{\mathrm{T}}, 1, ivx_1 + i\boldsymbol{v} \cdot \boldsymbol{x}_1)$ with $\nu_{\mathrm{T}} = (Z_{\mathrm{T}} - 1)/v$, whereas the potential $1/x_1$ is joined together with V_{12} to give the short-range potential ΔV_{12} in (17.3).

The third terms in (17.5) and (17.6) are the non-local potential operators

$$V_i^{\rm grad} = -\boldsymbol{\nabla}_{x_1} \ln \varphi_i(\boldsymbol{x}_1, \boldsymbol{x}_2) \cdot \boldsymbol{\nabla}_{s_1} \tag{17.7}$$

$$V_f^{\rm grad} = -\boldsymbol{\nabla}_{s_1} \ln \varphi_{f_1}^*(\boldsymbol{s}_1) \cdot \boldsymbol{\nabla}_{x_1}. \tag{17.8}$$

Due to $V_{i,f}^{\rm grad}$, the transition amplitudes $T_{if}^{\rm (CDW)\mp}$ contain their respective contributions from the two-center functions

$$\boldsymbol{\nabla}_{x_1}\varphi_i(\boldsymbol{x}_1, \boldsymbol{x}_2) \cdot \boldsymbol{\nabla}_{s_1} {}_1F_1(i\nu_{\rm P}, 1, ivs_1 + i\boldsymbol{v} \cdot \boldsymbol{s}_1) \tag{17.9}$$

$$\boldsymbol{\nabla}_{s_1}\varphi_{f_1}^*(\boldsymbol{s}_1) \cdot \boldsymbol{\nabla}_{x_1} {}_1F_1(i\nu_{\rm T}, 1, ivx_1 + i\boldsymbol{v} \cdot \boldsymbol{x}_1). \tag{17.10}$$

For example, the expression (17.10) is seen to couple the final bound state of e_1 on $Z_{\rm P}$ with its simultaneous continuum state in the field of $Z_{\rm T}$. In comparison with the full prior perturbation V_i from (17.5), the choice $V_i \approx V_i^{\rm grad}$ would correspond to neglect of the terms $\Delta V_{\rm P2} + O_i$. This is expected to produce small errors for process (17.14). However, keeping only the scalar product of the gradient operator in the interaction V_f from (17.6) via the approximation $V_f \approx V_f^{\rm grad}$ would result in a more severe error in the post form through omission of $\Delta V_{\rm P2} + \Delta V_{12}$. This will be illustrated for process (11.3).

The physical meaning of the perturbation potentials $V_i^{\rm grad}$ and $V_f^{\rm grad}$ from (17.7) and (17.8) can be seen by first recalling the well-known direct momentum matching mechanism for electron capture studied by means of the first-order methods. In this latter mechanism, capture occurs by way of interaction between the active electron e_1 and the projectile nucleus $V_{\rm P1} = -Z_{\rm P}/s_1$. To see the details, it is convenient to conceive the transition probability for capture in terms of the initial $\tilde{\varphi}_{n_i l_i m_i}$ and final $\tilde{\varphi}_{n_{f_1} l_{f_1} m_{f_1}}$ bound state wave functions in momentum space. When a fast impinging nucleus $Z_{\rm P}$ of mass $M_{\rm P}$ passes by a helium-like target with a large momentum $M_{\rm P}\boldsymbol{v}$ and velocity $\boldsymbol{v}$, the electron e_1 of mass $m_e \ll M_{\rm P}$, which is orbiting about the nucleus $Z_{\rm T}$ with the velocity $v_e \ll v$, could be captured via a single binary collision $Z_{\rm P} - e_1$ only if considerable momentum of the order of $m_e\boldsymbol{v}$ is imparted to e_1. Since this displacement $m_e\boldsymbol{v}$ in the momentum components of $\tilde{\varphi}_{n_{f_1} l_{f_1} m_{f_1}}$ will increase with augmentation of v, it is clear that only the largest components of the momentum space wave functions would be able to provide the direct momentum matching which, in turn, would yield a non-vanishing overlap of the initial and final orbitals. A large momentum $m_e\boldsymbol{v}$ could be transferred to e_1 only if $Z_{\rm P}$ comes close to the electron e_1, which possesses a very small initial momentum in the target ($m_e v_e \ll m_e v$). Hence, the direct momentum matching mechanism is expected to be operative mainly at small impact parameters. This is a common feature of the BK1-4B and JS1-4B methods for single electron capture from helium-like targets by fast nuclei.

Alternatively, capture can occur by considering the potential $V_{\rm P1}$ as a perturbation which distorts the asymptotic state Φ_i in entrance channel. As usual, Φ_i is given by the product of the two-electron target wave function $\varphi_i(\boldsymbol{x}_1, \boldsymbol{x}_2)$ and the plane wave for the relative motion of the projectile $\Phi_i = \varphi_i \exp{(i\boldsymbol{k}_i \cdot \boldsymbol{r}_i)}$. The said distortion of Φ_i is accomplished through the standard multiplication of the channel state Φ_i by the appropriate electronic Coulomb wave for electron e_1 in the field of $Z_{\rm P}$, and an automatic cancellation of $V_{\rm P1}$ from the initial perturbation $Z_{\rm P}(2/R - 1/s_1 - 1/s_2) = \Delta V_{\rm P2} - Z_{\rm P}/s_1 = \Delta V_{\rm P2} + V_{\rm P1}$ in the prior transition amplitude. This leads to the non-local two-center operator $V_i^{\rm grad}$ from (17.7) as typical for the CDW method. In this framework, the projectile obviously does not directly impart a large momentum of the order $m_e\boldsymbol{v}$ to the active electron e_1. Instead, this needed momentum is transferred indirectly via the long-range distortion effects and, hence, close encounters between $Z_{\rm P}$ and e_1 are no longer mandatory, as opposed to the first-order theories. This illustrates that the small size of the impact parameter cannot be used as an unambiguous signature of the direct momentum matching mechanism. In addition, an impact parameter is not an observable in the sense of being a physical quantity which could be directly measured in a collision experiment. However, a clearer situation emerges from the analysis of recoiled particle $(Z_{\rm T}, e_2)_{f_2}$ which is the target remainder in process (11.3). Namely, in order to conserve the total momentum of the whole four-body collision system, the residual $(Z_{\rm T}, e_2)_{f_2}$ ion in the exit channel must recoil in the backward direction. This effect can readily be observed by means of the COLTRIMS technique.

For the two-electron initial state $\varphi_i(\boldsymbol{x}_1, \boldsymbol{x}_2)$, we employ the CI wave function $(1s1s')$ of Silverman *et al* [71]

$$\varphi_i(\boldsymbol{x}_1, \boldsymbol{x}_2) = \frac{\mathcal{N}}{\pi}({\rm e}^{-ax_1 - bx_2} + {\rm e}^{-bx_1 - ax_2}) \tag{17.11}$$

where $\mathcal{N}$ is the normalization constant

$$\mathcal{N}^{-2} = 2[(ab)^{-3} + (a/2 + b/2)^{-6}]. \tag{17.12}$$

Despite its simplicity, this open-shell orbital of e.g. the ground-state wave function of helium includes radial correlations to a large extent, within approximately 95%. It is also instructive to consider the non-correlated initial bound-state wave function $\varphi_i(\boldsymbol{x}_1, \boldsymbol{x}_2)$ which can be chosen in the one-parameter form given by Hylleraas [58]

$$\varphi_i(\boldsymbol{x}_1, \boldsymbol{x}_2) = N_{\rm eff}^2 {\rm e}^{-Z_{\rm eff}(x_1 + x_2)} \qquad N_{\rm eff} = \left(\frac{Z_{\rm eff}^3}{\pi}\right)^{1/2}. \tag{17.13}$$

After an analytical calculation using the Nordsieck technique [311], the expressions for these two transition amplitudes can be reduced to a three-dimensional integral, which is evaluated numerically e.g. by means of the adaptive Gauss-Legendre quadratures, as in the initial study of Belkić *et al.* [189] and in a subsequent related work of Mančev [190]. The explicit computations for illustrations include only the ground-to-ground state transition of the captured electron. The contribution from excited states of hydrogen is taken into account approximately via the standard n^{-3} scaling law of Oppenheimer [247] with an overall multiplicative factor 1.202.

For illustrations, we shall first consider single electron capture from helium by fast protons in process

$$\mathrm{H}^+ + \mathrm{He}(1s^2) \longrightarrow \mathrm{H}(1s) + \mathrm{He}^+(1s). \tag{17.14}$$

As mentioned, except for the work of Belkić *et al.* [189] followed by Mančev [130, 190], all the previous computations in the CDW method on single electron capture from helium-like targets by fast nuclei used the perturbation interactions only in the form of the gradient potential operators V_i^{grad} and V_f^{grad} from (17.7) and (17.8), respectively.

17.1.1 Differential cross sections: The Thomas double scattering at all energies

Differential cross sections in the CDW-4B method for process (17.14) without the term $(\mu v \rho)^{2iZ_{\mathrm{P}}(Z_{\mathrm{T}}-1)/v}$ are shown in Fig. 17.1 highlighting the two Thomas peaks of the 1st and 2nd kind, which are mediated by the double scatterings $\mathrm{H}^+ - e_1 - \mathrm{He}^{2+}$ and $\mathrm{H}^+ - e_1 - e_2$, respectively [7, 49]. It can be seen that the Thomas peak for the $\mathrm{H}^+ - e_1 - e_2$ mechanism is systematically and clearly present at all impact energies and without any splitting. At sufficiently high energies, this latter peak is located at the Thomas critical angle $\theta_{\mathrm{c.m.}} \approx 0.027$ deg. At lower energies e.g. 30, 50 and 100 keV (not shown), the Thomas critical angles for the $\mathrm{H}^+ - e_1 - e_2$ peak are near 0.040, 0.035 and 0.031 deg, respectively. On the other hand, the Thomas peak for the $\mathrm{H}^+ - e_1 - \mathrm{He}^{2+}$ mechanism appears clearly only at sufficiently high energies in the CDW-4B method, where it is always split in the middle, but this has not been confirmed experimentally. To check these predictions of the CDW-4B theory, especially on the $\mathrm{H}^+ - e_1 - e_2$ mechanism, it would be important to have the experimental data on $\mathrm{d}Q/\mathrm{d}\Omega$ for process (17.14) with the two separate contributions from the Thomas double scattering of the 1st and 2nd kind. At present, such measured data are unavailable.

Detailed computations on $\mathrm{d}Q^+/\mathrm{d}\Omega \equiv \mathrm{d}Q/\mathrm{d}\Omega$ in the post CDW-4B method reveal that a contribution from ΔV_{12} increases with augmentation of the impact energy, as is also clear from Fig. 17.1 at $E = 0.293$, 2.0 and 7.4 MeV. The potential ΔV_{12} itself yields two maximae in $\mathrm{d}Q/\mathrm{d}\Omega$, one of which is in the forward direction ($\theta \approx 0$), whereas the other is in the vicinity of the critical Thomas angle $\theta_{\mathrm{c.m.}} \approx 0.027$ deg. The Thomas peak at $\theta_{\mathrm{c.m.}} \approx 0.027$ deg is due to a double elastic collision of the electron e_1, which first scatters on Z_{P} and then on e_2 before it finally becomes bound to the projectile, thus forming the $(Z_P, e_1)_{f_1}$ system. The difference between the heights of the peaks at zero angle and $\theta_{\rm c.m.} \approx 0.027$ deg decreases with increasing E, such that the relative importance of the Thomas double scattering increases as compared with the forward collision. For example, at $E = 2$ MeV the height of the forward peak is larger than the one at $\theta_{\rm c.m.} \approx 0.027$ deg by about a factor of 12 and this difference is reduced to only 3.5 at $E = 50$ MeV (not shown in Fig. 17.1 in order to avoid clutter).

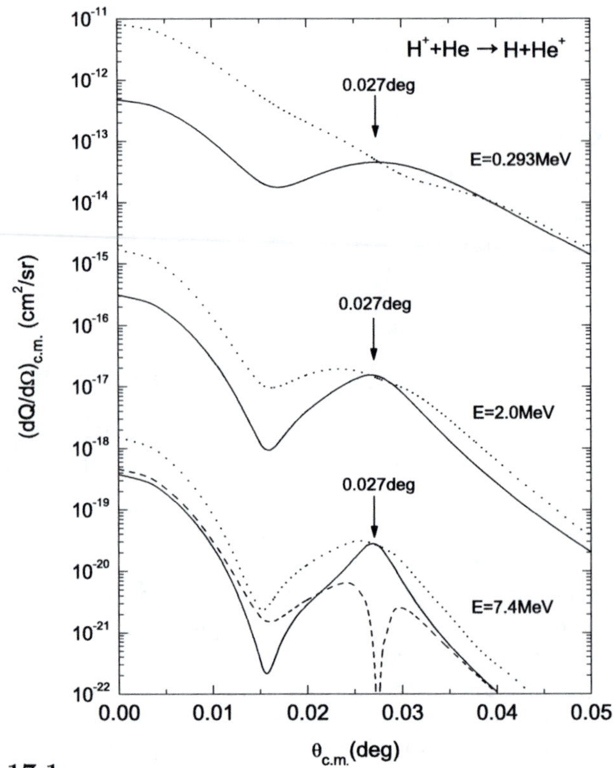

FIGURE 17.1

Post CDW-4B method [7, 49] for differential cross sections $dQ/d\Omega$ as a function of scattering angle θ at the incident energies $E = 0.293$, 2 and 7.4 MeV for process (17.14). Dotted lines: the results from the complete perturbation V_f summing the three contributions from (i) $-\boldsymbol{\nabla}_{s_1} \ln \varphi_f^*(\boldsymbol{s}_1) \cdot \boldsymbol{\nabla}_{x_1}$, (ii) $\Delta V_{12} = 1/r_{12} - 1/x_1$ and (iii) $\Delta V_{P2} = 1/R - 1/s_2$. Dashed and full lines: the separate contributions from (i) and (ii), respectively. To avoid clutter, the contribution from the gradient-gradient potential operator listed above as (i) is shown only at 7.4 MeV.

The Thomas peak also exists at energies lower than those shown in Fig. 17.1, but it is shifted to angles larger than 0.027 deg. For example, in the case of process (17.14), the peak at the critical angle due to the double binary collision $H^+ - e_1 - e_2$ becomes comparable to the forward maximum at $E = 50$ keV. Moreover, we have checked, but not displayed on Fig. 17.1 that near $\theta_{\text{c.m.}} \approx 0.054$ deg at $E = 25$ keV the same $H^+ - e_1 - e_2$ mechanism yields a peak which is an order of magnitude larger than the one at zero degree. The critical angle of the double binary scattering of e_1 depends only on the ratio m_e/M_P of the masses of the electron and the projectile nucleus. Thus, as seen in Fig. 17.1, the angle $\theta_{\text{c.m.}} \approx 0.027$ deg is the same irrespective of whether e_1 scatters on the target nucleus $Z_P \equiv He^{2+}$ or on the passive electron e_2 in the second collision.

In the CDW-4B method for process (17.14), the $H^+ - e_1 - He^{2+}$ double binary collision is included through the terms $V_{i,f}^{\text{grad}}$. For example, in the post CDW-4B method, the contribution of V_f^{grad} to $\text{d}Q/\text{d}\Omega$ near $\theta_{\text{c.m.}} \approx 0.027$ deg is very different from the one due to the ΔV_{12} potential for the $H^+ - e_1 - e_2$ Thomas peak. As conveyed by Fig. 17.1, while the $H^+ - e_1 - e_2$ mechanism produces an unsplit Thomas peak near the critical angle $\theta_{\text{c.m.}} \approx 0.027$ deg, the $H^+ - e_1 - He^{2+}$ double collision yields a narrow dip at the same angle. The splitting of the $H^+ - e_1 - He^{2+}$ Thomas peak is caused by a destructive interference around $\theta_{\text{c.m.}} \approx 0.027$ deg in the part of the integrand containing the two Kummer functions responsible for double scattering, ${}_1F_1(i\nu_P, 1, ivs_1 + i\boldsymbol{v}\cdot\boldsymbol{s}_1)\boldsymbol{\nabla}_{s_1}\varphi_{1s}^*(\boldsymbol{s}_1)\cdot\boldsymbol{\nabla}_{x_1}{}_1F_1(i\nu_T, 1, ivx_1 + i\boldsymbol{v}\cdot\boldsymbol{x}_1)$. However, a splitting of the Thomas peak has no physical significance for a positively charged projectile impinging on a helium-like target, and indeed it has never been observed experimentally. At present, there are no experimental data that could unambiguously disentangle the two competing Thomas mechanisms, but such measurements should be feasible using the COLTRIMS technique.

17.1.2 Total cross sections

Computations of the post and prior total cross sections for process (17.14) studied using the CDW-4B method have been carried out at incident energies ranging from 20 keV to 20000 keV by Belkić *et al.* [49, 189]. These results are summarized in Figs. 17.2–17.6. In our discussion, emphasis will be placed upon the relative role of various terms in the full prior V_i and post V_f perturbations from (17.5) and (17.6). This will be compared with the corresponding results obtained with the abridged perturbations in the forms of the gradient potential operators V_i^{grad} and V_f^{grad} from (17.7) and (17.8) that have predominantly been used in the past due to their computational simplicity.

In Figs. 17.2 and 17.3 we show the results obtained with and without the contributions from the potential ΔV_{P2}, which involves capture of e_1 through the screened $Z_P - e_2$ interaction, contained both in the prior and post CDW-4B methods from (17.1) and (17.2). It is seen that above 100 keV, the cross sections accounting for ΔV_{P2} in V_i and V_f are smaller by some 10% than those neglecting this term. The difference between the results with and without

ΔV_{P2} in both prior and post CDW-4B method remains nearly constant with increasing impact energy E.

The potential ΔV_{P2} acts within the so-called indirect velocity matching mechanism in process (17.14). The electron e_2 receives a large momentum $m_e \boldsymbol{v}$ through its interaction with the projectile nucleus $V_{\mathrm{P2}} = -Z_{\mathrm{P}}/s_2$. This latter momentum is afterwards transferred to e_1. The mediator of such a transfer is the static correlation of the two electrons in the helium target.

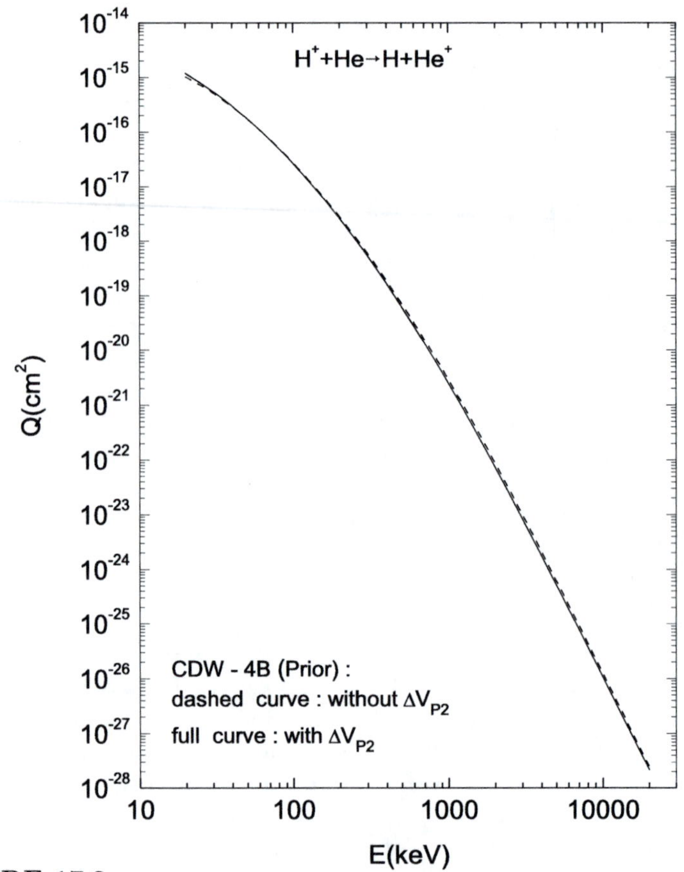

FIGURE 17.2

Total cross sections Q as a function of the incident energy E for for process (17.14). Theory: dashed curve (prior CDW-4B method without the perturbation ΔV_{P2} [127]) and full curve (prior CDW-4B method with the perturbation ΔV_{P2} [127]).

In this indirect way, electron e_1 acquires the necessary large momentum to be matched with the scattered projectile Z_{P} to create the $(Z_{\mathrm{P}}, e_1)_{f_1}$ system.

The relative importance of the indirect momentum matching mechanism has largely been overlooked in the past, with the exception of Refs. [189, 190]. Recall that ΔV_{P2} must be a short-range potential appearing in the transition amplitudes, as a consequence of fulfilling the correct boundary condition in the exit channel [41, 44, 49]. These conditions require that a perturbation potential which can cause a transition must be of short range, in the sense of vanishing faster than $1/R$ as R becomes infinitely large.

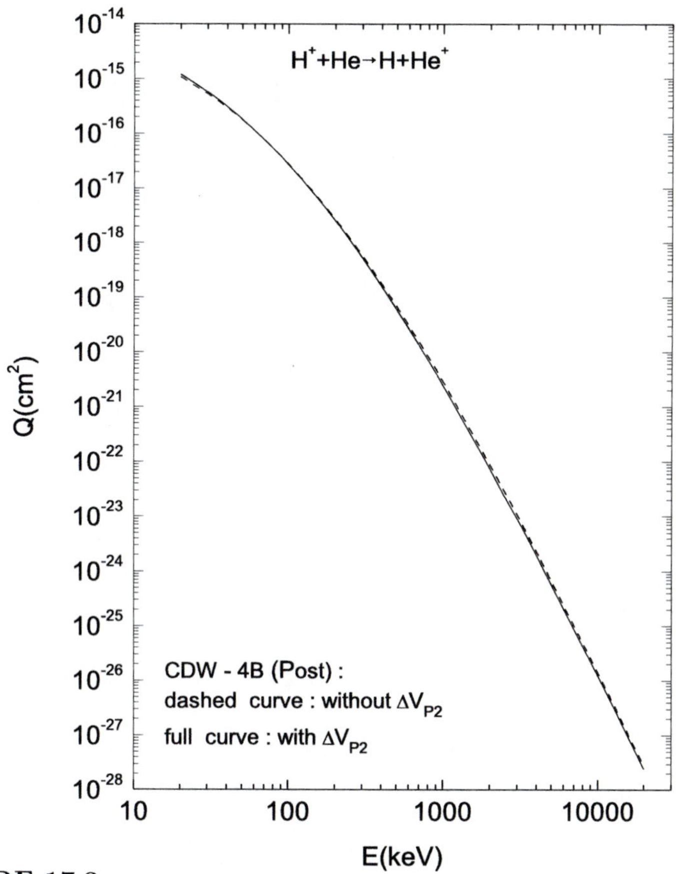

FIGURE 17.3
Total cross sections Q as a function of the incident energy E for for process (17.14). Theory: dashed curve (post CDW-4B method without the perturbation ΔV_{P2} [127]) and full curve (post CDW-4B method with the perturbation ΔV_{P2} [127]).

Only short-range potentials can lead to free particles in the asymptotic region. Free particles are necessary to unequivocally establish the two key situations, one in the remote past before collision, and the other in the distant future when scattering is completed. This is demanded by the concept of asymptotic freedom [41] according to which the interactive dynamics of the entire system with all the external perturbations must reduce to the unperturbed (free) dynamics at infinitely large times.

Post-prior discrepancy for process (17.14) is examined in Figs. 17.4 and 17.5. using the two sets of the perturbation potentials $V_{i,f}^{\rm grad}$ and $V_{i,f}$. As

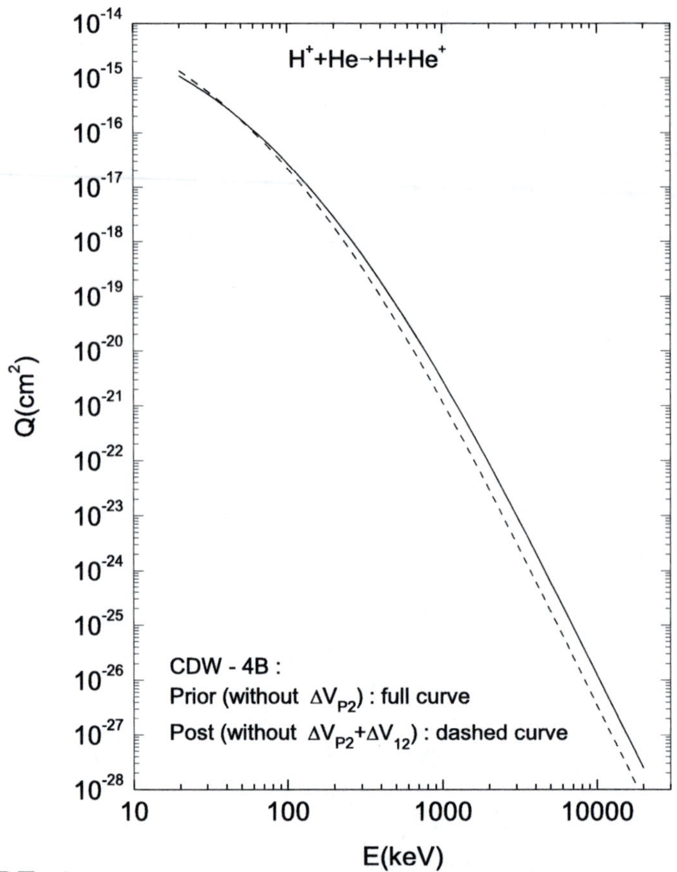

FIGURE 17.4
Total cross sections Q as a function of the incident energy E for for process (17.14). Theory: full curve (prior CDW-4B method without the perturbation $\Delta V_{\rm P2}$ [127]) and dashed curve (post CDW-4B method without the perturbations $\Delta V_{\rm P2}$ and ΔV_{12} [127]). Neglect of the term O_i is not explicitly mentioned regarding the dashed curve for the prior CDW-4B method, since this contribution is small and, as such, ignored in all the illustrations on single electron capture from helium-like targets.

mentioned, $V_i^{\rm grad}$ and $V_f^{\rm grad}$ from (17.7) and (17.8), respectively, were used in nearly all the previous studies on the CDW method with the exception of Refs. [189, 190]. Both $V_i^{\rm grad}$ and $V_f^{\rm grad}$ disregard the mediator $\Delta V_{\rm P2}$ of the static correlations. Additionally, $V_f^{\rm grad}$ neglects the dynamic correlations by ignoring the potential ΔV_{12}. We saw in Figs. 17.2 and 17.3 that neglect of $\Delta V_{\rm P2}$ leads to relatively small errors in total cross sections for process (17.14). However, the situation is radically different for the neglected term

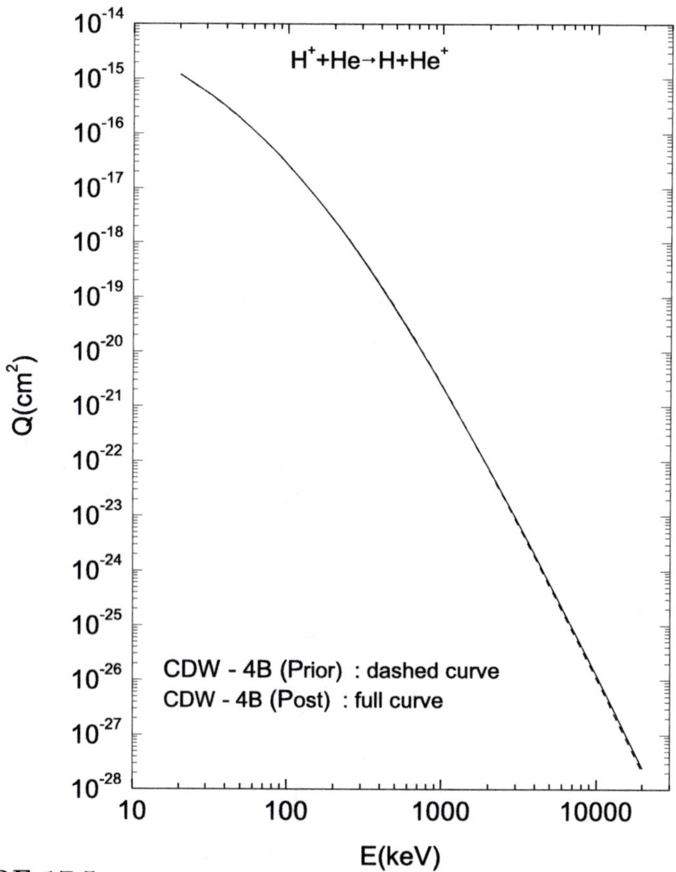

FIGURE 17.5
Total cross sections Q as a function of the incident energy E for for process (17.14). Theory: dashed curve (prior CDW-4B method with the complete perturbation V_i [127]) and full curve (post CDW-4B method with the complete perturbation V_f [127]).

ΔV_{12} in the post cross sections, as seen in Fig. 17.4. Thus, above 100 keV,

where the CDW model is expected to be valid, a huge post-prior discrepancy exists when using (17.7) and (17.8). In this region, the post cross sections are always smaller than those in the prior form. The discrepancy between the post and prior cross sections increases with augmentation of the impact energy. This indicates that the dynamic electron correlation effects ΔV_{12} become more prominent at higher incident energies. However, Fig. 17.5 shows that once the complete perturbations V_i and V_f are retained in the computations, according to (17.5) and (17.6), the full CDW-4B emerges with a very appealing feature of a dramatically reduced post-prior discrepancy. This is an excellent property of the method, since basically the same physical assumptions are involved in both the prior and post forms that, therefore, are expected to yield very similar results. Although the difference between the post and prior results with the complete V_i and V_f is small, it should be noticed that the two curves in Fig. 17.5 are not parallel to each other. The difference between these two curves becomes more significant at higher energies, where the post results are larger than the prior ones, as opposed to the situation in Fig. 17.4. Despite the fact that the post-prior discrepancy with the complete V_i and V_f from (17.5) and (17.6) does not exceed 15%, the trend of this discrepancy is significant, indicating again a greater role of the dynamic electron correlations at higher impact energies.

In Figs. 17.2–17.5, we examined a number of intrinsic reliability tests of the CDW-4B method. This is accomplished by studying the importance of various terms in the full perturbations V_i and V_f, as well as by investigating the post-prior discrepancy. The sensitivity of cross sections for process (17.14) to static inter-electron correlations in the helium target was examined by Belkić *et al.* [189] through the initial wave functions of Hylleraas [58] and Silverman *et al.* [71]. The common conclusion from these tests was that the CDW-4B method is internally consistent and reliable, with a negligible post-prior discrepancy, insofar as the full perturbations V_i and V_f from (17.5) and (17.6) are used. However, intrinsic reliability does not necessarily imply adequacy and validity in practice. Therefore, in order to critically assess the overall validity of the CDW-4B method, a comparison with measurements is necessary.

The total cross sections in the post version as a function of impact energy are displayed in Fig. 17.6. The full curve obtained by means of the CDW-4B method corresponds to the case where the complete perturbation $V_f = \Delta V_{\mathrm{P}2} + \Delta V_{12} - \boldsymbol{\nabla}_{s_1}\varphi_f^*(\boldsymbol{s}_1)\cdot\boldsymbol{\nabla}_{x_1}$ is used. A comparison with a number of experimental data is also shown in Fig. 17.6. As can be seen from this figure, the CDW-4B method [49, 189] is in excellent agreement with the available experimental findings. It should be emphasized that this comparison extends over the three orders of magnitude of the impact energy (20–20000) keV, for which the cross sections vary within twelve orders of magnitude $(10^{-27} - 10^{-15})\mathrm{cm}^2$. With such a stringent test, the CDW-4B method is seen in Fig. 17.6 to establish its reliability at energies $E \geq 70\,\mathrm{keV}$ in accordance with (14.69).

In order to determine the relative importance of dynamic correlations, the post total cross sections in the CDW-4B method used without the potential

ΔV_{12} from the complete perturbation V_f are plotted in Fig. 17.6 (dashed curve). It can be observed from this figure that dynamic electron correlations are essential, since exclusion of the relevant term ΔV_{12} yields results that significantly underestimate the experimental data at all energies above 100 keV. These theoretical findings provide evidence for the importance of dynamic electron correlations, especially at high impact energies.

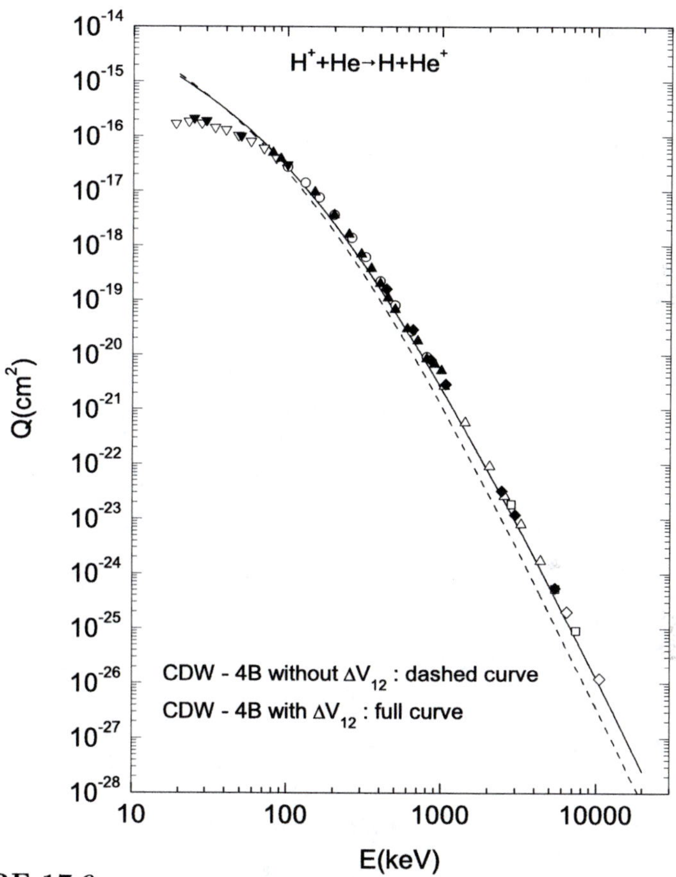

FIGURE 17.6
Total cross sections Q as a function of the incident energy E for for process (17.14). Theory: dashed curve (post CDW-4B method without the perturbation ΔV_{12} [127]) and full curve (post CDW-4B method with the perturbation ΔV_{12} [127]). Experiment: Δ [297], $\circ$ [309], ∇ [352], $\Box$ [393], $\Diamond$ [394], ▲ [395] ♦ [396] and ▼ [400].

Overall, as seen in Fig. 17.5, the difference between the post and prior cross sections with the full perturbations V_i and V_f from (17.5) and (17.6) is small and does not exceed 15%. By contrast, Fig. 17.4 shows that the usual approximations $V_i \approx V_i^{\rm grad}$ and $V_f \approx V_f^{\rm grad}$ with $V_i^{\rm grad}$ and $V_f^{\rm grad}$ given by (17.7) and (17.8) yield a very large post-prior discrepancy. Otherwise, the post cross sections from the CDW-4B method are found to be in excellent agreement with measurements (see Fig. 17.6).

The CDW-4B method has also been successfully applied by Mančev [130, 190] to other processes, such as the He^{2+} – He, Li^{3+} – He and H^+ – Li^+ collisions, as illustrated in Figs. 17.7–17.9. The main goal of these latter studies was to determine whether dynamic electronic correlations also remain important for the investigated collisions at high energies.

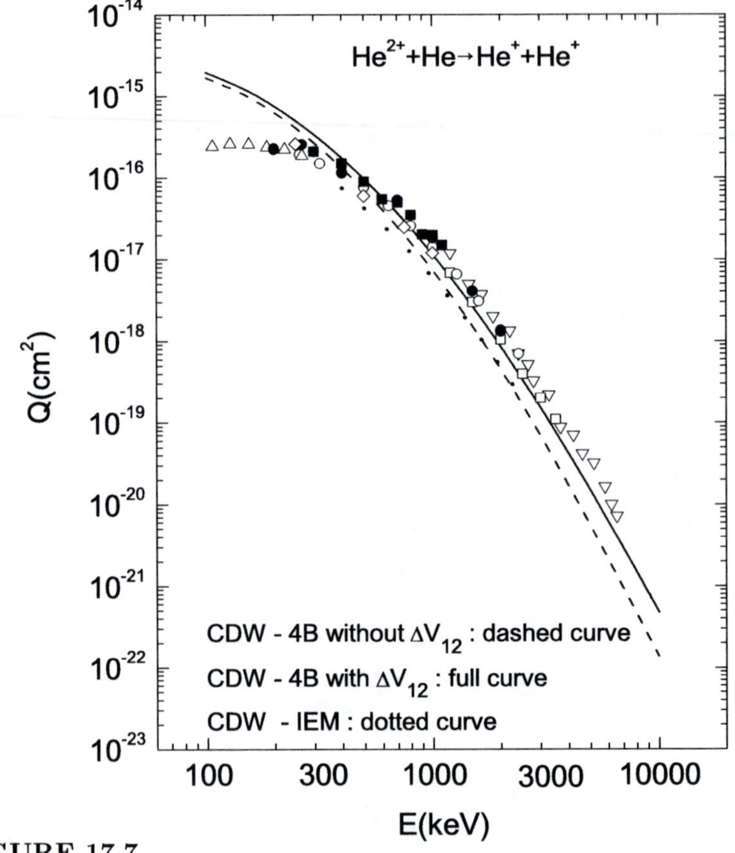

FIGURE 17.7
Total cross sections Q as a function of the incident energy E for for process (17.14). Theory: dashed curve (post CDW-4B method without the perturbation ΔV_{12} [190]), full curve (post CDW-4B method with the perturbation ΔV_{12} [190]) and dotted curve (CDW-IEM [350]). Experiment: • [304], □ [305], ○ [309], △ [352], ▽ [397], ◇ [398] and ■ [399].

The total cross sections obtained for the He^{2+} – He collision by means of the one-parameter wave function of Hylleraas [58] and the two-parameter radially correlated function of Silverman *et al.* [71] were found to be close to each other [130, 190]. Although the Hylleraas wave function is less accurate, it includes some form of the radial correlations through the presence of the Slater-screened effective charge of the target nucleus (e.g. $Z_{\rm T}^{\rm eff} = 1.6875$ for $Z_{\rm T} = 2$), as stated earlier. The prior form (17.1) does not contain the inter-electronic term $1/r_{12}$, which explicitly accounts for the dynamic correlations. As a result, the prior amplitude and, accordingly, the prior cross sections, are expected to be more sensitive to the accuracy of the initial wave function than the corresponding results from the post form. This has indeed been verified to be the case (see Ref. [190]). Nevertheless, this latter effect is less important for the Li^+ target, due to a higher nuclear charge [130].

As a check of the computations, when the term $\Delta V_{\rm P2}$ is neglected in the prior form, the earlier results [269, 270, 273] were reproduced exactly for capture into the ground state in terms of the wave function from Refs. [58] and [71]. More recently [189, 190], the prior total cross sections have been computed with the complete perturbation potential V_f. Here, the difference between the contributions from the usual gradient operators $-\boldsymbol{\nabla} \cdot \boldsymbol{\nabla}$ and the term $\Delta V_{\rm P2}$ does not exceed 15% above 30 keV/amu. The term $\Delta V_{\rm P2}$ also has a similar influence on the results obtained in the case of the post formalism. As stated, the potential $-Z_{\rm P}/s_2$ has the asymptotic value $-Z_{\rm P}/R$ at large distances between $Z_{\rm P}$ and e_2. A relatively small contribution of the term $\Delta V_{\rm P2} = Z_{\rm P}(1/R - 1/s_2)$ suggests that for single electron capture at intermediate and high energies, the potential $-Z_{\rm P}/s_2$ is nearly cancelled by $Z_{\rm P}/R$. Therefore, ignoring the term $\Delta V_{\rm P2}$ does not represent a severe approximation, and this is what has previously been done [269, 270, 273].

The post total cross sections derived with the full perturbation potential according to (17.2), as well as without the terms ΔV_{12}, are displayed in Fig. 17.7, where the existing experimental data are also plotted for single capture in the He^{2+} – He collisions. The CDW-4B method with the complete perturbation (full curve in Fig. 17.7) is in good agreement with the available measurements above 150 keV/amu. Neglect of the relevant term for the dynamic electron correlation $\Delta V_{12} = 1/r_{12} - 1/x_1$ from (17.2) leads to the results that are shown with the dashed curve in Fig. 17.7 and they are seen to underestimate the experimental findings. Similar to the case of the H^+ – He collision, the difference between the two curves for the He^{2+} – He scattering (with and without ΔV_{12}) becomes more significant at higher impact energies. This reaffirms the importance of dynamic electron correlations for single electron capture, especially at higher impact energies. In the same figure, comparison is made between the results from the CDW-4B method [190] and CDW-IEM [350] (the latter uses the wave function of Pluvinage [60] for the target). The formulation of Dunseath and Crothers [350] ignores dynamic correlations altogether, and this is the main reason for disagreement between the CDW-IEM and the experimental data, as is clear from Fig. 17.7. Moreover, this also indicates that the electron dynamic correlations in the perturbation potential which causes the transition are more important than the static ones in the target bound state wave function.

As discussed, the discrepancy between the post and prior cross sections in the CDW-4B method depends essentially on the choice of the ground state wave function of helium. This discrepancy is larger for the wave function of Hylleraas [58] than for the one due to Silverman *et al.* [71]. In the case of the wave functions of Hylleraas [58] and Silverman *et al.* [71], the post-prior discrepancy does not exceed 20% and 5%, respectively, for the He^{2+} – He collisions at energies considered in Fig. 17.7. For instance, at impact energies 600, 6000 and 10000 keV, the post-prior discrepancy for the Hylleraas wave function is 6.2%, 10.8% and 17.9%, respectively, whereas at the same energies for the orbital of Silverman *et al.* [71], the discrepancy is 2.2%, 1.2% and 4.9%, respectively. Of course, the post-prior discrepancy would not exist if the exact wave function were known for helium.

Next, we consider single electron capture from helium by the Li^{3+} ion

$$Li^{3+} + He \longrightarrow Li^{2+} + He^{+}. \tag{17.15}$$

This process has been treated previously using various theoretical methods as reported e.g. in Refs. [44, 198] and [401]–[404]. All these studies consider process (17.15) as an equivalent three-body problem by ignoring correlation effects from the outset. The contribution from the electron-electron interaction via the DEC within the CDW-4B method for the Li^{3+} – He collisions has been examined in Ref. [130].

The post and the prior total cross sections of the CDW-4B method obtained using the one-parameter orbital of Hylleraas [58] and the two-parameter wave function of Silverman *et al.* [71] are close to each other for process (17.15), similarly to the other related collisional systems examined in the present work. More specifically, in the case of the post form, the difference between the two results for these two wave functions is less than 7%, whereas for the prior form, the post-prior discrepancy is within 16% for process (17.15) [130]. As stated, the prior cross sections are more sensitive to the accuracy of the initial state wave function than the post ones. This is due to the fact that the expression for the prior amplitude in (17.1) does not contain the inter-electron term $1/r_{12}$, which is the signature of dynamic correlations.

The theoretical post total cross sections obtained with the CDW-4B method for single electron capture to the ground state in the Li^{3+} – He collisions are displayed in Fig. 17.8 (full curve) together with the experimental data from Refs. [309, 354]. Shape-wise, above 200 keV/amu, the computed cross sections are found to be in qualitative agreement with the experimental data. Quantitatively, however, the full curve from the CDW-4B method lies below the measured cross sections. This happens because the theoretical results include only capture into the ground state, while the contribution from exited states is roughly taken into account by an overall multiplying factor 1.202

from the scaling rule of Oppenheimer [247], as usual. When the relevant term for dynamic correlations ΔV_{12} is ignored in (17.2), the resulting cross sections are found to markedly underestimate the experimental data (see the dashed curve in Fig. 17.8). This provides direct evidence that dynamic correlations play an important role for electron capture in the ground state, especially at higher impact energies.

By increasing the charge of the projectile, the contribution from capture into excited states becomes more important. This has been investigated via the

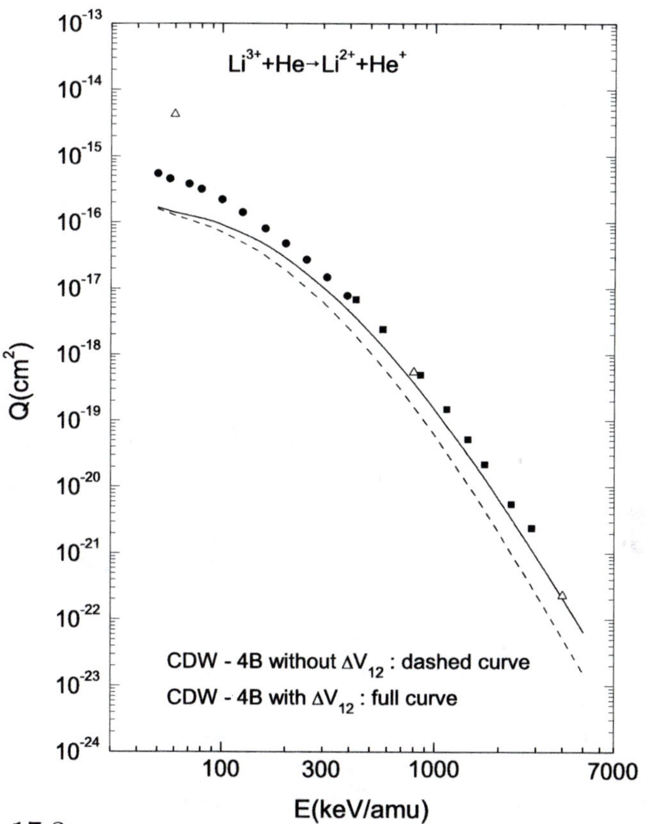

FIGURE 17.8

Total cross sections Q as a function of the incident energy E for for process (17.15). Theory: dashed curve (post CDW-4B method without the perturbation ΔV_{12} [130]) and full curve (post CDW-4B method with the perturbation ΔV_{12} [130]). Both curves include only the final ground state of Li^{2+} and He^{+} (excited states are taken into account roughly through the overall multiplying factor 1.202). The symbol ($\triangle$) refers to the post CDW-4B method for the cross sections Q_{Σ}^{+} obtained from (17.17). Experiment: • [309] and ■ [354].

extension of the CDW-4B method to encompass capture to the final excited states [130]. In these computations of total cross sections for single electron capture from He by Li^{3+} via process

$$Li^{3+} + He(1s^2) \longrightarrow Li^{2+}(n_f l_f) + He^+(1s) \tag{17.16}$$

the following values of the quantum numbers n_f and l_f were considered: $1s$, $2s$, $2p$, $3s$, $3p$ and $3d$. As an illustration, Mančev [130] gave the results at energies 60, 800 and 4000 keV/amu. The total cross sections $Q_\Sigma^\pm$ are obtained via the n_f^{-3} scaling law [247]

$$Q_\Sigma^\pm = Q_1^\pm + Q_2^\pm + 2.081 Q_3^\pm \tag{17.17}$$

where

$$Q_{n_f}^\pm = \sum_{l_f=0}^{n_f-1} \sum_{m_f=-l_f}^{+l_f} Q_{n_f l_f m_f}^\pm. \tag{17.18}$$

Here, $Q_{n_f l_f m_f}^\pm$ are the state-selective partial cross sections for capture into the state determined by the quantum numbers n_f, l_f and m_f. The results for Q_Σ^+ given in Fig. 17.8 show that the electronic correlations remain important for excited states [130]. The contribution from the term ΔV_{12} to the total cross section for excited states retains a similar trend relative to that for capture into the ground state. This can be demonstrated if we compare the post total cross sections computed with and without the term ΔV_{12} and denote the corresponding results by $Q_{nl,12}^+$ and $\bar{Q}_{nl,0}^+$. Then for the $1s$, $2s$ and $3s$ states at 4000 keV/amu, the following values are obtained: $Q_{nl,12}^+/\bar{Q}_{nl,0}^+ = 4.05$, 4.10, and 4.10, respectively. These ratios at 800 keV/amu become: 2.28, 2.24 and 2.21, whereas at 60 keV/amu they are 1.10, 1.66 and 1.48, respectively. The ratios for $Q_{nl,12}^+/\bar{Q}_{nl,0}^+$ also exhibit a similar behavior for the other excited states [130]. The values of the total cross sections Q_Σ^+ computed using (17.17) are displayed in Fig. 17.8. As expected, the contribution from excited states becomes less important as the impact energy increases. However, it is seen in Fig. 17.8 that at lower energies, the theoretical total cross sections Q_Σ^+ (denoted by open triangles) notably overestimate the experimental data. An evaluation of the contribution from the perturbation term ΔV_{P2} has been performed in Ref. [130] for the Li^{3+} – He collisions in process (17.15). The conclusion was that this contribution is significant (up to 40%).

It is also important to analyze collisions between two positively charged ions. An example is single electron capture from the Li^+ ion by fast protons, as in process

$$H^+ + {}^7Li^+(1s^2) \longrightarrow H(\Sigma) + {}^7Li^{2+}(1s). \tag{17.19}$$

The total cross sections from the CDW-4B method for this reaction are shown in Fig. 17.9, together with the experimental findings of Sewell *et al.* [405].

Unfortunately, their measured data have been reported only up to 250 keV. As can be seen, the CDW-4B method [190] is in fair agreement with the experimental data from Ref. [405].

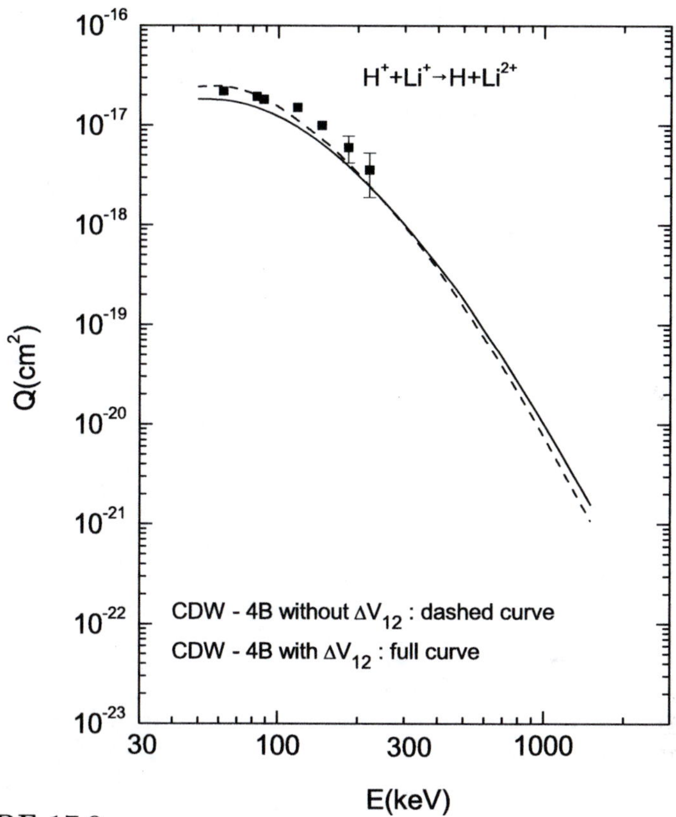

FIGURE 17.9
Total cross sections Q as a function of the incident energy E for process (17.19). Theory: dashed curve (post CDW-4B method without the perturbation ΔV_{12} [190]) and full curve (post CDW-4B method with the perturbation ΔV_{12} [190]). Experiment: ■ [405].

The classical trajectory Monte Carlo (CTMC) model was used by Wetmore and Olson [406] to describe reaction (17.19), along with a classical model for helium with the two active electrons. The CTMC approach treats all the participants in a collision (i.e. the projectile and target nuclei as well as two-target electrons) as classical point particles that interact through Coulomb potentials and move according to the Newton law. However, such a helium model removes the electron-electron force and allows each electron to interact independently with the target nucleus through a separate Coulomb potential. Hence, this version of the CTMC model from Ref. [406] cannot provide any information on electron correlation effects.

Charge exchange process (17.19) has also been studied theoretically by Ford *et al.* [407] by computing total cross sections at impact energies 50–250 keV. They employed the so-called "perturbative one-and-a-half-centered expansion" (POHCE) approximation. Here, a single-particle model was adopted for the target described through a local, exponentially screened potential of the type $V_{\rm T} = -2/r - \exp(-3.3954r)/r$. We note that there is good agreement between the POHCE and CDW-4B methods. By contrast, the cross sections of the CTMC model exhibit a different trend via a peak around 175 keV (not shown). However, such a peak is not present in the corresponding results of the POHCE and CDW-4B methods, or in the experimental data.

17.2 The CDW-BFS (prior BDW-4B) and CDW-BIS (post BDW-4B method)

In this section we analyze certain distorted wave methods that include continuum intermediate states of the captured electron only in one channel (see Refs. [89, 90] and [193]–[196]). We consider the prior form of the transition amplitude in the distorted wave formalism [89]

$$T_{if}^{-} = \langle \chi_f^{-} | U_i | \chi_i^{+} \rangle. \tag{17.20}$$

Here, we choose the perturbation potential U_i and the corresponding distorted wave of the initial state in the same manner as in the CDW-4B method

$$U_i = Z_{\rm P}\left(\frac{1}{R} - \frac{1}{s_2}\right) - \boldsymbol{\nabla}_{x_1} \ln \varphi_i \cdot \boldsymbol{\nabla}_{x_1} - \boldsymbol{\nabla}_{x_2} \ln \varphi_i \cdot \boldsymbol{\nabla}_{x_2} \tag{17.21}$$

$$\begin{aligned}\chi_i^{+} &= N^{+}(\nu_{\rm P})\mathcal{N}^{+}(\nu)\mathrm{e}^{i\boldsymbol{k}_i\cdot\boldsymbol{r}_i}\varphi_i(\boldsymbol{x}_1,\boldsymbol{x}_2)\\ &\quad\times\, {}_1F_1(i\nu_{\rm P},1,ivs_1+i\boldsymbol{v}\cdot\boldsymbol{s}_1)\, {}_1F_1(-i\nu,1,ik_ir_i-i\boldsymbol{k}_i\cdot\boldsymbol{r}_i)\end{aligned} \tag{17.22}$$

where $\nu = Z_{\rm P}(Z_{\rm T}-1)/v$ and $\nu_{\rm P} = Z_{\rm P}/v$. For the final state, we use the wave function χ_f^{-} from the CB1-4B method with $\chi_f^{-} \longrightarrow \Phi_f^{-}$ at $r_f \to \infty$

$$\chi_f^{-} = \mathcal{N}^{-}(\nu_f)\mathrm{e}^{-i\boldsymbol{k}_f\cdot\boldsymbol{r}_f}\varphi_{f_1}(\boldsymbol{s}_1)\varphi_{f_2}(\boldsymbol{x}_2)\, {}_1F_1(i\nu_f,1,-ik_fr_f-i\boldsymbol{k}_f\cdot\boldsymbol{r}_f) \tag{17.23}$$

where $\nu_f = (Z_{\rm P}-1)(Z_{\rm T}-1)/v$. With the choices (17.21) and (17.22), the transition amplitude (17.20) defines the prior BDW-4B method of Belkić [89]. The prior BDW-4B method has been used in Refs. [193, 194], and was subsequently renamed as the CDW-BFS method [196]. The explicit expression for the transition amplitude in the prior BDW-4B, or equivalently, CDW-BFS method is [193, 194]

$$T_{if}^{-} = N^{+}(\nu_{\mathrm{P}}) \iiint \mathrm{d}\boldsymbol{x}_1 \mathrm{d}\boldsymbol{x}_2 \mathrm{d}\boldsymbol{R}\, \mathrm{e}^{i\boldsymbol{\alpha}\cdot\boldsymbol{s}_1 + i\boldsymbol{\beta}\cdot\boldsymbol{x}_1} \mathcal{L}(\boldsymbol{R}) \varphi_{f_1}^{*}(\boldsymbol{s}_1) \varphi_{f_2}^{*}(\boldsymbol{x}_2)$$
$$\times [\Delta V_{\mathrm{P}2} \varphi_i(\boldsymbol{x}_1, \boldsymbol{x}_2)\, {}_1F_1(i\nu_{\mathrm{P}}, 1, ivs_1 + i\boldsymbol{v}\cdot\boldsymbol{s}_1)$$
$$- \boldsymbol{\nabla}_{x_1} \varphi_i(\boldsymbol{x}_1, \boldsymbol{x}_2) \cdot \boldsymbol{\nabla}_{s_1}\, {}_1F_1(i\nu_{\mathrm{P}}, 1, ivs_1 + i\boldsymbol{v}\cdot\boldsymbol{s}_1)]. \tag{17.24}$$

The function $\mathcal{L}(\boldsymbol{R})$ is

$$\mathcal{L}(\boldsymbol{R}) = \mathcal{N}^{-*}(\nu_f)\mathcal{N}^{+}(\nu)$$
$$\times\, {}_1F_1(-i\nu_f, 1, ik_f r_i + i\boldsymbol{k}_f\cdot\boldsymbol{r}_i)\, {}_1F_1(-i\nu, 1, ik_i r_i - i\boldsymbol{k}_i\cdot\boldsymbol{r}_i)$$
$$\simeq (k_f r_i + \boldsymbol{k}_f\cdot\boldsymbol{r}_i)^{i\nu_f} (k_i r_i - \boldsymbol{k}_i\cdot\boldsymbol{r}_i)^{i\nu} \qquad (r_i \to \infty). \tag{17.25}$$

Replacement of Coulomb scattering waves for the relative motion of heavy particles by their asymptotic forms has been made previously in Refs. [44, 49, 408]. Here, we use the well-known eikonal hypothesis, which assumes that the initial momentum k_i acquires large values. For heavy particle collisions, the reduced mass of the whole system in the entrance channel is very large ($\mu_i \gg 1$) and, hence, k_i is large even at very small incident velocities v_i (say, of the order of 0.01 a.u.). Due to their large mass, heavy projectiles are only slightly deflected from their initial direction and, therefore, scattering takes place predominantly in a narrow forward cone, which implies that $\hat{\boldsymbol{k}}_i \simeq \hat{\boldsymbol{k}}_f$. Thus, in the mass limit $M_{\mathrm{P,T}} \gg 1$, for the initial and final heavy particle velocities $\boldsymbol{v}_i \approx \boldsymbol{k}_i/\mu_i$ and $\boldsymbol{v}_f \approx \boldsymbol{k}_f/\mu_f$, we can write $\boldsymbol{v}_i \simeq \boldsymbol{v}_f \equiv \boldsymbol{v}$. This mass limit also justifies the replacement of $\boldsymbol{r}_i$ by $\boldsymbol{R}$. Hence, such an eikonal hypothesis permits a consistent reduction of the function $\mathcal{L}(\boldsymbol{R})$ to the following simplified form with the Coulomb logarithmic phase factors

$$\mathcal{L}(\boldsymbol{R}) \simeq (vR - \boldsymbol{v}\cdot\boldsymbol{R})^{iZ_{\mathrm{P}}(Z_{\mathrm{T}}-1)/v} (vR + \boldsymbol{v}\cdot\boldsymbol{R})^{i(Z_{\mathrm{P}}-1)(Z_{\mathrm{T}}-1)/v}$$
$$= \rho^{2iZ_{\mathrm{P}}(Z_{\mathrm{T}}-1)/v} (vR + \boldsymbol{v}\cdot\boldsymbol{R})^{-i\xi} \tag{17.26}$$

where $\xi = (Z_{\mathrm{T}} - 1)/v$. Hereafter, the unimportant phase factors of the unit moduli are ignored. The factor $\rho^{2iZ_{\mathrm{P}}(Z_{\mathrm{T}}-1)/v}$ has no influence on total cross sections and may be omitted [44]. The matrix elements in the prior BDW-4B method can be reduced to a two-dimensional numerical quadrature [193].

Further, we can employ the post form of the transition amplitude [89]

$$T_{if}^{+} = \langle \chi_f^{-} | U_f | \chi_i^{+} \rangle. \tag{17.27}$$

Here, we use χ_i^{+} from the CB1-4B method with $\chi_i^{+} \longrightarrow \Phi_i^{+}$ at $r_i \to \infty$

$$\chi_i^{+} = \mathcal{N}^{+}(\nu_i) \mathrm{e}^{i\boldsymbol{k}_i\cdot\boldsymbol{r}_i} \varphi_i(\boldsymbol{x}_1, \boldsymbol{x}_2)\, {}_1F_1(-i\nu_i, 1, ik_i r_i - i\boldsymbol{k}_i\cdot\boldsymbol{r}_i) \tag{17.28}$$

where $\nu_i = Z_{\mathrm{P}}(Z_{\mathrm{T}} - 2)/v$. On the other hand, we choose U_f and χ_f^{-} in the same forms as in the CDW-4B method

$$U_f \chi_f^{-} = \left[Z_{\mathrm{P}} \left(\frac{1}{R} - \frac{1}{s_2} \right) + \left(\frac{1}{r_{12}} - \frac{1}{x_1} \right) \right] \chi_f^{-}$$
$$- \varphi_{f_2}(\boldsymbol{x}_2) \boldsymbol{\nabla}_{s_1} \varphi_{f_1}(\boldsymbol{s}_1) \cdot \boldsymbol{\nabla}_{s_1} \mathcal{G}_f^{-} - \varphi_{f_1}(\boldsymbol{s}_1) \boldsymbol{\nabla}_{x_2} \varphi_{f_2}(\boldsymbol{x}_2) \cdot \boldsymbol{\nabla}_{x_2} \mathcal{G}_f^{-} \tag{17.29}$$

with

$$\chi_f^- = \varphi_{f_1}(\boldsymbol{s}_1)\varphi_{f_2}(\boldsymbol{x}_2)\mathrm{e}^{-i\boldsymbol{k}_f\cdot\boldsymbol{r}_f}\mathcal{N}^-(\nu)N^-(\nu_\mathrm{T}) \times {}_1F_1(-i\nu_\mathrm{T},1,-ivx_1-i\boldsymbol{v}\cdot\boldsymbol{x}_1)\,{}_1F_1(i\nu,1,-ik_fr_f+i\boldsymbol{k}_f\cdot\boldsymbol{r}_f) \qquad (17.30)$$

where $\nu_\mathrm{T} = (Z_\mathrm{T}-1)/v$. The pair (17.29) and (17.30), together with the transition amplitude (17.27) lead to the post BDW-4B method of Belkić [89]. The post BDW-4B method has been employed in Refs. [193, 194] under the original acronym and was subsequently renamed as the CDW-BIS method [196]. The explicit expression for the transition amplitude in the post BDW-4B, or equivalently, CDW-BIS method is [193, 194, 196]

$$T_{if}^+ = N^{-*}(\nu_\mathrm{T})\iiint \mathrm{d}\boldsymbol{x}_1\mathrm{d}\boldsymbol{x}_2\mathrm{d}\boldsymbol{R}\,\mathrm{e}^{i\boldsymbol{k}_i\cdot\boldsymbol{r}_i+i\boldsymbol{k}_f\cdot\boldsymbol{r}_f}\varphi_i(\boldsymbol{x}_1,\boldsymbol{x}_2)\mathcal{R} \times \left\{{}_1F_1(i\nu_\mathrm{T},1,ivx_1+i\boldsymbol{v}\cdot\boldsymbol{x}_1)[\Delta V_{\mathrm{P}2}+\Delta V_{12}]\varphi_{f_1}^*(\boldsymbol{s}_1)\varphi_{f_2}^*(\boldsymbol{x}_2) - \varphi_{f_2}^*(\boldsymbol{x}_2)\boldsymbol{\nabla}_{s_1}\varphi_{f_1}^*(\boldsymbol{s}_1)\cdot\boldsymbol{\nabla}_{x_1}\,{}_1F_1(i\nu_\mathrm{T},1,ivx_1+i\boldsymbol{v}\cdot\boldsymbol{x}_1)\right\} \qquad (17.31)$$

with

$$\mathcal{R} = \mathcal{N}^{-*}(\nu)\mathcal{N}^+(\nu_i)\,{}_1F_1(-i\nu,1,ik_fr_f-i\boldsymbol{k}_f\cdot\boldsymbol{r}_f)\,{}_1F_1(-i\nu_i,1,ik_ir_i-i\boldsymbol{k}_i\cdot\boldsymbol{r}_i) \simeq (vR+\boldsymbol{v}\cdot\boldsymbol{R})^{i\nu}(vR-\boldsymbol{v}\cdot\boldsymbol{R})^{i\nu_i} = (v\rho)^{2i\nu_i}(vR+\boldsymbol{v}\cdot\boldsymbol{R})^{-i\xi} \qquad (17.32)$$

where $\Delta V_{\mathrm{P}2} = Z_\mathrm{P}(1/R-1/s_2)$, $\Delta V_{12} = 1/r_{12}-1/x_1$ and $\xi = \nu_i-\nu = -Z_\mathrm{P}/v$.

For single capture (11.3) in the prior BDW-4B method for differential cross sections, a considerable simplification occurs for certain special values of the nuclear charges Z_P and Z_T. Specifically, for the H^- target ($Z_\mathrm{T}=1$) and an arbitrary Z_P, both Sommerfeld parameters $\nu = Z_\mathrm{P}(Z_\mathrm{T}-1)/v$ and $\nu_f = (Z_\mathrm{P}-1)(Z_\mathrm{T}-1)/v$ are equal to zero, so that

$$\mathcal{L}(\boldsymbol{R}) = 1 \qquad (17.33)$$

according to the eikonal expression (17.25) or (17.26). In this case, no $\boldsymbol{R}-$ or $\rho-$dependent distorted wave function occurs in T_{if}^- from (17.31) which, therefore, becomes identical to the prior form of the $T-$matrix element in the CDW-4B method for (11.3) [189]

$$T_{if}^- = N^+(\nu_\mathrm{P})\iiint \mathrm{d}\boldsymbol{x}_1\mathrm{d}\boldsymbol{x}_2\mathrm{d}\boldsymbol{R}\,\mathrm{e}^{i\boldsymbol{\alpha}\cdot\boldsymbol{s}_1+i\boldsymbol{\beta}\cdot\boldsymbol{x}_1}\varphi_{f_1}^*(\boldsymbol{s}_1)\varphi_{f_2}^*(\boldsymbol{x}_2) \times [\Delta V_{\mathrm{P}2}\varphi_i(\boldsymbol{x}_1,\boldsymbol{x}_2)\,{}_1F_1(i\nu_\mathrm{P},1,ivs_1+i\boldsymbol{v}\cdot\boldsymbol{s}_1) - \boldsymbol{\nabla}_{x_1}\varphi_i(\boldsymbol{x}_1,\boldsymbol{x}_2)\cdot\boldsymbol{\nabla}_{s_1}\,{}_1F_1(i\nu_\mathrm{P},1,ivs_1+i\boldsymbol{v}\cdot\boldsymbol{s}_1)] \equiv T_{if}^{(\mathrm{CDW-4B})-} \qquad (Z_\mathrm{T}=1\ ,\ \text{any } Z_\mathrm{P}). \qquad (17.34)$$

Here, the matrix element with the usual $\boldsymbol{\nabla}\cdot\boldsymbol{\nabla}$ potential operator can be calculated analytically, as in the CDW-3B method. Thus for process $Z_\mathrm{P}+\mathrm{H}^- \longrightarrow (Z_\mathrm{P},e_1)+\mathrm{H}$ where Z_P is arbitrary, the differential cross sections in prior

BDW-4B i.e. CDW-BFS method and CDW-4B method can be computed directly from (17.34) without the additional quadrature which necessitates the Fourier-Bessel transform

$$\frac{\mathrm{d}Q^-}{\mathrm{d}\Omega} = \left|\frac{\mu}{2\pi}T_{if}^-\right|^2 \qquad (Z_\mathrm{T} = 1, \text{ any } Z_\mathrm{P}) \tag{17.35}$$

where $\mu = M_\mathrm{P}M_\mathrm{T}/(M_\mathrm{P}+M_\mathrm{T})$. Another simplification occurs still in the prior BDW-4B method for $Z_\mathrm{P} = 1\,(\mathrm{H}^+)$ and arbitrary Z_T via

$$\begin{aligned}\mathcal{L}(\boldsymbol{R}) &= \mathcal{N}^+(\xi)\,{}_1F_1(-i\xi, 1, ik_ir_i - i\boldsymbol{k}_i\cdot\boldsymbol{r}_i)\\ &\simeq (vR - \boldsymbol{v}\cdot\boldsymbol{R})^{i\xi}\\ \xi &= \nu_\mathrm{T} = \frac{Z_\mathrm{T}-1}{v} \qquad (Z_\mathrm{P} = 1\ ,\ \text{any } Z_\mathrm{T}).\end{aligned} \tag{17.36}$$

In this case, there is no need to use the extra Fourier-Bessel transform either, since the same algorithm which computes total cross sections can also be used for obtaining differential cross sections in the prior BDW-4B method.

A similar simplification is also possible in the post BDW-4B method. This time, for any Z_P and $Z_\mathrm{T} = 1\ (\mathrm{H}^-)$, we have $\nu_i = Z_\mathrm{P}(Z_\mathrm{T}-2)/v = -Z_\mathrm{P}/v = -\nu_\mathrm{P}$ and $\nu = Z_\mathrm{P}(Z_\mathrm{T}-1)/v = 0$, so that

$$\begin{aligned}\mathcal{R} &= \mathcal{N}^{-*}(\nu_P)\,{}_1F_1(i\nu_P, 1, ik_ir_i - i\boldsymbol{k}_i\cdot\boldsymbol{r}_i)\\ &\simeq (vR - \boldsymbol{v}\cdot\boldsymbol{R})^{-i\nu_P}\\ \nu_\mathrm{P} &= \frac{Z_\mathrm{P}}{v} \qquad (Z_\mathrm{T} = 1\ ,\ \text{any } Z_\mathrm{P}).\end{aligned} \tag{17.37}$$

Thus, the same code for the post BDW-4B method can be used for computations of both total and differential cross sections without the need to use the supplementary Fourier-Bessel numerical integration. In the post BDW-4B method, a similar circumstance is also encountered for any Z_P and $Z_\mathrm{T} = 2$ (He), in which case $\nu_i = Z_\mathrm{P}(Z_\mathrm{T}-2)/v = 0$ and $\nu = Z_\mathrm{P}/v = \nu_\mathrm{P}$, yielding the eikonal simplification

$$\begin{aligned}\mathcal{R} &= \mathcal{N}^{-*}(\nu_P)\,{}_1F_1(-i\nu_P, 1, ik_fr_f - i\boldsymbol{k}_f\cdot\boldsymbol{r}_f)\\ &\simeq (vR + \boldsymbol{v}\cdot\boldsymbol{R})^{i\nu_P}\\ \nu_\mathrm{P} &= \frac{Z_\mathrm{P}}{v} \qquad (Z_\mathrm{T} = 2\ ,\ \text{any } Z_\mathrm{P}).\end{aligned} \tag{17.38}$$

Notice that relative to the transition amplitude from (17.24) in the prior BDW-4B method, the corresponding $T-$matrix element in the post BDW-4B method from (17.31) contains the extra term ΔV_{12}, which carries information about electronic correlations. Due to the presence of ΔV_{12}, the computation in the post form is more difficult than in its prior counterpart within the BDW-4B method. After an analytical calculation the expression for T_{if}^+ from (17.31) can be written in terms of a five-dimensional integral [196].

According to the CDW-BIS method, the following approximation is utilized for the distortion of initial scattering state in the entrance channel [195, 196]

$$F_i = \mathcal{N}^+(\nu_i)\mathrm{e}^{i\boldsymbol{k}_i\cdot\boldsymbol{r}_i}\,{}_1F_1(-i\nu_i,1,ik_ir_i-i\boldsymbol{k}_i\cdot\boldsymbol{r}_i)$$
$$\simeq \mathrm{e}^{i\boldsymbol{k}_i\cdot\boldsymbol{r}_i+i\nu_i\ln(k_ir_i-\boldsymbol{k}_i\cdot\boldsymbol{r}_i)} \simeq \mathrm{e}^{i\boldsymbol{k}_i\cdot\boldsymbol{r}_i}(vR-\boldsymbol{v}\cdot\boldsymbol{R})^{-i\nu_i}. \tag{17.39}$$

A similar eikonal approximation is chosen in the exit channel within the CDW-BFS method [193, 194]

$$F_f = \mathcal{N}^+(\nu_f)\mathrm{e}^{-i\boldsymbol{k}_f\cdot\boldsymbol{r}_f}\,{}_1F_1(i\nu_f,1,-ik_fr_f+i\boldsymbol{k}_f\cdot\boldsymbol{r}_f)$$
$$\simeq \mathrm{e}^{-i\boldsymbol{k}_f\cdot\boldsymbol{r}_f-i\nu_f\ln(k_fr_f-\boldsymbol{k}_f\cdot\boldsymbol{r}_f)} \simeq \mathrm{e}^{-i\boldsymbol{k}_f\cdot\boldsymbol{r}_f}(vR+\boldsymbol{v}\cdot\boldsymbol{R})^{-i\nu_f}. \tag{17.40}$$

In these expressions for the $\boldsymbol{R}$–dependent eikonal phases from both F_i and F_f, the unimportant phases of the unit moduli are ignored. The factor F_f comes from the long-range Coulomb repulsion between the screened positively charged nuclei $(Z_\mathrm{P}-1)$ and $(Z_\mathrm{T}-1)$ in the exit channel, whereas F_i originates from repulsion between a bare projectile Z_P and the screened target nucleus $(Z_\mathrm{T}-2)$ in the entrance channel. We recall that an eikonal phase is a good approximation to the full continuum wave function only when the third argument of the investigated confluent hyper-geometric function is sufficiently large. This is justified for heavy particle collisions even at quite low impact energies, due to large values of reduced masses.

The high-energy behavior of the cross sections in the CDW-BFS method for the $1s-n_fl_f$ capture is given by the asymptote $Q_{if}\sim v^{-11-2l_f}$. Thus, for capture into excited states with $l_f>0$, at very high energies, the CDW-BFS method does not give the correct v^{-11} behavior. Nevertheless, this deficiency in the asymptotic region is not of crucial importance in the intermediate and some high energies, prior to the outset of the Thomas double scattering mechanism. A similar deficiency in the asymptotic velocity dependence has also been found in the case of the other one-channel distorted wave models, such as the so-named target continuum distorted wave (TCDW) method [409]–[411] as well as in the CDW-EIS and CDW-EFS methods [197, 198]. Note that the prior form of the CIS-3B method of Belkić [275] has been relabelled as TCDW by Crothers and Dunseath [409, 410]. These latter authors studied asymmetric collisions ($Z_\mathrm{T}\gg Z_\mathrm{P}$) within the TCDW method, which uses the wave function of the CDW-3B method in the final state and the undistorted channel wave function (with the inadequate asymptotic behavior) for the initial state of the system for the general process (12.26). As such, the TCDW method is unsatisfactory, since it disregards the correct boundary condition in the entrance channel.

It should be reemphasized that the introduction of these hybrid-type second-order approximations was motivated mainly by the idea of approximating the exact wave function in one of the channels, by a simple analytical function which can provide an adequate description of the principal interaction region. Importantly, the CDW-BIS and CDW-BFS methods preserve the correct boundary conditions in both channels.

In order to illustrate the adequacy of the prior BDW-4B i.e. CDW-BFS method and post BDW-4B i.e. CDW-BIS method, we shall consider several collisional processes. The results obtained by the CDW-BFS method for the total cross sections in the case of the following rearrangement process

$$^4\mathrm{He}^{2+} + {}^4\mathrm{He}(1s^2) \longrightarrow {}^4\mathrm{He}^+(\Sigma) + {}^4\mathrm{He}^+(1s) \qquad (17.41)$$

are shown in Fig. 17.10 at impact energies E between 100 keV and 10 MeV. Explicit computations are carried out only for capture to the ground state ($1s$)

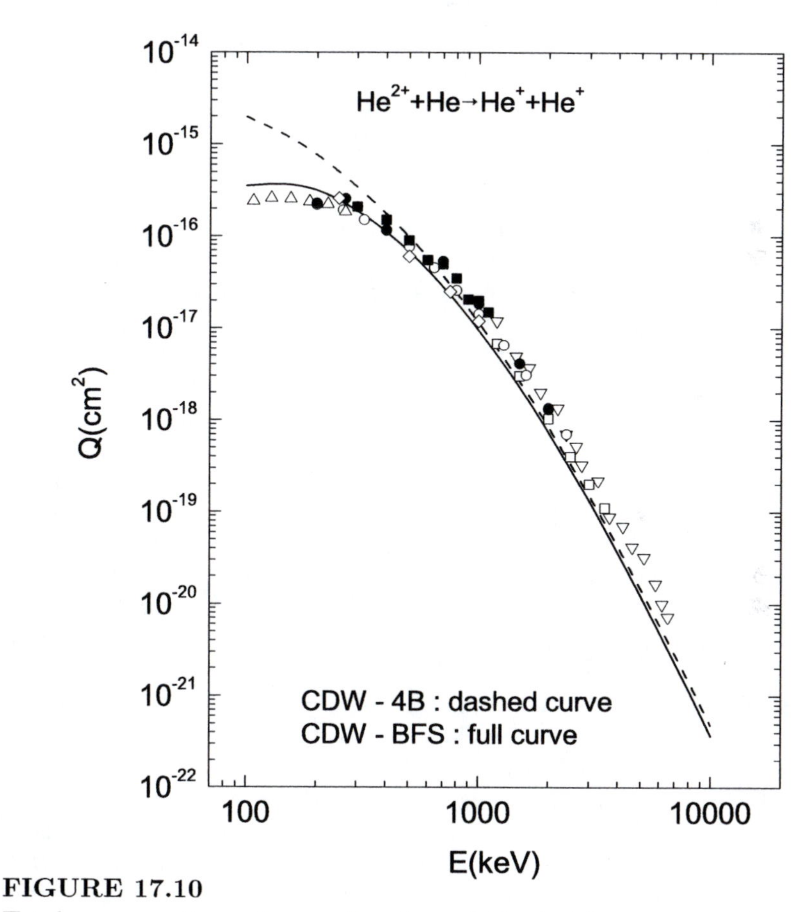

FIGURE 17.10
Total cross sections Q as a function of the incident energy E for process (17.14). Theory: dashed curve (prior CDW-4B method [190]) and full curve (CDW-BFS method [193]). Experiment: • [304], □ [305], ∘ [309], △ [352], ▽ [397], ◊ [398] and ■ [399].

of the He$^+$ ion. The symbol Σ in He$^+(\Sigma)$ in process (17.41) means that the cross sections for the $Q_{1s,1s}$ state are multiplied additionally by a factor 1.202 in order to approximately include the contribution from all excited states. In Fig. 17.10, the results of the CDW-BFS method are compared with those from the CDW-4B method [190]. We recall that the distorting potential U_i is the same in the prior CDW-4B and CDW-BFS methods. Both approximations satisfy the correct boundary conditions in the entrance and exit channels. However, unlike the CDW-4B with the initial and final electronic distortions, the CDW-BFS method includes Coulomb continuum intermediate state of the captured electron only in the entrance channel. Hence, by comparing these two theories, we can learn about the relative importance of the intermediate ionization electronic continua. As can be seen from Fig. 17.10, the CDW-BFS method provides similar cross sections as the CDW-4B method at higher impact energies. However, at lower energies, the results in the CDW-BFS method are smaller than the corresponding results of the CDW-4B method. This illustrates a remarkable sensitivity of second-order theories for single-charge exchange to the role of the electronic ionization continua. Also included in Fig. 17.10 are the experimental data for comparison with the theory. The CDW-BFS method is found to be in good agreement with the majority of measurements throughout the energy range of overlap. The theoretical results underestimate the total cross sections measured by Hvelplund *et al.* [397].

The CDW-BFS method also gives good results for other collisions such as single electron capture in the $H^+ - He$ and $H^+ - Li^+$ collisions [193]. The results of the computations for the $H^+ - He$ collisions in an energy interval from 40 keV to 15 MeV are shown in Fig. 17.11, where we compare the CDW-BFS method with experimental data. As can be seen, the CDW-BFS method is found to be consistently in excellent agreement with the available experimental data for the total cross sections from intermediate to high energies. Experimental data do not relate to the process $H^+ + {}^4He(1s^2) \longrightarrow H(\Sigma) + {}^4He^+(1s)$, but rather to the $H^+ + {}^4He(1s^2) \longrightarrow H(\Sigma) + {}^4He^+$ collisions, where the symbol ${}^4He^+$ indicates that no information is available on the post-collisional state of the target remainder He^+. This means that if a strict comparison with measurement is desired, the theory must allow for all possible contributions arising from transitions of the non-captured electron e_2 in the He^+ ion. In practice, however, only excitation and ionization of the He^+ ion can play a non-negligible role at high impact energies. In the same figure, a comparison with the theoretical results of Winter [412] is made. The computations of the cross sections from Ref. [412] were carried out at proton energies 50, 100 and 200 keV using the close coupling framework with the basis set expansion functions in terms of atomic orbitals. Specifically, these latter computations [412] use some 50 Sturmian functions and, moreover, the inter-electron interaction has been taken into account. It is seen from Fig. 17.11 that agreement between the results from the CDW-BFS and AO methods is quite good. The cross sections from the CDW-4B method obtained by Belkić *et al.* [189] (not shown in Fig. 17.11 to avoid clutter, but displayed earlier in Fig. 17.6) are also in good agreement with the findings of the CDW-BFS method. Further, it is found that the interaction ΔV_{P2} contributes about 9%–30% to the total cross section (dashed curve in Fig. 17.11).

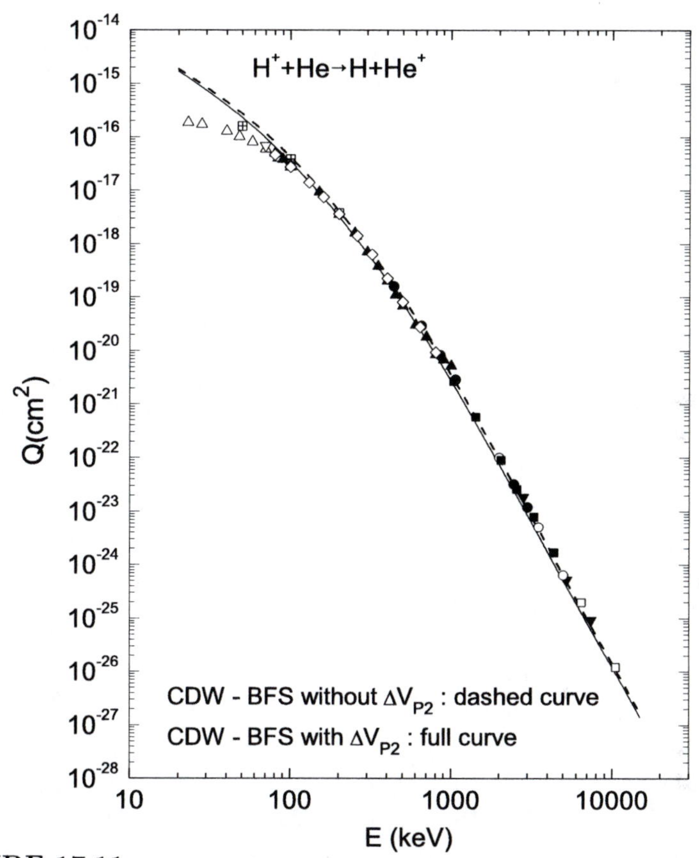

FIGURE 17.11
Total cross sections Q as a function of the incident energy E for process (17.14). Theory: dashed curve (CDW-BFS method without the perturbation ΔV_{P2} [193]) and full curve (CDW-BFS method with the perturbation ΔV_{P2} [193]). The symbol ⊞ represents the result from the AO method [412]. Experiment: ■ [297], ◊ [309], △ [352], ▼ [393], □ [394], • [396], ◦ [413], ▲ [414] and ▽ [415].

As to total cross sections in the CDW-BIS method, the corresponding computations have recently been performed by Mančev [196] in the energy interval 20–10000 keV. A comparison between the CDW-BIS method and numerous experimental data reveals excellent agreement. The CDW-BIS method provides very similar cross sections as the CDW-BFS method at higher impact energies. However, at lower energies, the results of the CDW-BIS method are smaller than the corresponding results of the CDW-BFS method. Moreover, the CDW-BIS method shows better agreement with measurements than the CDW-BFS method [196].

A more refined test of the validity of theoretical models is provided by differential cross sections (DCS) as can be seen in detail in Figs. 17.12–17.17.

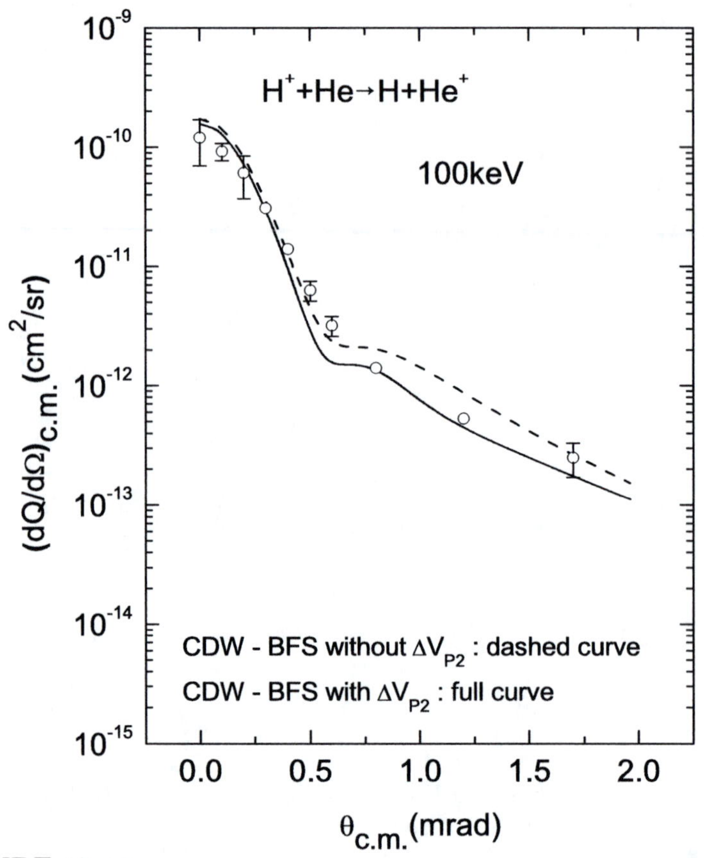

FIGURE 17.12
Differential cross sections $\mathrm{d}Q/\mathrm{d}\Omega$ as a function of the scattering angle θ at the incident energy $E = 100$ keV for process (17.14). Theory: dashed curve (CDW-BFS method without the potential ΔV_{P2} [194]) and full curve (CDW-BFS method with potential ΔV_{P2} [194]). Experiment: $\circ$ [400].

The DCS data from the CDW-BFS method for the $\mathrm{H}^+ - \mathrm{He}$ collisions at $E = 100$ keV are depicted in Fig. 17.12, together with the experimental data of Martin *et al.* [400]. It is seen from this figure that there is good agreement between the theory and measurements. Bross *et al.* [416] have also measured

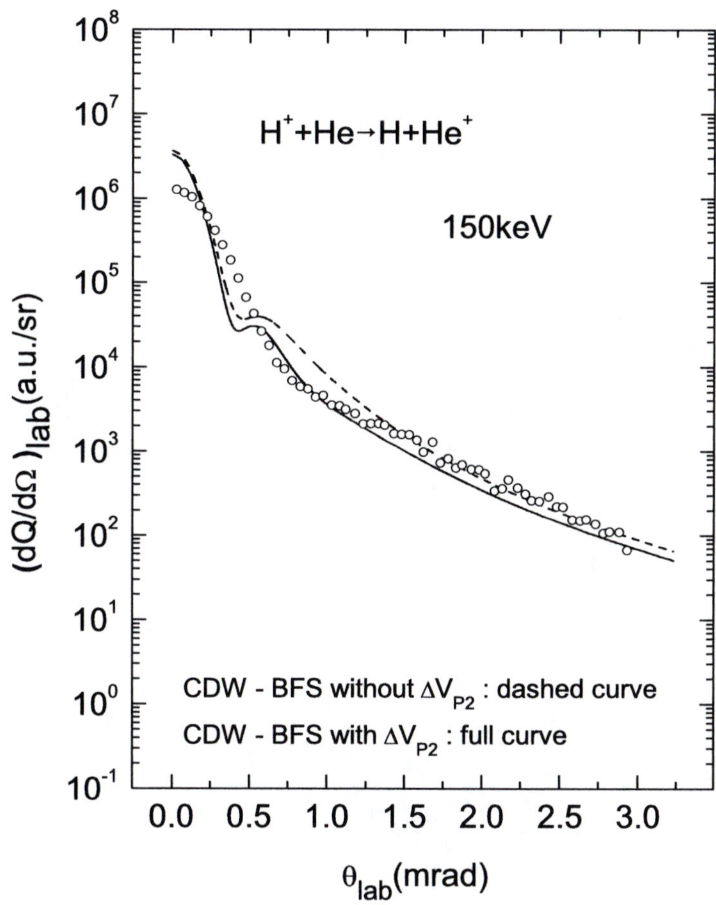

FIGURE 17.13
Differential cross sections $\mathrm{d}Q/\mathrm{d}\Omega$ as a function of the scattering angle θ at the incident energy $E = 150$ keV for process (17.14). Theory: dashed curve (CDW-BFS method without the potential ΔV_{P2} [194]) and full curve (CDW-BFS method with the potential ΔV_{P2} [194]). Experiment: $\circ$ [361].

differential cross sections for single electron capture in the $\mathrm{H}^+ - \mathrm{He}$ collisions at 100 keV, and their results are very close to those of Martin *et al.* [400].

In Figs. 17.13–17.15, more of the available DCS data are presented at proton energies 150, 200 and 300 keV. A comparison is made between the theory [193] and measurements [361]. Figures 17.13 and 17.14 show good agreement between theory and experiment at 150 and 200 keV. At 300 keV satisfactory agreement is also obtained, as seen in Fig. 17.15. Here, at larger scattering angles, the theoretical results underestimate the experimental data, but otherwise both theory and experiment exhibit quite similar behavior.

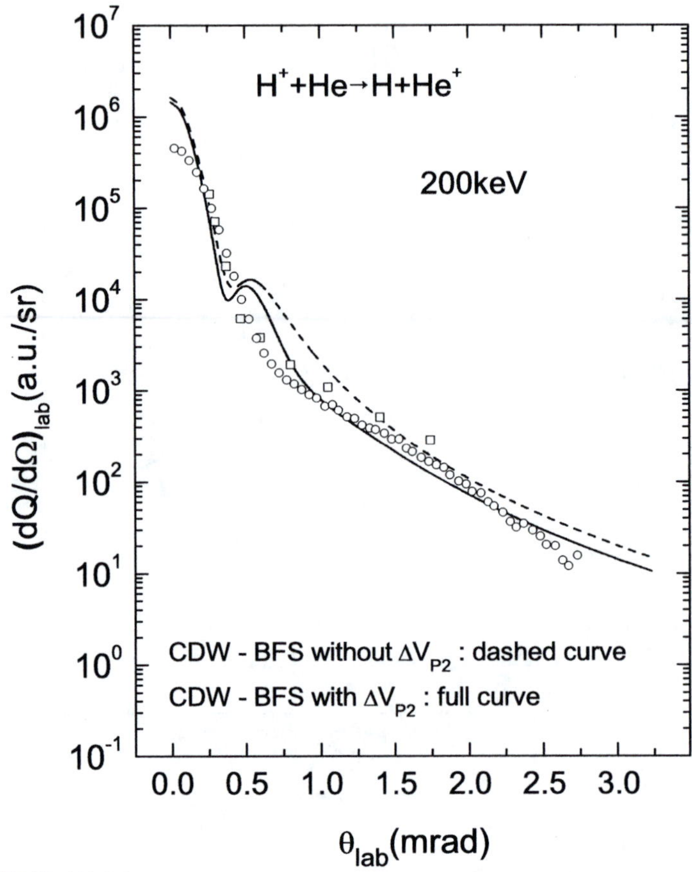

FIGURE 17.14
Differential cross sections $\mathrm{d}Q/\mathrm{d}\Omega$ as a function of the scattering angle θ at the incident energy $E = 200$ keV for process (17.14). Theory: dashed curve (CDW-BFS method without the potential ΔV_{P2} [194]) and full curve (CDW-BFS method with the potential ΔV_{P2} [194]). Experiment: $\circ$ [361] and $\square$ [417].

The results from the CDW-BIS method for differential cross sections at 400 and 630 keV for process (17.14) are depicted in Figs. 17.16 and 17.17. These figures show that at smaller scattering angles, the CDW-BIS and CDW-BFS methods give similar results and, moreover, these theoretical findings are in satisfactory agreement with the experimental data. In Fig. 17.16, the two sets of experimental data [361, 417] are displayed. Mergel *et al.* [361] measured the differential cross sections using COLTRIMS.

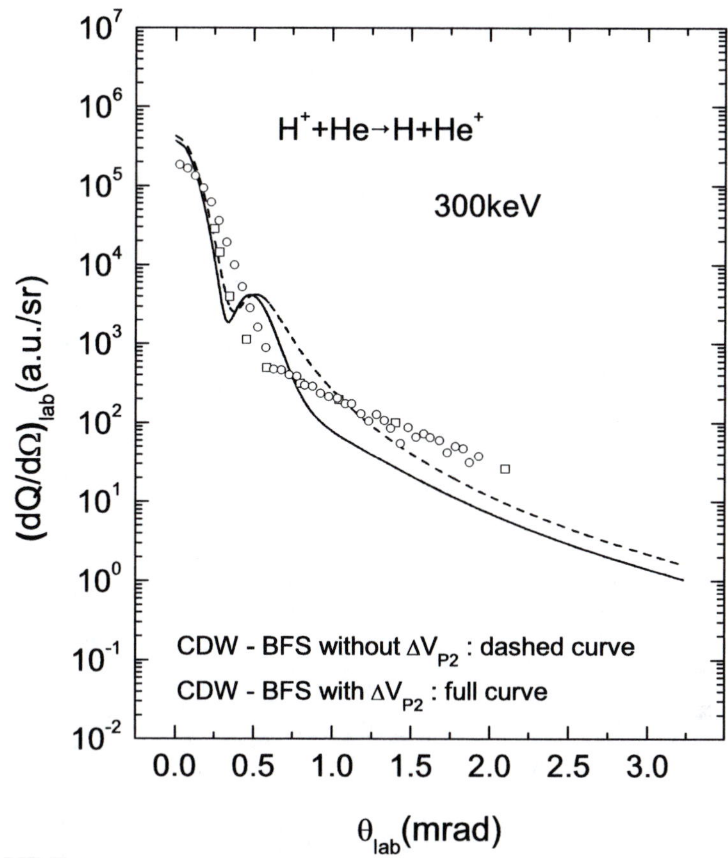

FIGURE 17.15
Differential cross sections $\mathrm{d}Q/\mathrm{d}\Omega$ as a function of the scattering angle θ at the incident energy $E = 300$ keV for process (17.14). Theory: dashed curve (CDW-BFS method without the potential ΔV_{P2} [194]) and full curve (CDW-BFS method with the potential ΔV_{P2} [194]). Experiment: $\circ$ [361] and $\square$ [417].

Different behaviors are obtained in the CDW-BIS and CDW-BFS methods around the Thomas peak region. When the impact energy increases, the CDW-BIS method shows the Thomas peak at the same angle as the CDW-BFS method. However, the presently considered impact energies are not sufficiently large for the cited experiments to detect the Thomas peak in an unambiguous way. The CDW-BIS method exhibits an unphysical and experimentally unobserved dip located after the Thomas peak region. This additional peak is due to a mutual cancellation among the individual terms in the perturbation potential U_f.

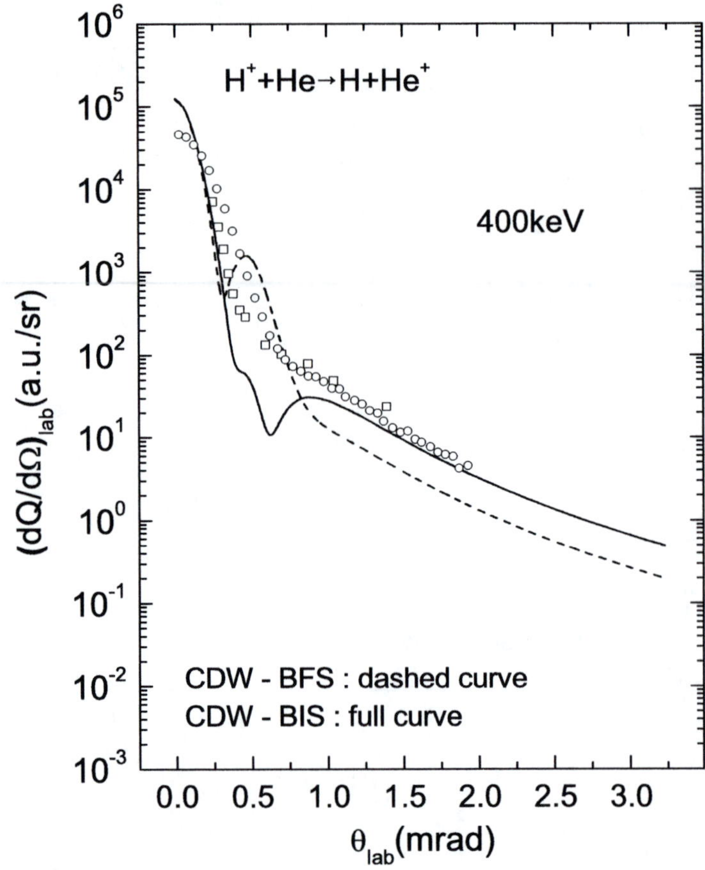

FIGURE 17.16
Differential cross sections $\mathrm{d}Q/\mathrm{d}\Omega$ as a function of the scattering angle θ at the incident energy $E = 400$ keV for process (17.14). Theory: dashed curve (CDW-BFS method without the potential ΔV_{P2} [193, 194]) and full curve (CDW-BFS method with the potential ΔV_{P2} [195, 196]). Experiment: $\circ$ [361] and $\square$ [417].

At larger scattering angles, the CDW-BIS method shows good agreement with the experimental data. This can be attributed to the inclusion of the interaction potential ΔV_{12} which describes dynamic electron correlations. We recall that the differential cross sections in the CDW-BFS method are obtained after performing a two-dimensional numerical integration. This is advantageous relative to the CDW-4B method in which a five-dimensional quadrature is required for the same purpose when (14.20) is used. Thus far, no differential cross sections obtained by means of the CDW-4B method have been reported in the literature.

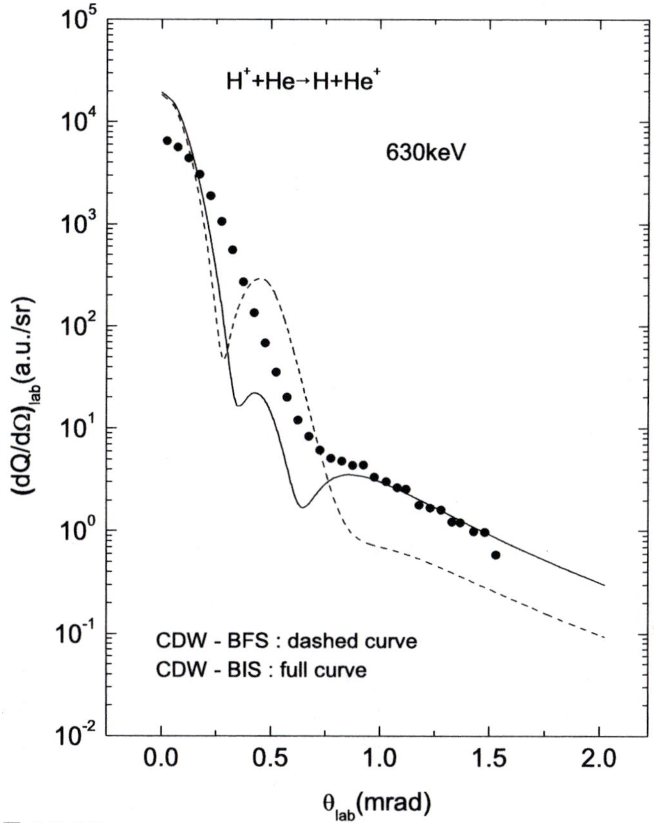

FIGURE 17.17
Differential cross sections $\mathrm{d}Q/\mathrm{d}\Omega$ as a function of the scattering angle θ at the incident energy $E = 630$ keV for process (17.14). Theory: dashed curve (CDW-BFS method without the potential $\Delta V_{\mathrm{P}2}$ [193, 194]) and full curve (CDW-BFS method with the potential $\Delta V_{\mathrm{P}2}$ [195, 196]). Experiment: • [361].

The theoretical DCS data for the $\mathrm{H}^+ - \mathrm{He}$ collision at intermediate and high impact energies have recently been reported in Refs. [244, 418]. Abufager *et al.* [418] employed an improved version of the three-body CDW-EIS method regarding the representation of the bound and continuum target states. This method gives a good description of single electron capture in the $H^+ - He$ collisions at intermediate and high energies, when an appropriate atomic model for the target potential is employed.

In Ref. [244] an artificial 'regularization' via the peaking and near-shell approximations (NSA) was used for an otherwise divergent second-order term in a perturbation series expansion of the Faddeev-Watson-Lovelace (FWL) equations[1] with the incorrect boundary conditions for Coulomb potentials [239]–[243]. If evaluated exactly, the FWL method to any order beyond the first term in the perturbation series would yield infinite results due to the presence of non-integrable i.e. non-regularizable branch-point singularities in the transition amplitude. Therefore, irrespective of its status relative to any conceivable comparison with experimental data, the method used in Refs. [239]–[246] is mathematically and physically unfounded for similar reasons that have discarded a once popular, but now defunct model called the strong potential Born (SPB) approximation [217]–[232].

Regarding the Thomas double scattering, despite considerable interest by theoreticians, there are only a few experimental investigations of electron capture in the fast $H^+ - He$ collisions. The Thomas peak in the DCS data at $\theta_{\text{lab}} = 0.47$ mrad has been observed by Horsdal-Pedersen *et al.* [393] at impact energies 2.82 MeV, 5.42 MeV and 7.4 MeV. Using the COLTRIMS technique, Fischer *et al.* [419] have recently reported the new DCS data for electron capture at 7.5 MeV and 12.5 MeV in the $H^+ - He$ collisions. In their experiment, the Thomas peak has clearly been identified. Moreover, the angle at which the dip occurs (and which is located between the Thomas and kinematic single electron capture processes) is determined with a higher precision than in the study of Horsdal-Pedersen *et al.* [393].

In the Faddeev equations for a three-body collisional problem, when the mass of one particle is either much smaller or much larger than that of the other two (as encountered in many examples in nuclear and atomic physics), an important practical simplification occurs as shown by Dodd and Greider [261]. This mass condition yields: (i) factorable three-particle Green's resolvent operators with the spectra solvable solely in terms of the known two-particle Green's propagators, (ii) a single integral equation for the transition $T-$operator, and (iii) elimination of all disconnected diagrams. Of course, the original Faddeev equations [51] solve the three-body problem with arbitrary masses in terms of the known solution of two-body problems and do not have disconnected diagrams. But they deal with a system of three coupled integral equations that are less tractable for practical computations.

[1]The original Faddeev equations [51] are often employed under the names of the Faddeev-Lovelace [235, 261] and Faddeev-Watson-Lovelace [239]–[246] equations.

Dodd and Greider [261] used the distorted wave representation and introduced an intermediate channel of scattering. With an adequate choice of the potential of this intermediate channel, the Faddeev equations in the Dodd-Greider formulation can fulfill the correct boundary conditions, including the most critical case when there are residual Coulomb potentials at large inter-aggregate separations in both channels. In practice, the Dodd-Greider variant of the Faddeev integral equations is solved perturbatively. It should be noted that the distorted wave formalism by itself cannot eliminate disconnected diagrams from any perturbation series expansion for rearranging collisions. However, this becomes possible in the version of the distorted wave methodology of Dodd and Greider [261].

The reason for which all disconnected diagrams must be avoided for rearranging collisions is that they lead to divergence of perturbation series of the transition $T-$operator, as actually happens in the corresponding Born expansion. Clearly, when a series is divergent, its lowest order cannot be interpreted as a mathematically and physically meaningful first-order approximation to an expansion of the transition amplitude. In such a case, the overall validity of theoretical models is judged solely upon the success of fitting experimental data and this is often accompanied by some qualitative and/or intuitive arguments. When the fit is good, the employed model is declared as satisfactory or reasonably correct, but this could be quite erroneous. If the fit is bad, then the model is hard to improve in any systematic and controllable manner, since no conclusion could be drawn about which part of the theoretical scheme is responsible for the failure to reproduce experimental data. Obviously, the success or failure in these phenomenologic models cannot be taken as the only criterion for the adequacy or inadequacy of the theory. Rather, certain more rigorous mathematical arguments should be used to judge the theory. One such basis is the establishment of a perturbation expansion series for the transition amplitude without any disconnected diagrams, as in the work of Dodd and Greider [261].

Hence, when properly implemented with a flexible intermediate channel along the lines envisaged by Dodd and Greider [261], the distorted wave formalism can appear as a synonym for a versatile regularization procedure of ill-defined and/or divergent $T-$matrix elements in scattering theories. The associated perturbation series provides the lowest-order approximation, which then represents a mathematically meaningful first-order term in the expansion. This is because such a first-order term stems from a divergence-free series (considered either as a whole, or term-by-term).

In two important studies, the rigorous first-order approximations to the perturbation series expansions of the Faddeev integral equations in the Dodd-Greider distorted wave formalism have been introduced by Gayet [43] for charge exchange and by Belkić [134] for ionization. Both processes belong to rearranging collisions between nuclei and hydrogen-like atomic systems. Moreover, the latter two studies prove the equivalence between the full quantum-mechanical and IPM versions of the said first-order terms. This equivalence is not limited to the first-order terms alone, but rather is valid for the entire Faddeev perturbation series in the Dodd-Greider formalism, as well as for the exact transition amplitude in the eikonal limit of two heavy and one light particles as shown by Belkić, Gayet and Salin [44].

The single remaining Faddeev equation in the Dodd-Greider distorted wave formalism invokes two-body perturbation potentials alongside the initial and final scattering states in the transition amplitudes. On the other hand, in the FWL formalism for three-body collisions, the three coupled Faddeev integral equations are given in terms of the corresponding two-body off-shell transition operators in the $T-$matrix elements with the initial and final scattering states. However, when the latter asymptotic states are taken to be the unperturbed channels states (i.e. the products of bound states and plane waves for relative motions of scattering aggregates), as has been done in all the applications of the FWL method [239]–[246], the correct boundary conditions are disregarded from the outset. As a consequence, non-regularizable singularities in momentum space are encountered. As is well-known, the off-shell two-body $T-$matrices for Coulomb potentials do not exist. They possess non-integrable branch-point singularities that lead to divergence in the physical on-shell limit. This occurs for both perturbative and non-perturbative analyses in the FWL equations despite the absence of disconnected diagrams. A compact kernel from the FWL integral equations does not salvage the situation.

Therefore, if one could compute the cross sections of the exact $T-$matrix from the full FWL equations, one would obtain the infinite result. Furthermore, infinity would also be obtained from each term separately in the perturbation expansion of the $T-$matrix in the FWL series expansion. The only reason for which this has not happened thus far in the literature, while using e.g. the second-order term in the FWL method [239]–[246] is that some additional approximations were introduced with the purpose of artificially avoiding the divergent terms.

The past inadequate treatment of the Coulomb problem within the FWL method is similar to previous mishandling of non-integrable logarithmic singularities in the SPB method [217]–[233]. Ironically, as initially pointed out by Belkić [208], the same logarithmic divergence in the same $T-$matrix element was previously encountered in 1968 by Mapleton [233] i.e. long before the appearance of the SPB method. Nearly two decades later, Dewangan and Eichler [234] rediscovered the finding of Mapleton [233], but failed to cite him. This was eventually rectified in 1988 by Bransden and Dewangan [45], but misinterpreted again in 1994 by Dewangan and Eichler [48].

In order to artificially avoid infinite cross sections, the literature on ion-atom collisions witnessed the emergence of a veritable menu of recipes. These artifices ranged from a straight neglect of the divergent term to the peaking approximation, the NSA, etc. None of these *ad hoc* prescriptions has any mathematical justification nor physical backing. Therefore, trying *a posteriori* to redeem the FWL and SPB models (or any such approximations with incorrect boundary conditions) by resorting to comparisons with experimental data is indeed misleading. At best, an eventually good agreement would only indicate how confusing comparisons with measurements could be if the design of a theoretical study is inadequate. However, the irony is that numerous explicit computations have established that the FWL and SPB methods do not agree quantitatively with experimental data at impact energies around the Massey peak ($\sim$ 50 keV/amu). Yet, quantitative improvement of theoretical predictions in comparison with measurements precisely within this latter energy region was the primary motivation for the introduction of the FWL and SPB methods.

The origin of the SPB method was in an earlier formulation of the IA by Briggs [171] who claimed that the $T-$matrices for all asymmetric collisions ($Z_{\rm T} \ll Z_{\rm P}$) should be treated as perturbation expansions in terms of the weaker nuclear charge which appears in the perturbation potential[2]. At the same time, the other constituents of these perturbative transition amplitudes, such as all bound states for the weaker potential, are treated non-perturbatively. This is mathematically unfounded. For consistency, each bound state supported by the weaker potential should also be subjected to the same perturbative expansion. This is because the whole $T-$matrix, as a given mathematical function, is developed in a perturbation series in terms of the weaker nuclear charge. Therefore, one cannot pick-and-choose only a part of the $T-$matrix for a perturbative treatment in terms of a weaker nuclear charge, while simultaneously treating this charge as an intact and fixed parameter in another part of the same transition amplitude. However, a severe drawback of this perturbative formalism is that it cannot be consistent in the requested sense, since bound states would cease to be bound if they are treated perturbatively to the lowest order in the field of the weaker potential. This vicious circle invalidates such perturbation developments of transition amplitudes in terms of the weaker potential.

Even if these fundamental obstacles were absent, such perturbative developments that eventually yielded the SPB method would still be inadequate from the standpoint of the main physics principles of theory of scattering, since they disregard the Dollard's asymptotic convergence problem. The only way to regularize this line of thought is to properly acknowledge the long-range nature of Coulomb perturbation interaction from the very beginning by consistently modifying the perturbation potential and the unperturbed channel states. In such a case, the boundary corrected second Born method extracted trivially from the exact eikonal $T-$matrix of Belkić, Gayet and Salin [44] would be rediscovered, as paraphrased by Dewangan and Eichler [234], thus making the introduction of the SPB approximation indeed unnecessary. The moral of this story is educational: no artificial handling of Coulomb perturbation potentials can avoid the ensuing fundamental inconsistencies stemming from

[2]Later, this was pursued further in a series of investigations by Macek, Taulbjerg, Alston, McGuire, etc [217]–[232] and eventually abandoned altogether, as expected.

the failure to properly address Dollard's asymptotic convergence problem.

As an alternative to the failed search for a particular perturbation development aimed at treating asymmetric collisions alone [171], the Dodd-Greider [261] formalism is available, together with its rigorous and consistent first-order term, the CDW method [42, 44, 134]. This method is known to be applicable equally well to symmetric and asymmetric collisions without encountering any of the discussed drawbacks of the SPB [217]–[232] and FWL [239]–[246] methods.

18

Electron capture by hydrogen-like projectiles

In this chapter, we consider single electron capture in two types of collisions, such as collisions between hydrogen-like projectiles and hydrogen-like targets as well as multi-electron targets

$$(Z_{\rm P}, e_1)_{i_1} + (Z_{\rm T}, e_2)_{i_2} \longrightarrow (Z_{\rm P}; e_1, e_2)_f + Z_{\rm T} \tag{18.1}$$

$$(Z_{\rm P}, e_1)_{i_1} + (Z_{\rm T}, e_2; \{e_3, e_4, \ldots, e_{N+2}\})_i \longrightarrow (Z_{\rm P}; e_1, e_2)_f \\ +(Z_{\rm T}; \{e_3, e_4, \ldots, e_{N+2}\})_{f'} \tag{18.2}$$

where the set $\{e_3, e_4, \ldots, e_{N+2}\}$ denotes the N non-captured electrons. In the case of a multi-electron target, we introduce the following assumptions. All the N non-captured electrons are viewed as being passive, such that their interactions with the active electrons e_1 and e_2 do not contribute to the capture process. We also suppose that the passive electrons occupy the same orbitals before and after the collisions [44]. In such a frozen-core approximation, the final state of the target is ignored. In this model, the passive electrons do not participate individually in capture of the active electron, and this permits the use of an effective local target potential $V_{\rm T}$.

18.1 The CB1-4B method

For the above simplified description of a multi-electron target, process (18.2) is effectively reduced to a four-body problem with a model target potential $V_{\rm T}$. In such a case, within the CB1-4B method for the ensuing equivalent process of a pure four-body collision, the transition amplitudes in either their post or prior form read as

$$T_{if}^{+}(\boldsymbol{\eta}) = \iiint {\rm d}\boldsymbol{s}_1 {\rm d}\boldsymbol{s}_2 {\rm d}\boldsymbol{R}\, \Phi_f^{-*}[V_{\rm T}(x_1) + V_{\rm T}(x_2) - 2V_{\rm T}(R)]\Phi_i^{+} \tag{18.3}$$

and

$$T_{if}^{-}(\boldsymbol{\eta}) = \iiint {\rm d}\boldsymbol{s}_1 {\rm d}\boldsymbol{x}_2 {\rm d}\boldsymbol{R}\, \Phi_f^{-*}\left[V_{\rm T}(x_1) - V_{\rm T}(R) + \frac{Z_{\rm P}-1}{R} - \frac{Z_{\rm P}}{s_2} + \frac{1}{r_{12}}\right]\Phi_i^{+} \tag{18.4}$$

with

$$\Phi_f^- = \varphi_f(\boldsymbol{s}_1, \boldsymbol{s}_2) \mathrm{e}^{-i\boldsymbol{k}_f \cdot \boldsymbol{r}_f - i\frac{Z_\mathrm{P}-2}{v} \int_Z^\infty \mathrm{d}Z' V_\mathrm{T}(R')} \tag{18.5}$$

$$\Phi_i^+ = \varphi_{i_1}(\boldsymbol{s}_1) \varphi_{i_2}(\boldsymbol{x}_2) \mathrm{e}^{i\boldsymbol{k}_i \cdot \boldsymbol{r}_i - i\frac{Z_\mathrm{P}-1}{v} \ln(vR - \boldsymbol{v} \cdot \boldsymbol{R}) - i\frac{Z_\mathrm{P}-1}{v} \int_{-\infty}^Z \mathrm{d}Z' V_\mathrm{T}(R')} \tag{18.6}$$

where $\boldsymbol{R}' = \boldsymbol{\rho} + Z'\hat{\boldsymbol{v}}$ and $\varphi_f(\boldsymbol{s}_1, \boldsymbol{s}_2)$ is the bound state wave function of the helium-like atomic system $(Z_\mathrm{P}; e_1, e_2)$, whose binding energy is E_f. The hydrogen-like wave function of the $(Z_\mathrm{P}, e_1)_{i_1}$ system is denoted as $\varphi_{i_1}(\boldsymbol{s}_1)$ and the corresponding binding energy is E_{i_1}. The initial orbital $\varphi_{i_2}(\boldsymbol{x}_2)$ of the active electron in a multi-electron target satisfies the following equation

$$\left[-\frac{1}{2} \nabla_{x_2}^2 - V_\mathrm{T}(x_2) - E_{i_2} \right] \varphi_{i_2}(\boldsymbol{x}_2) = 0. \tag{18.7}$$

The post and prior amplitudes (18.3) and (18.4), respectively, are identical to each other on the energy shell if (a) $\varphi_{i_2}(\boldsymbol{x}_2)$ is a solution of the eigen-value problem (18.7) for a selected atomic model potential $V_\mathrm{T}(x_2)$, and (b) the final bound state wave function $\varphi_f(\boldsymbol{s}_1, \boldsymbol{s}_2)$ is exact. Whenever at least one of these two conditions is not fulfilled, there will be a post-prior discrepancy.

For a given central field potential $V_\mathrm{T}(x_2)$, it is possible to obtain the numerical solution of (18.7). Certain realistically screened potentials for a neutral target atom are appealing, such as those of e.g. Green, Sellin and Zachor (GSZ) [420] or Herman and Skillman [421] (HS), since they exhibit two correct asymptotic behaviors $(-Z_\mathrm{T}/x_2)$ and $(-1/x_2)$ at small and large values of x_2, respectively. Nevertheless, these atomic model potentials are not practical within the present distorted wave formalism, since they preclude analytical calculation of the Coulomb phase integrals. However, the standard analytical calculation of the whole $T-$matrix e.g. within the CDW method can be reestablished by choosing a pure Coulomb potential $V_\mathrm{T}(x_2) = -Z_\mathrm{T}^\mathrm{eff}/x_2$ for a multi-electron target, where $Z_\mathrm{T}^\mathrm{eff}$ is an effective nuclear charge. We proceed in this way and choose the value of $Z_\mathrm{T}^\mathrm{eff}$ following Belkić *et al.* [44]. This is done by requiring that the energy due to $V_\mathrm{T}(x_2) = -Z_\mathrm{T}^\mathrm{eff}/x_2$, for an electron occupying the orbital with the principal quantum number n_i, is equal to the RHF orbital energy $E_{i_2}^\mathrm{RHF}$ [68] i.e. $Z_\mathrm{T}^\mathrm{eff} = n_{i_2}(-2E_{i_2}^\mathrm{RHF})^{1/2}$. For the initial state of the active electron in multi-electron target, we employ the RHF wave function given as a linear combination of the normalized STOs via

$$\varphi_{i_2}^\mathrm{RHF}(\boldsymbol{x}_2) = \sum_{k=1}^{N_{i_2}} C_k \chi_{n_k l_{i_2} m_{i_2}}^{(\alpha_k)}(\boldsymbol{x}_2) \tag{18.8}$$

$$\chi_{n_k l_{i_2} m_{i_2}}(\boldsymbol{x}_2) = \sqrt{\frac{(2\alpha_k)^{1+2n_k}}{(2n_k)!}}\, x_2^{n_k - 1} \mathrm{e}^{-\alpha_k x_2} Y_{l_{i_2} m_{i_2}}(\hat{\boldsymbol{x}}_2) \tag{18.9}$$

where C_k and α_k are the parameters obtained variationally by Clementi and Roetti [68] and n_k is the orbital number.

Over the years, various models for atomic potentials were encountered in studying electron capture from a multi-electron atomic target by either a bare or a dressed projectile (i.e. a nucleus or a hydrogen-like ion). The first suggestion was undertaken by Belkić *et al.* [44] who used the Coulomb potential $V_{\rm T}^{\rm eff}(r) \equiv -\lambda/r$ with an effective constant charge $Z_{\rm T}^{\rm eff} \equiv \lambda$ within the CDW method. Such a choice is practical, since it preserves the hydrogen-like Coulomb wave functions $\psi_{\boldsymbol{v}}^{\pm}(\boldsymbol{r})$ for continuum intermediate states. This circumstance and the usage of the RHF wave functions of Clementi and Roetti [68] for multi-electron bound states enable analytical calculations of the transition amplitudes in the CDW method. There are many ways to select the effective charge, such as the already mentioned value $\lambda = n(-2E^{\rm RHF})^{1/2}$ from Ref. [44], where $E^{\rm RHF}$ and n are the RHF binding energy and the corresponding orbital number [68]. This selection is reasonable, and it was extensively used in applications by many authors. However, a Coulombic potential with a fixed and constant effective nuclear charge cannot have the required asymptotic behaviors simultaneously at large and small distances for a multi-electron atomic system. Moreover, considering the same nuclear center, the RHF orbital for a bound state and a hydrogen-like Coulomb wave function for continuum are not consistent with each other. This is due to the fact that the RHF atomic model potential $V_{\rm T}^{\rm RHF}(r)$ is not equal to the Coulomb interaction $V_{\rm T}^{\rm eff}(r)$ for any fixed λ. In Ref. [44] an approximate link between the RHF bound orbital $\varphi(\boldsymbol{r})$ and the hydrogen-like continuum state is established by imposing a dependence of λ upon the parameters $\{n, E^{\rm RHF}\}$ of the RHF wave function via $\lambda = n(-2E^{\rm RHF})^{1/2}$. This link is not established directly through $V_{\rm T}^{\rm RHF}(r)$. An alternative choice of λ from $V_{\rm T}^{\rm eff}(r)$ is also possible by reliance upon $V_{\rm T}^{\rm RHF}(r)$ by imposing the condition

$$\langle\varphi|V_{\rm T}^{\rm eff}|\varphi\rangle = \langle\varphi|V_{\rm T}^{\rm RHF}|\varphi\rangle. \tag{18.10}$$

The rhs of this equation can be calculated using the virial theorem

$$\langle\varphi|V_{\rm T}^{\rm RHF}|\varphi\rangle = -2E^{\rm RHF}. \tag{18.11}$$

This procedure gives the effective charge in the following simple form

$$\lambda = -\frac{2E^{\rm RHF}}{\langle\varphi|1/r|\varphi\rangle} \equiv -\frac{2E^{\rm RHF}}{D}. \tag{18.12}$$

The remaining matrix element D in the denominator of this quotient for λ can be calculated analytically using the Clementi-Roetti wave function φ as the STO from (18.8) with the result

$$\left\langle\varphi\left|\frac{1}{r}\right|\varphi\right\rangle = \sum_{k=1}^{N}\sum_{k'=1}^{N}\tilde{C}_k\tilde{C}_{k'}(2-\delta_{k,k'})\delta_{l_k,l_{k'}}\delta_{m_k,m_{k'}} \times \frac{(n_k+n_{k'}-1)!}{(\alpha_k+\alpha_{k'})^{n_k+n_{k'}}} \qquad (k \geq k') \tag{18.13}$$

where $\delta_{k,k'}$ is the Kronecker δ−symbol and $\tilde{C}_j = C_j\sqrt{(2\alpha_j)^{1+2n_j}/[(2n_j)!]}\,(j = k, k')$. Upon inserting the RHF parameters for e.g. helium from the tables of Ref. [68], it follows $D \approx 1.684$ and $\lambda \approx 1.07$. The obtained effective charge $\lambda \approx 1.07$ is close to the fully screened charge $Z_{\rm T} - 1 = 1\,(Z_{\rm T} = 2)$ for He used previously in many computations within the CDW method [269, 270, 273].

Although the effective Coulomb potential has been shown to be useful in practice, a more adequate model is needed which could simultaneously satisfy the required behavior at short and large distances. Such improvements have been proposed in Refs. [44] and [422]–[426]. These studies introduce certain model short-range potentials alongside the long-range interactions using various atomic potential models (RHF, HS, GSZ) within different distorted wave theories. Thus, within the CDW-3B method used in Refs. [425, 426], the mentioned limitations of the effective Coulomb potential were lifted by obtaining both bound and continuum wave functions for the same interaction field. The ensuing results [425, 426] show noticeable improvement over the atomic model with the effective Coulomb potential [44]. Also the GSZ model potential has been used in the CB1-3B method [424]. A direct comparison between e.g. the GSZ and the RHF models shows that total cross sections in the CB1-3B method are not overly sensitive to the choice of the target potential. For example, using the GSZ and RHF models in the CB1-3B method [423], it was found that the magnitude of the difference of the corresponding two sets of charge exchange total cross sections for the $\rm H^+ - C$ collisions does not exceed 15% at intermediate and high energies (the same also applies to the $\rm He^{2+} - Li$ collisions). A test computation in the CB1-3B method [205] on the $\rm H^+ - He$ charge exchange demonstrated that the HS potential for the target gave total cross sections that differ from the RHF model by at most 12%. Assuming that a similar conclusion would also hold true for the problems considered in the present study, and bearing in mind the enormous advantages of analytical calculations, we consider a model with the Coulomb potential $V_{\rm T}$ and the target STOs. As mentioned, the related effective charge $Z_{\rm T}^{\rm eff}$ will be determined as in Ref. [44]. A direct and very useful consequence of these simplifications is a reduction of the complicated multi-electron process (18.2) to a more manageable equivalent collision of a pure four-body problem of the following type (18.1)

$$(Z_{\rm P}, e_1)_{i_1} + (Z_{\rm T}^{\rm eff}, e_2)_{i_2} \longrightarrow (Z_{\rm P}; e_1, e_2)_f + Z_{\rm T}^{\rm eff}. \tag{18.14}$$

Under these circumstances, the transition amplitudes (18.3) and (18.4) in the CB1-4B method for process (18.14) become [192]

$$T_{if}^{+}(\boldsymbol{\eta}) = Z_{\rm T}^{\rm eff} \iiint {\rm d}\boldsymbol{s}_1 {\rm d}\boldsymbol{s}_2 {\rm d}\boldsymbol{R}\,(v\rho)^{2i\nu_f}\,{\rm e}^{-i\boldsymbol{v}\cdot\boldsymbol{s}_2 + i\boldsymbol{\beta}\cdot\boldsymbol{R}} \varphi_f^*(\boldsymbol{s}_1, \boldsymbol{s}_2)$$
$$\times\left(\frac{2}{R} - \frac{1}{x_1} - \frac{1}{x_2}\right)\varphi_{i_1}(\boldsymbol{s}_1)\varphi_{i_2}^{\rm RHF}(\boldsymbol{x}_2)(vR - \boldsymbol{v}\cdot\boldsymbol{R})^{i\xi} \tag{18.15}$$

$$T_{if}^{-}(\boldsymbol{\eta}) = \iiint \mathrm{d}\boldsymbol{s}_1 \mathrm{d}\boldsymbol{x}_2 \mathrm{d}\boldsymbol{R}\, (v\rho)^{2i\nu_f}\, \mathrm{e}^{-i\boldsymbol{v}\cdot\boldsymbol{s}_2 + i\boldsymbol{\beta}\cdot\boldsymbol{R}} \varphi_f^*(\boldsymbol{s}_1, \boldsymbol{s}_2)(vR - \boldsymbol{v}\cdot\boldsymbol{R})^{i\xi}$$
$$\times \left(\frac{Z_{\mathrm{T}}^{\mathrm{eff}} + Z_{\mathrm{P}} - 1}{R} - \frac{Z_{\mathrm{T}}^{\mathrm{eff}}}{x_1} - \frac{Z_{\mathrm{P}}}{s_2} + \frac{1}{r_{12}} \right) \varphi_{i_1}(\boldsymbol{s}_1)\varphi_{i_2}^{\mathrm{RHF}}(\boldsymbol{x}_2) \qquad (18.16)$$

where $\xi = (Z_{\mathrm{T}}^{\mathrm{eff}} - Z_{\mathrm{P}} + 1)/v$, $\nu_f = Z_{\mathrm{T}}^{\mathrm{eff}}(Z_{\mathrm{P}} - 2)/v$ and

$$\boldsymbol{k}_i \cdot \boldsymbol{r}_i + \boldsymbol{k}_f \cdot \boldsymbol{r}_f = -\boldsymbol{v}\cdot\boldsymbol{s}_2 + \boldsymbol{\beta}\cdot\boldsymbol{R} \qquad \boldsymbol{\beta} = -\boldsymbol{\eta} - \beta_z \hat{\boldsymbol{v}}$$
$$\beta_z = \frac{v}{2} + \frac{\Delta E}{v} \qquad \Delta E = E_{i_1} + E_{i_2}^{\mathrm{RHF}} - E_f. \qquad (18.17)$$

Using the Nordsieck technique [311], the post amplitude T_{if}^{+} can be calculated as two-dimensional real integrals [192]. Alternatively, following Belkić [89], a different method can be employed to obtain T_{if}^{+} as a one-dimensional integral [192]. Calculations of the prior amplitude T_{if}^{-} are more difficult. This is due to the term $1/r_{12}$ in the perturbation potential from (18.16) which requires an additional three-dimensional numerical quadrature. In this way, T_{if}^{-} is reduced to a five-dimensional numerical integration.

18.2 Comparison between theories and experiments

In this section, we shall use two and three examples for processes (18.1) and (18.2), respectively. They are given by the following rearrangement collisions

$$\mathrm{H} + \mathrm{H} \longrightarrow \mathrm{H}^- + \mathrm{H}^+ \qquad (18.18)$$

$$^4\mathrm{He}^+ + \mathrm{H} \longrightarrow {}^4\mathrm{He} + \mathrm{H}^+ \qquad (18.19)$$

$$^4\mathrm{He}^+ + {}^4\mathrm{He} \longrightarrow {}^4\mathrm{He} + {}^4\mathrm{He}^+ \qquad (18.20)$$

$$\mathrm{H} + {}^4\mathrm{He} \longrightarrow \mathrm{H}^- + {}^4\mathrm{He}^+ \qquad (18.21)$$

$$^7\mathrm{Li}^{2+} + {}^4\mathrm{He} \longrightarrow {}^7\mathrm{Li}^+ + {}^4\mathrm{He}^+. \qquad (18.22)$$

The total cross sections from the CB1-4B method [191] for reaction (18.18) are shown in Fig. 18.1. The wave function for the H^- ion is described by means of the two-parameter orbital of Silverman *et al.* [71].

Comparison of these theoretical results with the experimental data is limited to the measurement of McClure alone [429]. His results are the only

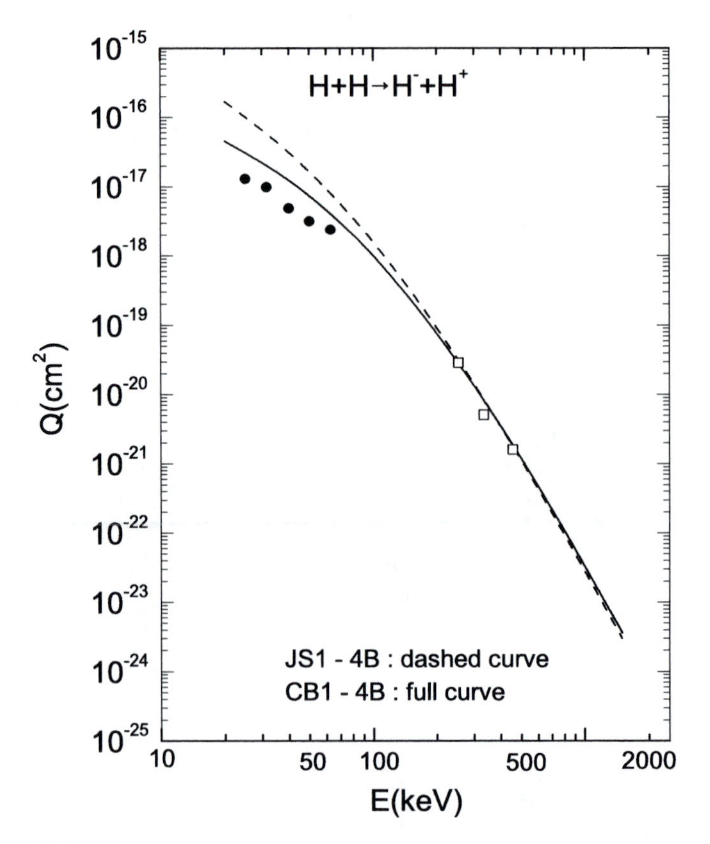

FIGURE 18.1
Total cross sections Q as a function of the incident energy E for process (18.18). Theory: dashed curve (JS1-4B method [427, 428]) and full curve (CB1-4B method [191]). Experiment: □ [297] and • [429].

experimental data available for (18.18) for the atomic hydrogen target and, moreover, they are restricted to impact energies $E \leq 63$ keV. However, in the same figure, the experimental results of Schryber [297] for an equivalent process with a molecular hydrogen target $\mathrm{H}+\mathrm{H}_2 \rightarrow \mathrm{H}^- + \mathrm{H}_2^+$ are also plotted. Here, the values for the experimentally-determined cross sections are divided by two. Namely, following the well-known Bragg sum rule [430]–[432], we assumed that the hydrogen molecule could be treated as two isolated and independent H atoms. It can be seen in Fig. 18.1 that the agreement between the CB1-4B method and the experimental data is good. The theoretical curve of Mapleton [427, 428] in the JS1-4B approximation is also displayed in Fig. 18.1. The difference between the CB1-4B [191] and JS1-4B [427, 428] approximations is in treating the Coulomb phase factors. These phases are completely ignored in Refs. [427, 428]. Such a comparison should provide valuable information about the importance of the correct boundary conditions for the problem under study. Figure 18.1 shows that the CB1-4B method [191] yields a significant improvement over the cross sections from the JS1-4B approximation [427, 428], when compared with the experimental data. As expected, Fig. 18.1 demonstrates that the logarithmic phase factors play a more prominent role at lower than at higher impact energies.

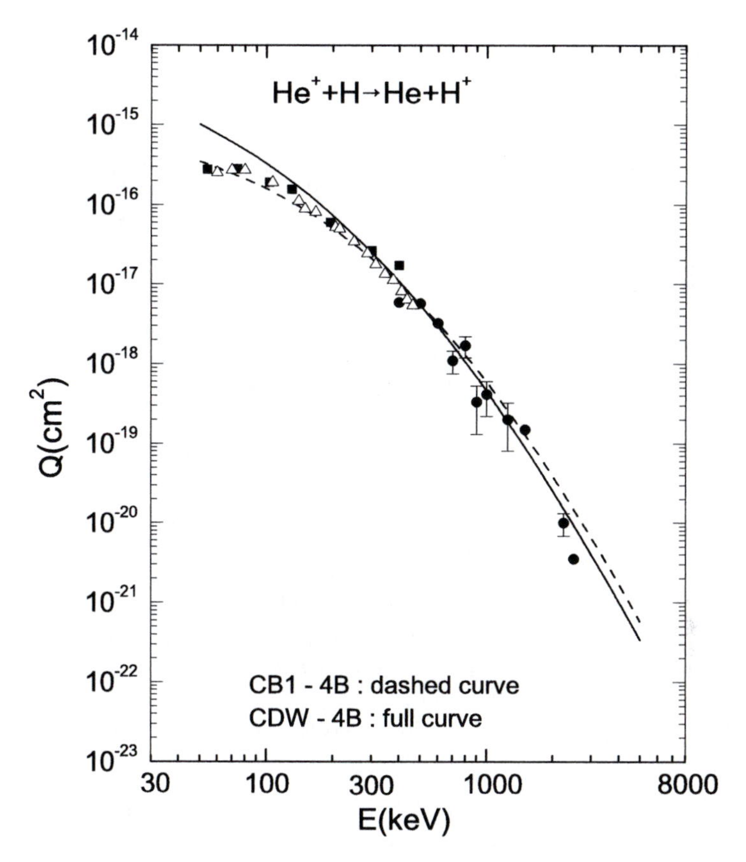

FIGURE 18.2
Total cross sections Q as a function of the incident energy E for process (18.19). Theory: dashed curve (post CB1 method [191]) and full curve (post CDW-4B method [433]). Experiment: ■ [434], • [435] and △ [436].

In Fig. 18.2 a comparison is made between the theoretical results from the CB1-4B as well as CDW-4B methods [192, 433] for process (18.19) with the experimental data [434]–[436]. The dashed and full curves refer, respectively, to the CB1-4B and CDW-4B methods implemented using the wave function of Silverman *et al.* [71] for helium. The wave function of Löwdin [69] for helium yields very similar cross sections [192]. It can be seen from Fig. 18.2 that both the CB1-4B and CDW-4B methods are in quite satisfactory agreement with the available experimental data.

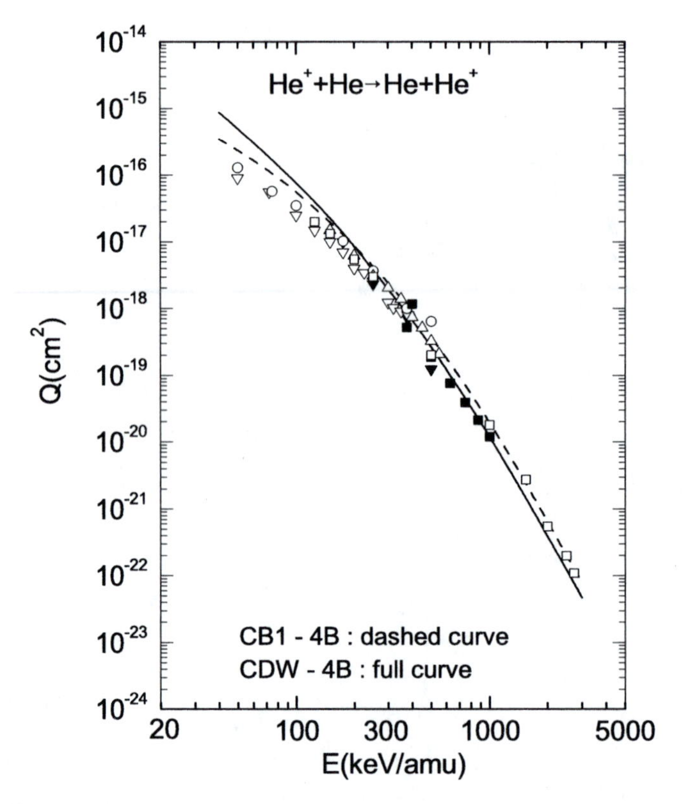

FIGURE 18.3
Total cross sections Q as a function of the incident energy E for process (18.20). Theory: dashed curve (post CB1 method [433]) and full curve (post CDW-4B method [433]). Experiment: • [305], ▽ [399], □ [437], ∘ [438], △ [439] and ▼ [440].

The cross sections for charge exchange in process (18.20) are depicted in Fig. 18.3. Despite the fact that there are many experimental data for reaction (18.20), theoretical studies are scarce, because of the difficulties which arise in treating collision systems that are more complex than a three-body problem. The CB1-4B and CDW-4B methods extended to process (18.20) by Mančev

[192, 433] provide the first quantum mechanical descriptions of this problem. The CB1-4B and CDW-4B methods are seen in Fig. 18.3 to be in close agreement with the available measurements, except at lower impact energies where the CB1-4B and CDW-4B methods overestimate the experimental data, as anticipated from the validity assessment (14.69).

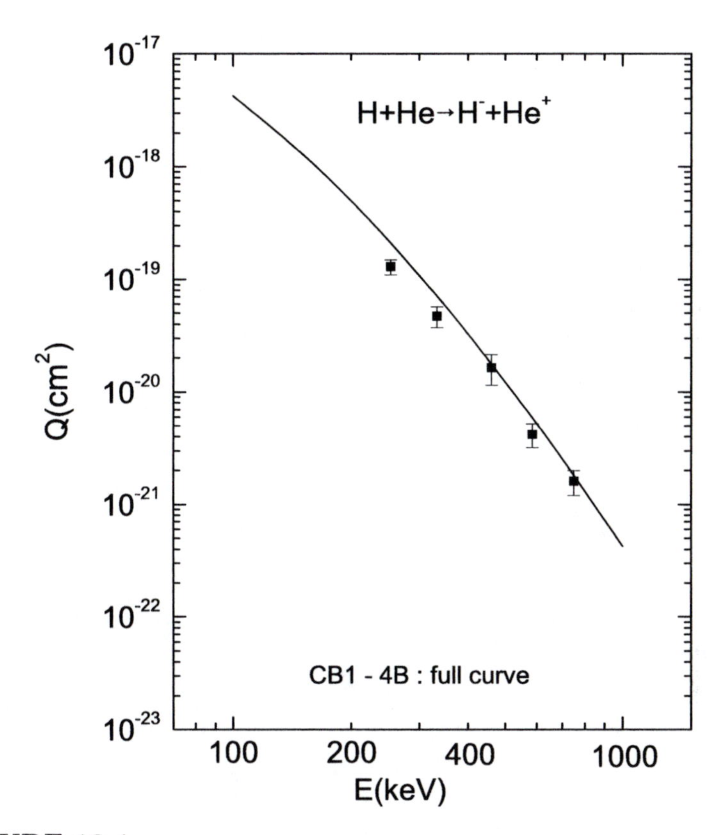

FIGURE 18.4
Total cross sections Q as a function of the incident energy E for process (18.21). Theory: full curve (post CB1 method [192]). Experiment: ■ [297].

The result of computations using the CB1-4B method [192] for the formation of $H^-(1s^2)$ ion in the H – He collisions i.e. for process (18.21), is plotted in Fig. 18.4. The ground state of the H^- ion is described by the two-parameter wave function of Silverman *et al.* [71]. It should be noted that the two-electron orbital of Silverman *et al.* [71] gives a bound energy for $H^-(1s^2)$ below -0.5 ($E_f = -0.5133289$) which ensures a stable bound state of the H^-

ion. In Fig. 18.4, the cross sections from the CB1-4B method are compared with the experimental data of Schryber [297]. Agreement between the theory and the measurement is very good.

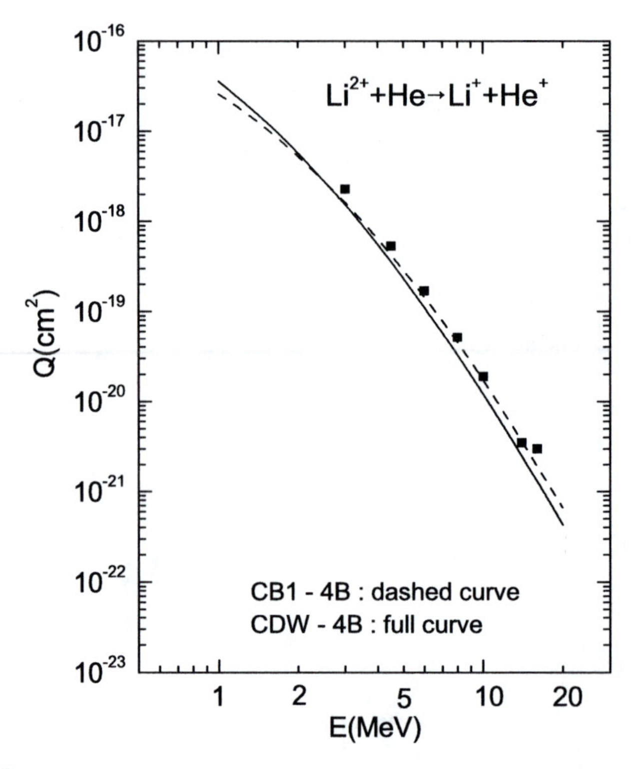

FIGURE 18.5
Total cross sections Q as a function of the incident energy E for process (18.22). Theory: dashed curve (post CB1 method [191]) and full curve (post CDW-4B method [433]). Experiment: ■ [354].

The total cross sections from the CB1-4B [191] and CDW-4B [433] methods for single electron capture in process (18.22) are given in Fig. 18.5. A comparison between these theoretical results and the experimental data of Woitke *et al.* [354] shows quite satisfactory agreement.

Note that it is also possible to compute the cross sections for direct processes (18.18)–(18.22) by using the principle of detailed balancing (quantum mechanical micro-reversibility) and the corresponding results obtained for the associated inverse reactions [317]. In these related inverse processes, the entrance and exit channels of reactions (18.18)–(18.22) exchange places. Electron statistics and the principle of detailed balancing can be illustrated on the following examples

$$H(1s) + H(1s) \longrightarrow H^-(1s^2) + H^+ \tag{18.23}$$

$$H^+ + H^-(1s^2) \longrightarrow H(1s) + H(1s). \tag{18.24}$$

Let us assume that we have computed only the cross sections $Q_{i;f_1,f_2} = Q_{1s^2;1s,1s}$ for process (18.24). We want to deduce the cross sections for the corresponding process (18.23) without any detailed computations i.e. solely in terms of the results $Q_{1s^2;1s,1s}$ for reaction (18.24).

With this goal in mind, the first thing which needs to be recalled is that the negative atomic hydrogen ion H^- has only one bound state. More specifically, in the infinite nuclear mass approximation ($M_{H^+} = \infty$), Hill [80] used the so-called comparison theorem to show that the H^- ion exists exclusively in its singlet state, $H^-(^1S)$.

The usual Rayleigh-Ritz variational principle with the one-parameter uncorrelated wave function of Hylleraas [58], $\propto \exp(-ax_1 - ax_2)$, with $a = 0.6875$ for the negative atomic hydrogen ion does not lead to a stable ground-state structure of H^-. This is because the ensuing energy $E(H^-) = -0.47265625$ of the closed $(1s)^2$ shell of H^- is larger than the energy $E(H) = -0.5$ of the neutral hydrogen atom H, thus yielding the unphysical negative ionization potential, $E(H^-) - E(H) = -0.0273 < 0$. Stability of permanent structure could only be secured if the ionization energy of the negative hydrogen ion (or equivalently, the affinity of the neutral atomic hydrogen) is positive $E(H^-) - E(H) > 0$. For example, using the same variational principle with the radially correlated two-parameter $(1s1s')$ open shell wave function of Silverman *et al.* [71], $\propto [\exp(-ax_1 - bx_2) + \exp(-bx_1 - ax_2)]$, as in (16.10), stability of the H^- ion is obtained with the variational parameters $a = 1.039230$ as well as $b = 0.2832215$ and energy $E(H^-) = -0.51330289$, yielding a positive ionization potential, $E(H^-) - E(H) = 0.0133 > 0$. A further improvement in the energy $E(H^-) = -0.52457430$ is obtained with the 3-parameter CI wave function of Silverman *et al.* [71], as in (16.11) which includes the angular electron correlation via the $\{(1s1s'), \lambda(2p)^2\}$ structure of the ground state of the $H^-(^1S)$ ion. The optimized constant λ multiplying the $2p$ hydrogen-like radial function has the value $\lambda = -0.112426$, whereas the corresponding orbiting parameters in this case are given by $a = 1.03556$ and $b = 0.323563$. The improved energy $E(H^-) = -0.52457430$ for the simple $\{(1s1s'), \lambda(2p)^2\}$ configuration from (16.11) is substantially better than the previous value $E(H^-) = -0.51330289$ for the $(1s1s')$ shell (3- versus 2-parameter CI orbitals, respectively). The energy with the 3-parameter orbital compares well with the corresponding estimate $E(H^-) = -0.5272225$ from the much more involved 61-parameter CI radially- and angularly-correlated wave function of Joachain and Terao [77] from (16.12). Recall also that the practically 'exact' energy $E(H^-) = -0.52775101654420$ for the ground state of the $H^-(^1S)$ ion was obtained by Drake [82] using the Hylleraas r_{12}−dependent wave function with some 616 parameters.

This illustrates the key role of static electronic correlation in predicting the stable structure of the H^- ion. In H^- the static electronic correlation consists of a tendency of the two electrons to avoid each (through their Coulomb repulsion) in order to decrease the total energy of the system and, thus, secure stability. Therefore, the inability of the one-parameter wave function of Hylleraas [58] for H^- to achieve the needed lowering of $E(H^-)$ can be attributed directly to the missing static electron correlation effect. In the one-parameter uncorrelated wave function [58] for H^-, both electrons are screened symmetrically to the same extent of about 31%, as implied by the variational value $a = 0.6875$ relative to the bare nuclear charge $Z = 1$ $(1-0.6875 = 0.3125)$. However, when the electron radial correlation is included via the two-parameter wave function [71], the two electrons are screened very differently in an atomic configuration with an open structure of the H^- ion. Here, the inner electron is screened only slightly to about 3.9% $(1.039230 - 1 = 0.0392)$, whereas screening of the outer electron is enormous, attaining some 72% $(1 - 0.2832215 = 0.7168)$. Relative to the radially correlated 2-parameter wave function, inclusion of the radial correlation in the 3-parameter orbital reduces the screening of the outer electron from 72% to 68% $(1 - 0.323563 = 0.6764)$. Simultaneously, the corresponding screening of the inner electron is altered only slightly i.e. from 3.9% to 3.6% $(1.03556 - 1 = 0.0356)$ when passing from the $(1s1s')$ to $(1s1s') + \lambda(2p)^2$ configurations.

The unstable structure of the H^- ion is prone to auto-ionization, which is a process of spontaneous dissociation into H and a free electron of positive energy, $H^- \longrightarrow H + e$. Alternatively, an unstable (or meta-stable) H^- ion can be viewed as existing only temporarily in a state with a transitory decaying character. This is the same type of auto-ionizing states as e.g. the doubly excited state $(2s)^2$ of helium, which has a larger energy than that of $He^+(1s)$, thus resulting in a negative ionization potential, and pointing to instability. In general, bound-state spectra are qualitatively different for negative, positive and neutral multi-electron atomic systems. All such negative ions (other than H^-) have only finite numbers of bound states. By contrast, all positive ions and neutral atoms have infinitely many bound states. Kato [63] was the first to prove (within the infinite nuclear mass limit) that helium has an infinite number of bound states. A similar proof also exists for general atoms.

Since the H^- ion exists only in the singlet state, it is possible to distinguish the two electrons in this system. Therefore, in the absence of spin-dependent forces that are, at any rate, not considered in this book, it is sufficient to compute cross sections for capture of electron e_1, and subsequently to multiply the result by 2, because either electron has the same probability to be transfered to the impinging proton. Note also that according to spin statistics (i.e. the law of conservation of the total spin of the considered system) in process (18.23), formation of the negative atomic hydrogen ion in the exit channel is possible only if the total spin state of the two constituent electrons of H^- leads to a singlet state, since the two colliding atomic hydrogen atoms are initially in their ground states ($1s$).

When the cross sections for process (18.23) are measured with two beams of unpolarized hydrogen atoms, the probability 1/4 is obtained for the chance of finding two electrons in a singlet state. Hence, the cross section for formation of the H^- ion by a collision of two atomic hydrogen atoms in their ground states, as in process (18.23), is equal to $(1/4)Q_{1s^2;1s,1s}$. Therefore, if the cross sections $Q_{1s^2;1s,1s}$ are available for single electron capture from $H^-(1s^2)$ by H^+, they can also be used for the inverse process (18.23). In order to deduce the cross sections for (18.23), it is necessary to divide the results $Q_{1s^2;1s,1s}$ for (18.24) by four[1]. Otherwise, the quantity $(1/4)Q_{1s^2;1s,1s}$ itself represents the cross sections for the following two processes

$$\mathrm{H_P}(1s) + \mathrm{H_T}(1s) \longrightarrow \mathrm{H_P^-}(1s^2) + \mathrm{H_T^+} \tag{18.25}$$

$$\mathrm{H_P}(1s) + \mathrm{H_T}(1s) \longrightarrow \mathrm{H_P^+} + \mathrm{H_T^-}(1s^2) \tag{18.26}$$

where $\mathrm{H_P}$ and $\mathrm{H_T}$ are the projectile and target atomic hydrogen. Process (18.25) is the usual electron capture from the target hydrogen $\mathrm{H_T}(1s)$ by the projectile $\mathrm{H_P}(1s)$. Similarly, process (18.26) is the equivalent electron capture from the projectile hydrogen $\mathrm{H_P}(1s)$ by the target $\mathrm{H_T}(1s)$. The cross sections for either of these two processes (18.25) or (18.26) are the same and equal to $(1/8)Q_{1s^2;1s,1s}$. This reasoning can be used in comparisons between theory and experiment. For example, if a measurement was performed for process (18.23), as done by McClure [429], but we have already computed the cross sections $Q_{1s^2;1s,1s}$ for the inverse reaction (18.24), then the theoretical findings $Q_{1s^2;1s,1s}/8$ should be compared with these experimental data.

Process (18.23), or equivalently, its inverse (18.24) are important for applications of the CB1-4B and CDW-4B methods, due to the fact that in one of the channels the two scattering aggregates attract each other. In nearly all other ion-atom collisions involving bound-free transitions, one aggregate is charged and the other neutral, or both positively charged with the center of ionic charges supposed to be placed on the nuclei. For example, the principal idea of the CB1-4B method is to allow for the asymptotic distortion of the unperturbed states in the entrance and exit channels. This is done through

[1]The CDW-4B method for single capture in general process (11.3) has been developed by Belkić *et al.* [189], using the helium-like wave functions of Hylleraas [58], Silverman et al. [71], Green *et al.* [70] and Löwdin [69]. The algorithm from Ref. [189] together with the principle of detailed balancing can be employed directly without any modification to obtain prior and post cross sections in the CDW-4B method for the inverse processes (18.18) and (18.19). The same program from Ref. [189] can also be applied to the remaining three processes (18.20)–(18.22) by using the frozen core approximation for the third non-captured electron as done in Ref. [433]. This should give identical results to the corresponding cross sections recently reported by Mančev [433] in the CDW-4B method for processes (18.19), (18.20) and (18.22).

the correct boundary conditions in the initial and final states by including the long-range Coulomb effects, due to the relative motion of the two scattering aggregates, and involving the appropriate screening of the bare nuclear charges by the electrons. In particular, in the exit channel of process (18.23), this distortion includes the attraction between the reaction products that are the two oppositely charged ions i.e. the newly-formed negative hydrogen ion H^- and the target rest H^+. Additionally, the negative charge of H^- is not centered on the the nucleus (proton) of this ion, and this differs from the usual assumption in collisions involving positive ions, such as those in process (18.22). This special feature as well as full respect of the correct boundary conditions are taken into account in the CB1-4B and CDW-4B methods[2]. Such features were seen in Fig. 18.1 to be very important for process (18.18). In particular, the Coulomb effect for the products $H^- - H^+$ in the exit channel of this reaction illustrated in Fig. 18.1 was noted to significantly improve agreement between the CB1-method and the experimental data, relative to the JS1-4B approximation, which uses the undistorted channel states Φ_i and Φ_f. A similar Coulomb repulsion also exists between the products H^- and He^+ in a more complicated five-body $H - He$ collision with single electron capture via process (18.21) for which good agreement is again found between the CB1-4B method and the available experimental data.

It should be noted that process (18.26) is not the only way to generate protons H_P^+ from atomic hydrogen beams $H_P(1s)$ during the passage through $H_T(1s)$. This could also be achieved via several other competitive channels such as stripping reactions that are alternatively called electron loss or projectile ionization [441]–[468]

$$H_P(1s) + H_T(1s) \longrightarrow H_P^+ + e + H_T(1s) \tag{18.27}$$

$$H_P(1s) + H_T(1s) \longrightarrow H_P^+ + e + H_T(n) \tag{18.28}$$

$$H_P(1s) + H_T(1s) \longrightarrow H_P^+ + e + H_T^+ + e. \tag{18.29}$$

In process (18.27), the target is left unaffected by the collision, and this is the simplest channel in electron loss involving only a one-electron transition. By contrast, processes (18.28) and (18.29) invoke simultaneous transitions of two electrons where the target can be excited leading to loss-excitation (LE) or ionized yielding loss-ionization (LI), respectively. In the B1-4B and CB1-4B methods for these problems [448]–[454], electron loss can occur through two distinct mechanisms: (i) interaction between the target proton and the projectile electron, as well as (ii) the electron-electron potential. The mechanisms

[2]Of course, the overall description is more adequate in the CDW-4B method than in the CB1-4B approximation, since the former theory incorporates the additional Coulomb distortions for electronic motions at all distances.

(i) and (ii) lead to the mentioned single transition (projectile ionization and target unaltered) and double transition (projectile ionization and target excitation or ionization) in electron loss, respectively. Importantly, computations using the B1-4B [448, 449] and CB1-4B [450]–[454] methods show that the LI mode represents the dominant channel already at the outset of intermediate energies (above 200 keV). This is another example of the critical importance of dynamic electron correlations by reference to the stated fact that the LI channel is produced solely by the $1/r_{12}$ perturbation potential.

19

Simultaneous Transfer and Excitation

In this chapter, we consider resonant collisions involving e.g. two hydrogen-like atomic systems

$$(Z_{\rm P}, e_1)_{i_1} + (Z_{\rm T}, e_2)_{i_2} \longrightarrow (Z_{\rm P}; e_1, e_2)_f^{**} + Z_{\rm T}. \qquad (19.1)$$

This is simultaneous transfer and excitation i.e. the TE process, where a doubly excited (auto-ionizing) state is produced in the projectile after capture of the target electron e_2. Here, there are two mechanisms that interfere through the resonant and non-resonant modes i.e. the RTE and NTE modes, respectively. The RTE mode occurs via capture of the target electron and simultaneous excitation of the projectile electron by means of the interaction $1/r_{12}$ between the two electrons. The NTE mode appears when the target electron is transferred by its interaction with the projectile nucleus P of charge $Z_{\rm P}$, whereas the excitation of the helium-like projectile formed in the exit channel comes from the interaction of the projectile electron e_1 with the target nucleus T of charge $Z_{\rm T}$ [187]. Process (19.1) is the prototype of the simplest example of the TE collision. Of course, the TE process can also occur with more complicated colliding particles, involving e.g. a multi-electron target. An example is the first work on TE by Tanis *et al.* [469] who measured the total cross sections for the S^{13+} – Ar colliding system.

The doubly excited state of the projectile $(Z_{\rm P}; e_1, e_2)_f^{**}$ in the exit channel of process (19.1) relaxes either by radiative decay via X-ray emission (TEX) or through the Auger mechanism (TEA), thus providing two different and complementary experimental approaches for understanding the TE phenomenon [470, 471]. Hereafter, the resonant and non-resonant TEA are denoted by RTEA and NTEA, respectively. The first experimental evidence of the resonant TEX mode (RTEX) was reported by Tanis *et al.* [469]. A similar measurement on the TE process via the RTEX mode was subsequently made with the H_2 target by Schulz *et al.* [472]. In the theoretical studies by Brandt [473] and Feagin *et al.* [474], the RTE and NTE modes were considered as independent. However, the basic features of the CDW-4B method could obviously provide a more adequate description of the TE by a natural introduction of the critical interference effects between the RTE and NTE modes [184, 185]. Furthermore, the CDW-4B method can be of help in interpreting the experiment of Justiniano *et al.* [471]. Here, a very asymmetric collisional system $S^{15+} - H_2$ was investigated, where the state $\left(S^{14+}\right)^{**}$ formed in the exit channel decays via the radiative emission lines $K\alpha - K\alpha$ and $K\alpha - K\beta$ that are dominated by the RTEX mode. The CDW-4B method has been found [184] to be in good agreement with the experimental data of Justiniano *et al.* [471], as well as with the results of Brandt [473]. Furthermore, it has been reported in Ref. [184] that the interference between the RTE and NTE modes can be important if the TE process occurs in a nearly symmetrical collisional system such as the He^+-H or He^+-He encounters. These two collisions involving the TE process were studied experimentally [470] and theoretically [185, 188]. The latter two theoretical studies used the CDW-4B method. Here, the TE process is observed experimentally through the TEA mode. Agreement between the experimental data obtained using the 0° electron spectroscopy technique and the theoretical cross sections computed by the CDW-4B method for the TEA mode is not satisfactory. For certain auto-ionizing states, the total cross sections from the CDW-4B method underestimate the corresponding experimental data of Itoh *et al.* [470]. This could be due to competition between the direct and indirect transfer excitation (ITE). The direct TE is the customary TE, which we have already defined. In the ITE, forward emitted electrons are generated through two intermediate channels (a) target ionization and (b) simultaneous capture of the target electron and projectile electron loss. The direct and indirect TE cannot be distinguished if the electron ejected from the target is not detected by a coincident measurement. Ourdane *et al.* [188] revisited this problem using the CDW-4B method, but this time with a more consistent description of the final auto-ionizing state, by including the adjacent continuum components in addition to the discrete ones. Specifically, the final state used in Ref. [188] is described within the atomic resonant structures from the formalism of Fano [475] as a a linear superposition of bound and continuum orbitals.

19.1 The CDW-4B method for the TE process

As emphasized, the main advantageous feature of the CDW-4B method of key relevance for the TE process is the preservation of the phase relation between the NTE and RTE modes [185]. The transition amplitude in the prior form for the basic TE process (19.1) described by means of the CDW-4B method, can be written as follows

$$\begin{aligned}
T_{if}^{-} &= N^{+}(\nu_{\mathrm{P}})N^{-*}(\nu_{\mathrm{T}})\iiint \mathrm{d}\boldsymbol{s}_1\mathrm{d}\boldsymbol{x}_2\mathrm{d}\boldsymbol{R}\,\mathrm{e}^{i\boldsymbol{k}_i\cdot\boldsymbol{r}_i+i\boldsymbol{k}_f\cdot\boldsymbol{r}_f}(\mu v\rho)^{2i\nu}\varphi_f^{*}(\boldsymbol{s}_1,\boldsymbol{s}_2)\\
&\times\ {}_1F_1(i\nu_{\mathrm{T}},1,ivx_2+i\boldsymbol{v}\cdot\boldsymbol{x}_2)\left\{\left(\frac{1}{r_{12}}-\frac{1}{s_2}+\frac{Z_{\mathrm{T}}}{R}-\frac{Z_{\mathrm{T}}}{x_1}\right)\right.\\
&\times\ \varphi_{i_1}(\boldsymbol{s}_1)\varphi_{i_2}(\boldsymbol{x}_2)\,{}_1F_1(i\nu_{\mathrm{P}},1,ivs_2+i\boldsymbol{v}\cdot\boldsymbol{s}_2)\\
&-\left.\varphi_{i_1}(\boldsymbol{s}_1)\boldsymbol{\nabla}_{\boldsymbol{x}_2}\varphi_{i_2}(\boldsymbol{x}_2)\cdot\boldsymbol{\nabla}_{\boldsymbol{s}_2}\,{}_1F_1(i\nu_{\mathrm{P}},1,ivs_2+i\boldsymbol{v}\cdot\boldsymbol{s}_2)\right\}
\end{aligned}\tag{19.2}$$

where $N^+(\nu_P) = \Gamma(1 + i\nu_P)e^{\pi\nu_P/2}$, $N^-(\nu_T) = \Gamma(1 - i\nu_T)e^{\pi\nu_T/2}$, $\nu_P = (Z_P - 1)/v$, $\nu_T = Z_T/v$ and $\nu = Z_T(Z_P - 1)/v$. Here, the phase factor $(\mu v\rho)^{2i\nu}$ stems from the eikonal approximation for the relative motion of heavy particles. As before, this phase can be omitted for the computation of total cross sections [44]. Moreover, such an omission is not limited to total cross sections alone, since it can be made whenever the integration over $\boldsymbol{\eta}$ has been carried out. This also applies to cross sections that are differential in the energy and/or angle or ejected electrons generated by either direct or indirect ionization. Since only the electron e_2 is transferred, one has the usual kinematic relation involving the initial and final wave vectors (see Ref. [44])

$$\boldsymbol{k}_i \cdot \boldsymbol{r}_i + \boldsymbol{k}_f \cdot \boldsymbol{r}_f \simeq -\boldsymbol{\eta} \cdot \boldsymbol{\rho} - \left(\frac{1}{2} + \frac{E_i - E_f}{v^2}\right) \boldsymbol{v} \cdot \boldsymbol{R} - \boldsymbol{v} \cdot \boldsymbol{s}_2 \tag{19.3}$$

where

$$E_i = E_{i_1} + E_{i_2} \qquad E_{i_1} = -\frac{Z_P^2}{2n_{i_1}^2} \qquad E_{i_2} = -\frac{Z_T^2}{2n_{i_2}^2}. \tag{19.4}$$

Here, E_{i_1} and E_{i_2} are the initial electronic energies of the bound states in the hydrogen-like systems $(Z_P, e_1)_{i_1}$ and $(Z_T, e_2)_{i_2}$, respectively. Likewise, E_f is the final electronic energy of the doubly excited state $(Z_P; e_1, e_2)_f^{**}$. Moreover, in (19.2) for T_{if}^-, the level of approximation made for the excitation process is similar to the usual first Born approximation, which is well-known to give good total cross sections for the excitation process at high impact velocities. More importantly, the transition amplitude from (19.2) for the TE process contains a coherent contribution from both the RTE and the NTE modes.

The type of deexcitation (radiative or Auger) occurring after the formation of an auto-ionizing state via the TE process depends on the ratio between the two nuclear charges Z_P and Z_T. Specifically, there are two limiting cases: (i) if $Z_P/Z_T \gg 1$, the process is radiative via the TEX mode, and (ii) $Z_P/Z_T \simeq 1$, the deexcitation is an Auger process via the TEA mode.

19.2 The TEX mode for radiative decays of asymmetric systems

19.2.1 A model for the RTEX modes

The first model for the RTEX modes was proposed by Brandt [473] along the lines of the dielectronic recombination (DR) process [476]. In the DR process, a free electron moving with momentum $\boldsymbol{p}$ relative to an ion of nuclear charge Z_P is captured via

$$(Z_P, e_1)_i + e_2 \longrightarrow (Z_P; e_1, e_2)_f \tag{19.5}$$

where i and f are the usual sets of the quantum numbers for the initial and final bound states in the entrance and exit channel, respectively. Energy conservation for this process requires that $p^2/2 + E_i = E_f$, where E_i and E_f are the initial and final electronic binding energies, respectively. Therefore, the DR process is resonant whenever

$$p = \sqrt{2(E_f - E_i)} \equiv p_r \tag{19.6}$$

where the corresponding resonance energy E_r is defined by

$$E_r = \frac{1}{2}p_r^2 = E_f - E_i. \tag{19.7}$$

The difference between the DR process and the RTEX mode of the TE collision is that the electron e_2 captured by the projectile $(Z_{\rm P}, e_1)_i$ is not free, but rather it is bound to the target with the binding energy E_{i_2}. The model of Brandt [473] is especially adapted to the case where $E_{i_2} \ll E_i$, which occurs in highly asymmetric collisions with $Z_{\rm P} \gg Z_{\rm T}$. In such an approximation, E_{i_2} can be neglected so that the influence of the target is manifested merely through the associated momentum distribution $|\tilde{\varphi}_{i_2}(\boldsymbol{\kappa})|^2$, where $\tilde{\varphi}_{i_2}(\boldsymbol{\kappa})$ is the target wave function in the momentum space representation. Here, $\boldsymbol{\kappa}$ is the electron momentum in the target frame i.e. relative to $Z_{\rm T}$. In fact, in the projectile frame, the quasi-free electron e_2 has an energy distribution which exhibits a very large peak. Following McLaughlin and Hahn [476], the cross section for the RTEX mode can be written as the standard convolution

$$Q_{\rm RTEX} = \int {\rm d}\boldsymbol{\kappa} Q_{\rm DR}(\boldsymbol{p})\, |\tilde{\varphi}_{i_2}(\boldsymbol{\kappa})|^2 \tag{19.8}$$

where $Q_{\rm DR}(\boldsymbol{p})$ is the cross section for the DR process and $\boldsymbol{p}$ is the electron momentum in the projectile frame i.e. relative to $Z_{\rm P}$

$$\boldsymbol{p} = \boldsymbol{\kappa} - \boldsymbol{v}. \tag{19.9}$$

Here, $\boldsymbol{v}$ is the impact velocity in the laboratory reference system whose origin is placed at the target, which is presumed to be at rest.

While $Q_{\rm DR}(\boldsymbol{p})$ is strongly peaked around $p = p_r$, the distribution $|\tilde{\varphi}_{i_2}(\boldsymbol{\kappa})|^2$ is much broader exhibiting a maximum at the value $\kappa = \kappa_{\rm max}$. For a collision occurring at a high impact energy, we have $v \gg \kappa_{\rm max}$, in which case the relationship (19.9) simplifies as

$$p \simeq \sqrt{v^2 - 2v\kappa_z} \tag{19.10}$$

where κ_z is the projection of momentum $\boldsymbol{\kappa}$ to velocity vector $\boldsymbol{v}$, which is itself along the $Z-$axis. In this case, (19.8) becomes

$$Q_{\rm RTEX} \simeq \int_{-\infty}^{+\infty} {\rm d}\kappa_z P_{i_2}(\kappa_z) Q_{\rm DR}(\boldsymbol{p}) \tag{19.11}$$

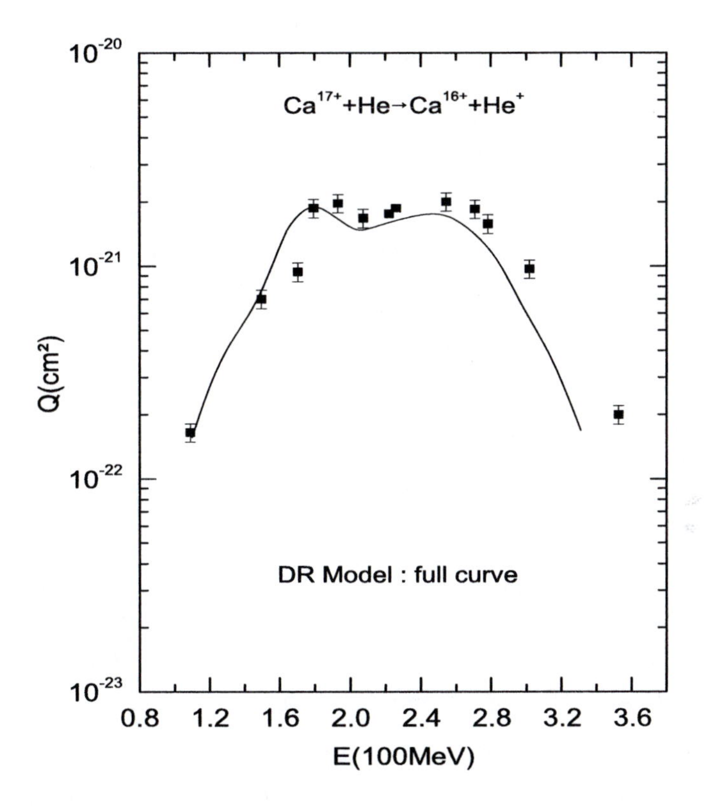

FIGURE 19.1
Total cross sections Q as a function of the incident energy E for process $\text{Ca}^{17+} + \text{He} \longrightarrow (\text{Ca}^{16+})^{**} + \text{He}^{+}$. Theory: full curve (DR method [476]). Experiment: ■ [469].

where $P_{i_2}(\kappa_z)$ is the Compton profile of the target

$$P_{i_2}(\kappa_z) = \int_{-\infty}^{+\infty} \mathrm{d}\kappa_x \int_{-\infty}^{+\infty} \mathrm{d}\kappa_y \, |\tilde{\varphi}_{i_2}(\boldsymbol{\kappa})|^2 . \tag{19.12}$$

As stated, the function $Q_{\text{DR}}(\boldsymbol{p})$ has a sharp and narrow peak of width Γ around p_r, whereas $P_{i_2}(\kappa_z)$ is a much more slowly varying function. Such a circumstance justifies the usage of the standard mean-value theorem [477] (also termed the impulse approximation by Brandt [473]), in which case the cross section Q_{RTEX} reads as

$$Q_{\text{RTEX}} \simeq \frac{1}{v} P_{i_2}(\kappa_{z_r}) \int_{E_r - \Delta E_r/2}^{E_r + \Delta E_r/2} \mathrm{d}\epsilon \, Q_{\text{DR}}(\epsilon) \tag{19.13}$$

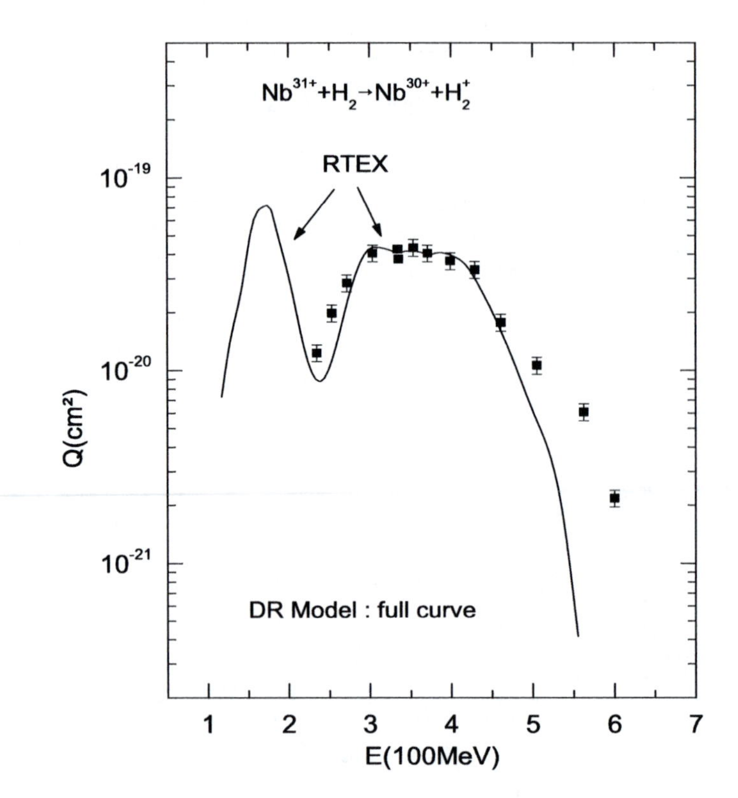

FIGURE 19.2
Total cross sections Q as a function of the incident energy E for process $Nb^{31+} + H_2 \longrightarrow (Nb^{30+})^{**} + H_2^+$. Theory: full curve (DR method [476]). Experiment: ■ [469].

where κ_{z_r} is the value of κ_z at the resonance, as per (19.6) and ΔE_r is an interval around E_r. Here, we used the relationship $\kappa_z = (p^2 - v^2)/(2v)$ which gives $\mathrm{d}\kappa_z = (1/v)p\mathrm{d}p \equiv (1/v)\mathrm{d}\epsilon$ with $\epsilon = p^2/2$, where p is given by (19.10). Thus, the cross section Q_{DR} depends upon ϵ via $\epsilon = p^2/2$. In (19.13) the integration over the whole resonance width Γ should not depend upon the interval ΔE_r. This will be the case if the parameter ΔE_r is taken to be sufficiently large relative to the resonance width ($\Delta E_r \gg \Gamma$). The model of Brandt [473] is expected to be successful in predicting the magnitude and shape of the RTEX cross section for collisions between multiply charged ions and light atoms or molecules. This is indeed the case, as illustrated in Figs. 19.1 and 19.2. The limitation of this approximation is the relationship (19.10), which entails the condition $v \gg \kappa_{\max}$. Overall, the model of Brandt [473] is restricted to asymmetric collisions ($Z_{\rm P} \gg Z_{\rm T}$) and, as such, cannot treat the non-resonant TEX mode (which is abbreviated by NTEX).

19.2.2 A model for the NTEX modes

In the NTEX mode, excitation is produced by the interaction between the projectile electron e_1 and the target nuclear charge $Z_{\rm T}$. Simultaneously, capture is due to interactions between the target electron e_2 and the projectile nuclear charge $Z_{\rm P}$. These two processes can be considered as being independent of each other only if the interaction between the two electrons e_1 and e_2 does not play a significant role in the collisional dynamics. In such a case the IPM applies, so that the total cross section for the NTEX mode has been defined by Brandt [473] via the integral

$$Q_{\rm NTEX}^{\rm IPM} \equiv Q_{\rm exc-cap} = 2\pi \int_0^\infty {\rm d}b\, b P_{\rm exc}(b) P_{\rm cap}(b). \tag{19.14}$$

Here, b is the impact parameter, whereas $P_{\rm exc}(b)$ and $P_{\rm cap}(b)$ are the excitation and capture probabilities, respectively. A refinement of this model, still within the IPM for the NTEX modes, has been made in Refs. [478]–[480]. This was achieved by considering that the formation of a doubly excited state (say, d) is followed by a stabilizing radiative decay of d with the so-called fluorescence yield, which is denoted by $\omega(d)$. Then the total cross section for the NTEX modes is given by the expression

$$Q_{\rm NTEX}^{\rm IPM} = \sum_d \omega(d) Q_{\rm NTE}^{\rm IPM}(d) \tag{19.15}$$

with

$$Q_{\rm NTE}^{\rm IPM}(d) = \frac{1}{2\pi v^2} \int_{q_{\rm min}}^{q_{\rm max}} q{\rm d}q \int {\rm d}\boldsymbol{\kappa} P_{i_2}(\kappa_z) \left|C(\boldsymbol{\kappa})\right|^2 \left|F(\boldsymbol{q} - \boldsymbol{\kappa})\right|^2 \tag{19.16}$$

where $\boldsymbol{q}$ is the transfer momentum $\boldsymbol{q} = \boldsymbol{k}_i - \boldsymbol{k}_f$ with $q_{\rm min} = \Delta E/v = (E_i - E_f)/v$ and $q_{\rm max} = k_i + k_f \simeq \infty$. Here, $|C(\boldsymbol{\kappa})|^2$ and $|F(\boldsymbol{q}-\boldsymbol{\kappa})|^2$ are capture and excitation probabilities, respectively, whereas $P_{i_2}(\kappa_z)$ is the target Compton profile from (19.12). Since $|C(\boldsymbol{\kappa})|^2$ decreases rapidly, and $|F(\boldsymbol{q}-\boldsymbol{\kappa})|^2$ increases with augmented energy, it follows that the product of these two probabilities should give a peak. Such a peak would occur at an energy which is lower than the one from the RTEX mode. Moreover, if the Compton profile is not too large, the NTEX and RTEX peaks are expected to be very well separated. Indeed, this is the case as illustrated in Fig. 19.3 for the S^{13+} – He system, for which two peaks are seen as being distinctly separated.

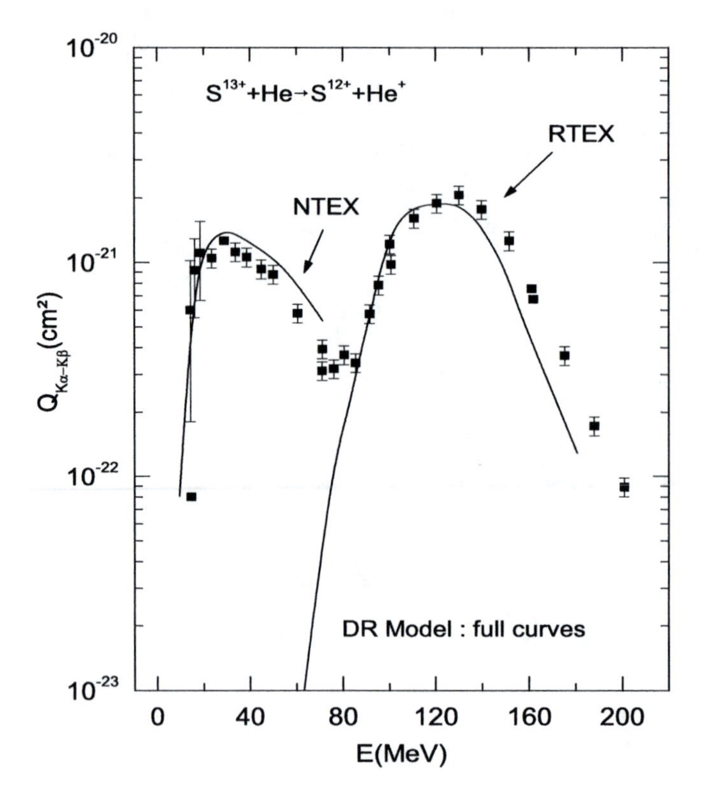

FIGURE 19.3
Total cross sections Q as a function of the incident energy E for process $S^{13+} + He \longrightarrow (S^{12+})^{**} + He^{+}$. Theory: full curve (DR method [478] with Q_{DR}^{NTEX} multiplied by 3). Experiment: ■ [479].

19.3 The CDW-4B method for the TEX modes

We consider the prototype reaction (19.1) by assuming that both hydrogen-like atomic systems in the entrance channel are in their ground states ($i_1 = i_2 = 1s$). For such a process, the prior transition amplitude in the CDW-4B method is given by (19.2) for arbitrary charges Z_P and Z_T. In (19.2), the auto-ionizing state $\varphi_f(\boldsymbol{s}_1, \boldsymbol{s}_2)$ can be defined within the well-known Feshbach formalism [481]. In this formalism of resonant scattering, a doubly excited state can be described using a set of hydrogen-like basis functions centered on the projectile nuclear charge Z_P [81]. Such a basis set should be adequate for doubly excited auto-ionizing states of helium-like atoms or ions. The formalism of Feshbach consists of discarding the lowest orbitals that could lead to an eigen-state with energy below that of the investigated doubly excited state. Surely, one such discarded basis function could be the ground state of the projectile $(Z_{\rm P}, e_1)_{i_1}$ if i_1 is chosen to be the lowest state of this hydrogen-like system in the TE process (19.1). Indeed, the illustrations for this latter process will be given only for the ground states ($i_1 = 1s, i_2 = 1s$) of the two colliding hydrogen-like atoms. Thus, whenever the wave function $\varphi_{i_1}(\boldsymbol{s}_1)$ describes the ground state of the $(Z_{\rm P}, e_1)_{i_1}$ system, this function will be absent from the CI basis set comprized of the purely hydrogen-like orbitals with the nuclear charge $Z_{\rm P}$ employed to construct $\varphi_f(\boldsymbol{s}_1, \boldsymbol{s}_2)$. As such, the overlap between the ground state $\varphi_{i_1}(\boldsymbol{s}_1)$ and $\varphi_f(\boldsymbol{s}_1, \boldsymbol{s}_2)$, i.e. the integral over $\boldsymbol{s}_1$, will be equal to zero. This implies that the interaction $Z_{\rm T}(1/R - 1/s_2)$ and the $\boldsymbol{\nabla} \cdot \boldsymbol{\nabla}$ potential operator will not contribute to the matrix element T_{if}^- in (19.2). Under these simplifying conditions [184], (19.2) is reduced to

$$\begin{aligned} T_{if}^- &= N^+(\nu_{\rm P})N^{-*}(\nu_{\rm T}) \iiint \mathrm{d}\boldsymbol{s}_1 \mathrm{d}\boldsymbol{x}_2 \mathrm{d}\boldsymbol{R}\, \mathrm{e}^{i\boldsymbol{k}_i\cdot\boldsymbol{r}_i + i\boldsymbol{k}_f\cdot\boldsymbol{r}_f}\, \varphi_f^*(\boldsymbol{s}_1, \boldsymbol{s}_2) \\ &\quad \times {}_1F_1(i\nu_{\rm T}, 1, ivx_2 + i\boldsymbol{v}\cdot\boldsymbol{x}_2) \left(\frac{1}{r_{12}} - \frac{Z_{\rm T}}{x_1}\right) \\ &\quad \times \varphi_{i_1}(\boldsymbol{s}_1)\varphi_{i_2}(\boldsymbol{x}_2)\, {}_1F_1(i\nu_{\rm P}, 1, ivs_2 + i\boldsymbol{v}\cdot\boldsymbol{s}_2) \equiv T_{if,12}^- + T_{if,1}^-. \qquad (19.17) \end{aligned}$$

As discussed, the phase $(\mu v\rho)^{2i\nu}$ is ignored whenever T_{if}^- is not used for cross sections that are differential in the angles of the scattered projectile. The matrix elements $T_{if,12}^-$ and $T_{if,1}^-$ denote the part of T_{if}^- associated with perturbations $1/r_{12}$ and $-Z_{\rm T}/x_1$, respectively. The remaining integrals in (19.17) are of the same type as those encountered previously by Belkić and Mančev [84] for double charge exchange in the CDW-4B method. As mentioned, for calculation of such integrals, the Nordsieck technique [311] can be used to reduce the transition amplitude to a triple quadrature which can be evaluated numerically by the procedure of Belkić and Mančev [84]. The same technique was subsequently used in Ref. [184] for computation of the transition amplitude T_{if}^- from (19.17).

19.4 The CDW-4B method for the TE process in asymmetric collisions

Here, the results obtained by means of the CDW-4B method are compared with the corresponding data from collisional experiments on the TE process with a molecular hydrogen target H_2 [471, 472]. As usual, we assumed that the molecule H_2 can be considered as two independent H atoms. Thus, an example of the equivalent four-body collisional system is

$$\mathrm{S}^{15+}(1s) + \mathrm{H}(1s) \rightarrow \mathrm{S}^{14+}(nl, n'l') + \mathrm{H}^{+}. \qquad (19.18)$$

The experimental data from Refs. [471, 472] show that the deexcitation of the auto-ionizing state $(nl, n'l')$ gives spectra with the radiative decay lines $K\alpha - K\alpha$, $K\alpha - K\beta$ and $K\beta - K\beta$. The computations from Ref. [184] within the CDW-4B method include the $K\alpha - K\alpha$ and $K\alpha - K\beta$ lines at impact energies ranging between 80 and 160 MeV, by covering most of the lowest resonances (LL, LM, LN, ...). This energy range corresponds to a peak in the RTEX mode as expected from (19.6).

19.4.1 The $K\alpha - K\alpha$ emission line from S^{14+}

The $K\alpha - K\alpha$ emission line is produced by the following three doubly excited states in a helium-like atomic system

$$(2s^2)\,^1S \qquad\qquad (2p^2)\,^1D \qquad\qquad (2p^2)\,^1S.$$

The wave functions of these latter states were obtained using the code from Ref. [81]. The configuration $(2p^2)$ appears to be the major component among these three states. The $(2p^2)$ state can decay radiatively through the following two electric dipoles transitions

$$(2p^2) \overset{\mathrm{Ly}\alpha}{\longrightarrow} (1s, 2p) \overset{\mathrm{Ly}\alpha}{\longrightarrow} (1s^2) \qquad (19.19)$$

where Lyα denotes the Lyman$-\alpha$ spectral line. Since the experiments of Justiniano *et al.* [471] and Schulz *et al.* [472] were performed with unpolarized colliding aggregates, there is a probability of 1/4 to find the system $\mathrm{S}^{15+} - \mathrm{H}$ in a definite singlet state. Therefore, a factor of 1/2 $(2 \times 1/4)$ must be introduced in the calculations of the total cross sections for the double $K\alpha - K\alpha$ line emission in addition to the fluorescence yield $\omega(d)$ which includes some cascade contributions. Here, the extra multiplying factor of 2 written inside the small parentheses stems from using the Bragg sum rule [430]–[432] to convert the theoretical results on the $\mathrm{S}^{15+} - \mathrm{H}$ system to the corresponding experimental data on the $\mathrm{S}^{15+} - \mathrm{H}_2$ system for the purpose of comparison. The contribution of cascades from higher doubly excited states (KLM, KLN, ...) has been ignored in these computations [184]. Both contributions from the RTEX and NTEX modes were coherently included in the CDW-4B method through the first and second terms $T^{-}_{if,12}$ and $T^{-}_{if,1}$ in the transition amplitude T^{-}_{if} from (19.17). However, it is expected that for a highly asymmetric collision with $Z_{\mathrm{P}} \gg Z_{\mathrm{T}}$ such as process (19.18), the contribution from the NTEX mode should always be negligible relative to the RTEX mode.

The total cross sections for reaction (19.18) are illustrated in Fig. 19.4 by comparing theory and experiment. The results for the RTEX mode treated by Brandt [473] within the IPM are shown by the dashed curve. The cross sections of a more complete description by the CDW-4B method used by Bachau *et al.* [184] are shown by the full curve. These two methods compare favorably to each other, as well as to the experimental data from Ref. [471]. The agreement between the cross sections $Q^{\rm IPM}_{\rm RTEX}$ [473] and $Q^{\rm CDW-4B}_{\rm RTEX+NTEX}$ [184] is good, as anticipated, due to a very small influence of the NTEX mode on process (19.18). In particular, the two theories are seen in Fig. 19.4 to be

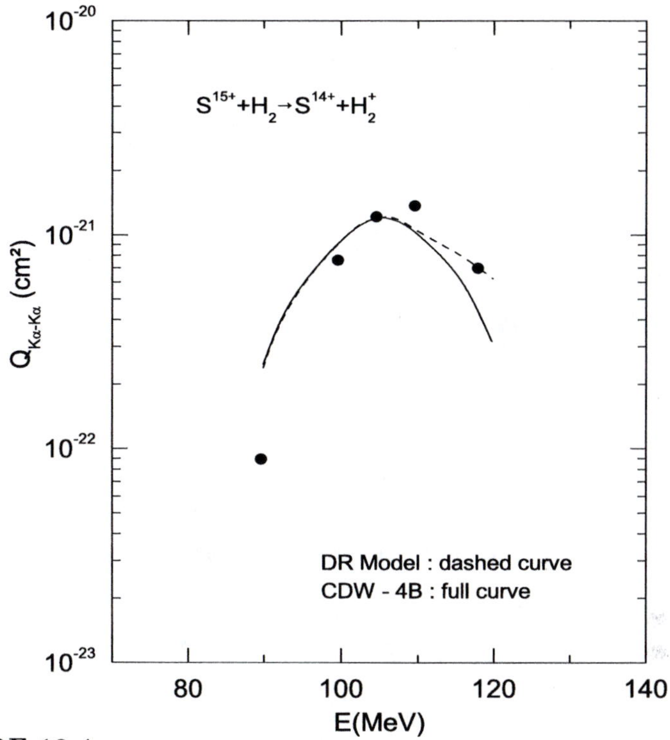

FIGURE 19.4
Total cross sections Q as a function of the incident energy E for process $S^{15+} + H_2 \longrightarrow (S^{14+})^{**} + H_2^+$ with the production of the $K\alpha - K\alpha$ lines from $(S^{14+})^{**}$. Theory: dashed curve (DR method [471]) and full curve (CDW-4B method [184]. Experiment: • [471].

in perfect accordance on the left wing of the resonance peak. On the right wing of the same peak, towards its high-energy tail, there is a discrepancy between the IPM and CDW-4B method. At these higher energies, the IPM is in better agreement with the experimental result than the CDW-4B method. This may be because the IPM includes some cascade effects, whereas the CDW-4B method does not. Such a conjecture is plausible, since Justiniano *et al.* [471] have shown that cascade-based contributions to the TE process are important at high impact energies like those considered in Fig. 19.4. A future application of the CDW-4B method should include cascades from higher excited states ($KLM, KLN, ...$) to improve the agreement with the experimental data [471] for the double Lyman α line above 150 MeV.

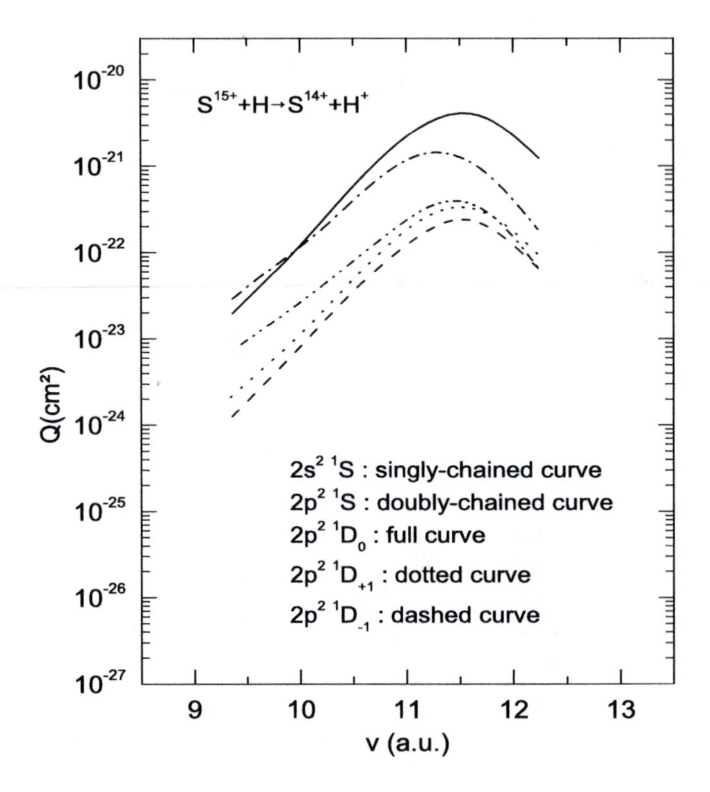

FIGURE 19.5
State-selective total cross sections Q as a function of the incident energy E for process $S^{15+} + H \longrightarrow (S^{14+})^{**} + H^+$ with the production of the $K\alpha - K\alpha$ lines from $(S^{14+})^{**}$. Theory (CDW-4B method [184]): $(2s^2)\,^1S$ (singly-chained curve), $(2p^2)\,^1S$ (doubly-chained curve), $(2p^2)\,^1D_0$ (full curve), $(2p^2)\,^1D_{+1}$ (dotted curve) and $(2p^2)\,^1D_{-1}$ (dashed curve).

In Fig. 19.5, we present the state-selective total cross sections for the TE process studied by means of the CDW-4B method for the sub-states $(\lambda_0\lambda_0')^{2S+1}L_M$, where M is the projection of the total momentum L. It is seen that the sub-state $(2p^2)\,^1D_0$ dominates at $v > 10$ a.u. ($E > 80$ MeV).

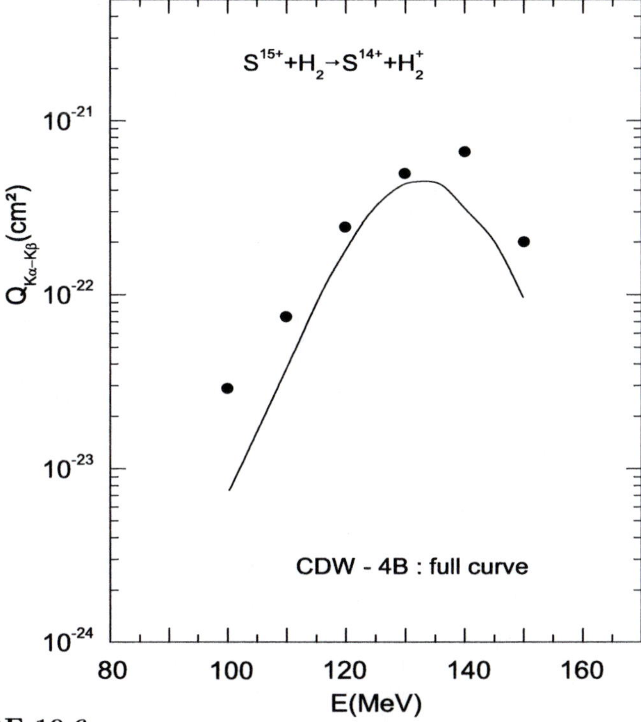

FIGURE 19.6
Total cross sections Q as a function of the incident energy E for process $S^{15+} + H_2 \longrightarrow (S^{14+})^{**} + H_2^+$ with the production of the $K\alpha - K\beta$ lines from $(S^{14+})^{**}$. Theory: full curve (CDW-4B method [184]). Experiment: • [471].

19.4.2 The $K\alpha - K\beta$ emission lines from S^{14+}

The $K\alpha - K\beta$ emission in process (19.18) is produced by five important configurations of the auto-ionizing state, and these are

$$(2s3s)\,^1S \qquad (2p3p)\,^1P \qquad (2p3p)\,^1D \qquad (2s3d)\,^1D \qquad (2p3p)\,^1S.$$

Following the preceeding case, the cascades (KLN, KLO, ...) have been ignored in Ref. [184]. In Fig. 19.6, the results of the CDW-4B method are displayed together with the experimental data from Ref. [471]. Here, we observe satisfactory agreement between theory and measurement up to about 130 MeV. The maximum of the experimentally detected $K\alpha - K\beta$ peak is located at an impact energy around 140 MeV. This peak corresponds to the RTEX mode induced by the KLM cascade transition. Both this peak and its high-energy slope are underestimated by the CDW-4B method due to neglect of all the cascade contributions.

Regarding the $K\alpha - K\alpha$ and $K\alpha - K\beta$ emission lines that are due primarily to the RTEX mode with the NTEX mode being negligible, it can generally be concluded that there is good agreement between the results from CDW-4B method and the available experimental data. This provides a strong motivation for extending the analysis to encompass the interference effects between the RTEX and NTEX modes.

19.5 Target charge $Z_{\rm T}$ and the interference between the RTEX and NTEX modes

In this section, we analyze the dependence of the cross sections upon interference effects between the RTEX and NTEX modes. This will be discussed for different charges $Z_{\rm T}$ of the target nucleus in the following process

$$\mathrm{S}^{15+}(1s) + (Z_{\rm T}, e_2)_i \to \mathrm{S}^{14+}[(2p^2)\,^1D_0] + Z_{\rm T} \tag{19.20}$$

where $i = 1s$. Here, the state $(2p^2)\,^1D_0$ is chosen for the analysis because this transition gives the largest cross section, as is clear from Fig. 19.5 (full curve). Furthermore, for each value of the selected charge $Z_{\rm T}$, the computations from Ref. [184] have been made in the energy range corresponding to the peak in the RTEX mode. Another restriction in these computations [184] is the standard condition for the validity of the general CDW method at impact energies that satisfy the inequality (14.69). As such, the lower limit of the validity of the CDW method is fulfilled at impact energies higher than (i) 78 MeV ($v = 9.9$ a.u.) for $Z_{\rm T} \leq 7$, (ii) 128 MeV ($v = 12.7$ a.u.) for $Z_{\rm T} = 10$ and (iii) 328 MeV ($v = 20.2$ a.u.) for $Z_{\rm T} = 16$.

The cross sections for process (19.20) treated by the CDW-4B method are presented in Figs. 19.7 and 19.8 for $Z_{\rm T} = 1$ and 10, respectively [184]. The results for the RTEX, NTEX and TEX modes are shown by the dashed, dotted and full curves, respectively. The complete description of the whole process includes the RTEX and NTEX nodes. Recall that, according to (19.17), the RTEX and NTEX modes are described by the terms $T^-_{if,12}$ and $T^-_{if,1}$, respectively. The computations have been carried out for $Z_{\rm T} = 1$, 4, 10 and 16 [184]. The obtained results show that the location of the maximum of the RTEX peak shifts to larger v with $Z_{\rm T}$ increased, as also implied by (19.6). For example, this latter peak moves from $v = 11.5$ a.u. for $Z_{\rm T} = 1$ (see Fig. 19.7) to $v = 12.6$ a.u. for $Z_{\rm T} = 16$ (not shown here; see Ref. [184]). It is seen in Fig. 19.7 for $Z_{\rm T} = 1$, that at the energies within the RTEX peak ($8 \leq v \leq 20$ a.u.), we have $Q_{\rm NTEX} \ll Q_{\rm RTEX}$. As a consequence, the interference between these two modes is totally negligible. Hence, there is no difference between the RTEX and TEX cross section, so that the corresponding dashed and full curves in Fig. 19.7 coincide with each other and only the full curve remains visible on the scale on this figure. Further, Fig. 19.7 confirms that the good agreement between the DR model (without NTEX) and CDW-4B methods (with NTEX), as seen earlier in Fig. 19.4, is not coincidental. In the case for $Z_{\rm T} = 4$ (not shown), although $Q_{\rm NTEX}$ is much smaller (by some 2 orders of magnitude) than $Q_{\rm RTEX}$, a destructive interference effect (to within $\simeq 15\%$) on the cross sections has been reported [184].

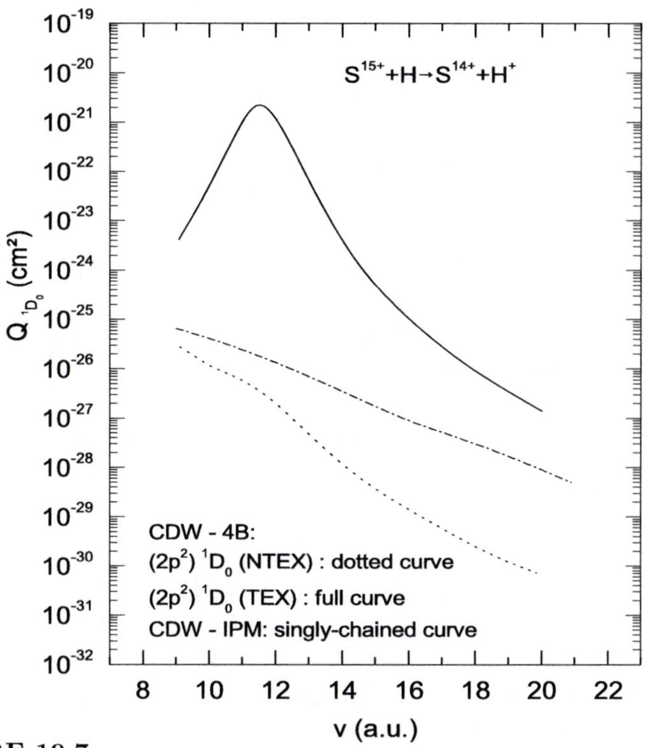

FIGURE 19.7
Total cross sections Q as a function of the incident velocity v for process $\mathrm{S}^{15+} + \mathrm{H} \longrightarrow (\mathrm{S}^{14+})^{**} + \mathrm{H}^+$ with the production of the $(2p^2)\,^1D_0$ state of $(\mathrm{S}^{14+})^{**}$. Here $Z_{\rm T} = 1$. Theory (CDW-4B method [184]): $Q_{1D_0}^{\rm NTEX}$ (dotted curve), $Q_{1D_0}^{\rm RTEX}$ (dashed curve) and $Q_{1D_0}^{\rm TEX}$) (full curve). Theory (CDW-IPM [482]): $Q_{1D_0}^{\rm TEX}$ (singly-chained curve). The results in the CDW-4B method for the RTEX and TEX modes coincide with each other due to negligible influence of the NTEX mode, yielding $Q_{1D_0}^{\rm TEX} \simeq Q_{1D_0}^{\rm RTEX}$. Thus, the full and dashed curves are indistinguishable on the scale of this figure. To alleviate overlapping of superscripts and subscripts, we write $1D_0 \equiv {}^1D_0$.

When the target nuclear charge $Z_{\rm T}$ is augmented, the contribution of the NTEX mode to the TE total cross section increases because its amplitude

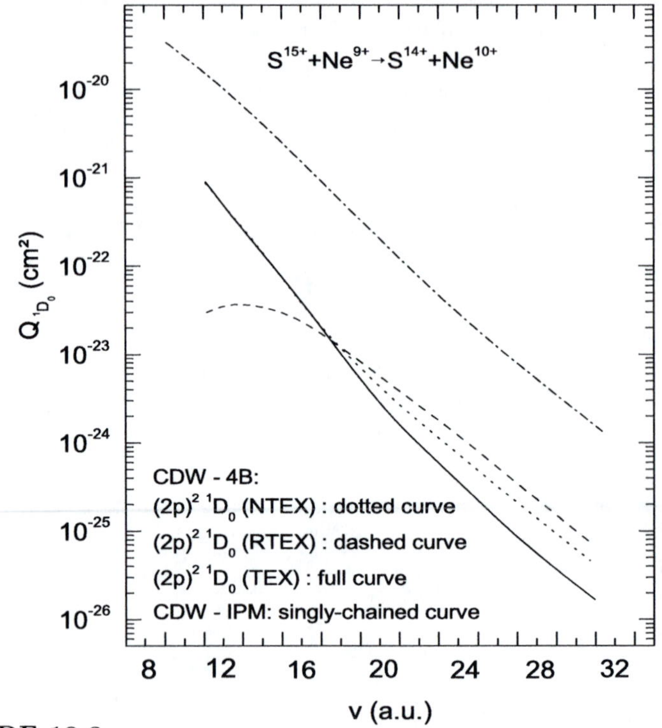

FIGURE 19.8
Total cross sections Q as a function of the incident velocity v for process $S^{15+} + Ne^{9+} \longrightarrow (S^{14+})^{**} + Ne^{10+}$ with the production of the $(2p^2)\,^1D_0$ state of $(S^{14+})^{**}$. Here $Z_T = 10$. Theory (CDW-4B method [184]): $Q_{1D_0}^{\rm NTEX}$ (dotted curve), $Q_{1D_0}^{\rm RTEX}$ (dashed curve) and $Q_{1D_0}^{\rm TEX}$) (full curve). Theory (CDW-IPM [482]): $Q_{1D_0}^{\rm TEX}$ (singly-chained curve). To alleviate overlapping of superscripts and subscripts, we write $1D_0 \equiv {}^1D_0$.

is nearly proportional to Z_T, as per $T_{if,1}^-$ from (19.17). As such, the cross section for the NTEX mode becomes dominant for $Z_T = 10$ (see Fig. 19.8) and $Z_T = 16$ [184]. In these cases, the shape of the RTEX peak disappears altogether. Thus, for production of the $(2p^2)\,^1D_0$ state, the RTEX and NTEX modes may lead to a destructive interference.

Overall, for $Z_P \gg Z_T$, the TEX process is dominated by the RTEX mode. For $Z_P \simeq Z_T$ interference effects can become important. Large discrepancies exist between the CDW-4B method and CDW-IPM. Thus, for the asymmetric $S^{15+} - H$ collisions, Fig. 19.7 shows that a resonance due to the RTEX mode (via perturbation $1/r_{12}$) is predicted by the CDW-4B method, but missed by the CDW-IPM. This reaffirms the importance of the dynamic correlations that are present in the CDW-4B method and absent from the CDW-IPM.

19.6 The TEA mode for nearly symmetrical systems: the Auger decay

In the preceeding section, the doubly excited state $(S^{14+})^{**}$ produced by the TE process in the S^{15+} – H collisions led to radiative decay. Alternatively, the same TE process could also be completed when this doubly excited state decays through the Auger mechanism i.e. via the TEA mode. This possibility within the TE processes has been investigated by Gayet *et al.* [186, 187] by applying the CDW-4B method to the He^+ – He and He^+ – H collisions. The obtained total cross sections from the CDW-4B method were not in quantitative agreement with the experimental data of Itoh *et al.* [470] and Zouros *et al.* [483], as can be seen in Figs. 19.9 and 19.10 for He^+ – H collisions. Figures 19.9 and 19.10 show that the interference between the RTEA and NTEA modes within the CDW-4B method can be important [186, 187]. However, the experimental data for the production of the $(2s2p)^1P_0$ state are seen in Fig. 19.9 as being markedly underestimated by the cross sections from the CDW-4B method. This is not understood at present[1]. By contrast, a reasonable agreement between the CDW-4B method and the experimental data is obtained in the case of the production of the $(2p^2)^1D_0$ state (Fig. 19.10).

It should be noted that the experimental total cross sections correspond to the integrated contribution of the two constituent overlapping peaks in the energy spectrum of the emitted electron [483]. The integrated sum of such two measured peaks yields good agreement between theory and experiment (Fig. 19.11). Nevertheless, such a comparison is questionable, because in the measurement, the $(2p^2)\,^1D_0$ cross section is smaller than the $(2s2p)\,^1P_0$ cross section, whereas precisely the reverse is true for the theory (full curve in Figs. 19.9 and 19.10). A similar remark is also pertinent to the case with the He target for the $(2s^2)\,^1S$ state, as well as for the $(2s2p)^3P_0$ state and their sum [186, 187]. The experimental data from Refs. [470] and [483] were measured using the technique of zero-degree electron spectroscopy by which the TEA and ITE mode were not separated. In other words, these experimentally determined cross sections do not correspond to the TEA alone, since they always contain an admixture from the ITE channel. This situation does not match the computations from Refs. [186, 187], where the CDW-4B method was employed with doubly excited states that include only the discrete orbitals. This assumes that all the ejected electrons are produced

[1] In general i.e. not only for formation of the $(2s2p)^1P_0$ state, it would be advisable to use both versions of the prior transition amplitude T_{if}^- reported in Ref. [185] for the same process (19.1) to assess the dependence of the CDW-4B method upon two different choices of the distorting potential. The computationally easier version for T_{if}^- has been chosen in Ref. [184] and this is the expression given by (19.2). Moreover, it would also be important to study the post-prior discrepancy for transfer-excitation.

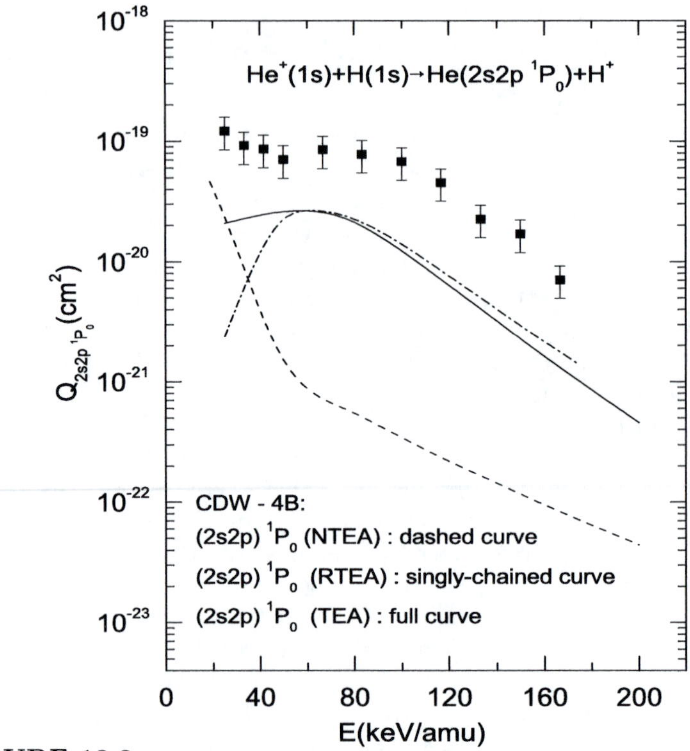

FIGURE 19.9
Total cross sections Q as a function of the incident energy E for process $\mathrm{He}^+(1s) + \mathrm{H}(1s) \longrightarrow \mathrm{He}^{**}(2s2p\,^1P_0) + \mathrm{H}^+$. Theory (CDW-4B method [186, 187]): Q_{NTEA} (dashed curve), Q_{RTEA} (singly-chained curve) and Q_{TEA} (full curve). Experiment: ■ [483].

exclusively by auto-ionizing decays i.e. via the Auger effect. However, the emitted electrons that are measured can be either the Auger electrons or the electrons from the ITE channel. Therefore, in order to properly interpret the experiments by Itoh *et al.* [470] and Zouros *et al.* [483], an improved model is needed using a more adequate description of doubly excited state, by taking into account the adjacent continuum orbitals in addition to the discrete ones.

This latter task has been undertaken by Ourdane *et al.* [188] who treated the TE process by means of the CDW-4B method improved by inclusion of the ITE channel within the theory of Fano [475] for a description of atomic resonant structures. Ourdane *et al.* [188] have investigated a nearly symmetrical four-body system by supposing that each line of the electron spectrum could be associated with a single isolated resonance. The additional continuum orbitals chosen in Ref. [188] were discretized via a basis set of the STOs,

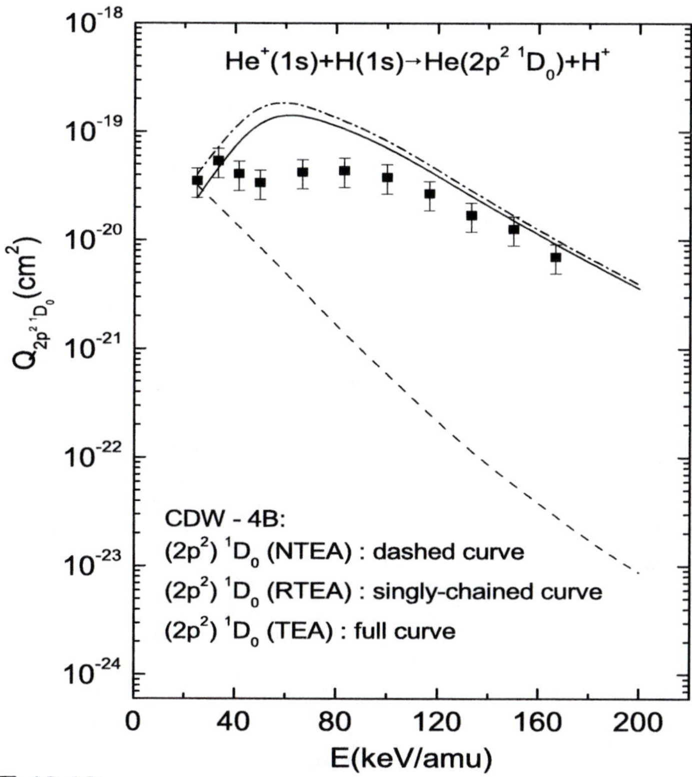

FIGURE 19.10
Total cross sections Q as a function of the incident energy E for process $\mathrm{He}^+(1s) + \mathrm{H}(1s) \longrightarrow \mathrm{He}^{**}(2p^2\,{}^1D_0) + \mathrm{H}^+$. Theory (CDW-4B method [186, 187]): Q_{NTEA} (dashed curve), Q_{RTEA} (singly-chained curve) and Q_{TEA} (full curve). Experiment: ■ [483].

following the procedure of Macías *et al.* [484], whereas pure discrete orbitals were the same as those from Gayet *et al.* [186, 187]. In this way, coherence effects between bound and continuum orbitals were included to a presumably sufficient extent. With this amelioration, it is reasonable to expect that the improved computation of Ourdane *et al.* [188] should give more adequate cross sections than those of Gayet *et al.* [186, 187]. We shall return to this point later on. In particular, it would be important to see whether the CDW-4B method with its improvement for the ITE effect could reproduce the asymmetric shape profiles in the spectra due to the $(2s2p)^3P$ doubly excited states. Such asymmetries have been observed experimentally in the measurements of Itoh *et al.* [470].

Specifically, it was suggested by Itoh *et al.* [470] that these asymmetric line shapes stem from the interference between the TE and ECC mechanisms. It

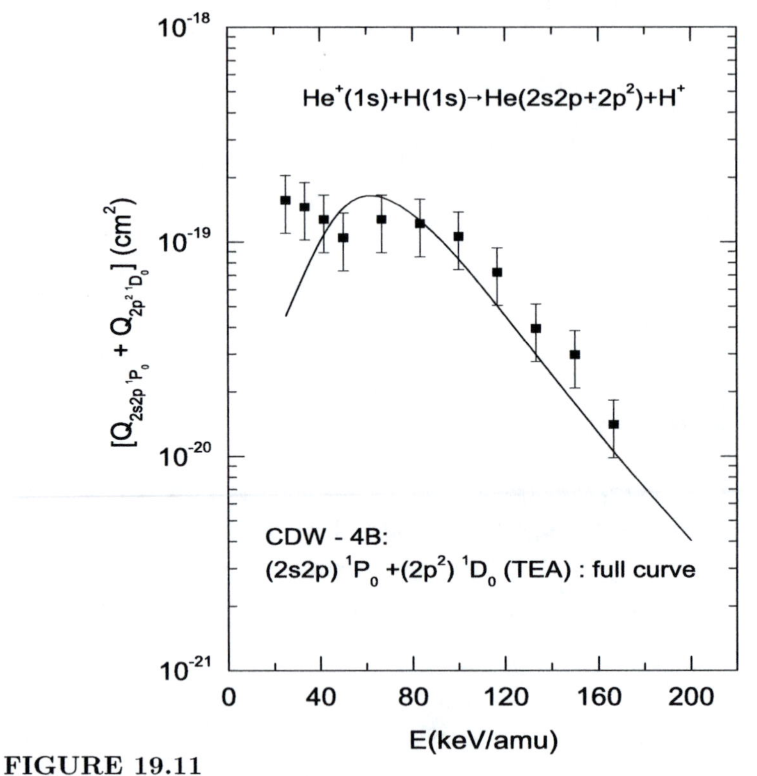

FIGURE 19.11
Total cross sections Q as a function of the incident energy E for process $\mathrm{He}^+(1s) + \mathrm{H}(1s) \longrightarrow \mathrm{He}^{**}(2s2p\,^1P_0 + 2p^2\,^1D_0) + \mathrm{H}^+$. Theory (CDW-4B method [186, 187]): Q_{TEA} (full curve). Experiment: ■ [483].

should also be mentioned that the role of the ITE channel has been assessed in the case of double excitation processes studied experimentally in Refs. [485] and [486]. An important conclusion from these experiments is that the role of interference among resonances with different origins can significantly alter the overall shape of the recorded spectral profiles.

19.7 The CDW-4B method for the TEA modes

We now return to the transition amplitude defined by (19.2), but without the term $(\mu v \rho)^{2i\nu}$. As mentioned, this is justified whenever the integration over the transverse momentum transfer η is performed, as in the total cross sections and the angular or energy distributions of ejected electrons [134]. Here, the main focus is on the establishment of the wave function $\varphi_f(\boldsymbol{s}_1, \boldsymbol{s}_2)$ for the doubly excited states.

19.8 Description of the final state

An isolated auto-ionizing state can be conceived as a superposition of certain bound and continuum states of an electron [475]. In this case, the final state wave function $\varphi_f(\boldsymbol{s}_1, \boldsymbol{s}_2) \equiv \varphi_f(E_e; \boldsymbol{s}_1, \boldsymbol{s}_2)$ becomes dependent on the electron energy E_e near the auto-ionizing threshold energy $E_f \equiv E_{\rm S}$ and this is needed for an adequate description of the complete TE process. Then, the final state can be written as

$$\varphi_f(E_e; \boldsymbol{s}_1, \boldsymbol{s}_2) = \alpha(E_e)[\Psi_{\rm S}(E_{\rm S}; \boldsymbol{s}_1, \boldsymbol{s}_2) + \mathcal{P}\int {\rm d}E'_e \Psi_{\rm C}(E'_e; \boldsymbol{s}_1, \boldsymbol{s}_2)\frac{1}{E_{\rm S} - E'_e}V(E'_e)] + \beta(E_e)\Psi_{\rm C}(E_{\rm S}; \boldsymbol{s}_1, \boldsymbol{s}_2) \tag{19.21}$$

where $\mathcal{P}$ is the standard symbol for the Cauchy principal value of the integral. Here, $\Psi_{\rm S}(E_{\rm S}; \boldsymbol{s}_1, \boldsymbol{s}_2)$ is the bound component of the resonant part of auto-ionizing state with the corresponding energy $E_{\rm S}$, which can be computed as in Ref. [81]. Likewise $\Psi_{\rm C}(E_e; \boldsymbol{s}_1, \boldsymbol{s}_2)$ is the adjacent continuum component. In (19.21), the function $V(E_e)$ represents the coupling between the bound and continuum components

$$V(E_e) = \left\langle \Psi_{\rm S}(E_{\rm S}; \boldsymbol{s}_1, \boldsymbol{s}_2) \left| \frac{1}{r_{12}} \right| \Psi_{\rm C}(E_e; \boldsymbol{s}_1, \boldsymbol{s}_2) \right\rangle \tag{19.22}$$

$$\alpha(E_e) = \frac{1}{\pi V(E_e)\sqrt{\epsilon_{\rm S}^2 + 1}} \qquad \beta(E_e) = \frac{\epsilon_{\rm S}}{\sqrt{\epsilon_{\rm S}^2 + 1}} \tag{19.23}$$

$$\epsilon_{\rm S} = 2\frac{E_e - E_{\rm S}}{\Gamma_{\rm S}} \qquad \Gamma_{\rm S} = 2\pi\left|V(E_{\rm S})\right|^2 \tag{19.24}$$

where $\Gamma_{\rm S}$ is the width of the auto-ionizing state of energy $E_{\rm S}$. The normalization of the continuum component $\Psi_{\rm C}(E_e; \boldsymbol{s}_1, \boldsymbol{s}_2)$ is given on the energy scale, with the accompanying static exchange approximation. Let the function $\Psi_{\rm C}(E_e; \boldsymbol{s}_1, \boldsymbol{s}_2)$ be represented via

$$\Psi_{\rm C}(E_e; \boldsymbol{s}_1, \boldsymbol{s}_2) = \mathcal{A}\left[\varphi_{i_1}(\boldsymbol{s}_1)\varphi^{\rm A}_{LM}(\boldsymbol{s}_2)\right] \tag{19.25}$$

where $i_1 = 1s$ and $\mathcal{A}$ is the usual anti-symmetrization operator. Here, $\varphi_{i_1}(\boldsymbol{s}_1)$ is the discrete orbital for the bound electron in the $\mathrm{He}^+(1s)$ ion with the binding energy $E_{i_1} = E_{1s}$. Likewise, $\varphi^{\rm A}_{LM}(\boldsymbol{s}_2)$ is the continuum orbital for the Auger electron with the corresponding energy $E_{\rm S} - E_{i_1}$. For the (L, M) configuration of an auto-ionizing state, the continuum orbital is given by

$$\varphi^{\rm A}_{LM}(\boldsymbol{s}_2) = \chi^{\rm A}_{LM}(s_2) Y_{LM}(\hat{\boldsymbol{s}}_2) \tag{19.26}$$

where $\chi^{\rm A}_{LM}(s_2)$ is the radial continuum wave function and $Y_{LM}(\hat{\boldsymbol{s}}_2)$ is the usual spherical harmonic. The radial function $\chi^{\rm A}_{LM}$ can be described by a linear combination of STOs

$$\chi^{\rm A}_{LM}(s_2) = \chi_j(s_2) = \sum_{\lambda=0}^{N_{\rm C}} b^{(j)}_\lambda S^{(j)}_\lambda(s_2)$$
$$S^{(j)}_\lambda(s_2) = s_2^{n_\lambda - 1} {\rm e}^{-\alpha^{(\gamma_j)}_\lambda s_2} \qquad j = (L, M) \tag{19.27}$$

where $N_{\rm C}$ is the total number of the retained continuum orbitals, and n_λ is the orbital number. Here, the expansion coefficients $b^{(j)}_\lambda$ are determined by a variational procedure, through a single diagonalization, which yields both bound and discretized continuum states. This can be achieved e.g. by the algorithm of Macías *et al.* [484]. Briefly, in this latter code, the damping coefficients $\alpha^{(\gamma_j)}_\lambda$ are chosen as a geometric sequence

$$\alpha^{(\gamma_j)}_\lambda = \alpha_0^{\gamma_j} \beta^{\lambda+\gamma} \qquad \lambda = 0, ..., N_{\rm C} \qquad j = (L, M) \tag{19.28}$$

where k_j, α_0, β and γ are the model parameters that are fixed for each studied system. Within the static exchange approximation, the continuum wave function $\chi_j(s_2)$, constructed with a set of the STOs from (19.27), represents a preparatory step for the subsequent diagonalization which gives the $(N_{\rm C}+1)$ eigen-functions $\Psi_{\rm C}(E_k; \boldsymbol{s}_1, \boldsymbol{s}_2)$ and the corresponding energies E_k. If the condition $E_k = 0$ is imposed for e.g. the double electron ionization threshold of helium, then the states with $E_k > -2$ a.u. would lie in the first continuum range. The non-linear parameter γ allows us to determine a sequence $\{\alpha^{(\gamma_j)}_\lambda\}$ such that one of the eigen-energies (say, E_m) matches the energy $E_{\rm S}$ of the auto-ionizing state. This procedure from Ref. [484] also permits an advantageous and convenient normalization for $\Psi_{\rm C}(E_m; \boldsymbol{s}_1, \boldsymbol{s}_2)$ via

$$N = \sqrt{\frac{2}{|E_{m-1} - E_{m+1}|}}. \tag{19.29}$$

Moreover, at $E_m = E_{\rm S}$, the principal value integral can be evaluated analytically via the spectral representation

$$\mathcal{P}\int {\rm d}E'_e \Psi_{\rm C}(E'_e; \boldsymbol{s}_1, \boldsymbol{s}_2) \frac{1}{E_{\rm S} - E'_e} V(E'_e) = \sum_{k \neq m} \bar{\Psi}_{\rm C}(E_k; \boldsymbol{s}_1, \boldsymbol{s}_2)$$
$$\times \frac{1}{E_m - E_k} \bar{V}(E_k) \tag{19.30}$$

$$\bar{V}(E_e) = \left\langle \Psi_S(E_S; \boldsymbol{s}_1, \boldsymbol{s}_2) \left| \frac{1}{r_{12}} \right| \bar{\Psi}_C(E_e; \boldsymbol{s}_1, \boldsymbol{s}_2) \right\rangle. \tag{19.31}$$

The bar sign over Ψ_C indicates that this continuum state is normalized to unity. Then finally, for E_e close to E_S, an isolated doubly excited state can be written as

$$\varphi_f(E_e; \boldsymbol{s}_1, \boldsymbol{s}_2) = \alpha(E_e)[\Psi_S(E_S; \boldsymbol{s}_1, \boldsymbol{s}_2) + \sum_{k \neq m} \bar{\Psi}_C(E_k; \boldsymbol{s}_1, \boldsymbol{s}_2) \frac{1}{E_m - E_k} \bar{V}(E_k)] + \beta(E_e)\Psi_C(E_S; \boldsymbol{s}_1, \boldsymbol{s}_2) \tag{19.32}$$

where $E_m = E_S$.

19.9 Cross sections for the TEA modes

The ionization transition amplitude $T_{if}^- \equiv T_{if}^-(E_e, \theta_e, \phi_e, \boldsymbol{\eta})$ is given by (19.2) for an electron of energy E_e ejected in the direction (θ_e, ϕ_e) at a fixed incident energy E. Note that when the wave function of the doubly excited state is described by (19.21) and (19.32), with the ansatz (19.25), then the interaction $Z_T/R - 1/s_2$ and the $\boldsymbol{\nabla} \cdot \boldsymbol{\nabla}$ potential operator give non-zero contributions. In such a case, (19.2) must be used because the simplified form (19.17) does not hold any longer. For the present purpose, E_e needs to be close to the resonance energy E_S. Then, the energy distribution of the emitted electron $Q_{if}^-(E_e, \theta_e, \phi_e)$ can be obtained through integration of $|T_{if}^-(E_e, \theta_e, \phi_e, \boldsymbol{\eta})|^2$ over the transverse momentum transfer $\boldsymbol{\eta}$

$$Q_{if}^-(E_e, \theta_e, \phi_e) \equiv \frac{\mathrm{d}^3 Q_{if}^-}{\mathrm{d}\Omega_e \mathrm{d}E_e} = \int \mathrm{d}\boldsymbol{\eta} \left| \frac{T_{if}^-(E_e, \theta_e, \phi_e, \boldsymbol{\eta})}{2\pi v} \right|^2 \tag{19.33}$$

where $\Omega_e = (\theta_e, \phi_e)$ and $\mathrm{d}\Omega_e = (\sin\theta_e)\mathrm{d}\theta_e \mathrm{d}\phi_e$. The quantity $Q_{if}^-(E_e, \theta_e, \phi_e)$ is a convenient notation for the triple differential cross sections $\mathrm{d}^3 Q_{if}^-/(\mathrm{d}\Omega_e \mathrm{d}E_e)$ in the solid angle and energy of the ejected electron. The so-named 'total' cross section $Q_{if}^-(E_S)$ for the TEA process in the case of an asymmetric profile resonance of total width Γ_S is defined by

$$Q_{if}^-(E_S) = \int \mathrm{d}\Omega_e \int_0^\infty \mathrm{d}E_e \left[Q_{if}^-(E_e, \theta_e, \phi_e) - Q_C^-(E_e, \theta_e, \phi_e) \right] \tag{19.34}$$

where $Q_C^-(E_e, \theta_e, \phi_e)$ is the ionization background determined by interpolation of the smooth electron spectrum which appears close to the given resonance.

It is clear that (19.34) will give a non-negligible total cross section only for the values of energy E_e located within the narrow range $[E_{\rm S}-\Gamma_{\rm S}, E_{\rm S}+\Gamma_{\rm S}]$. As mentioned earlier, the cross sections reported by Itoh *et al.* [470] have been measured using the 0° electron spectroscopy. This implies that $Q^-_{if}(E_e,\theta_e,\phi_e)$ must be computed with $\theta_e=0^\circ$ and $\phi_e=0^\circ$. As a consequence, the projectile is left in an $s-$state after auto-ionization, so that the non-zero contribution from the continuum will stem only from the case with $M=0$. Thus, with $\theta_e=0^\circ$, $\phi_e=0^\circ$ and $M=0$, the following simpler notation is more convenient

$$Q^-_{if}(E_e,\theta_e=0^\circ,\phi_e=0^\circ)\equiv Q^-_{if,0}(E_e)$$

$$Q^-_{\rm C}(E_e,\theta_e=0^\circ,\phi_e=0^\circ)\equiv Q^-_{\rm C,0}(E_e). \tag{19.35}$$

Under these conditions, (19.34) is reduced to

$$Q^-_{if}(E_{\rm S})=\frac{4\pi}{2L+1}\int_0^\infty {\rm d}E_e\left[Q^-_{if,0}(E_e)-Q^-_{\rm C,0}(E_e)\right] \tag{19.36}$$

where L is the total angular momentum of considered auto-ionizing state.

19.10 The CDW-4B method in the Feshbach resonance formalism

The concept discussed in the preceeding section for a physically more adequate description of the TEA process by means of the CDW-4B method, improved within the Feshbach formalism for resonances, has been tested by Ourdane *et al.* [188] against the experimental data of Itoh *et al.* [470] for the collision

$$^3{\rm He}^+(1s)+{}^4{\rm He}(1s^2)\to{}^3{\rm He}^{**}(nl,n'l')+{}^4{\rm He}^+(1s). \tag{19.37}$$

Four doubly excited states of $^3{\rm He}^{**}(nl,n',l')$ have been included in the computations and these were $(2s^2)\,^1S$, $(2s2p)\,^3P$, $(2p^2)\,^1D$ and $(2s2p)\,^1P$. Since the collisional system (19.37) is actually a five-body problem, an additional approximation has been made to account for the presence of the spectator electron in the target. The simplest way to proceed is to introduce the Slater screening of He. This amounts to first using $Z_{\rm T}^{\rm eff}=1.6875$ instead of the bare nuclear charge $Z_{\rm T}=2$ for the helium target. Further, in the interaction potential $Z_{\rm T}(1/R-1/x_1)$, the target charge $Z_{\rm T}$ is replaced by an effective charge $Z_{\rm NTE}$. In the computations from Ref. [186], $Z_{\rm NTE}$ has been taken to be equal to either 1 or 1.6875. All the parameters for the bound orbitals $\Psi_{\rm S}(E_{\rm S};\boldsymbol{s}_1,\boldsymbol{s}_2)$ of the doubly excited state of ${\rm He}^{**}(nl,n'l')$ were given in Ref. [188]. These parameters, that are resonant energies and the associated widths, compare favorably with the corresponding theoretical results of Bhatia and Temkin [487] based upon the projection-operator formalism. Writing the parameters from Ref. [487] in parentheses, we have e.g. for the auto-ionizing state $^1P^o$, $E_S(\text{a.u.}) = -0.6834(-0.6929)$, $\Gamma_S(\text{eV}) = 0.036(0.0363)$, whereas for the $^1D^e$ state it follows $E_S(\text{a.u.}) = -0.6918(-0.7028)$ and $\Gamma_S(\text{eV}) = 0.081(0.0729)$, where the superscripts o and e stand for odd and even parity of the state, respectively. Overall, a relatively good agreement (to within $\simeq 1\%$) exists between the computations from Refs. [188] and [487]. The continuum orbitals of the final state $\Psi_C(E_e; \boldsymbol{s}_1, \boldsymbol{s}_2)$ have been discretized by the mentioned procedure of Macías *et al.* [484]. The ensuing parameters appearing in the configuration $j = (L, 0)$ for a given index i ($i = 0, ..., N_C$) that have been used in the computations of Ourdane *et al.* [188] were $\alpha_i = 2\beta^{-i/2+\mu}$ ($n_i = L+1$, i even) and $\alpha_i = 2\beta^{-(i-1)/2+\mu}$ ($n_i = L+2$, i odd) with $\beta = 1.6$. The transition amplitude from (19.2) contains two different terms defined by the Coulomb interaction via the electrostatic potentials and by the gradient-gradient ($\boldsymbol{\nabla} \cdot \boldsymbol{\nabla}$) potential operator, which is the typical dynamic coupling occurring in the standard CDW method. This latter interaction term contains a contribution from the continuum component $\Psi_C(E_e; \boldsymbol{s}_1, \boldsymbol{s}_2)$ which describes the ECC channel. Therefore, the ECC effect is included in the CDW-4B method by the $\boldsymbol{\nabla} \cdot \boldsymbol{\nabla}$ interaction potential operator which couples the continuum state $\Psi_C(E_e; \boldsymbol{s}_1, \boldsymbol{s}_2)$ with the initial distorted wave function in the integral over $\boldsymbol{s}_2$ in (19.2). The matrix elements with the $\boldsymbol{\nabla} \cdot \boldsymbol{\nabla}$ term have caused some numerical instabilities in the computations from Ref. [188], especially regarding convergence of the discretized continuum states $\Psi_C(E_e; \boldsymbol{s}_1, \boldsymbol{s}_2)$. Ourdane *et al.* [188] have estimated that the contribution from the ECC process (via the said $\boldsymbol{\nabla} \cdot \boldsymbol{\nabla}$ term) is very small and, as such, was neglected.

19.11 Comparison between theories and experiments for electron spectra near Auger peaks

19.11.1 Electron energy spectral lines

We first discuss the energy distributions of the emitted electrons in the TE process (19.37). The theoretical electron spectra from the CDW-4B method [188] are plotted and compared with the corresponding experimental data [470] in Figs. 19.12–19.16 in the energy range $32 - 42$ eV at five impact energies of the $^3He^+$ ion, $E = 75$, 100, 200, 400 and 500 keV.

For a comparison between the experiment and the CDW-4B method, the theoretical data have been convoluted using a Gaussian function of a width equal to the experimental resolution of the spectrometer (about 0.2 eV) from Ref. [470]. The limitation of the validity of the standard CDW method for the $^3He^+$ ion in the case of reaction (19.37) is estimated using the empirical expression (14.69) to be about 330 keV (110 keV/amu). Although the impact energies 75 and 100 keV, considered in Figs. 19.12 and 19.13, are below

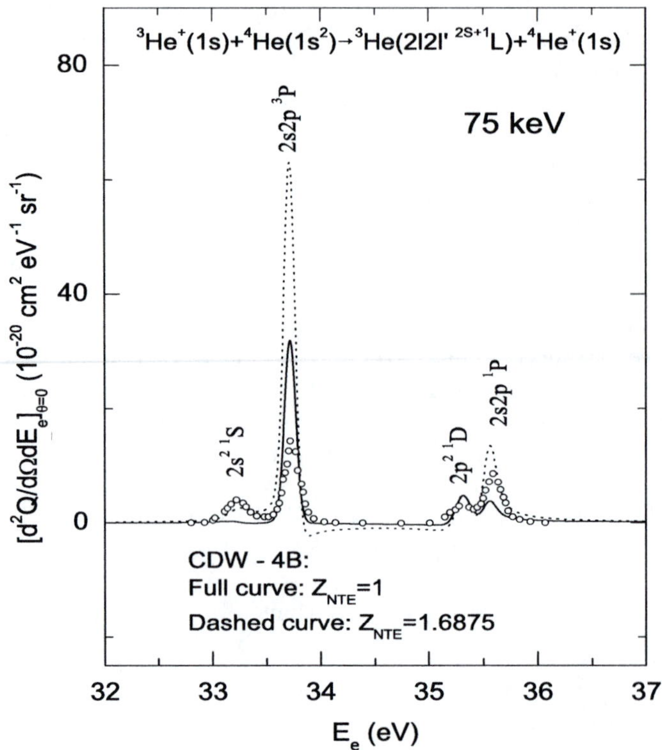

FIGURE 19.12
Zero degree electron energy spectra as a function of the emitted electron energy E_e at the impact energy $E = 75$ keV for process $^3\text{He}^+(1s) + {}^4\text{He}^+(1s) \to {}^3\text{He}(2l2l'\,{}^{2S+1}L) + {}^4\text{He}^+(1s)$. The $(2l2l')\,{}^{2S+1}L$ states are: $(2s^2)\,{}^1S$, $(2s2p)\,{}^3P$, $(2p^2)\,{}^1D$ and $(2s2p)\,{}^1P$. Theory (CDW-4B method [188]): full curve ($Z_{\text{NTE}} = 1$) and dashed curve ($Z_{\text{NTE}} = 1.6875$). Experiment: $\circ$ [470].

the said validity limit of the CDW method according to (14.69), a qualitative agreement between the theoretical and experimental results is still consistently obtained. Of course, such an agreement could be fortuitous.

Further, in Figs. 19.12 and 19.13, we display the results for the two effective charges $Z_{\text{NTE}} = 1$ and 1.6875 that were mentioned earlier. It is seen in these two figures that the choice of the effective charge Z_{NTE} can significantly influence the outcomes of the computations in the CDW-4B method at $E = 75$ and 100 keV. In this theoretical method, the initial and final states from

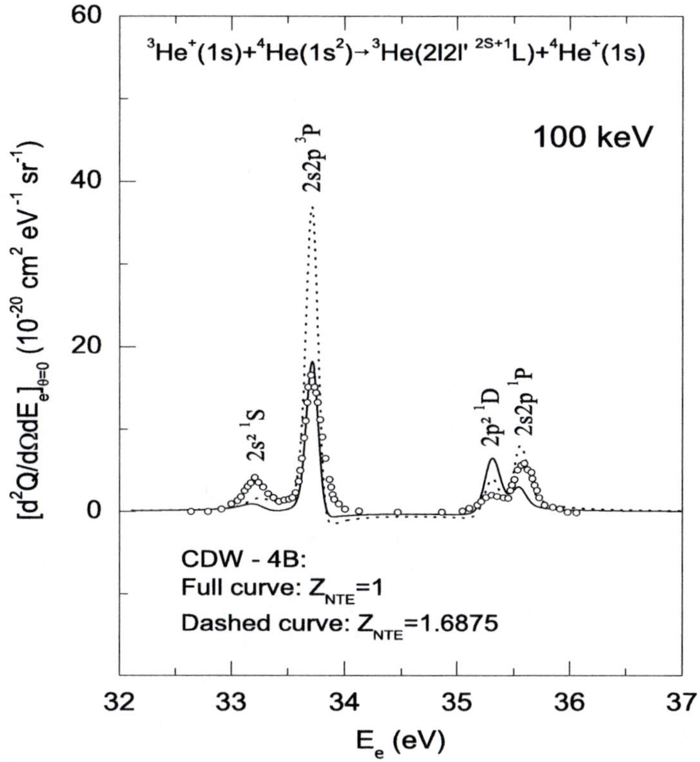

FIGURE 19.13
Zero degree electron energy spectra as a function of the emitted electron energy E_e at the impact energy $E = 100$ keV for process $^3\text{He}^+(1s)+^4\text{He}^+(1s) \to$ $^3\text{He}(2l2l'\,^{2S+1}L)+^4\text{He}^+(1s)$. The $(2l2l')\,^{2S+1}L$ states are: $(2s^2)\,^1S$, $(2s2p)\,^3P$, $(2p^2)\,^1D$ and $(2s2p)\,^1P$. Theory (CDW-4B method [188]): full curve ($Z_{\text{NTE}} = 1$) and dashed curve ($Z_{\text{NTE}} = 1.6875$). Experiment: ∘ [470].

the entrance and exit reaction channels are strongly coupled. As mentioned, this latter coupling, which is mediated by the usual $\boldsymbol{\nabla}\cdot\boldsymbol{\nabla}$ interaction potential operator, is responsible for a considerably enhanced influence of continuum intermediate states at lower impact energies. As a consequence, the total cross sections from the CDW-4B method systematically overestimate experimental data at lower energies i.e. below the usual Massey maximum. A varying degree of agreement or disagreement between theory and experiment also exists in comparisons of differential cross sections, especially when analyzing each individual peak, as is clear from Figs. 19.12–19.14. For example, in the case of the effective charge $Z_{\text{NTE}} = 1$, the strongest peak for the $2s2p\,^3P$ state predicted by the CDW-4B method (full line) overestimates or underestimates the corresponding measured maximum at 75 and 200 keV in Figs. 19.12

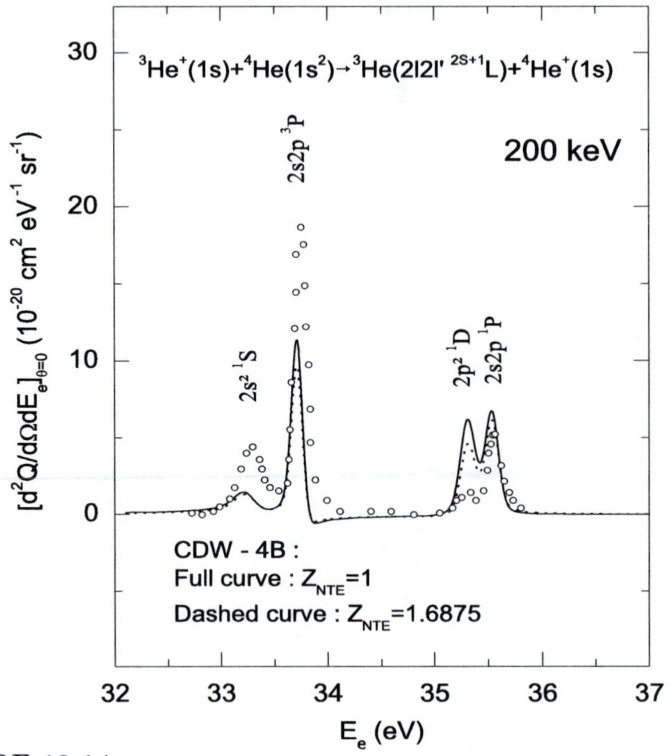

FIGURE 19.14
Zero degree electron energy spectra as a function of the emitted electron energy E_e at the impact energy $E = 200$ keV for process ${}^3\mathrm{He}^+(1s)+{}^4\mathrm{He}^+(1s) \rightarrow {}^3\mathrm{He}(2l2l'\,{}^{2S+1}L)+{}^4\mathrm{He}^+(1s)$. The $(2l2l')\,{}^{2S+1}L$ states are: $(2s^2)\,{}^1S$, $(2s2p)\,{}^3P$, $(2p^2)\,{}^1D$ and $(2s2p)\,{}^1P$. Theory (CDW-4B method [188]): full curve ($Z_{\mathrm{NTE}} = 1$) and dashed curve ($Z_{\mathrm{NTE}} = 1.6875$). Experiment: $\circ$ [470].

and 19.14, respectively. However, for the same peak considered at an impact energy of 100 keV, Fig. 19.13 shows very good agreement between theory (full line) and experiment. At higher impact energies (400 and 500 keV), good agreement persists between the theoretical and the experimental spectra, as displayed in Figs. 19.15 and 19.16, respectively. Moreover, it can be observed in Figs. 19.15 and 19.16 that at high impact energies, the influence of the choice of the effective charge Z_{NTE} becomes negligible.

Regarding the theory, it should be noted that the version of the CDW-4B method used in Ref. [188] is limited to isolated resonances alone, which amounts to ignoring overlapping resonances altogether. Therefore, the continuum of a given excited state has no simple relationship in magnitude and

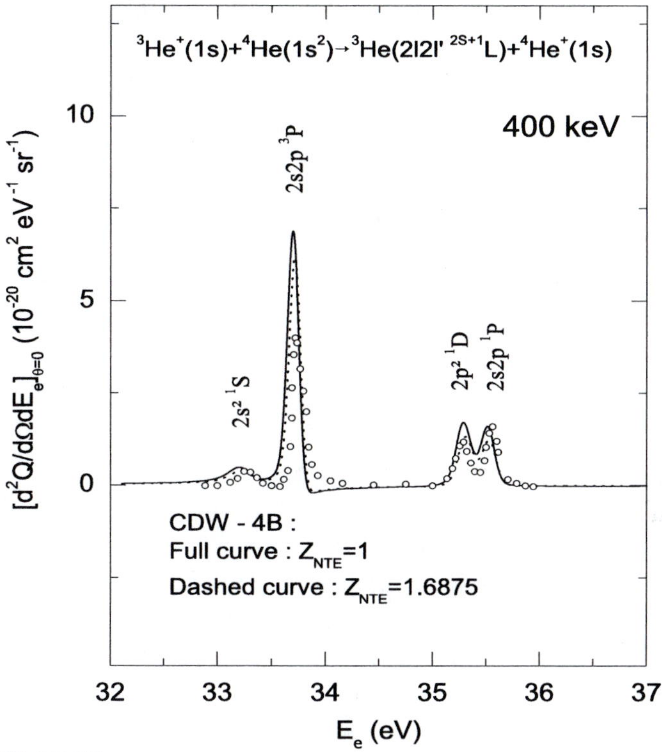

FIGURE 19.15
Zero degree electron energy spectra as a function of the emitted electron energy E_e at the impact energy $E = 400$ keV for process $^3\text{He}^+(1s)+^4\text{He}^+(1s) \rightarrow {^3\text{He}}(2l2l'\,^{2S+1}L)+^4\text{He}^+(1s)$. The $(2l2l')\,^{2S+1}L$ states are: $(2s^2)\,^1S$, $(2s2p)\,^3P$, $(2p^2)\,^1D$ and $(2s2p)\,^1P$. Theory (CDW-4B method [188]): full curve ($Z_{\text{NTE}} = 1$) and dashed curve ($Z_{\text{NTE}} = 1.6875$). Experiment: $\circ$ [470].

phase with the continuum adjacent to a doubly excited state. This can lead to certain problems in interference between transition amplitudes of any two consecutive auto-ionizing states. Moreover, neither interference between the transition amplitudes of contiguous doubly excited states, nor the effect due to the post-collisional interaction (PCI) were evaluated by Ourdane *et al.* [188] in the CDW-4B method. This is because the theoretical spectra from Ref. [188] are pure sums of contributions due to each line superimposed on top of its own background.

Using (19.37), the background contribution has been subtracted from the displayed theoretical cross sections. This was necessary because the estimate of the corresponding background has also been subtracted from the shown

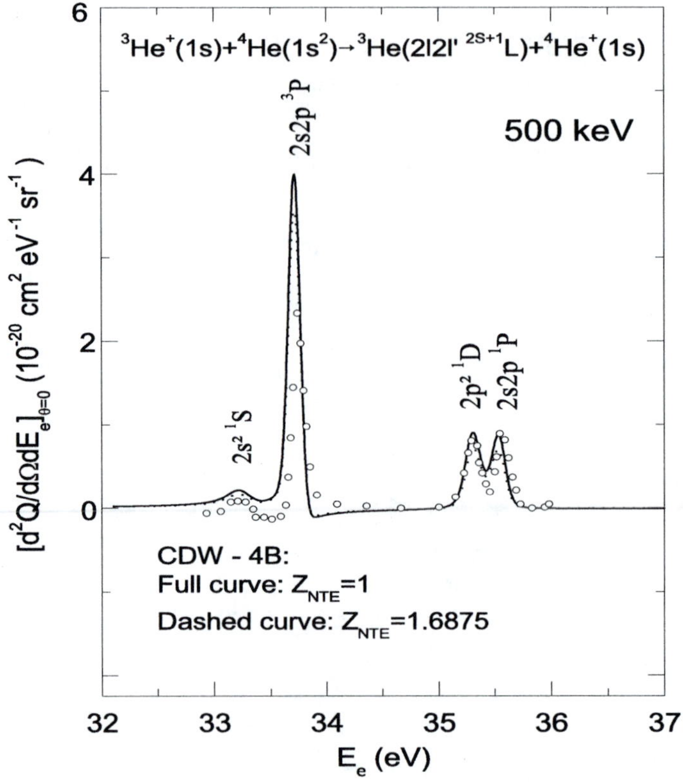

FIGURE 19.16
Zero degree electron energy spectra as a function of the emitted electron energy E_e at the impact energy $E = 500$ keV for process ${}^3\text{He}^+(1s)+{}^4\text{He}^+(1s) \to {}^3\text{He}(2l2l'\,{}^{2S+1}L)+{}^4\text{He}^+(1s)$. The $(2l2l')\,{}^{2S+1}L$ states are: $(2s^2)\,{}^1S$, $(2s2p)\,{}^3P$, $(2p^2)\,{}^1D$ and $(2s2p)\,{}^1P$. Theory (CDW-4B method [188]): full curve ($Z_{\text{NTE}} = 1$) and dashed curve ($Z_{\text{NTE}} = 1.6875$). Experiment: $\circ$ [470].

experimental data of Itoh *et al.* [470]. Of course, such a subtraction may lead to unphysical, negative cross sections in some parts of the spectrum. Indeed, this can be seen in Figs. 19.12–19.16 in both the theory and experiment.

Although overall good agreement between theory and experiment is obtained in Figs. 19.12–19.16, it should nevertheless be noted that the computed line shapes (especially the ${}^3P$ line shape) are very sensitive to the width which is used in the convolution. For the ${}^3P$ state, the code from Ref. [81] gives the width of 7.7 meV [188], which should be accurate. The corresponding experimental value is much larger than what is expected and this is due to the limitation of the spectrometer resolution (0.2 eV) in the measurement of Itoh *et al* [470]. Moreover, there is a slight shift between the theoretical and experimental resonant energies in Figs. 19.12–19.16.

19.11.2 Total cross section for the TEA mode

Here, we analyze the total cross sections for the TE reaction (19.37). In Figs. 19.17–19.20, the corresponding results of Ourdane *et al.* [188] from the CDW-4B method are compared with the experimental data of Itoh *et al.* [470] for the double excited states of helium $(2s2p)\,^1P$, $(2p^2)\,^1D$, $(2s^2)\,^1S$ and $(2s2p)\,^3P$. In these cases, for each doubly excited state, the cross sections in the CDW-4B method were obtained by performing numerical integration of the theoretical profiles over a sufficiently wide energy range around the resonance energy $E_{\rm S}$. It is clear from (19.32) that the final wave function for an energy E_e close to the auto-ionizing energy $E_{\rm S}$, exhibits its dependence upon E_e only through the coefficients $\alpha(E_e)$ and $\beta(E_e)$. Therefore, the general form of $Q^-_{if,0}(E_e)$ can be simplified as

$$Q^-_{if,0}(E_e) = \alpha^2(E_e)Q^-_{\rm D}(E_{\rm S}) + \alpha(E_e)\beta(E_e)Q^-_{\rm X}(E_{\rm S}) + \beta^2(E_e)Q^-_{\rm C}(E_{\rm S}) \quad (19.38)$$

where $Q^-_{\rm C}(E_{\rm S})$ is the local continuum contribution (background) and $Q^-_{\rm D}(E_{\rm S})$ is the total cross section from the discrete component $\Psi_{\rm S}(E_{\rm S};\boldsymbol{s}_1,\boldsymbol{s}_2)$ as well as from the resonant continuum, defined by the principal value integral in (19.30). The term $Q^-_{\rm X}(E_{\rm S})$ represents a cross section obtained using the product between transition amplitudes associated with $Q^-_{\rm C}(E_{\rm S})$ and $Q^-_{\rm D}(E_{\rm S})$. It can be shown that the TEA total cross section $Q^-_{if}(E_{\rm S})$ given by (19.36) may compactly be written as

$$Q^-_{if}(E_{\rm S}) = \frac{4\pi}{2L+1}\left[Q^-_{\rm D}(E_{\rm S}) - \frac{\pi\Gamma_{\rm S}}{2}Q^-_{\rm C}(E_{\rm S})\right]. \quad (19.39)$$

This formula hints at the two important arguments that run as follows. (i) The term $[4\pi/(2L+1)]Q^-_{\rm D}$ which defines the total TEA cross section is not equivalent to the cross section $Q^-_{\rm TE}$ from Gayet *et al.* [187] who employed only the discrete orbitals. As mentioned, $Q^-_{\rm D}(E_{\rm S})$ contains the same discrete orbitals as those from Ref. [187], but additionally it has the resonant continuum components. (ii) One could argue that the term $(\pi/2)\Gamma_{\rm S}Q^-_{\rm C}(E_{\rm S})$ could accidentally cancel the contribution from the resonant continuum in $Q^-_{\rm D}(E_{\rm S})$ due to the principal value integral. In such a case, $Q^-_{if}(E_{\rm S})$ from (19.39) would coincide with $Q^-_{\rm TE}$ from Ref. [187]. However, the explicit computations show that such a fortuitous and delicate cancellation does not occur. Therefore, the improvement of the CDW-4B method employed by Ourdane *et al.* [188] over the corresponding computations from Gayet *et al.* [187] is genuine (in figures, the former and the latter references are referred to as Variant 1 and Variant 2, respectively). Hence, it is pertinent to examine the results of Ourdane *et al.* [188] obtained by a numerical integration of (19.36). For the $(2s2p)\,^1P_0$ state, the result from the CDW-4B method without coupling between the doubly excited state and the continuum [187] is much smaller than the experimental data, as seen in Fig. 19.17. On the other hand, the results of Ourdane *et al.* [188] show good agreement with the measured cross sections of Itoh *et al.*

[470], even in the low-energy range $E < 110$ keV/amu i.e. below the validity limit for the standard CDW method. In this range, better agreement is obtained with $Z_{\rm NTE} = 1.6875$, which illustrates a previously mentioned feature i.e. an increasing influence of the spectator electron in the target at smaller impact energies. Complementary to this, the choice $Z_{\rm NTE} = 1$ gives better agreement with the experiment above 110 keV/amu, where the standard CDW method is expected to be most adequate.

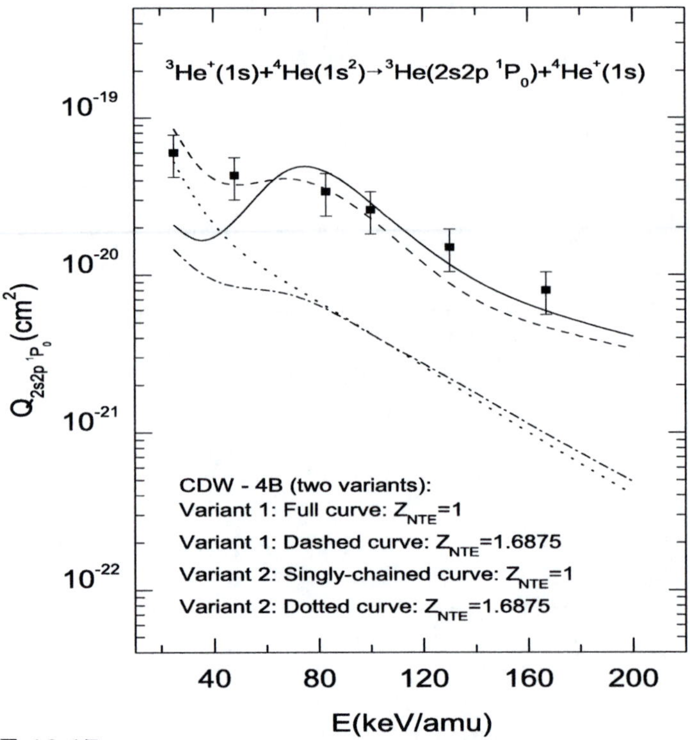

FIGURE 19.17
Total cross sections Q as a function of the incident energy E for process ${}^3\text{He}^+(1s) + {}^4\text{He}(1s^2) \rightarrow {}^3\text{He}^{**}(2s2p\,{}^1P_0) + {}^4\text{He}^+(1s)$. Theory (CDW-4B method – Variant 1 [188]): full curve ($Z_{\rm NTE} = 1$) and dashed curve ($Z_{\rm NTE} = 1.6875$). Theory (CDW-4B method – Variant 2 [186, 187]): singly-chained curve ($Z_{\rm NTE} = 1$) and dotted curve ($Z_{\rm NTE} = 1.6875$). Experiment: ■ [470].

In the case of the $(2p^2)\,{}^1D_0$ state, the theoretical results are seen in Fig. 19.18 to overestimate the measured cross sections below 110 keV/amu. Moreover, the CDW-4B method for $Z_{\rm NTE} = 1$ does not reproduce the experimen-

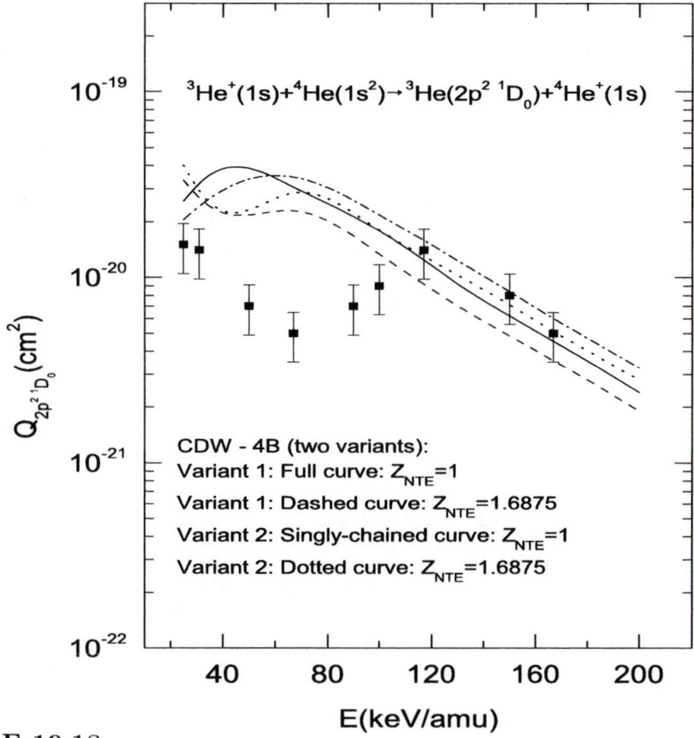

FIGURE 19.18
Total cross sections Q as a function of the incident energy E for process $^3\mathrm{He}^+(1s)+^4\mathrm{He}(1s^2) \to {}^3\mathrm{He}^{**}(2p^2\,{}^1D_0)+^4\mathrm{He}^+(1s)$. Theory (CDW-4B method – Variant 1 [188]): full curve ($Z_{\mathrm{NTE}} = 1$) and dashed curve ($Z_{\mathrm{NTE}} = 1.6875$). Theory (CDW-4B method – Variant 2 [186, 187]): singly-chained curve ($Z_{\mathrm{NTE}} = 1$) and dotted curve ($Z_{\mathrm{NTE}} = 1.6875$). Experiment: ■ [470].

tally observed dip at 70 keV/amu, but instead merely an oscillation appears at 50 keV/amu for $Z_{\mathrm{NTE}} = 1.6875$. However, above 110 keV/amu, the CDW-4B method for both values of $Z_{\mathrm{NTE}} = 1$ and 1.6875 [188] and the measurement [470] are in good agreement within the estimated experimental uncertainty. In Fig. 19.18 at $E > 60$ keV, the overall influence of coupling between the discrete component and adjacent continuum is such that the resulting cross sections are lowered, and this improves the agreement with the experiment, relative to the case when this coupling is ignored [187].

In an isolated resonance approach, the two above-mentioned states ($^1P_0$ and $^1D_0$) have been studied separately. The sum of the two theoretical cross sections for the formation of the states $^1P_0$ and $^1D_0$ is in good agreement with the corresponding sum of the experimental data above 110 keV/amu (not shown here, but the situation is similar to Fig. 19.11 for the $\mathrm{He}^+ - \mathrm{H}$

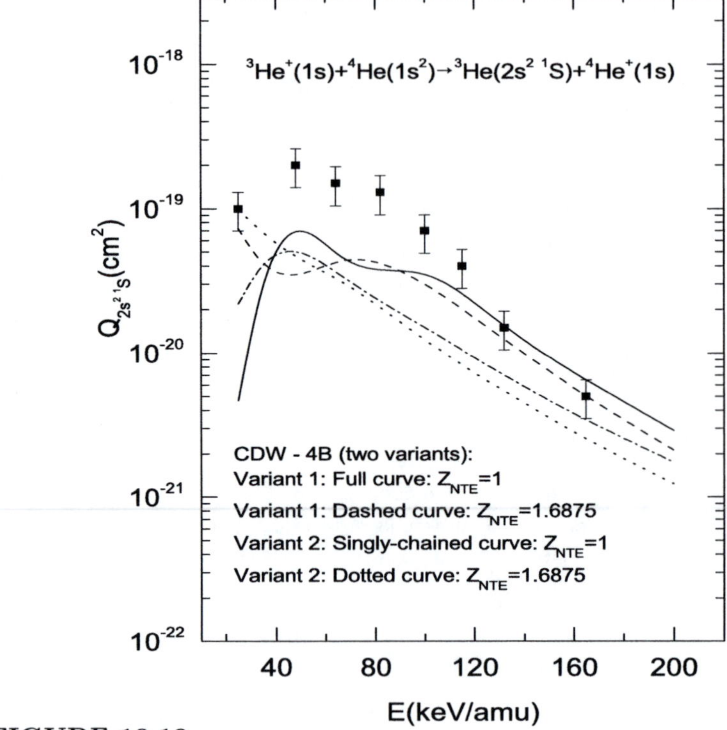

FIGURE 19.19
Total cross sections Q as a function of the incident energy E for process $^3\text{He}^+(1s) + {}^4\text{He}(1s^2) \rightarrow {}^3\text{He}^{**}(2s^2\,{}^1S) + {}^4\text{He}^+(1s)$. Theory (CDW-4B method – Variant 1 [188]): full curve ($Z_{\text{NTE}} = 1$) and dashed curve ($Z_{\text{NTE}} = 1.6875$). Theory (CDW-4B method – Variant 2 [186, 187]): singly-chained curve ($Z_{\text{NTE}} = 1$) and dotted curve ($Z_{\text{NTE}} = 1.6875$). Experiment: ■ [470].

collisions). Nevertheless, this theoretical sum is still unable to reproduce the oscillation from the experimental sum in the range 25–100 keV/amu. Such oscillatory structures might be due to the interference between the two states, $^1P_0$ and $^1D_0$, as argued by van der Straten and Morgenstern [486], who also pointed out that the dip in the cross sections for the production of the $^1D_0$ state could be the consequence of the effect of the PCI onto the $^1P_0$ state. As already mentioned, the CDW-4B method in the version used in Refs. [187, 188] does not take the PCI into account. A further improvement of this latter version of the CDW-4B method is possible with the PCI effect included, and this would be an important subject to study in the future. It is also possible that the discussed oscillatory structure may stem from a combining effect of both the discrete-continuum coupling and the PCI interference between the two investigated states.

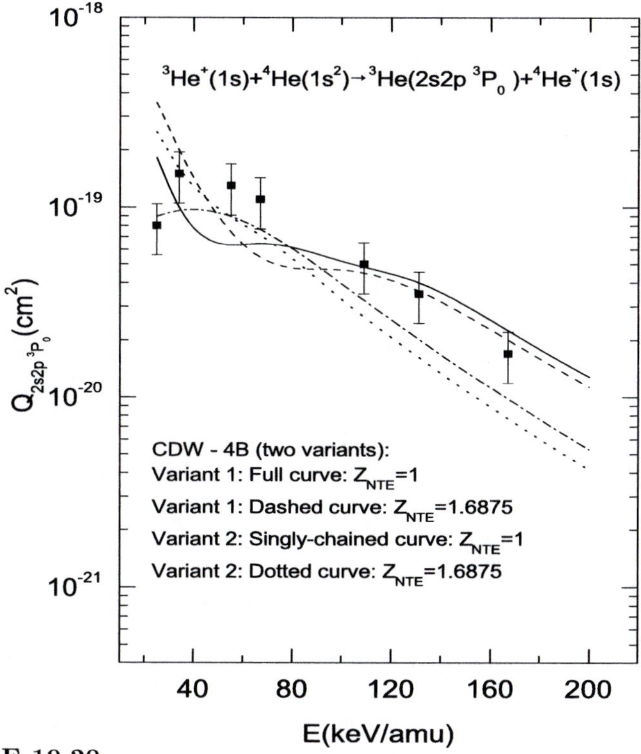

FIGURE 19.20

Total cross sections Q as a function of the incident energy E for process ${}^3\mathrm{He}^+(1s) + {}^4\mathrm{He}(1s^2) \rightarrow {}^3\mathrm{He}^{**}(2s2p\,{}^3P_0) + {}^4\mathrm{He}^+(1s)$. Theory (CDW-4B method – Variant 1 [188]): full curve ($Z_{\mathrm{NTE}} = 1$) and dashed curve ($Z_{\mathrm{NTE}} = 1.6875$). Theory (CDW-4B method – Variant 2 [186, 187]): singly-chained curve ($Z_{\mathrm{NTE}} = 1$) and dotted curve ($Z_{\mathrm{NTE}} = 1.6875$). Experiment: ■ [470].

Similar remarks to those for the ${}^1P_0$ and ${}^1D_0$ states could also be made for the formation of the ${}^1S$ state as illustrated in Fig. 19.19. In the range 110–200 keV/amu, the results of the CDW-4B method used by Ourdane *et al.* [188] are in much better agreement with the experiment than the ones from Gayet *et al.* [187]. Below the limit of the validity for the CDW-4B method, the theoretical results from Ref. [188] markedly underestimate the experimental data, that themselves do not exhibit any oscillatory structure. Since the ${}^1S$ state is well separated in energy from the ${}^3P$ state, the interference between the two states is expected to be negligible. Therefore, the oscillation in the theoretical cross sections is likely to be the result of coupling between discrete and continuum orbitals.

Finally, the situation for the ${}^3P_0$ state shown in Fig. 19.20 is comparable to that for the ^1S state from Fig. 19.19. As can be seen from Fig. 19.20, in the high-energy range, the CDW-4B method employed by Ourdane *et al.* [188] represents a major improvement over the results of Gayet *et al.* [187]. In addition, at energies smaller than 110 keV/amu, the discrepancy between the cross sections given by Ourdane *et al.* [188] and the experimental data is not so pronounced.

Overall, the study of the TE process in the $He^+ - He$ collisions carried out by Ourdane *et al.* [188] has shown that the adjacent continuum of a low-lying doubly excited state plays a significant role. This collisional system appears particularly interesting in two respects: (i) the target and the projectile nuclear charges (Z_T and Z_P) are either the same or comparable to each other, and this results in interference between competing contributions from the resonant and non-resonant modes[2], and (ii) the influence of the electron continuum is strong, because the final states of helium under consideration decay mainly through auto-ionization.

In the energy range where it applies (here above 110 keV/amu), the CDW-4B method gives reliable predictions for the TEA processes, provided that in the transition amplitude the wave function of the final doubly excited state includes both the discrete and the adjacent continuum components. For this case, the theoretical electron energy spectra near resonances (as well as the integrated energy profiles) are in much better agreement with the experimental data [470] than the previous version of the CDW-4B method which includes only discrete components [186, 187]. In fact, the results from the improved CDW-4B method [188] depart from the experimental spectra only at impact energies where the standard CDW method for single electron capture does not apply (below 110 keV/amu in this case). However, at higher collision energies, the CDW-4B method successfully reproduces the experimentally recorded electron spectra. Furthermore, good agreement of the integrated energy profiles from the CDW-4B method [188] with the experimentally measured cross section [470] for the TEA process represents a significant improvement over all the other predictions from earlier studies that considered the coupling with the adjacent continuum as the PCI effect. Nevertheless, it should be kept in mind that the CDW-4B method illustrated in Figs. 19.17–19.20 is restricted to isolated resonances alone. This could advantageously be overcome by using the non-perturbative Padé-based resonant scattering theory which treats both isolated and overlapping resonances on the same footing [34, 488]. Here, a non-perturbative theory is especially needed at intermediate impact energies E for total cross sections Q as well as for differential cross

[2]Of course, in a pure five-body treatment of the $He^+ - He$ collisions, we would have $Z_P = Z_T = 2$. However, in a simplified four-body description of the same $He^+ - He$ collisions, as done in Refs. [187, 188], the second (passive) electron in the target is treated approximately via e.g. the Slater screening of $Z_T = 2$ which then becomes $Z_T = 1.6875$. It is because of this screening that we may talk about the two comparable target and projectile nuclear charges even for the manifestly symmetric P and T such as those in the $He^+ - He$ collisions.

sections dQ/dE_e at intermediate energies E_e of the emitted electrons due to strong coupling between the initial and final channels. It should also be emphasized that the ignored continuum-continuum terms (arising from the $\boldsymbol{\nabla} \cdot \boldsymbol{\nabla}$ interaction operator), that Ourdane *et al.* [188] viewed as negligible, have caused convergence difficulties in the Feshbach-Fano spectral representation of auto-ionizing states. Moreover, the authors of Ref. [188] expect that these additional approximations introduced into the CDW-4B method should not seriously affect their main conclusions. Such an expectation needs to be verified by a new computation using the CDW-4B method without resorting to any of the mentioned simplifying approximations. This would be entirely feasible. Finally, it should be pointed out that the concept of the so-called 'pure transfer and excitation process' considered as a simple post-collisional Auger decay of the discrete component of a given doubly excited state might be quite misleading. In fact, it has been argued in Ref. [188] that this concept could make good sense only in the limit of a negligible auto-ionization width Γ_S. Of course, one may introduce the so-called 'transfer-excitation total cross section' by integrating an electron energy profile, after removal of the ionization background, as seen in (19.36). However, in so doing one must also remember that a hidden contribution from the continuum is necessarily included in the final result, unless $\Gamma_S = 0$.

20

Concluding remarks and outlooks

This book is comprized of two parts. In the first part through chapters 1–10, formal theory of scattering is presented. The second part via chapters 11–19 deals with a wide class of ion-atom collisions at high energies. The common thread which tightly binds together these two parts into one coherent whole is the necessary rigor of the mathematics of scattering theory, intertwined with the first principles of physics. Although theory is the central subject of this book, experiment is also in the main focus, as indeed pursuing one without the other in physics does not bear fruit.

The major theme which connects the first and second parts into a consistent scattering formalism is the following logical chain of key physical principles supported by a firm mathematical basis, as the necessity for experimentally relevant theoretical predictions: (i) the existence of the strong limit of the Schrödinger time-dependent full scattering state $\Psi(t)$ in the remote past ($t \rightarrow -\infty$) and distant future ($t \rightarrow +\infty$), (ii) the existence of the Møller wave operators $\Omega^{\pm}$, (iii) the existence of the scattering $\hat{\mathrm{S}}-$ as well as transition $\hat{\mathrm{T}}-$operators and (iv) the existence of isometry.

State vectors and operators are required to converge strongly in scattering theory. This implies the corresponding weak convergence by way of the Schwartz inequality. Adjectives 'strong' and 'weak' refer to convergence in terms of the norm and absolute value, respectively. Principle (i) is the requirement of the correct asymptotic behavior of the complete scattering states via the condition that the total dynamics is experimentally indistinguishable from the unperturbed (free) dynamics in the remote past and distant future. This is formulated by the strong limits $\mathrm{Lim}_{t\rightarrow\mp\infty}||\Psi(t)-\Psi_0(t)||=0$ by which the full states $\Psi(t)=\hat{\mathrm{U}}(t)\Psi_0$ associated with the complete evolution operator $\hat{\mathrm{U}}(t)=\exp(-i\hat{\mathrm{H}}t)$ of the entire system are reduced at $t\rightarrow\mp\infty$ to the state $\Psi_0(t)=\hat{\mathrm{U}}_0(t)\Psi_0$ corresponding to the unperturbed evolution operator $\hat{\mathrm{U}}_0(t)=\exp(-i\hat{\mathrm{H}}_0t)$, where $\Psi_0\equiv\Psi(0)$. Here, convergence in the norm is needed because the strong limit of the difference $\Psi(t)-\Psi_0(t)$ implies that both the difference $\Psi(t)-\Psi_0(t)$ itself is a zero state vector $\emptyset$ and that e.g. the spatial integral over $|\Psi^{\pm}(t)-\Psi_0(t)|^2$ tends to zero at every point in the configuration space. However, had it not been the case for the strong limit, each of the two states $\Psi(t)$ and $\Psi_0(t)$ could have independently approached $\emptyset$ at each spatial point when $t\rightarrow\mp\infty$. The most important physical significance of the scattering wave function $\Psi(t)$ is that it represents the particle state at infinitely large negative (and/or positive) times. Therefore, principle

(i) interprets scattering theory as the asymptotic agreement between the two descriptions, the one using the full dynamics (with H) and the other employing the unperturbed dynamics (with H_0). Principle (ii) necessitates that the strong limit of the product $\hat{U}^\dagger(t)\hat{U}_0(t) \equiv \Omega(t)$ of the two evolution operators, containing the full ($\hat{H}$) and unperturbed ($\hat{H}_0$) Hamiltonians, respectively, as $t \rightarrow \pm\infty$ must be the stationary Møller wave operators $\Omega^\mp$. Principle (iii) requires that the product of two stationary Møller operators via $\Omega^{-\dagger}\Omega^+$ is the $\hat{S}$−operator. Principle (iv) necessitates the relationship $||\hat{U}(t)\Psi_0|| = ||\Psi_0||$ which conserves the norm and probability.

None of the principles (i)–(iv) holds for Coulomb potentials. Nevertheless, in order to restore the usual interpretation of scattering theory for Coulomb potentials, the Hamiltonian must be appropriately modified so that the complete perturbation interaction is of short range. With such modifications, the principles (i)–(iv) could simultaneously hold, provided that the full Coulomb wave functions for electrons centered on both nuclei are used to distort the unperturbed channel states within theories that obey the correct boundary conditions, like the continuum distorted wave (CDW) method. However, when one or both of these Coulomb waves are replaced by their asymptotic forms that are logarithmic phase factors, the principle of isometry becomes invalid, since $||\hat{U}_{\rm eik}(t)\Psi_0|| \neq ||\Psi_0||$ where $\hat{U}_{\rm eik}(t)\Psi_0$ is the wave packet containing the Coulomb eikonal phase instead of the corresponding full Coulomb wave in $\hat{U}(t)\Psi_0$. In other words, the wave packet $\hat{U}_{\rm eik}(t)\Psi_0$ does not represent a physical state of a system of particles. Precisely this ansatz (i.e. the replacement of a full Coulomb wave by its asymptotic behavior at large distances) which violates isometry is used in a method called the continuum distorted wave eikonal initial 'state' (CDW-EIS) method. As such, the CDW-EIS method is obviously an approximate CDW method, irrespective of whether the former is established by reliance upon the latter or derived independently.

In the first part of this book, within formal scattering theory, a rigorous formalism is presented using the Abel and Cauchy strong limits. The concept of the strong limit, in general, is of critical importance to theory and experiment on scattering phenomena. The strong limits rely upon the correct boundary conditions through the requirement of the reduction of the complete dynamics under the full Hamiltonian $\hat{H}$ to the unperturbed dynamics governed by the unperturbed Hamiltonian $\hat{H}_0$. Without these limits it would be impossible to unequivocally distinguish the situations "before" and "after" the collision. And without this latter differentiation, it would be impossible to identify the real cause of the passage of the system from the initial to the final states. The wave function $\Psi(t)$ could tend to $\Psi_0(t)$ only if the interaction (as an external perturbation exerted onto the system) becomes negligibly small as $t \rightarrow \pm\infty$. On the other hand, at infinitely large times, the system has evolved to the asymptotic spatial region of scattering ($r \rightarrow \infty$). Thus the two limits $t \rightarrow +\infty$ and $r \rightarrow \infty$ consistently require that the interaction must vanish at infinitely large time-space variables. This concept coheres with the adiabatic theorem which treats the effect of screening and adiabatic switching of the interactions. These strategies via the existence of the strong limits, applications of the adiabatic theorem and other related aspects all hold in an adequate manner for short-range interactions from nuclear collisions. However, they fail for long-range Coulomb potentials from atomic collisions. For example, the rigorous work of Dollard proves that no screening whatsoever is applicable to Coulomb potentials. Yet, disregarding this fundamental objection, screened Coulomb potentials are regularly used throughout the literature on atomic collisions, as one of the inadequate attempts of avoiding to deal properly with the asymptotic convergence problem for long-range interactions.

Despite constituting a mathematically-oriented exposition, chapters 1–10 of the first part are presented in the typical physics setting with the relevant connections to the main observables, as experimentally measurable quantities, that can test and eventually validate the theory. Even the most rigorous concepts of strong topology can find their critical applications to typical situations encountered in scattering experiments as well as to associated numerical computations. Rigorous mathematical treatments of scattering phenomena, accompanied naturally with the corresponding robust algorithms, are necessary to secure two key achievements in modeling from first principles with no adjustable parameters: (i) enhancement of the predictive power of *ab initio* physics theories, and (ii) minimization as well as control of uncertainties in theoretical formulations and the ensuing results of computations.

Specifically, in this first part of the book, we begin with several fundamental notions and main observables in standard scattering problems. This theme is comprized of a number of interrelated topics, including: observables and elementary processes, energy as the most important physical property, classification of collisions, the role of wave packets, adiabatic switching of interaction potentials, collimation of beams of projectiles, general waves as well as quantum mechanical waves and probability character of quantum collisions.

Subsequently, we address the key issue of the requirements of the theory for the experiment. This encompasses elementary events versus multiple scatterings, average probabilities, differential and total cross sections, total probabilities, transmission phenomena, as well as quantum mechanical currents and cross sections. Next, we study continuous spectra and eigen-problems of resolvents. This subject is especially relevant to scattering where the continuum plays the dominant role. While in the corresponding bound-state spectrum, the Schrödinger eigen-value problem for Hamiltonians is at work, in scattering phenomena the Green resolvent operator proves to be much more appropriate. The topics analyzed here are: completeness and separability of the Hilbert spaces, the key realizations of abstract vector spaces, isomorphism of vector spaces, eigen-problems for continuous spectra, normal and Hermitean operators, strong and weak topology, compact operators for mapping of the weak to strong limits, strong differentiability and strong analyticity.

Further, we treat linear and bilinear functionals. In scattering theory, functionals play a very important role, as a special form of mapping between vector spaces and scalar fields. Therefore, it is necessary to present this subject, as well, and we do so by relying upon the Ries-Freshe theorem. We also focus on the definition of a quantum scattering event. This theme is addressed through Hamiltonian operators and boundedness, evolution operators and the Møller wave operators, as well as the Cauchy strong limit in non-stationary scattering theory, ending with three criteria for a quantum collisional system. In continuation, we elaborate on the adiabatic theorem and the Abel strong limit. The concept of screening of interaction potentials at asymptotically large distances is analyzed within the adiabatic theorem. The main idea behind this theorem is that all perturbation potentials must vanish at infinitely large times when the interacting particles are at asymptotic mutual distances. Special attention is paid to the application of the adiabatic theorem to scattering states that are improper, in the sense of being non-normalizable and, as such, not belonging to the Hilbert space of physical state vectors. Further, we inter-connect the adiabatic theorem with the Møller wave operators, the Abel strong limit in stationary scattering theory, the Green resolvent operators and the Lippmann-Schwinger equations.

Our next investigation is directed towards non-stationary and stationary scattering in the framework of the strong limits. The strong limits of the Møller wave operators are of central importance to scattering theory. For example, the Lippmann-Schwinger equations can be derived from the Abel limit, provided that the Møller operators $\Omega^{\pm}$ exist. The Cauchy and Abel limits deal with the processes $t \to \mp\infty$ and $\varepsilon \to 0^{\pm}$ that are typical in the non-stationary and stationary scattering theory. Here, ε is an infinitesimally small positive number added via its imaginary counterpart $i\varepsilon$ to the real energy E in the Green resolvent operator to avoid singularities (poles) at the eigen-energies of $\hat{\mathrm{H}}$. In the non-stationary {stationary} formalism, the limiting processes $t \to \pm\infty$ $\{\varepsilon \to 0^{\mp}\}$ imposed onto $\Psi(t)$ $\{\Psi(E)\}$ must lead to the total scattering states $\Psi^{\mp}$ with the incoming and outgoing boundary conditions, respectively. In the configuration space, the incoming {outgoing} asymptote is comprized of wavelets that are conceived as arriving {departing} to {from} the scattering center. It is possible to pass from the Cauchy to the Abel limit and *vice versa.* This passage effectively carries out the transformation from the time-dependent state vector $\Psi(t)$ to the corresponding stationary state $\Psi(E)$. Such a transformation is usually performed by means of the Fourier integral. The Cauchy-Abel mapping can totally replace the Fourier integral.

We expand the analysis to the scattering and transition matrices, as the two most important physical quantities that contain the entire information about any studied system. Here, we derive the $\hat{\mathrm{S}}$- and $\hat{\mathrm{T}}$-operators from the existence of the Abel strong limit. The squared absolute values of the $\hat{\mathrm{S}}$- and $\hat{\mathrm{T}}$-matrix elements are the key physical quantities that yield all the observables that could be measured experimentally, such as probabilities for transitions from the given initial to one and/or all final states, the corresponding cross sections, rate coefficients, density of states, spectral parameters, etc. To describe the characteristic features of the systems under investigation, we carry out spectral analysis of dynamical operators. As mentioned, the Cauchy and Abel limits can entirely replace the role of the Fourier integral. Moreover, these two limits do not necessarily need to rely upon each other. This implies that e.g. stationary scattering theory could be built with the sole reliance upon the Abel strong limit and with no recourse to the time-dependent formalism. Here, we deal with such issues as the proof of the existence of the Møller wave operators in an arbitrary representation, since this is one of the three conditions that define the so-called simple scattering system. Also addressed here are the themes of vital importance to scattering theory e.g. the spectral theorem, the link of the Møller operators with the Green resolvents, the relationships of the Abel strong limit with unitary operators and the Møller wave operators. We end the first part of this book with chapter 10 by considering the existence and completeness of Møller wave operators. The specific themes encompass linearity and isometry of wave operators, boundedness of wave operators throughout the Hilbert spaces, the spectral projection operators and the completeness of the Møller operators.

Overall, although the main emphasis in the first part of this book (chapters 1–10) is on the rigorous mathematical foundations of scattering theory, the basic general concepts of the related experiments are also analyzed. This is done on an intuitive level, as well as on deeper grounds that formulate a sequence of the requirements of the theory for the experiment.

The second part of the book (chapters 11–19) gives a thorough and systematic overview of the current status and a critical assessment of the existing quantum mechanical four-body methods for energetic ion-atom collisions. Proper descriptions of these collisions with two active electrons, such as the $Z_{\rm P} - (Z_{\rm T}; e_1, e_2)_i$ and $(Z_{\rm P}, e_1)_{i_1} - (Z_{\rm T}, e_2)_{i_2}$ scatterings require solutions of the pertinent four-body problems. We consider a number of inelastic collisions and special attention is focused on double electron capture, transfer ionization, transfer excitation, single electron detachment and single electron capture. We limit the scope of this book to intermediate and high non-relativistic impact energies.

A quantum mechanical treatment is adopted to set the formal theoretical framework for description of four-body rearrangement collisions. After establishing the basic notation, we present a succinct derivation of the Lippmann-Schwinger equations, the boundary-corrected Born as well as the Dodd-Greider perturbation expansions and the leading distorted-wave methods. All these theories satisfy the proper physical asymptotic conditions at large inter-particle distances in four-body collisions for which the CDW-4B method emerges as the most adequate. Subsequently, the particular one- and two-electron transitions in scatterings of completely stripped projectiles on helium-like atomic systems, as well as in collisions between two hydrogen-like atoms or ions, are analyzed in detail. Helium-like atoms or ions are the simplest many-electron systems where one can investigate the importance of electron correlation effects.

The reviewed problems indicate that for helium as a target, the dynamic electronic correlations in the given perturbation potential are much more important than the static ones stemming from the target bound state wave function. A substantial improvement of most of the examined four-body methods over e.g. the corresponding semi-classical impact parameter model (IPM) can be attributed exclusively to the role of dynamic electron correlation effects. The main drawback of the IPM and the related independent electron model (IEM) is in effectively reducing the initial four-body problem to the associated three-body problem. In this reduction, dynamic electron-electron correlations are completely ignored from the outset. The results of the correlated CDW method for double electron capture from helium by protons was the first proof of the complete break-down of the IPM. This is also the case for other two- and many-electron transitions involving complex-structured targets. All the presently analyzed boundary-correct four-body quantum mechanical methods are seen to be able to naturally incorporate both static and dynamic correlation effects of electrons.

In particular, static inter-electron correlations are shown to be very important for ionizing collisions involving H^- as a target, such as single electron detachment from H^- by H^+. Moreover, for these collisions, an even stronger emphasis is placed upon the proper connection between the distorted wave functions and the corresponding perturbations, as illustrated within the modified Coulomb Born (MCB) method, which is in excellent agreement with the available experimental data from the threshold through the Massey maximum to the Bethe region of high energies and beyond. By contrast, ignoring the said connection, as done in the eikonal Coulomb Born (ECB) method, leads to utterly unphysical total cross sections which overestimate the experimental data by 2–3 orders of magnitude, and tend to a peculiar constant value at high impact energies, instead of reaching the correct Bethe asymptotic limit.

In the present book, particular emphasis is placed upon the critical importance of preserving the proper Coulomb boundary conditions in formal four-body theory and in computational practice. This is guided by common past experience, which has shown that whenever such conditions (as one of the most basic requirements from scattering theory) are overlooked, severe and fundamental problems arose. As a consequence, models with incorrect boundary conditions are inadequate for describing experimental findings. In practice, the total scattering wave functions can satisfy the correct Coulomb boundary conditions in the initial and final asymptotic channels following the well-established procedure. For example, in addition to the long-range Coulomb distortions of the plane waves for relative motion of two charged heavy aggregates, account should be made for the intermediate ionization continua of the electrons in the entrance and exit channels for the CDW-4B method, or in either the entrance or exit channel for asymmetric distorted wave treatments, such as the boundary-corrected continuum intermediate state (BCIS-4B), Born distorted wave (BDW-4B), continuum distorted wave Born initial state (CDW-BIS) and continuum distorted wave Born final state (CDW-BFS) methods. As to the boundary-corrected first Born (CB1-4B) method, the pure electronic continuum intermediate states are not directly taken into account. Rather the presence of the electrons is felt here through a screening of the two nuclear charges in the Coulomb wave functions of the relative motion of the heavy scattering aggregates.

Double electron capture in the considered collisional systems is studied by means of the CB1-4B, CDW-4B, BDW-4B, BCIS-4B and CDW-EIS-4B methods. Unlike the well-documented success of the CB1-3B method for single-electron capture at intermediate and a wide range of high energies (all the way up to the outset of the Thomas double scattering), the CB1-4B method, as the prototype of four-body first-order theories, is satisfactory for double electron capture only at some intermediate energies, but flagrantly fails at higher energies. By contrast, as the prototype of four-body second-order theories, the CDW-4B method is successful for the majority of double electron capture processes, thus continuing with the excellent tradition of the corresponding three-body counterpart, which is the CDW-3B method. This is particularly true for two-electron capture from He by H^+ for which it is sufficient to include only the ground-to-ground state transition, due to the absence of the excited states of the H^- ion formed in the exit channel. For the same $H^+ - He$ double charge exchange, the cross sections from the CB1-4B method markedly overestimate all the experimental data by 1–3 orders of magnitude at all energies (10–1000 keV). Moreover, using the CDW-4B method for double electron capture in the $Z_P - He$ collisions with $Z_P \geq 3$, it is found that the contribution from excited states can be important compared to that from the corresponding ground states. As such, including excited states into computations can improve the agreement between the CDW-4B method and experimental data.

Specifically, within the distorted wave formalism, it is customary to refer to the first/second-order methods as those theories that exclude/include the electronic continuum intermediate states, respectively. On the other hand, any such second-order distorted wave method, may simultaneously be the lowest-order term in a consistent perturbation series expansion of the full transition amplitude. Then by a parallel nomenclature, this lowest-order term of a perturbation development (with or without distorted wave formalism) would be called a first-order approximation to the full transition amplitude. Thus, for example, the CDW-4B method is a second-order method (when viewed from the distorted wave perspective), as it includes the electronic continuum intermediate states. However, the same CDW-4B method is simultaneously the rigorous first-order term in the Dodd-Greider perturbation series. As such, the CDW-4B method is also the first-order approximation to the exact Dodd-Greider expansion. This latter fact should explicitly be indicated (whenever there is a chance for confusion) with a more specific acronym such as CDW-4B1. This is especially helpful whenever a reference should be made to the second term in the Dodd-Greider expansion, in which case the acronym CDW-4B2 is definitely needed.

Regarding double electron capture in the $He^{2+} - He(1s^2)$ collisions, excited states are expected to play a minor role due to the dominance of the resonant $1s^2 - 1s^2$ transition. The CDW-4B method confirms this anticipation, but does not quantitatively reproduce a part of the available experimental data at impact energies 200–3000 keV that are within the domain of the validity of this theory for the $He^{2+} - He(1s^2)$ collisions. Interestingly, for this symmetric scattering, the CB1-4B method substantially outperforms the CDW-4B method at energies 200–1500 keV. Such a surprising situation might seem to have been ameliorated in the past by using a crude approximation to the Green function from the second-order propagator of a perturbation expansion which, however, is not of the Dodd-Greider type. It is well-known that this latter circumstance could lead to certain serious difficulties.

All ordinary distorted wave perturbation expansions (non-Dodd-Greider), similar to the undistorted Born series, contain disconnected or dangerous diagrams that cause the transition operator to diverge for rearranging collisions. A seemingly improved agreement of the mentioned 'augmented' CDW-4B method for the two-electron transfer in the $He^{2+} - He(1s^2)$ collisions should therefore be taken with considerable caution, since the approximate Green function is merely off-shell. Moreover, only two hydrogen-like ground states centered on the projectile and target nucleus were taken into account from the sum over the discrete and continuous parts of the whole spectrum. More systematic work is needed for this particular colliding system, first by treating the on- and off-shell contributions on the same footing, and second by assessing the convergence rate in the spectral representation of the Green function from the second-order term of a chosen perturbation series. Needless to say, it would be important to use the second term in the Dodd-Greider perturbation series to obtain a relatively reasonable estimate of the proper CDW-4B2 method for double charge exchange. Comparing such an estimate to the contribution from the associated CDW-4B1 method would give an invaluable indication about convergence of the Dodd-Greider perturbation series which does not contain any disconnected diagrams.

As to the BCIS-4B and BDW-4B methods, they have been applied to double electron capture in the $He^{2+} - He$ collisions. At moderately high energies (1–3 MeV), good agreement with experiments is found using the BCIS-4B and BDW-4B methods. However, at still higher energies (4 and 6 MeV), the cross sections from the BCIS-4B and BDW-4B methods overestimate the experimental findings (that are the only two measured data points above 3 MeV). Below 1 MeV all the way up to 100 keV, the BCIS-4B and BDW-4B methods underestimate the available experimental data by a factor ranging from 2 to 10. Otherwise, throughout the range 100–7000 keV, the BCIS-4B and BDW-4B methods agree closely with each other. They both exhibit a broad Massey maximum near 175 keV. Such a behaviour is opposed to the CDW-4B method which gives the cross sections that continue to rise with decreasing impact energy, as usual, without any sign of the resonance peak. This latter pattern occurs because of an enhanced contribution from the discrete-continuum coupling, which is mediated by the typical $\boldsymbol{\nabla}\cdot\boldsymbol{\nabla}$ potential operator (for each of the two electrons) in the perturbation interaction from the transition amplitude of the CDW-4B method. Here, we have an example of discrete-continuum interference, because one gradient in the perturbation acts on e.g. the initial bound state centered on the target nucleus, whereas the other gradient is applied to the electronic full Coulomb wave centered on the projectile. The gradient-gradient perturbation describes the same electron being simultaneously bound to the target nucleus, and unbound in the field of the projectile nucleus. Therefore, this coupling of discrete and continuum states via $\boldsymbol{\nabla}\cdot\boldsymbol{\nabla}$ is a typical two-center effect. Once the underlining scalar product is carried out, at least two complex-valued terms are obtained in the transition amplitude T_{if}^{-}. The ensuing terms can have constructive or destructive interference in $|T_{if}^{-}|^2$, depending on the value of the impact energy. Specifically, when the impact energy decreases, constructive interference prevails between the two mentioned parts in $|T_{if}^{-}|^2$ and this causes the cross sections to increase in the CDW-4B method. Also, the Coulomb normalization constant for the full electronic continuum intermediate states increases with decreasing energy. These features are common to both the CDW-3B and CDW-4B methods. However, in the CDW-4B method, this constructive interference is further enhanced, since there are two gradient-gradient operators for each of the two actively participating electrons.

The CDW-EIS-4B method, as another hybrid method, was also applied to double capture from helium by alpha particles at impact energies ranging from 0.1 to 6 MeV. This method combines the CDW-4B method for the exit channel with the symmetric eikonal (SE-4B) method in the entrance channel. Unexpectedly, at energies 0.1–3 MeV, the CDW-EIS-4B method fails much more severely than the CDW-4B method for the same collision. Specifically, at energies 0.1–3 MeV the cross sections from the CDW-EIS-4B method underestimate all the experimental data by a factor ranging from 10 to 1000. Only at the two highest energies (4 and 6 MeV), the curve from the CDW-EIS-4B method passes through the estimated error bar limits of the measured cross sections. This breakdown of the CDW-EIS-4B method at energies 0.1–3 MeV is very surprising, especially given the success of the corresponding CDW-EIS-3B method for single electron capture at a wide range of intermediate and high energies. As an attempt to rescue this unsatisfactory situation, the 'augmented' CDW-EIS-4B method has been used in the past by including approximately a second-order term in a non-Dodd-Greider perturbation expansion, in precisely the same manner as done in the discussed 'augmented' CDW-4B method. However, this has not met with success at all and, therefore, further studies are needed to clarify the hidden drawbacks of the CDW-EIS-4B method for double electron capture. Such studies are needed in view of the similar inadequacy of the CDW-EIS-4B method for double electron capture in the Li^{3+} – He collisions. Crucially, no similar inadequacies are present in the BDW-4B and BCIS-4B methods.

Further, we conclude that for double charge exchange, the presented four-body methods are weakly dependent upon the choice of bound state wave functions. By implication, static correlations of two electrons bound to the target do not play a significant role in double electron capture.

The present work is also concerned with analyzing the role of continuum intermediate states of the electrons in the field of nuclei in the entrance and exit channels. The net effect of these latter states is observed to be striking, as illustrated in the case of symmetric resonant double-charge transfer in the He^{2+} – He collisions at high energies. For example, during comparisons of theory with measurements at high energies, it was found that the BCIS-4B method markedly improves (by 2 orders of magnitude) the predictions of the CB1-4B method. This startling effect occurs because the BCIS-4B method describes the motion of two electrons in the field of the projectile by two full Coulomb waves that are in the CB1-4B method approximated by their asymptotes (logarithmic phases) in terms of the variable $\boldsymbol{R}$, which is the distance between the two heavy scattering aggregates. Hence, a comparative analysis of the BCIS-4B and CB1-4B methods reveals the critical importance of the electronic continuum intermediate states for double charge exchange. While the CB1-3B and BCIS-3B methods for single electron transfer collisions give cross sections that are similar at all energies prior to the outset of the Thomas region, the corresponding CB1-4B and BCIS-4B methods depart from each other progressively more severely with increasing impact energies. *This indicates that continuum intermediate states are much more important for four- than for three-body fast collisions.*

Yet another instructive insight could be gained by examining the sensitivity of continuum intermediates states to the explicit form of the distortion of the wave function within e.g. the initial scattering state. The first invaluable hint is provided already by the discussed comparison between the CDW-4B and CDW-EIS-4B methods. Recall that the latter is a further approximation of the former. The additional approximation is in the replacement of the two electronic full Coulomb waves in the entrance channel by their logarithmic phase factors that are valid at asymptotically large distances. Everything else remains the same in the T-matrix elements from the CDW-4B and CDW-EIS-4B methods. An analogous replacement in the CDW-EIS-3B method gives total cross sections that typically bend down towards the experimental data as opposed to departing from them, as is the case with the CDW-3B method. The resulting agreement with experiments created enthusiasm about the CDW-EIS-3B method, although no rationale for the success has ever been reported. It is indeed counter-intuitive that an approximation to a given method works better than the exact version of that method itself (CDW-EIS-3B versus CDW-3B in the case under discussion). This remark cannot be countered by the argument that the CDW-EIS-3B method is derivable with no reference to the CDW-3B method. Namely, irrespective of the derivation, the final expression for the post transition amplitude in the CDW-EIS-3B method is immediately identified as the eikonal approximation of the corresponding exact CDW-3B method in which the full Coulomb wave in the entrance channel is replaced by its logarithmic phase asymptote.

An important question to ask here is: does the improvement of the CDW-

EIS-3B over CDW-3B method occur at the energies within the assessed validity domain of high-energy methods? Both the CDW-3B and CDW-EIS-3B methods are high-energy approximations as the lowest (first) orders in the Dodd-Greider perturbation expansion of the complete transition amplitude. This implies that they should be valid at impact velocities v exceeding (typically by several times) the classical orbital velocity v_0 of the electrons from the target state from which double capture takes place. At these latter impact velocities ($v \gg v_0$) from the domain of the joint validity, excellent agreement exists between the CDW-3B and CDW-EIS-3B methods that, in turn, compare favorably with measurements. However, improvement of the CDW-EIS-3B method over the CDW-3B method occurs outside the expected validity domain of both methods i.e. at those values of v that are close and smaller than v_0. As such, the answer to the raised question is in the negative. In other words, the improvement of the CDW-EIS-3B over CDW-3B method comes at energies at which it is not theoretically expected from the applicability criterion for the lowest-orders of perturbation series expansions. Technically, the reduction of large cross sections from the CDW-3B method in the region below its domain of applicability is achieved by simplifying this theory via eikonalization which yields the CDW-EIS-3B method. The eikonalization succeeds in weakening the intensity of bound-continuum coupling. This is possible by an admixture of destructive interference caused by approximating the full Coulomb wave by its asymptotic phase in the entrance channel. Such an outcome is plausible, since interference in $|T_{if}|^2$ is very sensitive to any change of phases in all the complex-valued constituents of the T-matrix element. Consequently, constructive interference from the CDW-3B method at lower energies can be mitigated or even converted into destructive interference when Coulomb waves are replaced by their asymptotes. And this is what happens in the CDW-EIS-3B method. Added to this destructive interference is the absence of the normalization constant in the Coulomb wave function in the eikonal initial state from the CDW-EIS-3B method. This Coulomb normalization increases with decreasing incident velocity.

In order not to view this switching (from constructive to destructive interference) as fortuitous and limited only to one active electron, it would be necessary that a similar phenomenon also occurs within eikonalization of electronic full Coulomb waves for two or more active electrons. This is indeed the case with the CDW-EIS-4B method for two-electron capture. The only problem is that this time the benefit of the CDW-EIS-3B method from destructive interference is not repeated at all by the CDW-EIS-4B method. Quite the contrary, the replacement of two electronic full Coulomb waves by the corresponding double phase factor in the entrance channel yields a markedly exaggerated destructive interference. Astoundingly, the CDW-EIS-4B method gives cross sections that grossly underestimate experimental data as well as the results from CDW-4B method by 1–3 orders of magnitude at the energies from the expected theoretical domain of validity (0.1–3 MeV). This effectively pushes the lower limit of the applicability domain of the CDW-

EIS-4B method to quite high energies ($E \geq 4$ MeV). On the other hand, the CDW-EIS-4B method is not adequate at energies where the Thomas double scattering becomes important, because the eikonalization of Coulomb waves destroys the proper velocity dependence for spherically non-symmetric states. As such, the domain of applicability of the CDW-EIS-4B method is restricted to a very narrow interval indeed (nearly void of experimental data) which is embedded in the high energy region. In practice, this excludes the CDW-EIS-4B method from the list of useful approximations for double electron capture.

Hence, there are no systematics in improving the CDW by CDW-EIS model when passing from single- to double-electron capture. Rather, quite the contrary happens. This indicates that the CDW-EIS method works for single capture by serendipity, but fails as soon as the physics and testings become more stringent, which is the case for double capture. This was bound to happen, since the CDW-EIS method is merely an eikonal approximation to the CDW method, as stated. Therefore, rather than resorting to comparisons with experiments, the genuine quality of this eikonalization must first and foremost be judged by the departure of the CDW-EIS from the CDW method within their joint domain of validity as high-energy theories. Experiments on single capture happened to favor the CDW-EIS over the CDW method at energies where neither method was expected to be adequate. However, if the course of the events had been otherwise in the past, with testings of the CDW-EIS method against the experiment being performed first on double capture, this eikonal model would be considered as utterly inadequate. At present, regarding capture processes alone, all we can say is that the CDW-EIS method is limited exclusively to single capture. Multiple capture is expected to be even more devastating for the CDW-EIS method than double capture, since probably with every additional electron becoming active, another higher order in the corresponding perturbation expansion would be necessary for a barely qualitative description, but on the expense of rendering the computations prohibitively impractical. This is an extrapolation of the current experience with a second-order CDW-EIS method which can hardly follow the shape of the line drawn through experimental data on double capture let alone quantitatively reproduce the measured cross sections. Whether these severe drawbacks and limitations also extend to double and multiple ionization remains to be seen. Thus far, the CDW-4B and CDW-EIS-4B methods have not been applied to double ionization in collisions of nuclei with helium-like atomic systems. Both the CDW-3B and CDW-EIS-3B methods are excellent for single ionization of hydrogen-like and multi-electron atomic targets (the latter within the frozen-core model) by fast nuclei. For highly charged projectiles, it has been demonstrated in the literature that the CDW-3B method outperforms the CDW-EIS-3B approximation.

Also illustrative is to juxtapose the BDW-4B and CDW-EIS-4B methods. This is interesting because the BDW-4B method differs from the CDW-EIS-4B method only in the inter-particle variables from the Coulomb logarithmic phase factors in the entrance channel. This variable is the inter-aggregate separation $\boldsymbol{R}$ in the BDW-4B method. In the CDW-EIS-4B method, a pair of two different variables ($\boldsymbol{s}_1$ and $\boldsymbol{s}_2$) appears as the distances of the two electrons from the projectile nucleus. This leads to two Coulomb logarithmic phase factors for the motion of two electrons in the field of the projectile nucleus. As stated, two such phases are the asymptotic forms of the corresponding full Coulomb wave functions for the electrons in the projectile-nucleus field. In the asymptotic region with large distances among all the particles, the product of the two electronic Coulomb phases in the CDW-EIS-4B method coincides with the corresponding logarithmic factor from the BDW-4B method. Such a high degree of similarity between these two methods also exists in the corresponding versions for one-electron transitions, such that the CDW-EIS-3B and BDW-3B methods give cross sections that are quite close to each other. However, this is not the case any longer for double capture, since the CDW-EIS-4B method underestimates the BDW-4B method by 1–2 orders of magnitude at all energies (0.1–6 MeV).

One wonders why the passage from the CDW-EIS-3B to CDW-EIS-4B method is so troublesome, as opposed to the extension of the BDW-3B to BDW-4B method? The answer is that the BDW-4B and CDW-EIS-4B methods have very different phase interference patterns away from large inter-particle separations. In the CDW-EIS-4B method, the Coulomb logarithmic phases for the eikonal initial state are always of a purely electronic origin (at all distances, finite and infinite), as is the discrete-continuum coupling via $\boldsymbol{\nabla} \cdot \boldsymbol{\nabla}$ in the perturbation potential, which causes the transition in the T-matrix element. For each of the two electrons, this latter two-center coupling is such that one of the gradient operators in the scalar product $\boldsymbol{\nabla} \cdot \boldsymbol{\nabla}$ acts on the final bound state in the field of the projectile nucleus, whereas the other applies to the electronic full Coulomb wave function centered on the target nucleus. As discussed, at finite distances the two electronic phase factors in the entrance channel introduce a strong destructive interference into the discrete-continuum coupling, thus yielding an enormous reduction of the ensuing cross sections in the CDW-EIS-4B versus CDW-4B method. This is one of the origins of the worsened agreement of the CDW-EIS-4B method with experiments. The other origin is the absence of the normalization constants of the two Coulomb wave functions, as a consequence of their eikonalization. In contradistinction, the Coulomb phases in the BDW-4B method are in terms of the inter-aggregate distance R and, as such, do not interfere significantly (at finite separations) with the electronic discrete-continuum coupling mediated by the same $\boldsymbol{\nabla} \cdot \boldsymbol{\nabla}$ potential operator which is common to both the BDW-4B and CDW-EIS-4B methods. More precisely, unlike the electronic logarithmic factors from the CDW-EIS-4B method, the scattering integrals involving the $\boldsymbol{R}$–dependent phases from T_{if} in the BDW-4B method are reduced to a folding- or convolution-type integral. As a result, even if destructive interference effects from the $\boldsymbol{R}$–dependent eikonal phases are present, they are effectively damped by this additional integration (folding). Recall that, by definition, every integral acts as a smoothening operator, which *de facto* averages over sharp phase-sensitive oscillations/undulations of the integrand. Additionally, the discrepancy between the total cross sections from the CDW-EIS-4B and BDW-4B methods increases with decreasing impact energy, attaining the largest values in the region of the Massey resonance peak. This is another independent confirmation of the critical relevance of phase interference effects that are radically different in these two methods. It should be recalled that the prominent role in any resonance phenomena is played by phases of functions whose interference can greatly influence the heights of resonant peaks. To recapitulate, as far as one is dealing with the continuum intermediate states, the common conclusion which emerges from the comparative analysis of the CDW-4B, BDW-4B and CDW-EIS-4B method is that the replacement of the full purely electronic Coulomb waves by their eikonal phases in the CDW-EIS-4B method is entirely unjustified.

Simultaneous electron transfer and ionization in the $He^{2+} - He$ and $Li^{3+} - He$ collisions are also extensively studied by means of the CDW-4B method. The theoretical results for the total cross sections for these processes show good agreement with the available experimental data at intermediate and high impact energies. A number of recent measurements of differential and total cross sections for transfer ionization in fast $H^+ - He$ collisions require additional theoretical considerations to achieve full agreement between theory and experiment.

Single electron capture has been the subject of intensive investigations since the early days of quantum mechanics, and interest in this fundamentally important process has remained steady. In this book, single charge exchange in collisions between completely stripped projectiles and helium-like atoms or ions is reviewed. Using the CDW-4B, CDW-BFS and CDW-BIS methods, we analyze cross sections for single electron capture in different processes involving a number of collisional particles such as $H^+ - He$, $He^{2+} - He$, $H^+ - Li^+$ and $Li^{3+} - He$. The CDW-4B method provides evidence that the dynamic correlations play a very important role for single electron capture, especially at higher impact energies for total and differential cross sections. This becomes clear from studying e.g. the post-prior discrepancy in total cross sections using the CDW-4B method for one electron transfer in the $H^+ - He$ collisions. If the dynamic correlations are ignored, the post-prior discrepancy is large and increases with impact energy. By contrast, when the dynamic correlations are included, the post-prior discrepancy is totally negligible at all energies. This is an excellent property of the CDW-4B method despite using very simple helium wave functions with 1–4 variational parameters.

For differential cross sections, dynamic inter-electron correlations lead to the Thomas peak of the 2nd kind, which is mediated by the $Z_P - e_1 - e_2$ double scattering. In the CDW-4B method, the Thomas peak of the 2nd kind appears at all impact energies without any splitting at the critical Thomas angle. Remarkably, at high energies, the strength of the Thomas $Z_P - e_1 - e_2$ peak remains very significant and comparable to that of the Thomas double scattering of the 1st kind ($Z_P - e_1 - Z_T$). The Thomas $Z_P - e_1 - Z_T$ peak, which is appreciable only at sufficiently high energies, is always split into two sub-peaks at the Thomas critical angle, but the ensuing dip is unphysical, as it has never been observed experimentally. To test these findings of the CDW-4B method for single electron capture involving helium-like targets, there is a need for experimental data that could provide two clearly separated contributions from the Thomas double scatterings of the 1st and 2nd kind.

We find that dynamic inter-electron correlations also remain important for one electron capture into excited states, as demonstrated in the Li^{3+} – He collisions. A more stringent test of theories is provided by comparisons with experimentally measured angular distributions of scattered projectiles. In the case of the H^+ – He collisions, good agreement of measurements with the CDW-BIS and CDW-BFS methods is also consistently found for differential cross sections. It should be noted that the CDW-BIS and CDW-BFS methods are the same as the post and prior BDW-4B methods, respectively.

Electron transfer in collisions between two hydrogen-like atoms or ions is presently analyzed by means of the CB1-4B and CDW-4B methods. These two methods can be adapted to investigate single electron capture from multi-electron targets by hydrogen-like projectiles. To this end, the initial state of the target active electron is described by the Roothan-Hartree-Fock wave function, which effectively reduces the original multi-electron problem to an equivalent four-body model problem. We can say that all the presented quantum mechanical boundary-corrected four-body methods are adequate for describing single electron capture in fast collisions of ions with helium-like atomic systems and show systematic agreement with experimental data at intermediate and high impact energies.

We also review a class of resonant collisions with a focus on simultaneous transfer and excitation. In transfer excitation, two modes have been highlighted as resonant transfer excitation and non-resonant transfer excitation. Interference between these two modes can be important, especially for nearly symmetric collisional systems (e.g. $He^+ - H$, $He^+ -$ He and the like). Doubly excited states, that are produced on the projectile after capture of one of the target electrons, can be relaxed either by radiative decay or through the Auger mechanism. These modes and their contributions are coherently included in the CDW-4B method. This is essential in order to preserve the importance of the interference phase. For highly asymmetric collisions ($Z_P \gg Z_T$), such as the $Si^{15+} -$H scattering, radiative decay in transfer excitation is dominated by the resonant mode relative to the corresponding non-resonant contribution. The influence of the target nuclear charge Z_T on the interference between these resonant and non-resonant radiative decays within transfer excitation is assessed. It follows that whenever $Z_P \simeq Z_T$, these latter interference effects can become important. Further, the CDW-4B method for transfer excitation in the He^+ – He collisions shows that the adjacent continuum of a low-lying doubly excited state can play a significant role.

Within the region of its validity, the CDW-4B method gives reliable predictions for transfer excitation via emission of Auger electrons. This is possible, provided that the transition amplitude in the CDW-4B method employs the wave function for the final doubly excited state with the discrete and adjacent continuum components. For this case, the theoretical electron energy spectra near resonances and also the integrated energy profiles are in much better agreement with the corresponding measurements than the associated predictions of the CDW-4B method based upon the discrete components alone. It is important to note that the CDW-4B method for transfer excitation is limited to isolated resonances because of the adapted Feshbach-Fano formalism. In reality, the corresponding experimental data contain isolated and overlapping resonances. Therefore, an improved version of the CDW-4B method is desired in the future by treating isolated and overlapping resonances on the same footing. This is feasible by using the Padé-based resonant scattering theory which has been shown to be remarkably successful for spectroscopy.

We can conclude that at least some of the presently analyzed quantum mechanical boundary-corrected four-body methods are able to provide adequate results for single as well as double electron transitions at intermediate and high energies. Interest in these methods remains steady, and further progress is expected in their extensions to pure five-body scattering problems without resorting to the customary frozen-core approximation, in order to adequately describe the existing coincidence experiments with three active electrons.

In addition to its fundamental importance within few-body quantum mechanics, the reviewed collisional problems also find significant applications in other neighboring research fields such as astrophysics, thermonuclear fusion, plasma physics and medical physics, through particle transport phenomena. This is because cross sections for the presently studied scattering problems are indispensable as entry data for accurate and reliable Monte Carlo simulations of the passage of energetic multiply charged light ions through matter including organic tissue. These energetic ions deposit nearly the total impact energy at the end of their track, via the Bragg peak, and they are neutralized by single or multiple electron capture. Therefore, stopping powers must be determined as precisely as possible to optimize modeling of the passage of these ions through matter. At present, electronic stopping powers used in heavy ion transport physics are exclusively based upon the empirically corrected Bethe-Bloch formula (as a high-energy asymptote of the plane wave Born approximation), which accounts only for excitation and ionization energy losses. The corresponding substantially improved data bases can be generated using the modern methodologies and cross sections from the quantum mechanical methods studied in the present book. This represents an added value of paramount importance for the reviewed methods that go beyond the field of atomic physics, where they were originally established. The most prominent example is hadron therapy by energetic heavy ions such as $H^+, He^{2+}, ..., C^{6+}$. Here, reliable energy deposition by beam particles in tissue is predicted by judicious intertwining of powerful Monte Carlo simulation algorithms with atomic physics data bases for cross sections of processes analyzed in the present work, as well as in several related previous books, monographs and reviews on extensively studied fast ion-atom three-body collisions.

From all the presented detailed illustrations, we can single out dynamic inter-electron correlation effects, as one of the main mechanisms in fast ion-atom collisions involving two or more electrons. Such a highlight is justified by reference to the progressively increasing importance of dynamic correlations with augmentation of impact energies. This automatically enhances the probabilities for double and multiple electron transitions. For example, emission of some 6–8 electrons from argon by heavy multiply charged ions of high impact energies 3–7 MeV/amu can reach a remarkable $\sim 40\%$ of the total ionization yield. Consequently, larger probabilities for multiple ionization at high energies automatically increase the importance of multiple electron continuum intermediate states in multiple capture. Dramatic evidence for this was provided by passing from single- to double-electron capture. Here, we have seen a complete breakdown of methods that either neglect these twofold continuum intermediate states or treat them asymptotically through their eikonal Coulomb phase factors.

Overall, fast heavy particle collisions are topical again, as greatly stimulated by the recent favorable settlement of the International Thermonuclear Reactor for fusion. Likewise, high-energy multiply charged ion beams, as a powerful part of hadron therapy, are increasingly in demand, and this motivates construction of medical accelerators worldwide. In these and other important practical applications, estimates of stopping powers of heavy nuclei traversing matter cannot be reliably made by taking into account single ionization/excitation events alone, as is currently the case in particle transport physics used in fusion research and hadron therapy. Inclusion of channels with two or more actively participating electrons substantially influences the overall energy balance, as well as stability of ion plasma, and considerably alters the energy deposition of ion beams in the traversed matter. Moreover a persistent sequence of charge-changing equilibrium and non-equilibrium phenomena along the incident beam track (predominantly near the Bragg peak) stemming from electron capture and projectile electron loss, must also be accurately incorporated, as these processes that are missing from the Bethe-Bloch formula can substantially change electronic stopping powers. All these critical conclusions demand a thorough and appropriate upgrade of the customary procedures for electronic stopping power data bases in fast heavy ion transport physics. This is deemed necessary because, at present, the atomic physics input into the existing major Monte Carlo algorithms for simulations of the passage of multiply charged ions through matter is based upon single ionization and single excitation alongside abundant empirical expressions with adjustable parameters determined by fitting experimental data. After nearly 80 years of intensive cross-disciplinary applications, the extremely important concept of stopping power is still awaiting to be incorporated into the well-established general body of theory on quantum mechanical heavy ion-atom collisions. This crucial task can confidently be accomplished by relying upon the general distorted wave methodologies expounded in the present book.

List of acronyms in the main text and bibliography

AO	Atomic orbitals
B1-4B	Four-body first Born
BCIS-3B	Three-body continuum intermediate state with correct boundary conditions
BCIS-4B	Four-body continuum intermediate state with correct boundary conditions
BDW-4B	Four-body Born distorted wave
BFS	Born final state
BIS	Born initial state
BK1-3B	Three-body first-order Brinkman-Kramers
BK2-3B	Three-body second-order Brinkman-Kramers
BK1-4B	Four-body first-order Brinkman-Kramers
C2 or 2C	Two Coulomb wave function
C3 or 3C	Three Coulomb wave function
CB	Coulomb-Born
CB1-3B	Three-body first Born with correct boundary conditions
CB1-4B	Four-body first Born with correct boundary conditions
CB2-3B	Three-body second Born with correct boundary conditions
CB2-4B	Four-body second Born with correct boundary conditions
CB3-4B	Four-body third Born with correct boundary conditions
CBn-4B	Four-body nth Born with correct boundary conditions
CDW-3B	Three-body continuum distorted wave
CDW-4B	Four-body continuum distorted wave
CDW-4B1	Four-body first-order continuum distorted wave
CDW-4B2	Four-body second-order continuum distorted wave
CDW-BFS	Continuum distorted wave Born final state
CDW-BIS	Continuum distorted wave Born initial state
CDW-CB1	Continuum distorted wave boundary-corrected first Born
CDW-EFS	Continuum distorted wave eikonal final state
CDW-EFS-3B	Three-body continuum distorted wave eikonal final state
CDW-EFS-4B	Four-body continuum distorted wave eikonal final state

CDW-EIS	Continuum distorted wave eikonal initial state
CDW-EIS-3B	Three-body continuum distorted wave eikonal initial state
CDW-EIS-4B	Four-body continuum distorted wave eikonal initial state
CDW-EIS-4B1	Four-body first-order continuum distorted wave eikonal initial state
CDW-EIS-4B2	Four-body second-order continuum distorted wave eikonal initial state
CDW-EIS-IPM	Continuum distorted wave eikonal initial state independent particle model
CDW-IEM	Continuum distorted wave independent event model
CDW-IPM	Continuum distorted wave independent particle model
CI	Configuration interaction
CIS-3B	Three-body continuum intermediate states
COLTRIMS	Cold-target recoil-ion momentum spectroscopy
CTMC	Classical trajectory Monte Carlo
DCS	Differential cross section
DEC	Dynamic electron correlations
DR	Dielectronic recombination
ECB-4B	Four-body eikonal Coulomb-Born
ECC	Electron capture to continuum (or CTC - capture to continuum)
EFS	Eikonal final state
EIS	Eikonal initial state
FDCS	Fully differential cross section
FWHM	Full width at half maximum
FWL	Faddeev-Watson-Lovelace
GSZ	Green-Sellin-Zachor
HS	Herman-Skillman
IA	Impulse approximation
IEM	Independent-event model
IPM	Independent-particle model
ITE	Indirect transfer excitation
JS1-3B	Three-body first-order Jackson-Schiff
JS2-3B	Three-body second-order Jackson-Schiff
JS1-4B	Four-body first-order Jackson-Schiff
KTI	Kinematic transfer ionization
LE	Loss excitation
LI	Loss ionization
Lyα	Lyman-alpha or Lyman-α
MCB-4B	Four-body modified Coulomb-Born
MO	Molecular orbitals
MS	Multiple scattering
NSA	Near-shell approximation

NTE	Non-resonant transfer excitation
NTEA	Non-resonant transfer excitation via the Auger electron emission
NTEX	Non-resonant transfer excitation via X-ray emission (radiative decays)
PCI	Post-collisional interaction
POHCE	Perturbative one-and-a-half-centered expansion
PWB-4B	Four-body plane wave Born
PWBO	Plane wave Born-Oppenheimer
RIA-3B	Three-body reformulated impulse approximation
RIA-4B	Four-body reformulated impulse approximation
RHF	Roothaan-Hartree-Fock
RTE	Resonant transfer excitation
RTEA	Resonant transfer excitation via the Auger electron emission
RTEX	Resonant transfer excitation via X-ray emission (radiative decays)
SC	Single capture
SEC	Static electron correlations
SE-3B	Three-body symmetric eikonal
SE-4B	Four-body symmetric eikonal
SPB	Strong potential Born
STO	Slater-type orbital
TCDW	Target continuum distorted wave
TE	Transfer excitation
TEA	Transfer excitation via the Auger electron emission
TEX	Transfer excitation via X-ray emission (radiative decays)
TI	Transfer ionization
TTI	Thomas transfer ionization
VPS	Vainstein-Presnyakov-Sobelman

References

[1] Rutherford, E., The scattering of alpha and beta particles by matter and the structure of the atom, *Phil. Mag.*, 21, 669 – 688, 1911.

[2] Rutherford, E., The capture and loss of electrons by alpha particles *Phil. Mag.*, 47, 277 – 303, 1924.

[3] Rutherford, E., Chadwick, J., and Ellis, C., *Radiation from Radioactive Substances*, McMillan, New York, 1930.

[4] Lattes, C.M.G., Muirhead, H., Occhialini, G.P.S., and Powell, C.F., Processes involving charged mesons, *Nature*, 159, 694 – 697, 1947.

[5] Lattes, C.M.G., Occhialini, G.P.S., and Powell, C.F., Observations on the tracks of slow mesons in photographic emulsions. Part 1, *Nature*, 160, 453 – 456, 1947.

[6] Yukawa, H., On the interaction of elementary particles, *Proc. Phys. Math. Soc. (Japan)*, 17, 48 – 57, 1935.

[7] Belkić, Dž., *Principles of Quantum Scattering Theory*, Institute of Physics Publishing, Bristol, 2004.

[8] Delves, L.M., Tertiary and general-order collisions, *Nucl. Phys.*, 9, 391 – 399, 1958/1959.

[9] Delves, L.M., Tertiary and general-order collisions (II), *Nucl. Phys.*, 20, 275 – 308, 1960.

[10] Smith, F.T., Generalized angular momentum in many-body collisions, *Phys. Rev.*, 120, 1058 – 1069, 1960.

[11] Parker, G.A., Walker, R.B., Kendrick, B.K., and Pack, R.T., Accurate quantum calculations on three-body collisions in recombination and collision-induced dissociation. I. Converged probabilities for the $H+Ne_2$ system, *J. Chem. Phys.*, 117, 6083 – 6102, 2002.

[12] Esry, B.D., Greene, C.H., and Burke, J.P. Jr., Recombination of three atoms in the ultracold limit, *Phys. Rev. Lett.*, 83, 1751 – 1754, 1999.

[13] Suno, H., Esry, B.D., Greene, C.H., and Burke, J.P. Jr., Three-body recombination of cold helium atoms, *Phys. Rev. A*, 65, 042725, 2002.

[14] Nielsen, E., and Macek, J.H., Low-energy recombination of identical bosons by three-body collisions, *Phys. Rev. Lett.*, 83, 1566 – 1569, 1999.

[15] Bedaque, P.F., Braaten, E., and Hammer, H.-W., Three-body recombination in Bose gases with large scattering length, *Phys. Rev. Lett.*, 85, 908 – 911, 2000.

[16] Quinteros, T., Gao, H., DeWitt, D.R., Schuch, R., Pajek, M., Asp, S., and Belkić, Dž., Recombination of D^+ and He^+ ions with low-energy free electrons, *Phys. Rev. A*, 51, 1340 – 1346, 1995.

[17] Schuch, R., Belkić, Dž., Justiniano, E., Zong, W., and Gao, H., Formation of D^- by double radiative recombination of D^+, *Hyperf. Inter.*, 108, 195 – 203, 1997.

[18] Melchert, F., Krüdener, S., Huber, K., and Salzborn, E., Electron detachment in $H^+ - H^-$ collisions, *J. Phys. B*, 32, L139 – L144, 1999.

[19] Voigt, W., Über das Gesetz der Intensitätsverteilung innerhalb der Linien eines Gasspektrums, *Münch. Ber.*, 603 – 620, 1912.

[20] Lorentz, H.A., The width of spectral lines, *Proc. Roy. Acad. Amsterdam*, 13, 134 – 150, 1914.

[21] Unsöld, A., *Physik der Sternatmosphären*, Springer, Berlin, 1968.

[22] Kielkopf, J.F., New approximation to the Voigt function with applications to spectral-line profile analysis, *J. Opt. Soc. Am.*, 63, 987 – 995, 1973.

[23] Belkić, Dž., Asymptotic convergence in quantum scattering theory, *J. Comput. Meth. Sci. Eng.*, 1, 353 – 496, 2001.

[24] Gel'fand, I.M., and Vilenkin, N. Ya., *Generalized Functions*, Academic Press, New York, 1964.

[25] Vladimirov, V.S., *Equations of Mathematical Physics*, Marcel Dekker Inc., New York, 1979.

[26] Vladimirov, V.S., *Generalized Functions in Mathematical Physics* (in English), Mir Publisher, Moscow, 1979.

[27] McDowell, M.R.C., and Coleman, J.P., *Introduction to the Theory of Ion-Atom Scatterings*, North-Holland, Amsterdam, 1970.

[28] Schuch, R., Storage rings for investigation of ion-atom collisions, in: *Lecture Notes in Physics: High-Energy Ion-Atom Collisions* (Eds. D. Berényi and G. Hock, Springer-Verlag, Berlin), 294, 509 – 524, 1988.

[29] Schuch, R., *Cooler storage rings: new tool in atomic physics*, pp. 169 – 200, in: *Review of Fundamental Processes and Applications of Atoms and Ions*, Ed. C.D. Lin, World Scientific Publishing Company, Singapore, 1993.

[30] Bohr, N., Theory of the decrease of velocity of moving electrified particles on passing through matter, *Phil. Mag.*, 25, 10 – 31, 1913.

[31] Davisson, C., and Germer, L.H., Diffraction of electrons by a crystal of Nickel, *Phys. Rev.*, 30, 705 – 740, 1927.

[32] Franck, J., and Hertz, G., Über Zusammenstöße zwischen Elektronen und Moleklen des Quecksilberdampfes und die Ionisierungsspannung desselben, *Verh. Dtsch. Phys. Ges.*, 16, 457 – 467, 1914.

[33] Mejaddem, Y., Belkić, Dž., Hyödynmaa, S., and Brahme, A., Calculations of electron energy loss straggling, *Nucl. Instr. Meth. Phys. Res. B*, 187, 499 – 524, 2002.

[34] Belkić, Dž., *Quantum-Mechanical Signal Processing and Spectral Analysis*, Institute of Physics Publishing, Bristol, 2005.

[35] Rotenberg, M., Application of Sturmian functions to the Schrödinger three-body problem: elastic $e^+ - \mathrm{H}$ scattering, *Ann. Phys.*, 19, 262 – 278, 1962.

[36] Dirac, P.A.M., *The Principles of Quantum Mechanics*, The international series of monographs in physics, 4th Ed., Clarendon, Oxford, 1947.

[37] Taylor, J.R., *Scattering Theory*, John Wiley & Sons, New York, 1972.

[38] Dollard, J.D., Adiabatic switching in the Schrödinger theory of scattering, *J. Math. Phys.*, 7, 802 – 810, 1966.

[39] Dollard, J.D., Screening in the Schrödinger theory of scattering, *J. Math. Phys.*, 7, 620 – 624, 1968.

[40] Dollard, J.D., Quantum-mechanical scattering theory for short-range and Coulomb interactions, *Rocky Mount. J. Math.*, 1, 5 – 88, 1971.

[41] Dollard, J.D., Asymptotic convergence and the Coulomb interaction, *J. Math. Phys.*, 5, 729 – 738, 1964.

[42] Cheshire, I.M., Continuum distorted wave approximation; resonant charge transfer by fast protons in atomic hydrogen, *Proc. Phys. Soc.*, 84, 89 – 98, 1964.

[43] Gayet, R., Charge exchange scattering amplitude to first order of a three body expansion, *J. Phys. B*, 5, 483 – 491, 1972.

[44] Belkić, Dž., Gayet, R., and Salin, A., Electron capture in high-energy ion-atom collisions, *Phys. Rep.*, 56, 279 – 369, 1979.

[45] Bransden, B.H., and Dewangan, D.P., High energy charge transfer, *Adv. At. Mol. Opt. Phys.*, 25, 343 – 374, 1988.

[46] Bransden, B.H., and McDowell, M.R.C., *Charge Exchange and the Theory of Ion-Atom Collisions*, The international series of monographs in physics, Clarendon, Oxford, 1992.

[47] Crothers, D.S.F., and Dubé, L.J., Continuum distorted wave methods in ion-atom collisions, *Adv. At. Mol. Opt. Phys.*, 30, 287 – 337, 1993.

[48] Dewangan, D.P., and Eichler, J., Charge exchange in energetic ion-atom collisions, *Phys. Rep.*, 247, 59 – 219, 1994.

[49] Belkić, Dž., Leading distorted wave theories and computational methods for fast ion-atom collisions, *J. Comput. Meth. Sci. Eng.*, 1, 1 – 74, 2001.

[50] Bransden, B.H., and Joachain, C.J., *Physics of Atoms and Molecules*, 2nd Ed. Prentice Hall, New York, 2003.

[51] Faddeev, L.D., Scattering theory for a system of three particles, *J. Exp. Theor. Phys. JETP*, 12, 1011 – 1014, 1961. [*Zh. Eksper. Teor. Fiz.*, 39, 1459 – 1467, 1960.]

[52] Faddeev, L.D., *Mathematical Problems of the Quantum Theory of Scattering for a Three-Particle System*, Steklov Mathematical Institute, Leningrad, 1963, No. 69 [H.M. Stationary Office, Harwell, 1964].

[53] Faddeev, L.D., *Mathematical Aspects of the Three-Body Problem in the Quantum Scattering Theory*, Israel Program of Scientific Translations, Jerusalem, 1965.

[54] Faddeev, L.D., and Merkuriev, S.P., *Quantum Scattering Theory for Several Particle Systems*, Kluwer Academic Publishers, Dordrecht, 1993. [H.M. Stationary Office, Harwell, 1964].

[55] Jauch, J.M., On the relation between scattering phase and bound states, *Helv. Phys. Acta*, 30, 143 – 156, 1957.

[56] Jauch, J.M., Theory of the scattering operator, *Helv. Phys. Acta*, 31, 127 – 158, 1958.

[57] Jauch, J.M., Theory of the scattering operator II. Multichannel scattering, *Helv. Phys. Acta*, 31, 661 – 684, 1958.

[58] Hylleraas, E.A., Neue Berechtnung der Energie des Heliums im Grundzustande, sowie tiefsten Terms von Ortho-Helium, *Z. Phys.*, 54, 347 – 366, 1929.

[59] Hylleraas, E.A., Über den Grundterm der Zweielektronenprobleme von H^-, He, Li^+, Be^{++} usw, *Z. Phys.*, 65, 209 – 225, 1930.

[60] Pluvinage, P., Fonction d'onde approchée à un parametre pour l'état fondamental de l'hélium, *Ann. Phys. NY*, 5, 145 – 152, 1950.

[61] Pluvinage, P., Nouvelle famille de solutions approchées pour certaines équations de Schrödinger non séparables. Application à l'état fondamental de l'hélium, *J. Phys. Radium*, 12, 789 – 792, 1951.

[62] Kato, T., Upper and lower bounds of eigenvalues, *Phys. Rev.*, 77, 413, 1950.

[63] Kato, T., On the existence of solutions of the helium wave equations, *Trans. Amer. Math. Soc.*, 70, 212 – 218, 1951.

[64] Bazley, N.W., Lower bounds for eigenvalues with application to the helium atom, *Proc. Natl. Acad. Sci. USA*, 40, 850 – 853, 1959.

[65] Roothaan, C.C.J., New developments in molecular orbital theory, *Rev. Mod. Phys.*, 23, 69 – 89, 1951.

[66] Roothaan, C.C.J., Self-consistent field theory for open shells of electronic systems, *Rev. Mod. Phys.*, 32, 179 – 185, 1960.

[67] Huzinaga, S., Analytical methods in Hartree-Fock self-consistent field theory, *Rev. Phys.*, 122, 131 – 138, 1961.

[68] Clementi, E., and Roetti, C., Roothaan-Hartree-Fock atomic wavefunctions: Basis functions and their coefficients for ground and certain excited states of neutral and ionized atoms, $Z = 54$, *At. Data Nucl. Data Tables*, 14, 177 – 478, 1974.

[69] Löwdin, P.-O., Studies of atomic self-consistent fields. I. Calculation of Slater functions, *Phys. Rev.*, 90, 120 – 125, 1953.

[70] Green, L.C., Mulder, M.M., Lewis, M.N., and Woll, J.W. Jr., A discussion of analytic and Hartree-Fock wave functions for $1s^2$ configurations from H^- to C v, *Phys. Rev.*, 93, 757 – 761, 1954.

[71] Silverman, J., Platas, O., and Matsen, F.A., Simple configuration-interaction wave functions. I. Two-electron ions: A numerical study, *J. Chem. Phys.*, 32, 1402 – 1406, 1960.

[72] Weiss, A.W., Superposition of configurations and atomic oscillator strengths – Carbon I and II, *Phys. Rev.*, 162, 71 – 80, 1969.

[73] Weiss, A.W., Correlations in excited states of atoms, *Adv. At. Mol. Phys.*, 9, 1 – 46, 1973.

[74] Byron, F.W. Jr., and Joachain, C.J., Importance of correlation effects in the ionization of helium by electron impact, *Phys. Rev. Lett.*, 16, 1139 – 1142, 1966.

[75] Joachain, C.J., and Vanderpoorten R., Configuration-interaction wave functions for two-electron systems, *Physica*, 46, 333 – 343, 1970.

[76] Tweed, R.J., Correlated wavefunctions for helium-like atomic systems, *J. Phys. B*, 5, 810 – 819, 1972.

[77] Joachain, C.J., and Terao, M., *Private communication*, 1991.

[78] Rotenberg, M., and Stein, J., Use of asymptotically correct wave function for three-body Rayleigh-Ritz calculations, *Phys. Rev.*, 182, 1 – 7, 1969.

[79] Bethe, H., and Salpeter, E., *Quantum Mechanics of One- and Two-Electron Atoms*, Plenum, New York, 1977.

[80] Hill, R.N., Proof that the H^- ion has only one bound state, *Phys. Rev. Lett.*, 38, 643 – 646, 1977.

[81] Bachau, H., Position and widths of auto-ionizing states in the helium isoelectronic sequence above the $N = 2$ continuum, *J. Phys. B*, 17, 1771 – 1784, 1984.

[82] Drake, G.W.F., High precision variational calculations for the $1s^2\,{}^1S$ state of H^- and the $1s^2\,{}^1S$, $1s2s\,{}^1S$ and $1s2s\,{}^3S$ states of helium, *Nucl. Inst. Meth. Phys. Res. B*, 31, 7 – 13, 1988.

[83] Tanner, G., Richter, K., and Rost, J.-M., The theory of two-electron atoms: between ground state and complete fragmentation, *Rev. Mod. Phys.*, 72, 497 – 544, 2000.

[84] Belkić, Dž., and Mančev, I., Formation of H^- by double charge exchange in fast proton-helium collisions, *Phys. Scr.*, 45, 35 – 42, 1992.

[85] Belkić, Dž., and Mančev, I., Four-body CDW approximation: dependence of prior and post total cross sections for double charge exchange upon bound-state wave-functions, *Phys. Scr.*, 46, 18 – 23, 1993.

[86] Belkić, Dž., Importance of intermediate ionization continua for double charge exchange at high energies, *Phys. Rev. A*, 47, 3824 – 3844, 1993.

[87] Belkić, Dž., Symmetric double charge exchange in fast collisions of bare nuclei with helium-like atomic systems, *Phys. Rev. A*, 47, 189 – 200, 1993.

[88] Belkić, Dž., Two-electron capture from helium-like atomic systems by completely stripped projectiles, *J. Phys. B*, 26, 497 – 508, 1993.

[89] Belkić, Dž., Double charge exchange at high impact energies, *Nucl. Inst. Meth. Phys. Res. B*, 86, 62 – 81, 1994.

[90] Belkić, Dž., Mančev, I., and Mudrinić, M., Two-electron capture from helium by fast alpha particles, *Phys. Rev. A*, 49, 3646 – 3658, 1994.

[91] Belkić, Dž., Double detachment in collisions between protons and negative hydrogen ions, *Nucl. Instr. Meth. Phys. Res. B*, 154, 365 – 376, 1999.

[92] Gayet, R., Hanssen, J., Martínez, A., and Rivarola, R., Double electron capture in ion-atom collisions at high impact velocities, *Nucl. Instr. Meth. Phys. Res. B*, 86, 158 – 160, 1994.

[93] Gayet, R., Hanssen, J., Jacqui, L., Martínez, A., and Rivarola, R., Double electron capture by fast bare ions in helium atoms: production of singly and doubly excited states, *Phys. Scr.*, 53, 549 – 556, 1996.

[94] Martínez, A.E., Rivarola, R.D., Gayet, R., and Hanssen, J., Double electron capture theories: second order contributions, *Phys. Scr.*, T80, 124 – 127, 1999.

[95] Gravielle, M.S., and Miraglia, J.E., Double-electron capture as a two-step process, *Phys. Rev. A*, 45, 2965 – 2973, 1992.

[96] Gayet, R., Hanssen, J., Martínez, A., and Rivarola, R., Status of two-electron processes in ion-atom collisions at intermediate and high impact energies, *Comments At. Mol. Phys.*, 30, 231 – 248, 1994.

[97] Belkić, Dž., Mančev, I., and Hanssen, J., Four-body methods for high-energy ion-atom collisions, *Rev. Mod. Phys.* 80, 249 – 314, 2008.

[98] Hansteen, J.M., and Mosebekk, O.P., Simultaneous Coulomb ejection of K- and L-shell electrons by heavy charged projectiles, *Phys. Rev. Lett.*, 29, 1361 – 1362, 1972.

[99] McGuire, J.H., and Weaver, L., Independent electron approximation for atomic scattering by heavy particles, *Phys. Rev. A*, 16, 41 – 47, 1977.

[100] Stolterfoht, N., Time ordering of two-step processes in energetic ion-atom collisions: basic formalism, *Phys. Rev. A*, 48, 2980 – 2985, 1993.

[101] Sidorovich, V.A., Nikolaev, V.S., and McGuire, J.H., Calculation of charge-changing cross sections in collisions of H^+, He^{2+} and Li^{3+} with He atoms, *Phys. Rev. A*, 31, 2193 – 2201, 1985.

[102] McCartney, M., The double ionization of helium by ion impact, *J. Phys. B*, 30, L155 – L160, 1997.

[103] Gayet, R., Rivarola, R.D., and Salin, A., Double electron capture by fast nuclei, *J. Phys. B*, 14, 2421 – 2427, 1981.

[104] Martínez, A.E., Gayet, R., Hanssen, J., and Rivarola, R.D., Thomas two-step mechanisms for double electron transfer, *J. Phys. B*, 27, L375 – L382, 1994.

[105] Martínez A.E., Busnengo, H.F., Gayet, R., Hanssen, J., and Rivarola, R.D., Double electron capture in atomic collisions at intermediate and high collision energies: contribution of capture into excited states, *Nucl. Instr. Meth. Phys. Res. B*, 132, 344 – 349, 1997.

[106] Gayet, R., Hanssen, J., Martínez, A., and Rivarola, R., CDW and CDW-EIS investigations in an independent electron approximation for resonant double electron capture by swift He^{2+} in helium, *Z. Phys. D*, 18, 345 – 350, 1991.

[107] Deco, G., and Grün, N., An approximate description of the double capture process in $He^{2+} + He$ collisions with static correlation, *Z. Phys. D*, 18, 339 – 343, 1991.

[108] Ghosh, M., Mandal, C.R., and Mukharjee, S.C., Single and double electron capture from lithium by fast a particles, *J. Phys. B*, 18, 3797 – 3803, 1985.

[109] Ghosh, M., Mandal, C.R., and Mukharjee, S.C., Double-electron capture from helium by ions of helium, lithium, carbon, and oxygen, *Phys. Rev. A*, 35, 5259 – 5261, 1987.

[110] Theisen, T.C., and McGuire, J.H., Single and double electron capture in the independent-electron approximation at high velocities, *Phys. Rev. A*, 20, 1406 – 1408, 1979.

[111] Biswas, S., Bhadra, K., and Basu, D., Double-electron capture by protons from helium, *Phys. Rev. A*, 15, 1900 – 1905, 1977.

[112] Shingal, R., and Lin, C.D., Calculations of two-electron transition cross sections between fully stripped ions and helium atoms, *J. Phys. B*, 24, 251 – 264, 1991.

[113] Gayet, R., and Salin, A., Simultaneous capture and ionization by fast ion impact on helium, *J. Phys. B*, 20, L571 – L576, 1987.

[114] Salin, A., Helium ionization by high-energy ions as a function of impact parameter and projectile scattering angle, *J. Phys. B*, 22, 3901 – 3914, 1989.

[115] Gayet, R., Multiple capture and ionization in high energy ion-atom collisions, *Journal de Physique*, Colloque C1, Supplément au n° 1, 50, 53 – 71, 1989.

[116] Olson, R.E., Electron capture and ionization in H^+, He^{2+}+Li collisions, *J. Phys. B*, 15, L163 – L167, 1982.

[117] Olson, R.E., Wetmore, A.E., and McKenzie, M.L., Double electron transitions in collisions between multiply charged ions and helium atoms, *J. Phys. B*, 19, L629 – L634, 1986.

[118] Mukherjee, S.C., Roy, K., and Sil, N.C., Electron capture and excitation in He^{++} – He collisions, *J. Phys. B*, 6, 467 – 476, 1973.

[119] Zerarka, A., Non-resonant transfer excitation mode in S^{15+} + H collisions, *Phys. Rev. A*, 55, 1976 – 1979, 1997.

[120] Posthumus, J.H., Lukey, P., Morgenstern, R., Double electron capture into highly charged ions: correlated or independent?, *Z. Phys. D. (Supplement)*, 21, S285 – S286, 1991.

[121] Ford, A.L., Wehrman, L.A., Hall, K.A., and Reading, J.F., Single and double electron removal from helium by protons, *J. Phys. B*, 30, 2889 – 2897, 1997.

[122] Jain, A., Shingal, R., and Zouros, T.J.M., State-selective non-resonant transfer excitation in 50-400 keV $^3He^+ + H_2$ and He collisions, *Phys. Rev. A*, 43, 1621 – 1624, 1991.

[123] McGuire, J.H., Correlation in atomic scattering, *Phys. Rev. A*, 36, 1114 – 1123, 1987.

[124] Stolterfoht, N., Dynamics of electron correlation processes in atoms and atomic collisions, *Phys. Scr.*, 42, 192 – 204, 1990.

[125] McGuire, J.H., Multiple-electron excitation, ionization, and transfer in high-velocity atomic and molecular collisions, *Adv. At. Mol. Opt. Phys.*, 29, 217 – 323, 1992.

[126] McGuire, J.H., *Electron Correlation Dynamics in Atomic Collisions*, Cambridge University Press, Cambridge, 1997.

[127] Belkić, Dž., Mančev, I., and Mergel, V., Four-body model for transfer ionization in fast ion-atom collisions, *Phys. Rev. A*, 55, 378 – 395, 1997.

[128] Belkić, Dž., Mančev I., and Mergel, V., Transfer ionization in energetic $\alpha-$He collisions, *Hyperf. Inter.*, 108, 141 – 146, 1997.

[129] Mančev, I., Transfer ionization in fast Li^{3+} – He collisions, *Nucl. Instr. Meth. Phys. Res. B*, 154, 291 – 294, 1999.

[130] Mančev, I., Single-electron capture and transfer ionization in collisions of Li^{3+} ions with helium, *Phys. Rev. A*, 64, 012708, 2001.

[131] Gayet, R., Janev, R., and Salin, A., Electron detachment from negative ions by charged particles: I. Proton impact, *J. Phys. B*, 6, 993 – 1002, 1973.

[132] Belkić, Dž., Single electron detachment from H^- by proton impact, *Nucl. Instr. Meth. Phys. Res. B*, 124, 365 – 376, 1997.

[133] Belkić, Dž., Electron detachment from the negative hydrogen ion by proton impact, *J. Phys. B*, 30, 1731 – 1745, 1997.

[134] Belkić, Dž., A quantum theory of ionization in fast collisions between ions and atomic systems, *J. Phys. B*, 11, 3529 – 3552, 1978.

[135] Garibotti, C.R., and Miraglia, J.E., Ionization and electron capture to continuum in the H^+-hydrogen-atom collisions, *Phys. Rev.*, 21, 572 – 580, 1980.

[136] Garibotti, C.R., and Miraglia, J.E., Asymmetry of the CTC peak on the forward cross section for ionization of a H atom by a fast stripped ion, *J. Phys. B*, 14, 863 – 868, 1981.

[137] Garibotti, C.R., and Miraglia, J.E., Single differential and total scattering cross sections for electrons ejected in collisions of fast bare ions with atomic hydrogen, *Phys. Rev. A*, 25, 1440 – 1444, 1982.

[138] Brauner, M., Briggs, J.S., and Klar, H., Triply-differential cross sections for ionization of hydrogen atoms by electrons and positrons, *J. Phys. B*, 22, 2265 – 2287, 1989.

[139] Brauner, M., Briggs, J.S., Klar, H., Broad, J.T., Rösel, T., Jung, K., and Erhradt, H., Triply differential cross sections for ionization of hydrogen atoms by electrons: the intermediate and threshold energy regions, *J. Phys. B*, 24, 657 – 673, 1991.

[140] Berakdar, J., Briggs, J.S., and Klar, H., Proton and anti-proton impact ionization of atomic hydrogen and helium, *Z. Phys. D*, 24, 351 – 364, 1992.

[141] Maulbetsch, F., and Briggs, J.S., Asymmetry parameter for double photoionization, *Phys. Rev. Lett.*, 68, 2004 – 2006, 1992.

[142] Maulbetsch, F., and Briggs, J.S., Angular distribution of electrons following double photoionization, *J. Phys. B*, 26, 1679 – 1696, 1993.

[143] Berakdar, J., Three-body Coulomb continuum problem, *Phys. Rev. Lett.*, 72, 3799 – 3802, 1994.

[144] Berakdar, J., Parabolic-hyperspherical approach to the fragmentation of three-particle Coulomb systems, *Phys. Rev. A*, 54, 1480 – 1486, 1996.

[145] Belkić, Dž., Charge dependence of ionization cross sections, *J. Phys. B*, 13, L589 – L593, 1980.

[146] Dubé, L.J., and Dewangan, D.P., Reinstating an ionization theory beyond reasonable doubt, *19th International Conference on the Physics of Electronic and Atomic Collisions*, Book of Abstracts, p. 62, Whistler, Canada, 1995.

[147] O'Rourke, S.F.C., and Crothers, D.S.F., Single ionization of He by 3.6 MeV amu^{-1} Ni^{24+} ions, *J. Phys. B*, 30, 2443 – 2454, 1997.

[148] Gulyás, L., and Fainstein, P.D., CDW theory of ionization by ion impact with a Hartree-Fock-Slater description of the target, *J. Phys. B*, 31, 3297 – 3305, 1998.

[149] Ciappina, M.F., Cravero, W.R., and Garibotti, C.R., Post-prior discrepancies in the continuum distorted wave-eikonal initial state approximation for ion-helium ionization, *J. Phys. B*, 36, 3775 – 3786, 2003.

[150] Ciappina, M.F., and Cravero, W.R., CDW and CDW-EIS calculations for FDCSs in highly charged ion impact ionization of helium, *Brazilian J. Phys. B*, 36, 524 – 528, 2006.

[151] Crothers, D.S.F., and McCann, J.F., Ionization of atoms by ion impact, *J. Phys. B*, 16, 3229 – 3242, 1983.

[152] Fainstein, P.D., Ponce, V.H., and Rivarola, R.D., Two-center effects in ionization by ion impact, *J. Phys. B*, 24, 3091 – 3119, 1991.

[153] Stolterfoht, N., DuBois, R., and Rivarola, R.D., *Electron Emission in Heavy Ion-Atom Collisions*, Springer, Berlin, 1997.

[154] Fainstein, P.D., Ponce, V.H., and Rivarola, R.D., Ionization of the first excited state of hydrogen by bare ions at intermediate and high velocities, *J. Phys. B*, 23, 1481 – 1489, 1990.

[155] Olivera, G.H., Fainstein, P.D., Ponce, V.H., and Rivarola, R.D., Ionization of excited states of hydrogen atoms by protons, antiprotons, electrons and multiply charged bare ions, *17th International Conference on the Physics of Electronic and Atomic Collisions*, Abstract of Contributed Papers, Eds., Andersen, T., Fastrup, B., Folkmann, F., and Knudsen, H., p. 449, Aarhus University, Aarhus, 1993.

[156] Igarashi, A., and Shirai, T., Ionization of excited states of atoms by collisions with bare ions, *Phys. Rev. A*, 50, 4945 – 4950, 1994.

[157] Olivera, G.H., Rivarola, R.D., and Fainstein, P.D., Ionization of the excited states of hydrogen by proton impact, *Phys. Rev. A*, 51, 847 – 849, 1995.

[158] McCartney, M., and Crothers, D.S.F., Comment on ionization of the first excited state of hydrogen by ion impact using the continuum distorted wave eikonal initial state approximation, *Z. Phys. D*, 35, 1 – 2, 1995.

[159] Crothers, D.S.F., and McCartney, M., The role of correlation in single ionization of helium, *J. Phys. B*, 30, 3211 – 3226, 1997.

[160] Crothers, D.S.F., and McCartney, M., Low energy ionization using continuum-distorted-wave eikonal-initial-state model, *J. Phys. B*, 25, L281 – L285, 1992.

[161] Tribedi, L.C., Richard, P., Ling, B., Wang, Y.D., Lin, C.D., Moshammer, R., Kerby, G.W. III., Gealy, M.W., and Rudd, M.E., Double differential cross sections of low-energy electrons emitted in ionization of molecular hydrogen by bare carbon ions, *Phys. Rev. A*, 54, 2154 – 2160, 1996.

[162] Tribedi, L.C., Richard, P., Ling, D., DePaola, B., Wang, Y.D., Lin, C.D., and Rudd, M.E., Double differential cross sections for soft electron emission in ionization of hydrogen by bare carbon ions, *Phys. Scr.*, T73, 233 – 234, 1997.

[163] O'Rourke, S.F.C., McSherry, D.M., and Crothers, D.S.F., Two-center effects in ionization by ion-impact in heavy-particle collision, *Adv. Chem. Phys.*, 121, 311 – 356, 2002.

[164] Jones, S., and Madison, D.H., Evidence of initial-state two-center effect in $(e, 2e)$ reactions, *Phys. Rev. Lett.*, 81, 2886 – 2889, 2000.

[165] Jones, S., and Madison, D.H., The ionization of atoms by electron impact, *Phys. Rev. A*, 62, 042701, 2000.

[166] Fainstein, P.D., Ponce, V.H., and Martínez, A.E., Distorted-wave calculation of stopping powers for light ions traversing H targets, *Phys. Rev. A*, 47, 3055 – 3061, 1993.

[167] Olivera, G.H., Martínez, A.E., and Rivarola, R.D., Electron-capture contributions to the stopping power of low-energy hydrogen beams passing through helium, *Phys. Rev. A*, 49, 603 – 606, 1994.

[168] Olivera, G.H., Fainstein, P.D., and Rivarola, R.D., Contribution from the inner shell of water to dose profiles under proton and alpha particle irradiation, *Phys. Med. Biol.*, 41, 1633 – 1647, 1996.

[169] Martínez, A.E., Rivarola, R.D., and Fainstein, P.D., Electronic stopping power of hydrogen beams traversing oxygen, *Nucl. Instr. Meth. B*, 11, 7 – 23, 1996.

[170] Belkić, Dž., Continuum distorted wave methodologies for energetic collisions of radio-therapeutic light ions, *J. Math. Chem.*, 51, 500 – 532, 2008.

[171] Briggs, J.S., Impact parameter formulation of the impulse approximation for charge exchange, *J. Phys. B*, 10, 3075 – 3089, 1977.

[172] Briggs, J.S., A symmetric impulse approximation for charge exchange *J. Phys. B*, 13, L717 – L722, 1980.

[173] Miraglia, J.E., and Macek, J., Quantum-mechanical impulse approximation for single ionization of hydrogen-like atoms by multi-charged ions, *Phys. Rev. A*, 43, 5919 – 5928, 1991.

[174] Brauner, M., and Macek, J.H., Ion-impact ionization of He targets, *Phys. Rev. A*, 46, 2519 – 2531, 1994.

[175] McCann, J.F., The distorted-wave impulse approximation for electron capture processes at intermediate collision energies, *J. Phys. B*, 25, 449 – 461, 1992.

[176] McCann, J.F., and Ng, Y.H., The distorted-wave impulse approximation for electron capture into excited states, *Phys. Scr.*, 61, 180 – 186, 2000.

[177] Belkić, Dž., Single and double charge exchange at high energies, *Nucl. Instr. Meth. Phys. Res. B*, 99, 218 – 224, 1995.

[178] Belkić, Dž., Reformulated impulse approximation (RIA) for charge exchange in fast ion-atom collisions, *Phys. Scr.*, 53, 414 – 430, 1996.

[179] Belkić, Dž., Critical validity assessment of theoretical models: charge exchange at intermediate and high energies, *Nucl. Instr. Meth. Phys. Res. B*, 154, 220 – 246, 1999.

[180] Mergel, V., Dörner, R., Achler, M., Khayyat, Kh., Lencinas, S., Euler, J., Jagutzki, O., Nüttgens, S., Unverzagt, M., Spielberger, L., Wu, W., Ali, R., Ullrich, J., Cederquist, H., Salin, A., Wood, C., Olson, R.E., Belkić, Dž., Cocke, C.L., and Schmidt-Böcking, H., Intra-atomic electron-electron scattering in p-He collisions (Thomas Process) investigated by cold target recoil ion momentum spectroscopy, *Phys. Rev. Lett.*, 79, 387 – 390, 1997.

[181] Thomas, L.H., On the capture of electrons by swiftly moving electrified particles, *Proc. Roy. Soc.*, 114, 561 – 576, 1927.

[182] Schmidt, H.T., Fardi, A., Schuch, R., Schwartz, S.H., Zettergren, H., Cederquist, H., Bagge, L., Danared, H., Källberg, A., Jensen, J., Rensfelt, K.-G., Mergel, V., Schmidt, L., Schmidt-Böcking, H., and Cocke, C.L., Double-to-single target ionization ratio for electron capture in fast p-He collisions, *Phys. Rev. Lett.*, 89, 163201, 2002.

[183] Schmidt, H.T., Jensen, J., Reinhed, P., Schuch, R., Støchkel, K., Zettergren, H., Cederquist, H., Bagge, L., Danared, H., Källberg, A., Schmidt-Böcking, H., and Cocke, C.L., Recoil-ion momentum distributions for transfer ionization in fast proton-He collisions, *Phys. Rev. A*, 72, 012713, 2005.

[184] Bachau, H., Gayet, R., Hanssen J., and Zerarka, A., Transfer and excitation in ion-atom collisions at high impact velocities: a unified continuum distorted wave treatment of resonant and non-resonant modes in a four-body approach. II. Application to the collision $S^{15+}(1s)+H(1s)$, *J. Phys. B*, 25, 839 – 852, 1992.

[185] Gayet, R., and Hanssen, J., Transfer and excitation in ion-atom collisions at high impact velocities: a unified continuum distorted wave treatment of resonant and non-resonant modes in a four-body approach. I. Theory, *J. Phys. B*, 25, 825 – 837, 1992.

[186] Gayet, R., Hanssen, J., and Jacqui, L., Transfer and excitation in ion-atom collisions at high impact velocities. III. Application of the CDW-4B theory to an almost symmetrical system: $He^{+}-He$, *J. Phys. B*, 28, 2193 – 2208, 1995. [Corrigendum: *J. Phys. B*, 30, 1619 – 1622, 1997.]

[187] Gayet, R., Hanssen, J., Jacqui, L., and Ourdane, M.A., Transfer and excitation in ion - atom collisions at high impact velocities: IV. Application of the CDW-4B theory to an almost symmetrical system: $He^{+}-H$, *J. Phys. B*, 30, 2209 – 2219, 1997.

[188] Ourdane, M., Bachau, H., Gayet, R., and Hanssen, J., Transfer and excitation in high-energy $He^{+}-He$ collisions: V. Electronic continuum

influence on ejected electron distributions and TE cross sections, *J. Phys. B*, 32, 2041 – 2055, 1999.

[189] Belkić, Dž., Gayet, R., Hanssen, J., Mančev, I., and Nuñez, A., Dynamic electron correlations in single capture from helium by fast protons, *Phys. Rev. A*, 56, 3675 – 3681, 1997.

[190] Mančev, I., Electron correlations in single-electron capture from helium-like atomic systems, *Phys. Rev. A*, 60, 351 – 358, 1999.

[191] Mančev I., Four-body corrected first Born approximation for single charge exchange at high impact energies, *Phys. Scr.*, 51, 762 – 768, 1995.

[192] Mančev, I., Single-electron capture by hydrogen atoms and helium ions from helium atoms, *Phys. Rev. A*, 54, 423 – 431, 1996.

[193] Mančev, I., Single charge exchange in fast collisions of alpha particles with helium, *J. Phys. B*, 36, 93 – 104, 2003.

[194] Mančev, I., Mergel, V., and Schmidt, L., Electron capture from helium atoms by fast protons, *J. Phys. B*, 36, 2733 – 2746, 2003.

[195] Mančev, I., Continuum distorted wave - Born initial state (CDW-BIS) model for single charge exchange, *J. Comput. Meth. Sci. Eng.*, 5, 73 – 89, 2005.

[196] Mančev, I., Single-electron capture from helium-like atomic systems by bare projectiles, *Europhys. Lett.*, 69, 200 – 206, 2005.

[197] Busnengo, H.F., Martínez, A.E., and Rivarola, R.D., Distorted wave models for electron capture in asymmetric collisions, *Phys. Scr.*, 51, 190 – 195, 1995.

[198] Busnengo, H.F., Martínez, A.E., and Rivarola, R.D., Single electron capture from He targets, *J. Phys. B*, 29, 4193 – 4205, 1996.

[199] Belkić, Dž., Gayet, R., Hanssen, J., and Salin, A., The first Born approximation for charge transfer collisions, *J. Phys. B*, 19, 2945 – 2953, 1986.

[200] Belkić, Dž., Saini, S., and Taylor, H.S., Electron capture by protons from the K-shell of H and Ar, *Z. Phys. D*, 3, 59 – 76, 1986.

[201] Belkić, Dž., Saini, S., and Taylor, H.S., A critical test of first-order theories for electron transfer in collisions between multi-charged ions and atomic hydrogen, *Phys. Rev. A*, 36, 1601 – 1617, 1987.

[202] Belkić, Dž., and Taylor, H.S., First-order theory for charge exchange with correct boundary conditions: general results for hydrogen-like and multi-electron targets, *Phys. Rev. A*, 35, 1991 – 2006, 1987.

[203] Belkić, Dž., Electron capture by fast protons from helium, nitrogen and oxygen: the corrected first Born approximation, *Phys. Rev. A*, 37, 55 – 67, 1988.

[204] Dewangan, D.P., and Eichler, J., A first-order Born approximation for charge exchange with Coulomb boundary conditions, *J. Phys. B*, 19, 2939 – 2944, 1986.

[205] Belkić, Dž., State-selective capture cross sections in proton-hydrogen and proton-helium collisions at intermediate and high energies, *Phys. Scr.*, T28, 106 – 111, 1989.

[206] Grozdanov, T.P., and Krstić, P.S., On a first order theory for charge exchange with Coulomb boundary conditions, *Phys. Scr.*, 38, 32 – 36, 1988.

[207] Alston, S., Nuclear-charge screening in electron capture, *Phys. Rev. A*, 41, 1705 – 1708, 1990.

[208] Belkić, Dž., Second Born approximation for charge exchange with correct boundary conditions, *Europhys. Lett.*, 7, 323 – 327, 1988.

[209] Belkić, Dž., Exact second Born approximation with correct boundary conditions for symmetric charge exchange, *Phys. Rev. A*, 43, 4751 – 4770, 1991.

[210] Belkić, Dž., and Taylor, H.S., Non-perturbative treatments of charge exchange at arbitrary energies: an alternative variational principle, *Phys. Rev. A*, 39, 6134 – 6147, 1989.

[211] Decker, F., and Eichler, J., Exact second-order Born calculations for charge exchange with Coulomb boundary conditions, *J. Phys. B*, 22, L95 – L100, 1989.

[212] Decker, F., and Eichler, J., Comparative study of the distorted-wave Born and boundary-corrected Born approximation for charge exchange up to the second order, *J. Phys. B*, 22, 3023 – 3036, 1989.

[213] Toshima, N., and Igarashi, A., Second Born approximation differential cross sections for p+H and p+He charge-exchange collisions, *Phys. Rev. A*, 45, 6313 – 6317, 1992.

[214] Landau, L.D., and Lifshitz, E.M., *Quantum Mechanics* (*Non-relativistic Theory*), 3rd Ed., Pergamon Press, Oxford, 1985 [3rd revised edition reprinted by Butterworth-Heinemann, London, 1998.]

[215] McGuire, J.H., Straton, J.C., Axmann, W.J., Ishihara, T., and Horsdal, E., Recoil distributions in particle transfer, *Phys. Rev. Lett.*, 62, 2933 – 2936, 1989.

[216] Crothers, D.S.F., and Todd, N.R., One-electron capture by fast multiply charged ions in H: q^3 scaling, *J. Phys. B*, 13, 2277 – 2294, 1980.

[217] Macek, J.H., and Shakeshaft, R., Second Born approximation with Coulomb Green's function: electron capture from a hydrogen-like ion by a bare ion, *Phys. Rev. A*, 22, 1441 – 1446, 1980.

[218] Briggs, J.S., Macek, J.H., and Taulbjerg, K., Theory of asymmetric electron capture collisions, *Comments. At. Mol. Phys.*, 12, 1 – 17, 1982.

[219] Macek, J.H. and Alston, S., Theory of electron capture from a hydrogen-like ion by a bare ion, *Phys. Rev. A*, 26, 250 – 270, 1982.

[220] Taulbjerg, K., and Briggs, J.S., Multiple scattering theory of electron capture in intermediate- to high-velocity collisions, *J. Phys. B*, 16, 3811 – 3824, 1983.

[221] Jakubaßa-Amundsen, D.H., On the effect of off-shell wavefunctions on K and L shell charge transfer in fast, asymmetric collisions, *Z. Phys.*, 316, 161 – 167, 1984.

[222] Burdörfer, J., and Taulbjerg, K., Distorted-wave methods for electron capture in ion-atom collisions, *Phys. Rev. A*, 39, 2959 – 2969, 1986.

[223] Macek, J.H., Treatment of divergent terms in the strong potential Born approximation, *J. Phys. B*, 18, L71 – L74, 1985.

[224] McGuire, J.H., Non-orthogonality in the strong potential Born approximation, *J. Phys. B*, 18, L75 – L77, 1985.

[225] Sil, N.C., and McGuire, J.H., A technique for the evaluation of the strong potential Born approximation for electron capture, *J. Math. Phys.*, 26, 845 – 853, 1985.

[226] Hsin, S.H., and Liber, M., Third-Born-approximation effects in electron capture, *Phys. Rev. A*, 35, 4833 – 4835, 1988.

[227] Marxer, H., and Briggs, J.S., The capture of inner-shell electrons in the strong potential Born (SPB) approximation, *Z. Phys. D*, 13, 75 – 76, 1989.

[228] Marxer, H., and Briggs, J.S., Total cross section for K-K electron transfer in fast ion-atom collisions: the impulse and strong potential Born approximations, *J. Phys. B*, 25, 3823 – 3848, 1992.

[229] Salin, A., Comments on 'Strong potential Born expansions for ion-atom collisions', *Comments At. Mol. Phys.*, 26, 1 – 10, 1991.

[230] Salin, A., Some remarks on theory of high energy electron capture in ion-atom collision, *J. Phys. B*, 25, L137 – L143, 1992.

[231] Macek, J.H., Reply to A Salin: Some remarks on theory of high energy electron capture in ion-atom collision, *J. Phys. B*, 24, 5121 – 5131, 1991.

[232] Alston, S., Inner-shell capture using atomic potentials: A distorted strong-potential Born treatment, *Phys. Rev. A*, 49, 310 – 320, 1994.

[233] Mapleton, R., Asymptotic form of the electron capture cross section in the second Born approximation, *Proc. Phys. Soc.*, 91, 868 – 880, 1968.

[234] Dewangan, D.P., and Eichler, J., Boundary conditions and the strong potential Born approximation for electron capture, *J. Phys. B*, 18, L65 – L69, 1985.

[235] Janev, R.K., and Salin, A., Faddeev-Lovelace equations in the eikonal approximation, *Ann. Phys. NY*, 73, 136 – 155, 1972.

[236] Alt, E.O., and Mukhamedzhanov, A.M., Influence of intermediate-state Coulomb scattering on the behaviour of the two-particle scattering amplitude at small momentum transfer, *J. Phys. B*, 27, 63 – 79, 1994.

[237] Alt, E.O., Levin, S.B., and Yakovlev, S.L., Coulomb Fourier transformation: A novel approach to three-body scattering with charged particles, *Phys. Rev. C*, 69, 034002, 2004.

[238] Avakov, G.V., Blokhintsev, L.D., Lago, R., and Poletayeva, M.V., Three-body integral equations and their applications to the reactions of muon transfer, *Hyperf. Inter.*, 101, 257 – 262, 2006.

[239] Alston, S., Limiting behaviours of off-shell scattering wave functions and T matrices for centrally modified Coulomb potentials, *Phys. Rev. A*, 38, 636 – 644, 1988.

[240] Alston, S., Closed-form expressions for $1s \rightarrow 1s$ electron-capture amplitude in a second-order Faddeev approximation, *Phys. Rev. A*, 40, 4907 – 4913, 1989.

[241] Alston, S., Unified Faddeev treatment of high-energy electron capture, *Phys. Rev. A*, 42, 331 – 350, 1990.

[242] Alston, S., Modified Faddeev treatment of electron capture, *Phys. Rev. A*, 54, 2011 – 2021, 1996.

[243] Roberts, M.J., A comparative study of the second-order Born and Faddeev-Watson approximations: II. Charge transfer, *J. Phys. B*, 20, 551 – 564, 1987.

[244] Adivi, E.G., and Bolorizadeh, M.A., Faddeev treatment of single-electron capture by protons in collision with many-electron atoms, *J. Phys. B*, 37, 3321 – 3338, 2004.

[245] Adivi, E.G., and Bolorizadeh, M.A., A symmetric-impulse approximation applied to Faddeev theory on the charge transfer amplitude, *Few-Body Systems*, 39, 11 – 25, 2006.

[246] Adivi, E.G., Brunger, M.J., Bolorizadeh, M.A., and Campbell, L., Closed-form expressions for state-to-state charge transfer differential cross sections in a modified Faddeev three-body approach, *Phys. Rev. A*, 38, 022704, 2007.

[247] Oppenheimer, J.R., On the quantum theory of the capture of electrons, *Phys. Rev.*, 31, 349 – 356, 1928.

[248] Brinkman, H.C., and Kramers, H.A., Zur Theorie der Eifangung von Elektronen durch α–Teilchen, *Proc. Acad. Sci. Amsterdam*, 33, 973 – 984, 1930.

[249] Miraglia, J.E., Piacentini, R.D., Rivarola, R.D., and Salin, A., Discussion of electron capture theories for ion-atom collisions at high energies, *J. Phys. B*, 14, L197 – L202, 1981.

[250] Simony, P.R., and McGuire, J.H., Exact second Born calculations of $1s - 1s$ electron capture in p+H, *J. Phys. B*, 14, L737 – L741, 1981.

[251] Wadehra, J.M., Shakeshaft, R., and Macek, J.H., Evaluation of the second Born amplitude as a two-dimensional integral, for $\mathrm{H}^+ + \mathrm{H}(1s) \longrightarrow \mathrm{H}(1s) + \mathrm{H}^+$, *J. Phys. B*, 14, L767 – L771, 1981.

[252] Simony, P.R., McGuire, J.H., and Eichler, J., Exact second Born electron capture for p+He, *Phys. Rev. A*, 26, 1337 – 1343, 1982.

[253] McGuire, J.H., Eichler, J., and Simony, P.R., Exact second Born calculations for electron capture for systems with various projectile and target charges, *Phys. Rev. A*, 28, 2104 – 2112, 1983.

[254] Bates, D.R., and Dalgarno, A., Electron capture – I: resonance capture from hydrogen atoms by fast protons, *Proc. Phys. Soc. A*, 65, 919 – 925, 1952.

[255] Bates, D.R., and Dalgarno, A., Electron capture – III: capture into excited states in encounters between hydrogen atoms and fast protons, *Proc. Phys. Soc. A*, 66, 972 – 976, 1953.

[256] Jackson, J.D., and Schiff, H., Electron capture by protons passing through hydrogen, *Phys. Rev.*, 89, 359 – 365, 1953.

[257] Schiff, H., Electron capture cross sections, *Canad. J. Phys.*, 32, 393 – 405, 1954.

[258] Jackson, J.D., On the use of the complete interaction Hamiltonian in atomic rearrangement collisions, *Proc. Phys. Soc. A*, 70, 26 – 33, 1957.

[259] Kramer, P.J., Exact calculations of the second-order Born terms for proton-hydrogen electron-transfer collisions, *Phys. Rev. A*, 6, 2125 – 2130, 1972.

[260] Greider, K.R., and Dodd, L.R., Divergence of the distorted-wave Born series for rearrangement scattering, *Phys. Rev.*, 146, 671 – 675, 1966.

[261] Dodd, L.R., and Greider, K.R., Rigorous solution of three-body scattering processes in the distorted-wave formalism, *Phys. Rev.*, 146, 675 – 686, 1966.

[262] Deco, G., Maidagan, J., and Rivarola, R., Do symmetric eikonal and continuum distorted wave models satisfy the correct boundary conditions?, *Phys. Scr.*, 51, 334 – 338, 1995.

[263] Pedlow, R.T., O'Rourke, S.F.C., and Crothers, D.S.F., Fully differential cross sections for 3.6 MeV u^{-1} Au^{Z_P+} +He collisions, *Phys. Rev. A*, 72, 062719, 2005.

[264] Belkić, Dž., and Salin, A., Coulomb Brinkman-Kramers approximation: differential cross section for electron capture, *J. Phys. B*, 9, L397 – L402, 1976.

[265] Belkić, Dž., and Salin, A., Differential cross sections for charge exchange at high energies, *J. Phys. B*, 11, 3905 – 3911, 1978.

[266] Rivarola, R.D., Piacentini, R.D., Salin, A., and Belkić, Dž., The influence of the static potential in high energy K-shell capture collisions, *J. Phys. B*, 13, 2601 – 2609, 1980.

[267] Abramowitz, M., and Stegun, I., *Handbook of Mathematical Functions*, Dover, New York, 1956.

[268] Press, W., Teukolsky, S., Vetterling, W., and Flannery, B., *Numerical Recipes in Fortran 77: The Art of Scientific Computing*, Cambridge University Press, Cambridge, Vol. 1, 1992.

[269] Salin, A., Calculations of high-energy charge transfer with the continuum distorted wave approximation, *J. Phys. B*, 3, 937 – 951, 1970.

[270] Belkić, Dž., and Janev, R., Electron capture from hydrogen and helium atoms by fast alpha particles, *J. Phys. B*, 6, 1020 – 1027, 1973.

[271] Belkić, Dž., and Janev, R.K., Formation of muonium in fast collisions of μ^+ mesons with atomic systems, *J. Phys. B*, 6, 2613 – 2617, 1973.

[272] Belkić, Dž., and Gayet, R., Electron capture from atomic hydrogen by fast protons and alpha particles, *J. Phys. B*, 10, 1911 – 1921, 1977.

[273] Belkić, Dž., and Gayet, R., Charge exchange in fast collisions of H^+ and He^{++} with helium, *J. Phys. B*, 10, 1923 – 1932, 1977.

[274] Dž. Belkić, Dž., and McCarroll, R., Projectile charge dependence of electron capture cross sections, *J. Phys. B*, 10, 1933 – 1943, 1977.

[275] Belkić, Dž., High energy behaviour of transition probability and total cross sections for charge exchange, *J. Phys. B*, 10, 3491 – 3510, 1977. [Corrigendum: Belkić, Dž., *J. Phys. B*, 12, 337 – 337, 1979.]

[276] Banyard, K.E., and Szuster, B., Electron capture cross sections for high-energy alpha particle-helium atom collisions, *J. Phys. B*, 10, L503 – L511, 1977.

[277] Banyard, K.E., and Szuster, B., Continuum distorted wave calculations for rearrangement cross sections and their sensitivity to improvements in the target wave function: proton-helium collisions, *Phys. Rev. A*, 16, 129 – 132, 1977.

[278] Moore, J.C., and Banyard, K.E., Continuum distorted wave calculations for electron capture from negative hydrogen ions by fast protons, *J. Phys. B*, 11, 1613 – 1621, 1978.

[279] Banyard, K.E., and Shirtcliffe, G.W., Electron capture from lithium and its ions by high-energy protons, *J. Phys. B*, 12, 3247 – 3256, 1979.

[280] Shirtcliffe, G.W., and Banyard, K.E., Ordering of cross sections for electron capture from He-like targets by fast projectiles, *Phys. Rev. A*, 21, 1197 – 1201, 1980.

[281] Banyard, K.E., and Shirtcliffe, G.W., Charge exchange between simple structured projectiles in high-energy collisions, *Phys. Rev. A*, 22, 1452 – 1454, 1980.

[282] Belkić, Dž., Gayet, R., and Salin, A., Computation of total cross sections for electron capture in high energy ion-atom collisions, *Comp. Phys. Commun.*, 23, 153 – 167, 1981.

[283] Belkić, Dž., Gayet, R., and Salin, A., Computation of total cross sections for electron capture in high energy ion-atom collisions. II, *Comp. Phys. Commun.*, 30, 193 – 205, 1983.

[284] Belkić, Dž., Gayet, R., and Salin, A., Computation of total cross sections for electron capture in high energy ion-atom collisions. III, *Comp. Phys. Commun.*, 32, 385 – 397, 1984.

[285] Chetioui, A., Wohrer, K., Rozet, J.P., Joly, A., Stephan, C., Belkić, Dž., Gayet, R., and Salin, A., State-to-state charge exchange cross sections in high-velocity asymmetric and near symmetric collisions at 400 MeV Fe^{+26} ions, *J. Phys. B*, 16, 3993 – 4003, 1983.

[286] Andriamonje, S., Chemin, J.F., Routier, J., Saboya, B., Scheurer, J.N., Belkić, Dž., Gayet, R., and Salin, A., Electron capture from the krypton M-shell by MeV protons, *J. de Physique*, 46, 349 – 353, 1985.

[287] Belkić, Dž., Electron transfer from hydrogen-like atoms to partially and completely stripped projectiles: CDW approximation, *Phys. Scr.*, 43, 561 – 571, 1991.

[288] Belkić, Dž., Gayet, R., and Salin, A., Cross sections for electron capture from atomic hydrogen by fully stripped ions, *Atom. Data Nucl. Data Tables*, 51, 59 – 150, 1992.

[289] Kunikeev, S.D., The CDW approximation and Coulomb boundary conditions: bound states *J. Phys. B*, 31, L849 – L853, 1998.

[290] Ferreira da Silva, M.F., Electron capture in $H^+ - H(1s)$ collisions: two-state approximation with a continuum distorted wave basis, *J. Phys. B*, 36, 2357 – 2370, 2003.

[291] Purkait, M., Double electron capture cross-sections of the ground state in the collisions of He^{2+} and Li^{3+} with He, *Eur. Phys. J. D*, 30, 11 – 14, 2004.

[292] Purkait, M., Sounda, S., Dhara, A., and Mandal, C.R., Double-charge-transfer cross sections in inelastic collisions of bare ions with helium atoms, *Phys. Rev. A*, 74, 042723, 2006.

[293] Gulyás, L., and Szabo, Gy., Resonant double electron capture by fast He^{2+} from helium: the first-order Born approximation with correct boundary condition, *Z. Phys. D*, 29, 115 – 119, 1994.

[294] Belkić, Dž., Double electron capture in fast ion-atom collisions, *J. Math. Chem.*, 51, 450 – 480, 2008.

[295] Lin, C.D., Double K-shell electron capture for ion-atom collisions at intermediate energies, *Phys. Rev. A*, 19, 1510 – 1516, 1979.

[296] Gerasimenko, V.I., The two-electron charge exchange of protons in helium during fast collisions, *J. Exp. Theor. Phys. JETP*, 14, 789 – 791, 1962. [*Zh. Eksp. Teor. Fiz.*, 41, 1104 – 1106, 1961.]

[297] Schryber, U., Elektroneneinfang schneller protonen in gasen, *Helv. Phys. Acta*, 40, 1023 – 1051, 1967.

[298] Toburen, L.H., and Nakai, M.Y., Double-electron-capture cross sections for incident protons in the energy range 75 to 250 keV, *Phys. Rev.*, 177, 191 – 196, 1969.

[299] Fogel', Ya. M., Mitin, R.V., Kozlov, V.F., and Romashko, N.D., On applicability of the Massey adiabatic hypothesis to double charge exchange, *J. Exp. Theor. Phys. JETP*, 8, 390 – 398, 1959. [*Zh. Eksp. Teor. Fiz.*, 35, 565 – 573, 1958.]

[300] Williams, J.F., Cross sections for double electron capture by 2-50 keV protons incident upon hydrogen and the inert gases, *Phys. Rev. A*, 150, 7 – 10, 1966.

[301] Schuch, R., Justiniano, E., Vogt, H., Deco, G., and Grün, N., Double electron capture by He^{2+} from He at high velocity, *J. Phys. B*, 24, L133 – L138, 1991.

[302] Pivovar, L.I., Novikov, M.T., and Tubaev, V.M., Electron capture by helium ions in various gases in the 300-1500 keV energy range, *J. Exp. Theor. Phys. JETP*, 15, 1035 – 1039, 1962. [*Zh. Eksp. Teor. Fiz.*, 42, 1490 – 1494, 1962.]

[303] McDaniel, E.W., Flannery, M.R., Ellis, H.W., Eisele, F.L., and Pope, W., *US army missile research and development command technical report*, No. H, 1, 78, 1977.

[304] DuBois, R.D., Ionization and charge transfer in He^{2+} – rare gas collisions. II, *Phys. Rev. A*, 36, 2585 – 2593, 1987.

[305] de Castro Faria, N.V., Freire, F.L. Jr., and de Pinho, A.G., Electron loss and capture by fast helium ions in noble gases, *Phys. Rev. A*, 37, 280 – 283 , 1988.

[306] Berkner, K.H., Pyle, R.V., Stearns, J.W., and Warren, J.C., Single- and double-electron capture by 7.2 to 181 keV $^3He^{++}$ ions in He, *Phys. Rev.* 166, 44 – 46 (1968)

[307] Gerasimenko, V.I., and Rosentsveig, L.N., Two-electron charge exchange of α–particles in helium, *J. Exp. Theor. Phys. JETP*, 4, 509 – 512, 1957. [*Zh. Eksp. Teor. Fiz.*, 31, 684 – 687, 1956.]

[308] Crothers, D.S.F., Refined orthogonal variational treatment of continuum distorted waves, *J. Phys. B*, 15, 2061 – 2074, 1982.

[309] Shah, M.B., and Gilbody, H.B., Single and double ionization of helium by H^+, He^{2+} and Li^{3+} ions, *J. Phys. B*, 18, 899 – 913, 1985.

[310] Crothers, D.S.F., and McCarroll, R., Correlated continuum distorted-wave resonant double electron capture in He^{2+} – He collisions, *J. Phys. B*, 20, 2835 – 2842, 1987.

[311] Nordsieck, A., Reduction of an integral in the theory of bremsstrahlung, *Phys. Rev.*, 93, 785 – 787, 1954.

[312] Belkić, Dž., Bound-free non-relativistic transition form factors in atomic hydrogen, *J. Phys. B*, 14, 1907 – 1914, 1981.

[313] Belkić, Dž., Analytical results for Coulomb integrals with hydrogenic and Slater-type orbitals, *J. Phys. B*, 16, 2773 – 2784, 1983.

[314] Belkić, Dž., Bound-free transition form factors in hydrogen-like and multi-electron atoms, *J. Phys. B*, 17, 3629 – 3636, 1984.

[315] Belkić, Dž., and Lazur, V.Yu., Sum rules for the bound-free transition form factors in hydrogen-like atoms, *Z. Phys. A*, 319, 261 – 267, 1984.

[316] Belkić, Dž., and Taylor, H.S., A unified formula for the Fourier transform of Slater-type orbitals, *Phys. Scr.*, 39, 226 – 229, 1989.

[317] Belkić, Dž., Vector spherical harmonics and scattering integrals, *Phys. Scr.*, 45, 9 – 17, 1992.

[318] Hippler, R., Datz, S., Miller, P., Pepmiler, P., and Dittner, P., Double- and single-electron capture and loss in collisions of 1-2 MeV/u boron,

oxygen, and silicon projectiles with helium atoms, *Phys. Rev. A*, 35, 585 – 590, 1987.

[319] Lepage, G.P., A new algorithm for adaptive multidimensional integration, *J. Comput. Phys.*, 27, 192 – 203, 1978.

[320] Kawabata, S., A new Monte Carlo event generator for high energy physics, *Comput. Phys. Commun.*, 41, 127 – 153, 1986.

[321] Kawabata, S., A new version of the multi-dimensional integration and event generation package BASES/SPRING, *Comput. Phys. Commun.*, 41, 309 – 326, 1995.

[322] Dörner, R., Mergel, V., Spielberger, L., Jagutzki, O., Schmidt-Böcking, H., and Ullrich, J., State-selective differential cross sections for double-electron capture in 0.25–0.75 MeV He^{2+} – He collisions, *Phys. Rev. A*, 57, 312 – 317, 1998.

[323] Dörner, R., Mergel, V., Jagutzki, O., Spielberger, L., Ullrich, J., Moshammer, R., and Schmidt-Böcking, H., Cold Target Recoil Ion Momentum Spectroscopy: a 'momentum microscope' to view atomic collision dynamics, *Phys. Rep.*, 330, 95 – 192, 2000.

[324] Dörner, R., *Private communication*, 2005.

[325] McCann, J.F., and Crothers, D.S.F., Classical and quantal perturbation approximations for continuum capture by fast multiply charged ions, *Nucl. Instr. Meth. Phys. Res. B*, 23, 164 – 166, 1987.

[326] Crothers, D.S.F., and McCann, J.F., Electron capture from helium to the continuum of fast cations, *J. Phys. B*, 20, 19 – 23, 1987.

[327] Gulyás, L., and Szabó, Gy., Electron capture to continuum and simultaneous target excitation or ionization, *Phys. Rev. A*, 43, 5133 – 5136, 1991.

[328] Kunikeev, S.D., ECC cusp and Coulomb boundary conditions, *J. Phys. B*, 31, 2649 – 2658, 1997.

[329] Pregliasco, R.G., Garibotti, C.R., and Barrachina, R., ECC cusp analysis, *Nucl. Instr. Meth. Phys. Res. B*, 86, 168 – 171, 1994.

[330] Barrachina, R.O., and Sarkadi, L., On the divergency problem of the ECC cusp: a final state interaction theory, *Nucl. Instr. Meth. Phys. Res. B*, 233, 260 – 265, 2005.

[331] Sarkadi, L., and Barrachina, R.O., Divergency problem of the electron capture to continuum cusp: classical trajectory Monte Carlo simulation and experimental data, *Phys. Rev. A*, 71, 062712, 2005.

[332] Fainstein, P.D., Ponce, V.H., and Rivarola, R.D., Z_P^3 effects in the ionization of helium by ion impact, *Phys. Rev. A*, 36, 3639 – 3641, 1987.

[333] Fainstein, P.D., Ponce, V.H., and Rivarola, R.D., Ionization of helium by antiproton and proton impact, *J. Phys. B*, 21, 2989 – 2998, 1988.

[334] Pregliasco, R.G., Garibotti, C.R., and Barrachina, R.O., Measurement and analysis of the ECC cusp structure, *J. Phys. B*, 27, 1151 – 1166, 1994.

[335] Shah, M.B., McGrath, C., Illescas, C., Pons, B., Riera, A., Luna, H., Crothers, D.S.F., O'Rourke, S.F.C., and Gilbody, H.B., Shifts in electron capture to the continuum at low collision energies: enhanced role of target post-collisional interaction, *Phys. Rev. A*, 67, 010704, 2003.

[336] Sarkadi, L., Lugosi, L., Tökési, K., Gulyás, L., and Kövér, Á., Study of the transfer ionization process by observing the electron cusp in 100-300 keV He^{2+} + He collisions, *J. Phys. B*, 34, 4901 – 4917, 2001.

[337] Rosenberg, L., Variational methods in charged-particle collision theory, *Phys. Rev. D*, 8, 1833 – 1843, 1973.

[338] Vainstein, L., Presnyakov, L., and Sobelman, I., On a model for computation of cross sections for excitation of atoms, *J. Exp. Theor. Phys. JETP*, 18, 1383 – 1385, 1964. [*Zh. Eksper. Teor. Fiz.*, 45, 2015 – 2021, 1964.]

[339] Presnyakov, L., The ionization of atoms by electron impact, *J. Exp. Theor. Phys. JETP*, 20, 760 – 761, 1965. [*Zh. Eksper. Teor. Fiz.*, 47, 1134 – 1135, 1964.]

[340] Dewangan, D.P., and Bransden, B.H., The Vainstein, Presnyakov and Sobelman approximation in heavy particle collisions, *J. Phys. B*, 15, 4561 – 4576, 1982.

[341] Peterkop, R.K., *Theory of Ionization of Atoms by Electron Impact*, Associated University Press, Boulder, Colorado, 1977.

[342] Rudge, M.R.H., and Seaton, M.J., Ionization of atomic hydrogen by electron impact, *Proc. Roy. Soc. A*, 283, 262 – 290, 1965.

[343] Rudge, M.R.H., Theory of the ionization of atoms by electron impact, *Rev. Mod. Phys.*, 40, 564 – 590, 1968.

[344] Salin, A., Ionization of atomic hydrogen by proton impact, *J. Phys. B*, 2, 631 – 639, 1969.

[345] Macek, J., Theory of forward peak in the angular distribution of ejected electrons by fast protons, *Phys. Rev. A*, 1, 235 – 241, 1970.

[346] Salin, A., Ionization of complex atoms by ion impact, *J. Phys. B*, 2, L1255 – L1256, 1969.

[347] Salin, A., Ionization of helium by proton impact, *J. Phys. B*, 5, 979 – 986, 1972.

[348] Salin, A., Three-body Coulomb continuum and ionization by ion impact, in: *Lecture Notes in Physics: High-Energy Ion-Atom Collisions*, (Eds. D. Berényi and G. Hock, Springer-Verlag, Berlin), 294, 245 – 261, 1988.

[349] Gulyás, L., Fainstein, P.D., and Shirai, T., Multiple-scattering analysis of triply differential cross sections for electron emission in energetic ion-atom collisions, *J. Phys. B*, 34, 1473 – 1483, 2001.

[350] Dunseath, K.M., and Crothers, D.S.F., Transfer and ionization processes during the collision of fast H^+, He^{2+} nuclei with helium, *J. Phys. B*, 24, 5003 – 5022, 1991.

[351] Rodriguez, K.V., and Gasaneo, G., Accurate Hylleraas-like functions for the He atom with correct cusp conditions, *J. Phys. B*, 38, L259 – L267, 2005.

[352] Shah, M.B., McCallion, P., and Gilbody, H.B, Electron capture and ionization in collisions of slow H^+ and He^{2+} ions with helium, *J. Phys. B*, 22, 3037 – 3045, 1989.

[353] Bhattacharyya, S., Rinn, K., Salzborn, E., and Chatterjee, L., High-energy electron capture by fully stripped ions from He atoms - a QED approach, *J. Phys. B*, 21, 111 – 118, 1998.

[354] Woitke, O., Závodsky, P.A., Ferguson, S.M., Houck J.H., and Tanis, J.A., Target ionization and projectile charge changing in 0.5 – 8 MeV/q Li^{q+} + He ($q = 1, 2, 3$) collisions, *Phys. Rev. A*, 57, 2692 – 2700, 1998.

[355] Schmidt, H.T., Fardi, A., Jensen, J., Reinhed, P., Schuch, R., Støchkel, K., Zettergren, H., Cederquist, H., and Cocke, C.L., Transfer ionization in p + He collisions, *Nucl. Instr. Meth. Phys. Res. B*, 233, 43 – 47, 2005.

[356] Tolmanov, S.G., and McGuire, J.H., Electron-electron Thomas peak in fast transfer ionization, *Phys. Rev. A*, 62, 032711, 2000.

[357] Schmidt-Böcking, H., Schöffler, M.S., Janke, T., Czasch, A., Mergel, V., Schmidt, L., Dörner, R., Jagutzki, O., Hattass, M., Weber, T., Weigold, E., Schmidt, H.T., Schuch, R., Cederquist, H., Demkov, Y., Whelan, C., Godunov, A., and Walters, J., Many-particle fragmentation processes in atomic and molecular physics – new insight into the world of correlation, *Nucl. Instr. Meth. Phys. Res. B*, 233, 3 – 11, 2005.

[358] Schöffler, M., Godunov, A.L., Whelan, C.T., Walters, H.R.J., Schipakov, V.S., Mergel, V., Dörner, R., Jagutzki, O., Schmidt, L.H., Titze, J., Weigold, E., and Schmidt-Böcking, H., Revealing the effect of angular correlation in the ground-state He wavefunction: a coincidence study of the transfer ionization process, *J. Phys. B*, 38, L123 – L128, 2005.

[359] Godunov, A.L., Whelan, C.T., and Walters, H.R.J., Fully differential cross sections for transfer ionization – a sensitive probe of high level correlation effects in atoms, *J. Phys. B*, 37, L201 – L208, 2004.

[360] Godunov, A.L., Whelan, C.T., Walers, H.R.J., Schipakov, V.S., Schöffler, M., Mergel, V., Dörner, R., Jagutzki, O., Schmidt, L.H., Titze, J., and Schmidt-Böcking, H., Transfer ionization process p+He $\rightarrow$ H + He^{2+} + e with the ejected electron detected in the plane perpendicular to the incident beam direction, *Phys. Rev. A*, 71, 052712, 2005.

[361] Mergel, V., Dörner, R., Khayyat, Kh., Achler, M., Weber, T., Jagutzki, O., Lüdde, H.J., Cocke, C.L., and Schmidt-Böcking, H., Strong correlations in the He ground state momentum wave function observed in the fully differential momentum distributions for the p + He transfer ionization process, *Phys. Rev. Lett.*, 86, 2257 – 2260, 2001.

[362] Schmidt-Böcking, H., Mergel, V., Dörner, R., Cocke, C.L., Jagutzki, O., Schmidt, L., Weber, T., Lüdde, H.J., Weigold, E., Barakdar, J., Cederquist, H., Schmidt, H.T., Schuch, R., and Kheifets, A.S., Revealing the non-s^2 contributions in the momentum wave function of ground-state He, *Europhys. Lett.*, 62, 477 – 483, 2003.

[363] Schmidt-Böcking, H., Mergel, V., Schmidt, L., Dörner, R., Jagutzki, O., Ullmann, K., Weber, T., Lüdde, H.J., Weigold, E., and Kheifets, A.S., Dynamics of ionization processes studied with the COLTRIMS method: New insight into e-e correlation, *Rad. Phys. Chem.*, 68, 41 – 50, 2003.

[364] Popov, Yu.V., Chuluunbaatar, O., Vinitsky, S.I., Ancarani, L.U., Dal Cappello, C., and Vinitsky, P.S., Theoretical Investigation of the p + He $\rightarrow$ H + He^{+} and p + He $\rightarrow$ H + He^{++} + e reactions at very small scattering angles of hydrogen, *J. Exp. Theor. Phys. JETP*, 95, 620 – 624, 2002. [*Zh. Eksper. Teor. Fiz.*, 122, 717 – 722, 2002.]

[365] Ishihara, T., and McGuire, J.H., Second-order singularities in transfer ionization, *Phys. Rev. A.*, 38, 3310 – 3318, 1988.

[366] Galassi, M.E., Abufager, P.N., Martínez, A.E., Rivarola, R.D., and Fainstein, P.D., The continuum distorted wave eikonal initial state model for transfer ionization in H^{+}, He^{2+} + He collisions, *J. Phys. B*, 35, 1727 – 1739, 2002.

[367] Belkić, Dž., Quantum-Mechanical Methods for Loss-Excitation and Loss-Ionization in Fast Ion-Atom Collisions, *Adv. Quantum Chem.*, 56, 150 – 216, 2008.

[368] Belkić, Dž., Four-body continuum distorted wave (CDW-4B) method for ionization in fast ion-atom collisions, *J. Math. Chem.*, 51, 610 – 642, 2008.

[369] Pálinkás, J., Schuch, R., Cederquist, H., and Gustafsson, O., Observation of electron-electron scattering in electron capture by fast protons from He, *Phys. Rev. Lett.*, 63, 2464 – 2467, 1989.

[370] Pálinkás, J., Schuch, R., Cederquist, H., and Gustafsson, O., Evidence for electron-electron scattering in simultaneous capture and ionization by 1 MeV protons in He, *Phys. Scr.*, 42, 175 – 179, 1990.

[371] Briggs, J.S., and Taulbjerg, K., Charge transfer by a double-scattering mechanism involving target electrons, *J. Phys. B*, 12, 2565 – 2573, 1979.

[372] Horsdal-Pedersen, E., Jensen, B., and Nielsen, K.O., Critical angle in electron capture, *Phys. Rev. Lett.*, 57, 1414 – 1416, 1986.

[373] Shi, T.Y., and Lin, C.D., Double photo-ionization and transfer ionization of He: shake-off theory revisited, *Phys. Rev. Lett.*, 89, 163202, 2002.

[374] Vinitsky, P.S., Popov, Yu.V., and Chuluunbaatar, O., Fast proton-hydrogen charge exchange reaction at small scattering angles, *Phys. Rev. A*, 71, 012706, 2005.

[375] Nath, B., and Sinha, C., Simultaneous transfer and ionization in a positron-helium atom system, *Phys. Rev. A*, 61, 062705, 2000.

[376] Fregenal, D., Fiol, J., Bernardi, G., Suárez, S., Focke, P., González, A.D., Muthig, A., Jalowy, T., Groeneveld, K.O., and Luna, H., Double capture with simultaneous ionization in He^{2+} on Ar collisions, *Phys. Rev. A*, 62, 012703, 2000.

[377] Massey, H.S.W., Collisions between atoms and molecules at ordinary temperatures, *Rep. Prog. Phys.*, 12, 248 – 269, 1949.

[378] Geltman, S., Electron detachment from the negative hydrogen ion by electron impact, *Proc. Phys. Soc.*, 75, 67 – 76, 1960.

[379] McDowell, M.R.C., and Williamson, J.H., Electron detachment from H^- by electrons, *Phys. Lett.*, 4, 159 – 161, 1963.

[380] Belly, O., and Schwartz, S.B., Electron detachment from H^- by electron impact, *J. Phys. B*, 2, 159 – 161, 1969.

[381] Bell, K.L., Kingston, A.E., and Madden, P.J., Electron detachment from H^- ions by proton impact, *J. Phys. B*, 11, 3977 – 3982, 1978.

[382] Sidis, V., Kubach, C., and Fussen, D., Ionic-covalent problem in the $H^+ + H^- \leftrightarrow H^* + H$ collisional system, *Phys. Rev. A*, 27, 2431 – 2446, 1983.

[383] Fussen D., and Claeys, W., Electron detachment in $H^+ - H^-$ collisions, *J. Phys. B*, 17, L89 – L93, 1984.

[384] Ermolaev, A.M., Neutralization and detachment in collisions between protons and negative hydrogen ions in the proton energy range from 0.62 to 80.0 keV lab, *J. Phys. B*, 21, 81 – 101, 1988.

[385] Lucey, S., Whelan, C.T., Allan, R.J., and Walters, H.R.J., $(e, 2e)$ on hydrogen minus, H^-, *J. Phys. B*, 29, L489 – L495, 1996.

[386] Kazansky, A.K., and Taulbjerg, K., Quantum wave packet study of electron detachment from H^- by electron impact, *J. Phys. B*, 29, 4465 – 4475, 1996.

[387] Peart, B., Walton, D.S., and Dolder, K.T., Electron detachment from H^- ions by electron impact, *J. Phys. B*, 3, 1346 – 1356, 1970.

[388] Walton, D.S., Peart, B., and Dolder, K.T., A measurement of cross sections for detachment from H^- by a method employing inclined ion and electron beams, *J. Phys. B*, 4, 1343 – 1348, 1971.

[389] Peart, B., Grey, R., and Dolder, K.T., Measurements of cross sections for electron detachment from H^- ions by proton impact, *J. Phys. B*, 9, 3047 – 3053, 1976.

[390] Andersen, L.H., Mathur, D., Schmidt, H.T., and Vejby-Christensen, J., Electron-impact detachment of D^-: near-threshold behaviour and non-existence of D^{2-} resonances, *Phys. Rev. Lett.*, 74, 892 – 895, 1995.

[391] Fritioff, K., Sandström, J., Andersson, P., Hanstorp, D., Hellberg, F., Thomas, R., Geppert, W., Larsson, M., Österdahl, F., Collins, G.F., Pegg, D.J., Danared, H., Källberg, A., and Gibson, N.D., Single and double detachment from H^-, *Phys. Rev. A*, 69, 042707, 2004.

[392] Gradshteyn, I.S., and Ryzhik, I.M., *Tables of Integrals, Series and Products*, Academic Press, New York, 1980.

[393] Horsdal-Pedersen, E., Cocke, C.L., and Stockli, M., Experimental observation of the Thomas peak in high-velocity electron capture by protons from He, *Phys. Rev. Lett.*, 50, 1910 – 1913, 1983.

[394] Berkner, K.H., Kaplan, S.N., Paulikas, G.A., and Pyle, R.V., Electron capture by high-energy deuterons in gases, *Phys. Rev.*, 140, A729 – A731, 1965.

[395] Williams, J.F., Measurement of charge-transfer cross sections for 0.25 to 2.5 MeV protons and hydrogen atoms incident upon hydrogen and helium gases, *Phys. Rev.*, 157, 97 – 100, 1967.

[396] Welsh, L.M., Berkner, K.H., Kaplan, S.N., and Pyle, R.V., Cross sections for electron capture by fast protons in H_2, He, N_2 and Ar, *Phys. Rev.*, 158, 85 – 92, 1967.

[397] Hvelplund, P., Heinemei, J., Horsdal-Pedersen, E., and Simpson, F.R., Electron capture by fast He^{2+} ions in gases, *J. Phys. B*, 9, 491 – 496, 1976.

[398] Mergel, V., Dörner, R., Ullrich, J., Jagutzki, O., Lencinas, S., Nüttgens, S., Spielberger, L., Unverzagt, M., Cocke, C.L., Olson, R.E., Schulz, M., Buck, U., Zanger, E., Theisinger, W., Isser, M., Geis, S., and Schmidt-Böcking, H., State selective scattering angle dependent capture cross sections measured by cold target recoil ion momentum spectroscopy, *Phys. Rev. Lett.*, 74, 2200 – 2203, 1995.

[399] Pivovar, L.I., Tubaev, V.M., and Novikov, M.T., Electron loss and capture by 200-1500 keV helium ions in various gases, *J. Exp. Theor. Phys. JETP*, 14, 20 – 25, 1962. [*Zh. Eksp. Teor. Fiz.*, 41, 26 – 31, 1961.]

[400] Martin, P.J., Arnett, K., Blankenship, D.M., Kvale, T.J., Peacher, J.L., Redd, I., Sutcliffe, V.C., Park, J.T., Lin, C.D., and McGuire, J.H., Differential cross sections for electron capture from helium by 25 to 100 keV incident protons, *Phys. Rev. A*, 23, 2858 – 2865, 1981.

[401] Suzuki, H., Kajikawa, Y., Toshima, N., Ryufuku, H., and Watanabe, T., Electron-capture cross sections from He in collision with bare nuclear ions, *Phys. Rev. A*, 29, 525 – 528, 1984.

[402] Saha, G.C., Datta, S., and Mukherjee, S.C., Electron capture from multi-electron atoms by fast ions in the continuum intermediate-state approximation, *Phys. Rev. A*, 31, 3633 – 3638, 1985.

[403] Saha, G.C., Datta, S., and Mukherjee, S.C., Electron capture in collisions of He^{2+} with Li atoms and of Li^{3+}, C^{6+}, and O^{8+} with He atoms in the high-energy region, *Phys. Rev. A*, 34, 2809 – 2821, 1986.

[404] Gravielle, M.S., and Miraglia, J., State-selective scaling in electron capture by multi-charged ions on light atoms, *Phys. Rev. A*, 51, 2131 – 2139, 1995.

[405] Sewell, E.C., Angel, G.C., Dunn, K.F., and Gilbody, H.B., Ionization and charge transfer in fast $H^+ - Li^+$ collisions, *J. Phys. B*, 13, 2269 – 2275, 1980.

[406] Wetmore, A.E., and Olson, R.E., Electron loss from helium atoms by collisions with fully stripped ions, *Phys. Rev. A*, 38, 5563 – 5570, 1988.

[407] Ford, A.L., Reading, J.F., and Becker, R.L., Coupled-channel calculations of ionization and charge transfer in $p + Li^{+,2+}$ and transfer in $Li^{2+,3+} + H(1s)$ collisions, *J. Phys. B*, 15, 3257 – 3273, 1982.

[408] McCarroll, R., and Salin, A., High-energy behaviour of the cross section for charge exchange, *Proc. Phys. Soc.*, 90, 63 – 72, 1967.

[409] Crothers, D.S.F., and Dunseath, K.M., Target continuum distorted-wave theory for capture of inner-shell electrons by fully stripped ions, *J. Phys. B*, 20, 4115 – 4128, 1987.

[410] Crothers, D.S.F., and Dunseath, K.M., Target continuum distorted-wave theory for collisions of fast protons with atomic hydrogen, *J. Phys. B*, 23, L365 – L371, 1987.

[411] Dubé, L.J., A note on the asymptotic behaviour of asymmetric charge transfer theories, *J. Phys. B*, 16, 1783 – 1791, 1983.

[412] Winter, T.G., Electron transfer and ionization in proton-helium collisions studied using a Sturmian basis, *Phys. Rev. A*, 44, 4353 – 4367, 1991.

[413] Schwab, W., Baptista, G.B., Justiniano, E., Schuch, R., Vogt, H., and Weber, E.W., Measurement of the total cross sections for electron capture of 2.0-7.5 MeV H^+ in H, H_2 and He, *J. Phys. B*, 20, 2825 – 2834, 1987.

[414] Allison, S.K., Experimental results on charge-changing collisions of hydrogen and helium atoms and ions at kinetic energies above 0.2 keV, *Rev. Mod. Phys.*, 30, 1137 – 1168, 1958.

[415] Rudd, M.E., DuBois, R.D., Toburen, L.H., Ratcliffe, C.A., and Goffe, T.V., Cross sections for ionization of gases by 5-4000 keV protons and for electron capture by 5-150 keV protons, *Phys. Rev. A*, 28, 3244 – 3257, 1983.

[416] Bross, S.W., Bonham, S., Gaus, A., Peacher, J., Vajnai, T., Schulz, M., and Schmidt-Böcking, H., Differential transfer ionization cross sections for 50-175 keV proton-helium collisions, *Phys. Rev. A*, 50, 337 – 342, 1994.

[417] Loftager, P., *Private communication*, 2002.

[418] Abufager, P.N., Fainstein, P.D., Martínez, A.E., and Rivarola, R.D., Single electron capture differential cross section in $H^+ + He$ collisions at intermediate and high collision energies, *J. Phys. B*, 38, 11 – 22, 2005.

[419] Fischer, D., Støchkel, K., Cederquist, H., Zettergren, H., Reinhed, P., Schuch, R., Källberg, A., Simonsson, A., and Schmidt, H.T., *Phys. Rev. A*, 73, 052713, 2006.

[420] Green, A.E.S., Sellin, D.L., and Zachor, A.S., Analytic independent-particle model for atoms, *Phys. Rev.*, 184, 1 – 9, 1969.

[421] Herman, F., and Skillman, S., *Atomic Structure Calculations*, Prentice-Hall, Englewood Cliffs, New Jarsay, 1963.

[422] Belkić, Dž., State-selective and total single-capture cross sections for fast collisions of multiply charged ions with helium and lithium, *Phys. Scr.*, 40, 610 – 624, 1989.

[423] Belkić, Dž., and Mančev, I., Single electron capture from carbon by completely stripped projectiles, *Phys. Scr.*, 42, 285 – 292, 1990.

[424] Decker, F., and Eichler, J., Consistent treatment of electron screening in charge transfer, *Phys. Rev. A*, 39, 1530 – 1533, 1989.

[425] Bachau, H., Deco, G., and Salin, A., Introduction of short-range interactions in continuum distorted-wave theory of electron capture for ion-atom collisions, *J. Phys. B*, 21, 1403 – 1410, 1988.

[426] Gulyás, L., Fainstein, P.D., and Shirai, T., Extended description for electron capture in ion-atom collisions: Application of model potentials within the framework of continuum distorted wave theory, *Phys. Rev. A*, 65, 052720, 2002.

[427] Mapleton, R., Production of $H^-(1s^2)$ by hydrogen atom collisions, *Phys. Rev.*, 117, 479 – 485, 1960.

[428] Mapleton, R.A., Production of $H^-(1s^2)$ by hydrogen atom collisions, *Proc. Phys. Soc.*, 85, 841 – 844, 1965.

[429] McClure, G.W., Ionization and electron transfer in collisions of two H atoms: 1.25-117 keV, *Phys. Rev.*, 166, 22 – 29, 1968.

[430] Bragg, W.H., On the absorption of α rays and on the classification of the α rays from radium, *Phil. Mag.*, 8, 719 – 725, 1904.

[431] Bragg, W.H., and Kleeman, R.D., On the ionization curves of radium, *Phil. Mag.*, 8, 726 – 738, 1904.

[432] Bragg, W.H., and Kleeman, R.D., On the $\alpha-$particles of radium and their loss of range in their passing through various atoms and molecules, *Phil. Mag.*, 10, 318 – 340, 1905.

[433] Mančev, I., Four-body continuum-distorted-wave model for charge exchange between hydrogen-like projectiles and atoms, *Phys. Rev. A*, 75, 052716, 2007.

[434] Olson, R.E., Salop, A., Phaneuf, R.A., and Mayer, F.W., Electron loss by atomic and molecular hydrogen in collisions with $^3He^{++}$ and $^4He^+$, *Phys. Rev. A*, 16, 1867 – 1872, 1977.

[435] Hvelplund, P., and Andersen, A., Electron capture by fast H^+, He^+, and He^{++} ions in collisions with atomic and molecular hydrogen, *Phys. Scr.*, 26, 375 – 380, 1982.

[436] Shah, M.B., and Gilbody, H.B., *Private communication*, 1995.

[437] Forest, J.L., Tanis, J.A., Ferguson, S.M., Haar, R.R., Lifrieri, K., and Plano, V.L., Single and double ionization of helium by intermediate-to-high-velocity He^+ projectiles, *Phys. Rev. A*, 52, 350 – 356, 1995.

[438] DuBois, R.D., Multiple ionization in He^+–rare-gas collisions, *Phys. Rev. A*, 39, 4440 – 4450, 1989.

[439] Atan, H., Steckelmacher, W., and Lukas, M.W., Single electron loss and single electron capture for 0.6-2.2 MeV He^+ colliding with rare gases, *J. Phys. B*, 24, 2559 – 2569, 1991.

[440] Itoh, A., Asari, M., and Fukuzawa, F., Charge-changing collisions of 0.7-2.0 MeV helium beams in various gases. I. Electron capture, *J. Phys. Soc. (Japan)*, 48, 943 – 950, 1980.

[441] Stedford, J.B.H., and Hasted, J.B., Further investigation of charge exchange and electron detachment. I. Ion energies 3 to 40 keV. II. Ion energies 100 to 4000 eV, *Proc. Roy. Soc. A*, 227, 466 – 486, 1955.

[442] Gilbody, H.B., and Hasted, J.B., Anomalies in the adiabatic interpretation of charge-transfer collisions, *Proc. Roy. Soc. A*, 238, 334 – 343, 1957.

[443] Stier, P.M., and Barnett, C.F., Charge exchange cross sections of hydrogen ions in gases, *Phys. Rev.*, 103, 896 – 907, 1956.

[444] Wittkower, A.B., Levy, G., and Gilbody, H.B., An experimental study of electron loss during the passage of fast hydrogen atoms through atomic hydrogen, *Proc. Phys. Soc.*, 91, 306 – 309, 1967.

[445] Shah, M.B., Goffe, T.V., and Gilbody, H.B., Electron loss by 35-1000 keV He^+ ions in collisions with atomic and molecular hydrogen, *J. Phys. B*, 10, L723 – L725, 1977.

[446] Hvelplund, P., and Andersen, A., Electron loss by fast He^+ H, He, H^- and He^- projectiles in collisions with atomic and molecular hydrogen, *Phys. Scr.*, 26, 370 – 374, 1982.

[447] Sant'Anna, M.M., Melo, W.S., Santos, A.C.F., Sigaud, G.M., Montenegro, E.C., Shah, M.B., and Meyerhof, W.E., Absolute measurements of electron-loss cross sections of He^+ and C^{3+} with atomic hydrogen at intermediate velocities, *Phys. Rev. A*, 58, 1204 – 1211, 1998.

[448] Bates, D.R., and Griffing, G.W., Inelastic collisions between heavy particles. I: Excitation and ionization of hydrogen atoms in fast encounters with protons and with other hydrogen atoms, *Proc. Phys. Soc. A*, 66, 961 – 971, 1953.

[449] Bates, D.R., and Griffing, G.W., Inelastic collisions between heavy particles. IV: Contributions of double transitions to certain cross sections including that associated with the ionization of hydrogen atoms in fast

encounters with other hydrogen atoms, *Proc. Phys. Soc. A*, 68, 90 – 96, 1955.

[450] Belkić, Dž., and Gayet, R., Corrected closure approximation for electron loss from fast hydrogen atoms passing through atomic hydrogen, *J. Phys. B*, 8, 442 – 447, 1975.

[451] Bates, D.R., Electron loss from fast hydrogen atoms passing through atomic hydrogen, *J. Phys. B*, 8, L117 – L118, 1975.

[452] Belkić, Dž., and Gayet, R., Corrected closure approximation for electron loss from fast hydrogen atoms passing through atomic hydrogen. II, *J. Phys. B*, 9, L111 – L116, 1976.

[453] Belkić, Dž., Electron loss in collisions between two hydrogen-like atoms/ions: four-body accelerated first Born approximation with correct boundary conditions, *J. Math. Chem.*, 51, 498 – 526, 2008.

[454] Belkić, Dž., Electron loss in collisions between two hydrogen-like atoms/ions: corrected closure approximation, *J. Math. Chem.*, 51, 527 – 558, 2008.

[455] Bates, D.R., and Griffing, G.W., Inelastic collisions between heavy particles. II: Contributions of double-transitions to the cross sections associated with excitation of hydrogen atoms in fast encounters with other hydrogen atoms, *Proc. Phys. Soc. A*, 66, 663 – 668, 1954.

[456] Moiseiwitsch, B.L., and Stewart, A.L., Inelastic collisions between heavy particles. III: Excitation of helium atoms in fast encounters with hydrogen atoms, protons and positive helium ions, *Proc. Phys. Soc. A*, 67, 1069 – 1074, 1954.

[457] Boyd, T.J.M., Moiseiwitsch, B.L., and Stewart, A.L., Inelastic collisions between heavy particles. V: Electron loss for fast He^+ ions passing through atomic hydrogen, and ionization of hydrogen atoms by fast He^+ ions, *Proc. Phys. Soc. A*, 70, 110 – 116, 1957.

[458] Bell, K.L., and Kingston, A.E., Excitation and ionization processes in $He^+(1s) + H(1s)$ collisions, *J. Phys. B*, 11, 1259 – 1265, 1978.

[459] Bethe, H., Zur Theorie des Durchgangs schneller Korpuskularstrahlen durch Materie, *Ann. Phys. Lpz.* 5, 325 – 400, 1930.

[460] Unsöld, A., Quantentheorie des Wasserstoffmolekülions und der Born-Landéschen Abstroßungskräfte, *Z. Phys.*, 43, 563 – 574, 1927.

[461] Massey, H.S.W., and Mohr, C.B.O., The collisions of slow electrons with atoms – IV, *Proc. Roy. Soc. A*, 146, 880 – 990, 1934.

[462] May, R.M., Sum rules for hydrogenic wave functions, with applications to charge-exchange and ionization processes, *Phys. Rev.*, 136, 669 – 674, 1964.

[463] Cheshire, I.M., and Kyle, H.L., Excitation of atomic hydrogen by lithium atoms, *Phys. Lett.*, 17, 115 – 116, 1965.

[464] Cheshire, I.M., and Kyle, H.L., Excitation of H atoms by fast protons, *Proc. Phys. Soc.*, 88, 785 – 786, 1966.

[465] Becker, R.L., and MacKellar, A.D., Classical four-body calculations of $^4He^+ + H$ and $H + H$ collisions, *J. Phys. B*, 12, L345 – L350, 1979.

[466] Zouros, T.J.M., Sulik, B., Gulyás, L., and Orbán, A., Production of $1s2s2p\,^4P$ states by transfer-loss cascaded in O^{5+} collisions with He and H_2 targets, *Brazilian J. Phys. B*, 36, 505 – 508, 2006.

[467] Barnett, C.F., and Reynolds, H.K., Charge exchange cross sections for hydrogen particles in gases at high energies, *Phys. Rev.*, 109, 355 – 359, 1958.

[468] Barnett, C.F., and Stier, P.M., Charge exchange cross sections for helium ions in gases, *Phys. Rev.*, 109, 385 – 390 1958.

[469] Tanis, J.A., Bernstein, E.M., Graham, W.G., Clark, M., Shafroth, S.M., Johnson, B.M., Jones, K.W., and Meron, M., Resonant behavior in the projectile X-ray yield associated with electron capture in S + Ar collisions, *Phys. Rev. Lett.*, 49, 1325 – 1328, 1982.

[470] Itoh, A., Zouros, T.J.M., Schneider, D., Stettner, U., Zeitz, W., and Stolterfoht, N., Transfer excitation in $He^+ + He$ collisions studied by 0 degrees electron spectroscopy, *J. Phys. B*, 18, 4581 – 4587, 1985.

[471] Justiniano, E., Schuch, R., Schulz, M., Reusch, S., Mokler, P.H., McLaughlin, D.J., and Hahn, Y., X-ray – X-ray coincidence measurements of resonant transfer and excitation, *Proceedings of the XVth International Conference on the Physics of Electronic and Atomic Collisions*, Brighton, Book of Invited Talks, Eds. H.B. Gilbody, W.R. Newell, F.H. Read, and C.H. Smith, Elsevier Science Publishers B.V., Amsterdam, 1988, 477 – 483.

[472] Schulz, M., Justiniano, E., Schuch, R., Mokler, P.H., and Reusch, S., Separated resonances in simultaneous capture and excitation of S^{15+} in H_2 observed by K-X-ray – K-X-ray coincidences, *Phys. Rev. Lett.*, 58, 1734 – 1737, 1987.

[473] Brandt, D., Resonant transfer and excitation in ion-atom collisions, *Phys. Rev. A*, 27, 1314 – 1318, 1983.

[474] Feagin, J.M., Briggs, J.S., and Reeves, T.M., Simultaneous charge transfer and excitation, *J. Phys. B*, 17, 1057 – 1068, 1984.

[475] Fano, U., Effects of configuration interaction on intensities and phase shifts, *Phys. Rev.*, 124, 1866 – 1878, 1961.

[476] McLaughlin, D.J., and Hahn, Y., Dielectronic recombination cross sections for Si^{11+} and S^{13+}, *Phys. Lett. A*, 88, 394 – 397, 1982.

[477] Prudnikov, A.P., Britchkov, Yu., and Marietchev, O., *Integrals and Series* (in Russian), Nauka, Moscow, Vol. 1, 1981.

[478] Hahn, Y., Transfer excitation processes in ion-atom collisions at high energies, *Phys. Rev. A*, 40, 2950 – 2957, 1989.

[479] Tanis, J., Electron transfer and projectile excitation in single collisions, *Nucl. Instr. Meth. Phys. Res. A*, 262, 52 – 61, 1987.

[480] Hahn, Y., and Ramadan, H., Uncorrelated transfer excitation collisions at high energies, *Phys. Rev. A*, 40, 6206 – 6209, 1989.

[481] Feshbach, H., A unified theory of nuclear reactions. II, *Ann. Phys. NY*, 19, 287 – 313, 1962.

[482] Gayet, R., and Hanssen, J., Resonant and non-resonant modes of transfer and excitation processes in ion-atom collisions, *Nucl. Instr. Meth. Phys. Res. B*, 86, 52 – 61, 1994.

[483] Zouros T.J.M., Schneider, D., and Stolterfoht, N., State-selective observation of resonant and non-resonant transfer-excitation in 50-500 keV $^3He^+ + H_2$ collisions, *J. Phys. B*, 21, L671 – L676, 1988.

[484] Macías, A., Martín, F., Riera, A., and Yáñez, N., A practical solution to the "unknown normalization" problem, *Int. J. Quantum Chem.*, 33, 279 – 300, 1988.

[485] Bordenave-Montesquieu, A., Gleizes, A., and Benoit-Cattin, P., Excitation of the helium auto-ionizing states in $He^+ + He$ collisions, between 3 and 140 keV, *Phys. Rev. A*, 25, 245 – 267, 1982.

[486] van der Straten, P., and Morgenstern, R., Interference of auto-ionizing transitions in fast ion-atom collisions, *J. Phys. B*, 19, 1361 – 1370, 1986.

[487] Bhatia A.K., and Temkin, A., Calculation of auto-ionization of He and H^- using the projection-operator formalism, *Phys. Rev. A*, 11, 2018 – 2024, 1975.

[488] Belkić, Dž., *Quantum Theory of Resonant Scattering, Spectroscopy and Signal Processing*, Taylor and Francis, London, 2009.

Index